# WAVELET THEORY
# APPROACH TO
# PATTERN RECOGNITION

## 3rd Edition

# SERIES IN MACHINE PERCEPTION AND ARTIFICIAL INTELLIGENCE*

ISSN: 1793-0839

*Editors:* **H. Bunke** (University of Bern, Switzerland)
**Cheng-Lin Liu** (Chinese Academy of Sciences, China)

This book series addresses all aspects of machine perception and artificial intelligence. Of particular interest are the areas of pattern recognition, image processing, computer vision, natural language understanding, speech processing, neural computing, machine learning, hardware architectures, software tools, and others. The series includes publications of various types, for example, textbooks, monographs, edited volumes, conference and workshop proceedings, PhD theses with significant impact, and special issues of the International Journal of Pattern Recognition and Artificial Intelligence.

*Published*

Vol. 90: *Wavelet Theory Approach to Pattern Recognition*
*Third Edition*
by Yuan Yan Tang and Lixiang Xu

Vol. 89: *Handwritten Historical Document Analysis, Recognition, and Retrieval —*
*State of the Art and Future Trends*
edited by Andreas Fischer, Marcus Liwicki and Rolf Ingold

Vol. 88: *The Lognormality Principle and its Applications in e-Security, e-Learning*
*and e-Health*
by Réjean Plamondon, Angelo Marcelli and Miguel Angel Ferrer

Vol. 87: *Fuzzy Systems to Quantum Mechanics*
by Hong-Xing Li

Vol. 86: *Graph-Based Keyword Spotting*
edited by Michael Stauffer, Andreas Fischerand Kaspar Riesen

Vol. 85: *Ensemble Learning: Pattern Classification Using Ensemble Methods*
*Second Edition*
by L. Rokach

Vol. 84: *Hybrid Metaheuristics: Research and Applications*
by Siddhartha Bhattacharyya

Vol. 83: *Data Mining in Time Series and Streaming Databases*
edited by Mark Last, Horst Bunke and Abraham Kandel

Vol. 82: *Document Analysis and Text Recognition:*
*Benchmarking State-of-the-Art Systems*
edited by Volker Märgner, Umapada Pal and Apostolos Antonacopoulos

Vol. 81: *Data Mining with Decision Trees: Theory and Applications*
*Second Edition*
by L. Rokach and O. Maimon

*The complete list of the published volumes in the series can be found at
https://www.worldscientific.com/series/smpai

Series in Machine Perception and Artificial Intelligence – Vol. 90

# WAVELET THEORY APPROACH TO PATTERN RECOGNITION

## 3rd Edition

Yuan Yan Tang

University Macau, China

Lixiang Xu

Hefei University, China

**World Scientific**

NEW JERSEY · LONDON · SINGAPORE · BEIJING · SHANGHAI · HONG KONG · TAIPEI · CHENNAI · TOKYO

*Published by*

World Scientific Publishing Co. Pte. Ltd.

5 Toh Tuck Link, Singapore 596224

*USA office:* 27 Warren Street, Suite 401-402, Hackensack, NJ 07601

*UK office:* 57 Shelton Street, Covent Garden, London WC2H 9HE

**Library of Congress Cataloging-in-Publication Data**

Names: Tang, Yuan Yan, 1943–    author. | Xu, Lixiang, author.

Title: Wavelet theory approach to pattern recognition / Yuan Yan Tang, University Macau, China,
 Lixiang Xu, Hefei University, China.

Description: 3rd edition. | New Jersey : World Scientific, 2025. | Series: Series in machine
 perceptionand artificial intelligence, 1793-0839 ; vol. 90 | Revised edition of:
 Wavelet theory approach to pattern recognition / Yuan Yan Tang. New Jersey :
 World Scientific, c2009. | Includes bibliographical references and index.

Identifiers: LCCN 2023057798 | ISBN 9789811284045 (hardcover) |
 ISBN 9789811284052 (ebook for institutions) | ISBN 9789811284069 (ebook for individuals)

Subjects: LCSH: Wavelets (Mathematics) | Optical pattern recognition--Mathematics. |
 Image processing--Digital techniques.

Classification: LCC QA403.3 .T36 2025 | DDC 515/.2433--dc23/eng/20240416

LC record available at https://lccn.loc.gov/2023057798

**British Library Cataloguing-in-Publication Data**

A catalogue record for this book is available from the British Library.

For any available supplementary material, please visit
https://www.worldscientific.com/worldscibooks/10.1142/13616#t=suppl

Desk Editors: Balasubramanian Shanmugam/Steven Patt

Typeset by Stallion Press
Email: enquiries@stallionpress.com

# Preface

This is the third edition of the book *Wavelet Theory Approach to Pattern Recognition*. The first edition was published in 2000 and the second one was in 2009. Since that time, many achievements in the area of pattern recognition with wavelet analysis have been made. Especially, deep learning has become a huge breakthrough in artificial intelligence including pattern recognition that can automatically learn the essential features in pattern recognition. It greatly increases the efficiency and accuracy of pattern recognition system. It is necessary to have a new book to involve these new developments including the basic principle of deep learning as well as the applications of combination of wavelet theory with deep learning to pattern recognition. The main part of the previous versions of the book is kept; it contains 12 chapters of the former versions. Meanwhile, five new chapters related to the combination of wavelet theory and deep learning approach to pattern recognition are added. The new version emphasizes the applications of wavelet theory and deep learning to pattern recognition rather than the wavelet theory and deep learning themselves.

Wavelet analysis with deep learning and its applications have become one of the fastest growing research areas in recent years. This is in part attributed to the pioneering work by the researchers as well as practitioners in the fields of mathematics and signal processing. Wavelet theory with deep learning has been employed in many fields and applications, such as signal and image processing, communication systems, biomedical imaging, radar, air acoustics, control systems, and endless other areas. However, the research on applying the wavelets with deep learning to pattern recognition is still too weak; only a few publications deal with this topic at present. This book focuses on this challenging research topic.

The most fascinating area of artificial intelligence with practical applications is pattern recognition. Making computer see and recognize objects like humans has captured the attention of many scientists in different disciplines. Indeed, machine recognition of different patterns, such as printed and handwritten characters, fingerprints, and biomedical images, has been intensively and extensively researched by scientists in different countries around the world. The area of pattern recognition, after over six decades of continued development, is now definitely playing a very major role in advanced automation in the 21st century. Although a lot of achievements have been made in the area of pattern recognition, many problems still have to be solved. The goal of this book is to, through mathematically sound derivations and experiments, develop some new application-oriented techniques in wavelet theory and, thereafter, apply these new techniques to solve some particular problems in the area of pattern recognition.

This book is organized into four parts:

**Part 1:** Chapter 1 presents a brief survey of the basic wavelet theory as well as its applications to pattern recognition.

**Part 2:** Chapters 2–5 extend the basic theories of wavelets as well as deep learning.

**Part 3:** Chapters 6–13 deal with the detailed presentation of the applications of the wavelet theory to pattern recognition.

**Part 4:** Chapter 14–17 provide the detailed presentation of the applications of combination of wavelet theory with deep learning to pattern recognition.

Parts 3 and 4 play the core role of this book.

Considering the major readers of this book are scientists and engineers, thus, in some chapters/sections, we give up the exactness in mathematics temporarily.

Initially, in Chapter 1, a brief description of wavelet theory is introduced, and a comparison between the wavelet and Fourier transforms is discussed along with several pictorial examples. This chapter reviews established applications of the wavelet theory to pattern recognition. The review is not detailed, since this book concentrates on the novel research results developed by the author. However, the references are cited if additional details are desired.

Throughout Chapters 2 and 3, both the wavelet transform and wavelet bases are of critical concern, which formulate the basic wavelet theory. In Chapter 2, the general theory of the continuous wavelet transform is addressed, and its major properties are investigated including the characterization of Lipschitz regularity of signals by the wavelet transform. Chapter 2 also primarily examines an important property by relating the processing to matched filtering concepts. Chapter 3 considers multiresolution analysis (MRA) and wavelet bases, where the basic concepts of the both are presented as well as the construction of them. As an important algorithm for implementing the discrete wavelet transform, Mallat algorithm is introduced.

After studying these basic concepts of the wavelet theory, some typical wavelet bases including the orthonormal and non-orthonormal bases are provided in Chapter 4. They benefit from the application of wavelet theory to engineering, such as pattern recognition and image processing.

Chapter 5 is used to introduce the basic principle of deep learning including the main models of deep learning, such as RNN-LSTM model, RNN-GRU model, GAN model, and transformer model.

By formulating the above wavelet theory and the general applications with wavelet theory, the third part of this book (Chapters 6–13) demonstrates more detailed applications, which become the core chapters. All of these applications were made by our research group.

Chapter 6 develops a method to identify different structures of the edges and design an algorithm to detect the step structure edges. This technique can be employed to contour extraction in document processing as well as 2D object recognition.

Chapter 7 aims at studying the characterization of Dirac structure edges with wavelet transform and selecting the suitable wavelet functions to detect the Dirac edges. A mapping technique is applied in this chapter to construct such a wavelet function. In this way, a low-pass function is mapped onto a wavelet function by a derivation operation. In this chapter, the quadratic spline wavelet is utilized to characterize the Dirac structure edges and an algorithm to extract the Dirac structure edges by wavelet transform is also developed.

Chapter 8 introduces a new wavelet function called Tang–Yang wavelet, which is constructed by our research group. The characteristics of the Tang–Yang wavelet with curves are discussed. They are gray-level invariant, slope invariant, and width invariant. The application of new wavelet function to curve analysis is presented.

In Chapter 9, skeletonization of ribbon-like shapes based on the Tang–Yang wavelet function is presented. Characterization of the ribbon-like shape with wavelet transform using the Tang–Yang wavelet is investigated. Some useful algorithms are also provided.

Chapter 10 presents an approach to feature extraction. In this way, the wavelet decomposition is used to produce wavelet sub-patterns and, thereafter, the fractal divider dimensions are utilized to find the numerical features from these sub-patterns.

Chapter 11 applies 2D MRA and Mallat algorithm to form document analysis. The HL and LH sub-images are utilized to find the reference lines in a form document, and furthermore, the useful information can be extracted in accordance with these reference lines. This application is verified by several concrete examples of bank checks.

Chapter 12 uses B-spline wavelet for Chinese computing, which consists of three operations: (1) compression of Chinese characters, (2) enlargement of type size with arbitrary scales, and (3) generation of type styles of Chinese fonts.

Chapter 13 deals with the classification of patterns with wavelet theory, where the orthogonal wavelet series are used for the probability density estimation in the classifier design.

The fourth part of this book (Chapters 14–17) as a new one joins to the third version of this book. It contains some applications of the combination of the wavelet and deep learning to pattern recognition, which become the core chapters for the new edition of the book.

Chapter 14 introduces a method that utilizes deep learning for texture classification via multiple wavelets fusion of the modified scattering transforms, which mix together two different wavelet basic functions (Morlet and Shannon). This hierarchical structure of deep networks can build a suitable representation of different texture categories.

Chapter 15 proposes a deep wavelet network architecture for image classification. It involves classifying all classes in the dataset by applying the stacked autoencoders and a linear classifier in the last layer.

Chapter 16 represents an application of combination of wavelet theory with deep learning to medical image processing. Precisely, a hybrid deep learning is utilized for brain tumor classification and segmentation. This approach comprises the stationary wavelet packet transform and adaptive kernel fuzzy $c$ means clustering. A CNN-LSTM technique with hybrid optimization is used.

Finally, Chapter 17 studies the hybrid combination of wavelets and deep learning for the enhancement of speech signals. It can improve the quality of speech signals degraded by noise, reverberation, or other artifacts that can affect the intelligibility, automatic recognition, or other attributes involved in speech technologies and telecommunications among others.

The main components of this book are the achievements of our research group as well as the research scholars from various universities:

Professor Seong-Whan Lee at Korea University, Korea;
Professor Lihua Yan at Zhongshan (Sun Yat-Sen) University, China;
Professor Ching Y. Suen at Concordia University, Canada;
Professor Xinge You at Huazhong University of Science and Technology, China;
Professors Hong Ma and Bing-Fa Li at Sichuan University, China;
Professor Feng Yang at the South Medical University, China;
Professors A. Dadashnialehi, A. Bab-Hadiashar and R. Hoseinnezhad at RMIT University, Australia [Dadashnialehi *et al.*, 2017];
Professors D. Blel, S. Hassairi and R. Ejbali at University of Gabes, Tunisia [Blel *et al.*, 2022];
Professors R. Sindhiya Devi, B. Perumal and M. Pallikonda Rajasekaran at Kalasalingam Academy of Research and Education, India [Devi *et al.*, 2022];
Professors M. Gutierrez-Murioz and M. Coto-Jimenez at University of Costa Rica, Casta Rica [Gutiérrez-Muñoz and Coto-Jiménez, 2022].

We would like to especially appreciate their contribution to this book.

To review the established applications of the wavelet theory to pattern recognition, in Chapter 1, some materials including figures from the published papers are quoted. We would like to thank the following authors for releasing the copyrights to this book: S. H. Yoon, J. H. Kim, W. E. Alexander, S. M. Park, and K. H. Sohn [Yoon *et al.*, 1998], F. Murtagh and J.-L. Starck [Murtagh and Starck, 1998], and K. H. Liang, F. Chang, T. M. Tan, and W. L. Hwang [Liang *et al.*, 1999].

A specific international journal called *International Journal on Wavelets, Multiresolution, and Information Processing* (*IJWMIP*) was founded by myself in 2003. In Chapter 1, the following papers from *IJWMIP* are quoted. We would like to record my appreciation to the authors for their contributions to this book: Daugman [2003], Yang *et al.* [2003b], Kouzani and Ong [2003], Kumar *et al.* [2003], Kumar and Kumar [2005], Sharnia *et al.* [2004], Kunte and Samuel [2007], Muneeswaran *et al.* [2005], Ksantini *et al.* [2006], and El-Khamy *et al.* [2006].

We would like to express our deep gratitude to the students, Y. W. Man, Limin Cui, Yan Tang, D. H. Xi, L. Feng, W. Xu, W. Ge, Q. Z. Cui, and H. F. Liu.

We would like to also thank the many people at the University of Macau, Hong Kong Baptist University, and Hefei University, for their strong support.

Many research projects involved in this book were supported by the research grants received from Macau FDCT and Research Grant Council (RGC) of Hong Kong, National Natural Science Foundation of China (62176085, 62172458), Industry University Research Cooperation Project (GP/026/2020 and HF-010-2021) Zhuhai City, Guangdong Province, Talent Fund of Hefei University under Grant 20RC25.

# About the Authors

**Yuan Yan Tang** is an IEEE Life Fellow, IAPR Fellow and AAIA Fellow. He currently is the Director of Smart City Research Center and a Chair Professor at the University of Macau. He was the Dean of Science Faculty at Australia UOW College Hong Kong/Community College of City University, a Chair Professor at Hong Kong Baptist University, the Dean of Computer Science at Chongqing University and an Honorary/Adjunct Professor at several universities. He is a Historical-Highly-Cited-Researcher. Prof. Tang published 13 monographs, 16 book chapters and about 400 *SCIE International Journal* Papers. Dr. Y. Y. Tang is the Founder and Editor-in-Chief of *International Journal of Wavelets, Multiresolution, and Information Processing (IJWMIP)* and served as the Chair for more than 50 international conferences.

**Lixiang Xu** (Member, IEEE) is a Professor at Hefei University of China. He received the B.Sc. degree and M.Sc. degree in Applied Mathematics in 2005 and 2008, respectively. He worked at Huawei Technologies Co., Ltd. in 2008 before joining Hefei University in the following year. He received his Ph.D. degree from the School of Computer Science and Technology, Anhui University, Hefei, China, in 2017. He has been awarded a scholarship to pursue his studies in Germany as a joint Ph.D. student from 2015 to 2017. He did postdoctoral research at the University of Science and Technology of China from 2019 to 2022. His current research interests include wavelet analysis, structural pattern recognition, data mining and machine learning, social network analysis, and recommender systems.

# Contents

# PART 1

# Chapter 1

# Introduction

## 1.1 Wavelet: A Novel Mathematical Tool for Pattern Recognition

Wavelet analysis is a relatively recent development of applied mathematics from the 1980s. Independent from its developments in harmonic analysis, Grossmann, Morlet and their coworkers studied the wavelet transform in its continuous form and initially applied it to analyze geological data [Grossmann and Morlet, 1984; Grossmann and Morlet, 1985; Grossmann *et al.*, 1985; Morlet *et al.*, 1982a,b]. However, at that time, the roughness of wavelets made the mathematicians suspect the existence of a "good" wavelet basis until two great events took place in 1988:

1. Daubechies, a female mathematician, constructed a class of wavelet bases, which are smooth, compactly supported and orthonormal. They are referred to as Daubechies bases and successfully applied to many fields today.
2. French signal analysis expert, Mallat, with mathematician, Meyer, proposed a general method to construct wavelet bases. It is termed multiresolution analysis (MRA) and is intrinsically consistent with sub-band coding in signal analysis.

The above achievements play important roles in mathematics as well as engineering. They established the theoretical frame for wavelet analysis in mathematics. On the other hand, they improved the traditional theory of frequency analysis in engineering. As a result, a novel method for time–frequency analysis has been formed. In this way, signals can be locally characterized in both the time domain and the frequency domain simultaneously and self-adaptively. According to this property, wavelet analysis can be efficiently applied to analyze and process the non-stationary signals. Engineers and scientists have paid ample attention to wavelet theory and carried out research of wavelets in very wide areas during the recent years. As a matter of fact, significant success of wavelet analysis has been achieved on a variety of disciplines, such as signal processing, image compressing and enhancement, pattern recognition, communication systems, control systems, biomedical imaging, air acoustic, radar, theoretical mathematics, and endless other signal processing fields. Moreover, the potentiality of wavelets in both research and development would be immeasurable.

What is a wavelet? The simplest answer is a "short" wave (wave + let ⇒ wavelet). The suffix "let" initially means "a small kind of" in a general English dictionary [Procter, 1993]. However, in the mathematical term "wavelet", the meaning of the suffix "let" comes "short", which indicates that the duration of the function is very limited. In other words, it is said that wavelets are localized. Look at the following simple examples in Fig. 1.1 to obtain a brief idea: What is a wave, and what is a wavelet?

We consider two typical mechanical movements:

(1) non-damped simple harmonic vibration,
(2) damped oscillation.

A wave can be produced by a non-damped simple harmonic vibration, which is illustrated in Fig. 1.1(a). It forms a sinusoidal wave with non-attenuated amplitude. By contrast, a wavelet can be formed by a damped oscillation, as shown in Fig. 1.1(b), its amplitudes quickly decay to zero in both the positive and negative directions. In other words, the wavelet is a special signal, in which two conditions have to be satisfied:

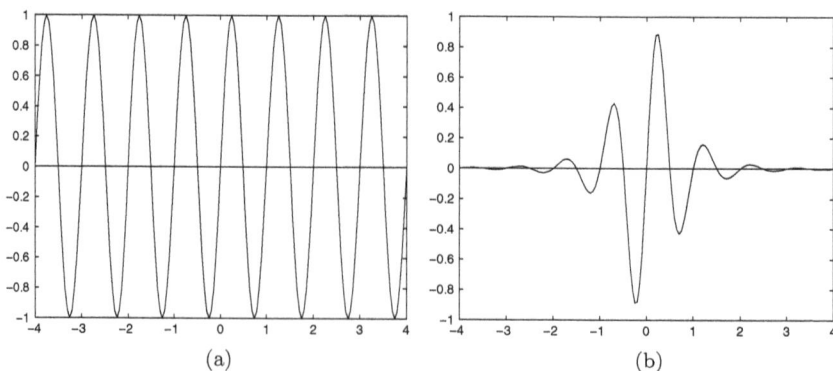

     (a)           (b)

Fig. 1.1 Simple examples of wave and wavelet: (a) a wave is produced by a non-damped simple harmonic vibration. (b) a wavelet is formed by a damped oscillation.

1. The wavelet must be oscillatory (wave).
2. Its amplitudes are non-zero only during a short interval (short).

The required oscillatory condition leads to sinusoids as the building blocks. The quick decay condition is a tapering or windowing operation. These two conditions must be simultaneously satisfied for the function to be a wavelet [Young, 1993]. According to these conditions, we can justify whether a function is a wavelet.

Consider the functions in Fig. 1.2. While applying these two conditions, it is clear that the functions in Figs. 1.2(a) and 1.2(b) are wavelets, while the ones in Figs. 1.2(c) and 1.2(d) are not. The functions in Figs. 1.2(a) and 1.2(b) are oscillatory and have amplitudes which quickly decay to zero with time. Thus, they are wavelets. The functions in Figs. 1.2(c) and 1.2(d) do not satisfy both wave and short simultaneously. Figure 1.2(c) decays quickly to zero but does not oscillate while Fig. 1.2(d) is a wave but not short. Therefore, they are not wavelets.

The next question is: What we can do with this simple wave? It is well known that any complex movement, which is described by $f(x)$, can be represented by the summation of several simple harmonic vibrations with

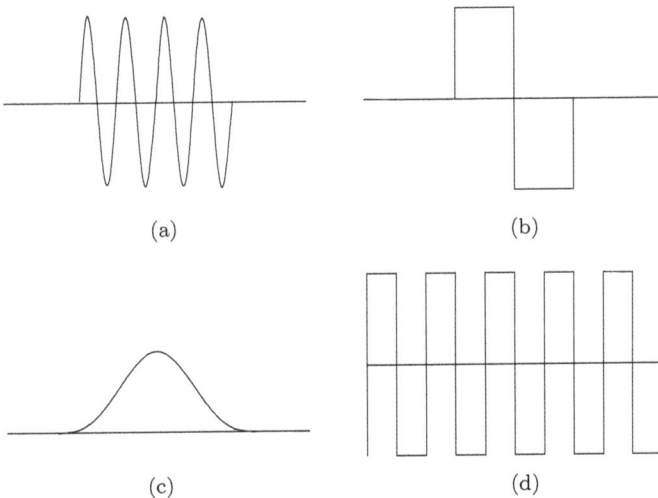

Fig. 1.2 Simple examples of wave and wavelet. (a) and (b) are wavelets, since they are oscillatory and in limited duration; (c) and (d) are not wavelets because the function in (c) decays quickly to zero but does not oscillate and the function in (d) is wave but not short.

different frequencies:

$$f(x) = \frac{a_0}{2} + \sum_{k=1}^{\infty}(a_k \cos kx + b_k \sin kx), \tag{1.1}$$

where $\cos kx$ and $\sin kx$ denote the simple harmonic vibrations.

Fourier first discovered this important fact in 1807, when he was studying the equation of heat conduction. This is why the above series expansion is called Fourier expansion. It infers that any complex wave $f(x)$ can be formed by the linear combination of the basic elements, which are produced by the dilations and translations of $\sin x$. $\sin x$ is the simplest basic wave and is viewed as a basic "atom". Therefore, through the investigation of rules, which indicate how a complex function $f(x)$ can be composed of such small atoms, we can understand function $f(x)$ itself well. It has been mathematically proved that the sum of the finite number of terms, which are the fore ones in the Fourier series, is the best approximation of $f(x)$ under the consideration of energy:

$$\int_0^{2\pi}\left|f(x) - \left[\frac{a_0}{2} + \sum_{k=1}^{N}(a_k \cos kx + b_k \sin kx)\right]\right|^2 dx$$

$$= \min_{c_0,c_1,d_1,\ldots,c_N,d_N}\int_0^{2\pi}\left|f(x) - \left[\frac{c_0}{2} + \sum_{k=1}^{N}(c_k \cos kx + d_k \sin kx)\right]\right|^2 dx \to 0$$

$$(N \to \infty). \tag{1.2}$$

For these reasons, Fourier analysis has been regarded as a traditional and efficient tool in many fields of science and engineering, during the past 200 years. It will also play an indispensable role in the future. However, the development of science and engineering has no limits. Fourier analysis is not a solution that will be good for all times and in all situations. Meanwhile, Fourier analysis has its own deficiency. In fact, it has two major problems:

(1) Fourier analysis cannot characterize the signals locally in the time domain.
(2) Fourier expansion can approximate the stationary signals well, but cannot do so for the non-stationary ones.

These drawbacks can be found in the following concrete examples. Consider two signals shown in Fig. 1.3.

According to the above analysis, it can be known that a complicated function $f(x)$ can be represented by a series of Fourier coefficients, say $\{a_0, a_1, b_1, a-2, b_2, \ldots\}$, i.e.,

$$f(x) \Longrightarrow \{a_0, a_1, b_1, \ldots, a_i, b_i, \ldots, a_j, b_j, \ldots\},$$

where $j > i$ and the Fourier coefficients $a_j$ and $b_j$ correspond to the components with higher frequency in signal $f(x)$. By contrast, the coefficients $a_i$ and $b_i$ correspond to the components with lower frequency in signal $f(x)$. For the given signals $f_1(x)$ and $f_2(x)$, we have

$$\begin{aligned} f_1(x) &\Longrightarrow \{a_0^1, a_1^1, b_1^1, \ldots, a_i^1, b_i^1, \ldots, a_j^1, b_j^1, \ldots\}, \\ f_2(x) &\Longrightarrow \{a_0^2, a_1^2, b_1^2, \ldots, a_i^2, b_i^2, \ldots, a_j^2, b_j^2, \ldots\}. \end{aligned} \tag{1.3}$$

It is easy to understand, from Figs. 1.3(c) and 1.3(d), that the values of the first several terms, which correspond to lower frequencies, in the Fourier coefficients are greater than those of the other terms, which are located in

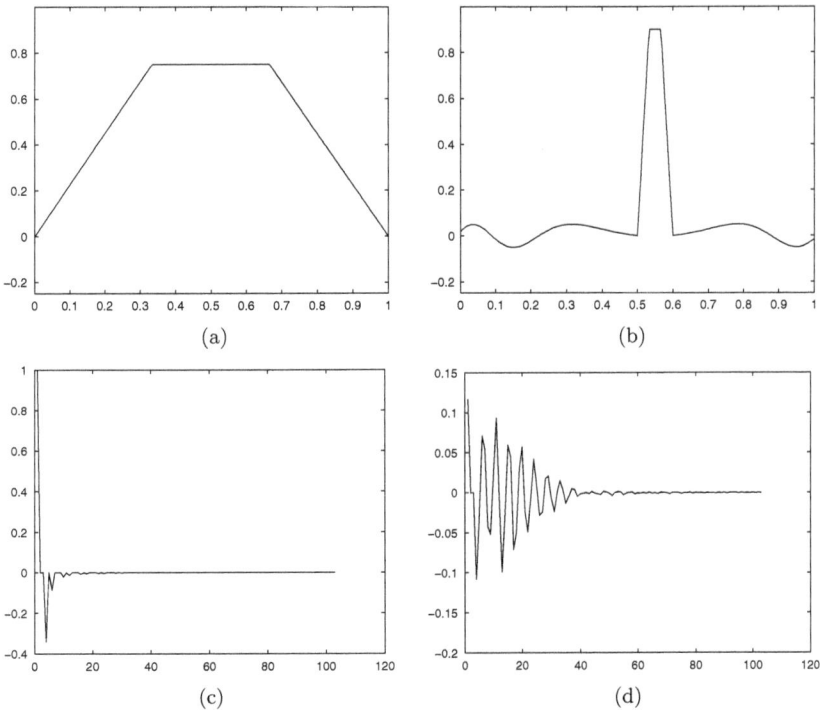

Fig. 1.3  (a) The original signal $f_1(x)$; (b) another original signal $f_2(x)$; (c) the Fourier coefficients corresponding to $f_1(x)$; (d) the Fourier coefficients corresponding to $f_2(x)$.

the tail of the series and have higher frequencies. The backer the coefficients are located, the closer to zero the values are. Thus, the energy of a signal is distributed on the front terms (lower frequencies) in the Fourier coefficients.

Consider the number of cycles in Figs. 1.3(c) and 1.3(d), it can be found that the number of cycles in Fig. 1.3(d) is greater than that in Fig. 1.3(c), which indicates that signal $f_2(x)$ has more non-zero components in the frequency domain than what $f_1(x)$ has. This situation is also described in Eq. (1.3), where the number of non-zero terms in the Fourier expansion for signal $f_2(x)$ is more than that in the expansion for signal $f_1(x)$. This property arrests the time–frequency analysis using the Fourier theory. For instance, signal $f_2(x)$ possesses a transient component with a very short interval. However, its corresponding Fourier expansion contains many terms which produce a long-duration vibration, as illustrated in Fig. 1.3(d). Thus, a short-duration transient in the time domain corresponds to a long-duration vibration in the frequency domain. The characteristic of the localization of the narrow transient signal in the time domain will disappear in the frequency domain. Therefore, it is impossible to locate and characterize the transient components in both the time and frequency domains using the Fourier coefficients. This makes the Fourier transforms less than optimal representations for analyzing signals, images and patterns containing transient or localized components. As a matter of fact, in pattern recognition, many important features are highly localized in spatial positions. The above weakness of the Fourier transform obstructs its application to pattern recognition.

Look at the following three examples, which deal with the field of pattern recognition, shown in Figs. 1.4 and 1.5. First, we consider an image of an aircraft with several drawing lines and character strings, which are displayed in Fig. 1.4(a). The contour of the aircraft is a localized image component, meanwhile the drawing lines and character strings are also the transient image components. To recognize the aircraft from other objects, its contour is a useful feature, which is required to extract from the image. On the other hand, the drawing lines and character strings pose as obstacles for recognition and have to be removed. For these two different image components, the Fourier analysis cannot treat them efficiently.

The second example is illustrated in Fig. 1.4(b), which shows an English letter "A: with noise. In pattern recognition, the boundary of the character should be extracted as a good feature, while the noise has to be deleted. To do so, the transient component, the boundary of letter "A", needs to be

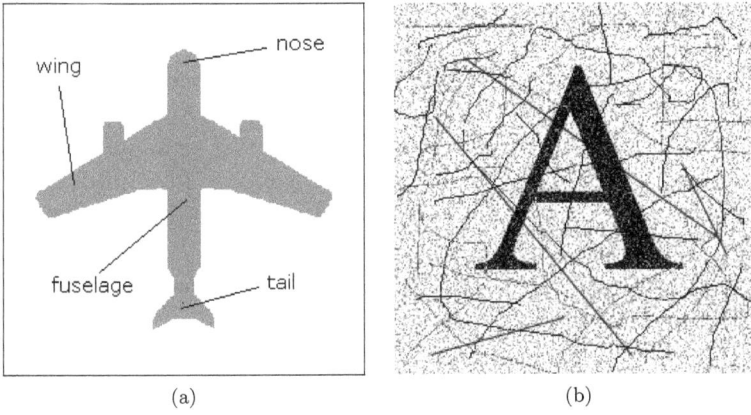

(a)            (b)

Fig. 1.4    The contours of objects, drawing lines and texts are transient components.

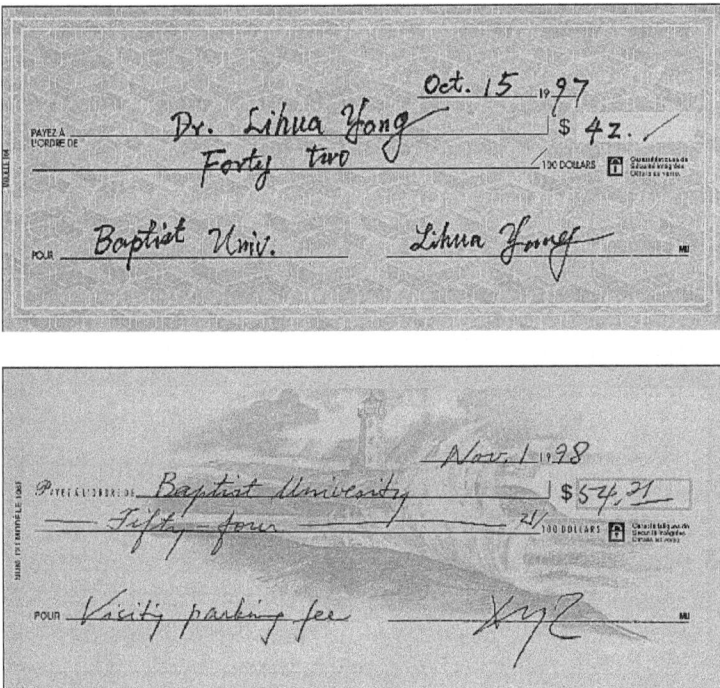

Fig. 1.5    The reference lines of bank check are transient components.

localized and processed. The deficiency of the Fourier analysis will obstruct this task.

The last example, as shown in Fig. 1.5, deals with document analysis, which is an active branch in pattern recognition. A bank check is very important and widely used in our daily life. In order to automatically process it, detection of reference lines plays a key role. However, these lines belong to the narrow transient components of the image. It is difficult to process them using Fourier analysis.

We now turn to discuss the approximation of signals $f_1(x)$ and $f_2(x)$ using basic wave function $\sin x$.

Figure 1.6 consists of two columns (a)–(j). The left column (a,c,e,g,i) describes the original non-transient signal $f_1(x)$ as shown in the top, and its approximations using the partial sums of the Fourier expansions with $N = 4, 8, 15, 30$ from the top to the bottom, respectively. The right column (b,d,f,h,j) presents the transient signal $f_2(x)$ and its approximations using the partial sums of the first $N$ coefficients of Fourier expansions, $N = 4, 8, 15, 30$ from the top to the bottom, respectively.

The errors between the original signals and the partial sums of the Fourier expansions are illustrated in Fig. 1.7. The left column (a,c,e,g,i) presents the original non-transient signal $f_1(x)$ and the errors. $f_1(x)$ is located on the top, and the errors with different numbers of terms of Fourier expansions are described from the next-to-the-top to the bottom, which correspond to $n = 4, 8, 15, 30$, respectively. The right column (b,d,f,h,j) presents the original signal $f_2(x)$ and the errors with $n = 4, 8, 15, 30$ from the top to the bottom, respectively.

From the left columns (a,c,e,g,i) in Figs. 1.6 and 1.7, it is clear that, for the non-transient signal $f_1(x)$, the fore 17 terms of the Fourier expansion can approximate the original signal very well. This corresponds to $n = 8$ in Eq. (1.3). It means that the series of $\{a_0, a_1, b_1, \ldots, a_8, b_8\}$ contain the most of information in $f_1(x)$. Therefore, the representation of the non-transient signal $f_1(x)$ by the basic function $\sin x$ is extremely well founded.

Unfortunately, the situation is not so good for the transient signal $f_2(x)$. It can be found, in the right columns (b,d,f,h,j) in Figs. 1.6 and 1.7, that in the case of $n = 8$, the approximation error is still large. Although the error is relatively small when $n = 30$, it is still visible. The reason for this occurrence is that for any transient or localized signals, like $f_2(x)$, the smooth basic wave $\sin x$ cannot fit them perfectly.

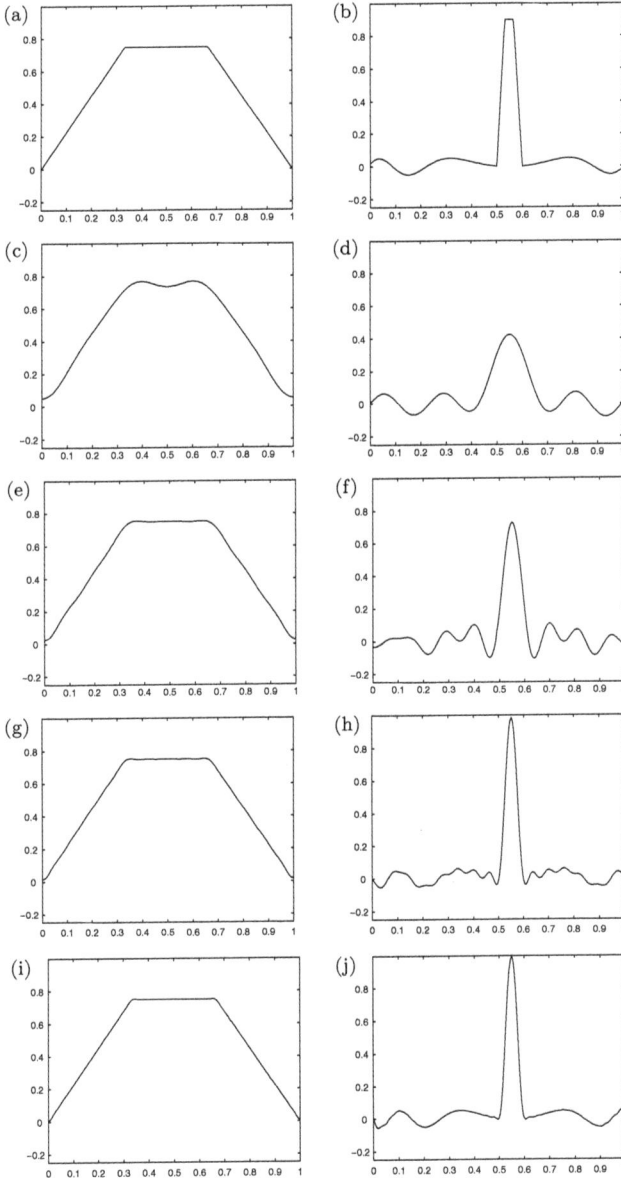

Fig. 1.6 The left column (a,c,e,g,i): original signal; $f_1(x)$ and its partial sums of the Fourier expansion with $N = 4, 8, 15, 30$ respectively. The right column (b,d,f,h,j): original signal $f_2(x)$ and its partial sums of the Fourier series of $f_2(x)$ with $N = 4, 8, 15, 30$, respectively.

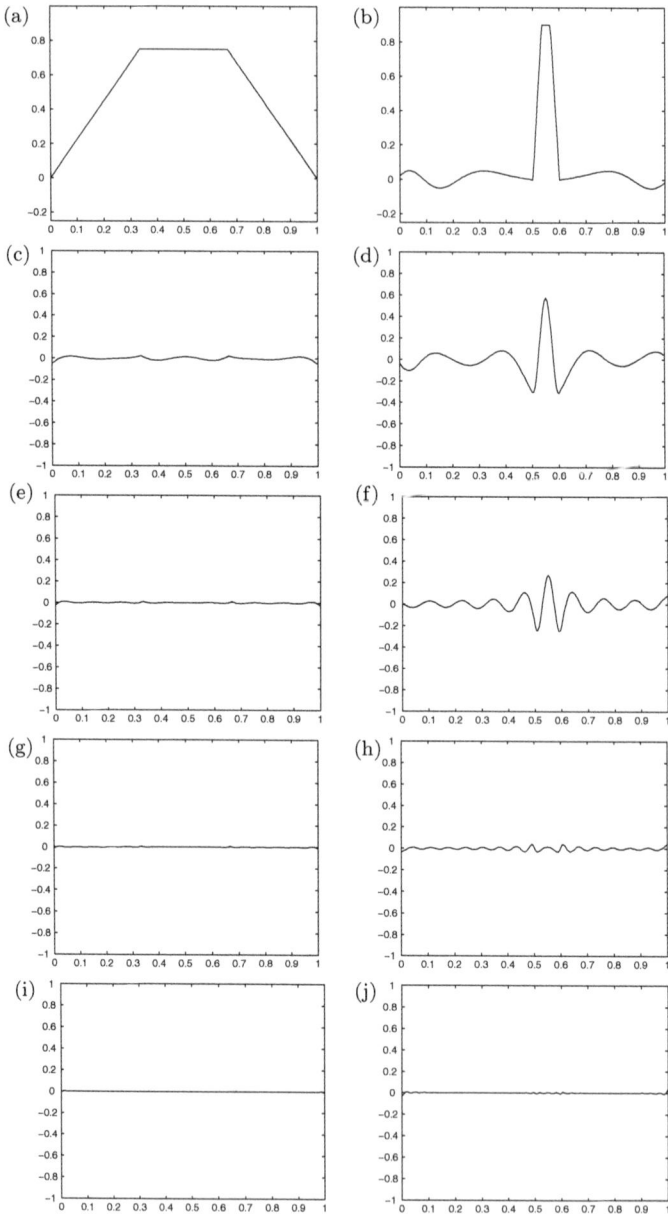

Fig. 1.7   The left column (a,c,e,g,i): original signal $f_1(x)$ and the errors with $N = 4$, 8, 15, 30, respectively; The second column (b,d,f,h,j): original signal $f_2(x)$ and the errors with $N = 4$, 8, 15, 30, respectively.

To contest the above deficiencies, we should find an other basic function $\psi(x)$ to replace the basic function $\sin x$. This new basic function should satisfy the following conditions:

- Similar to $\sin x$, any complicated signal $f(x)$ can be constructed by the linear combination of $\psi(jx - k)$ $(j, k \in \mathbb{Z})$, which are produced by the dilations and translations of the basic function $\psi(x)$.
- The expansion coefficients of a signal using the basic function $\psi(x)$ can reflect the locations of the transient or localized image components in the time domain.
- This new basic function $\psi(x)$ and its family can "fit" transient signal $f(x)$ much better than Fourier basic wave $\sin x$. In other words, they can minimize the error between the approximation of the signal $f(x)$ and $f(x)$ itself.

In fact, we already found such a basic function $\psi(x)$ as early as 1910. It is the well-known square wave function: Haar function. This basic function $\psi(x)$ can be written in the form of

$$\psi(x) = \begin{cases} 1 & x \in [0, \frac{1}{2}), \\ -1 & x \in [\frac{1}{2}, 1). \end{cases}$$

The graphic illustration of the Haar function is presented in Fig. 1.8.

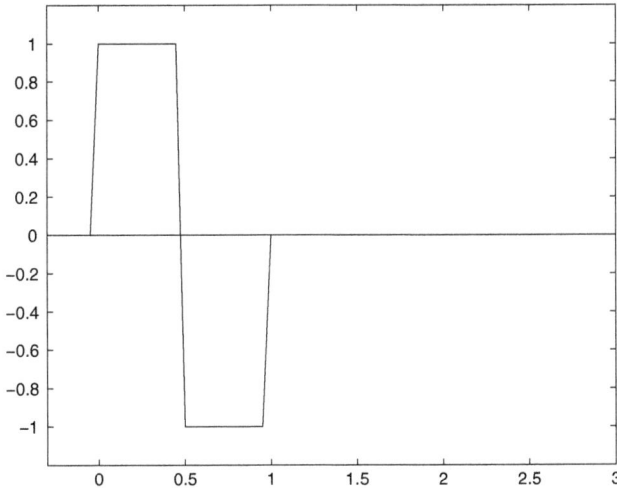

Fig. 1.8  Haar wavelet.

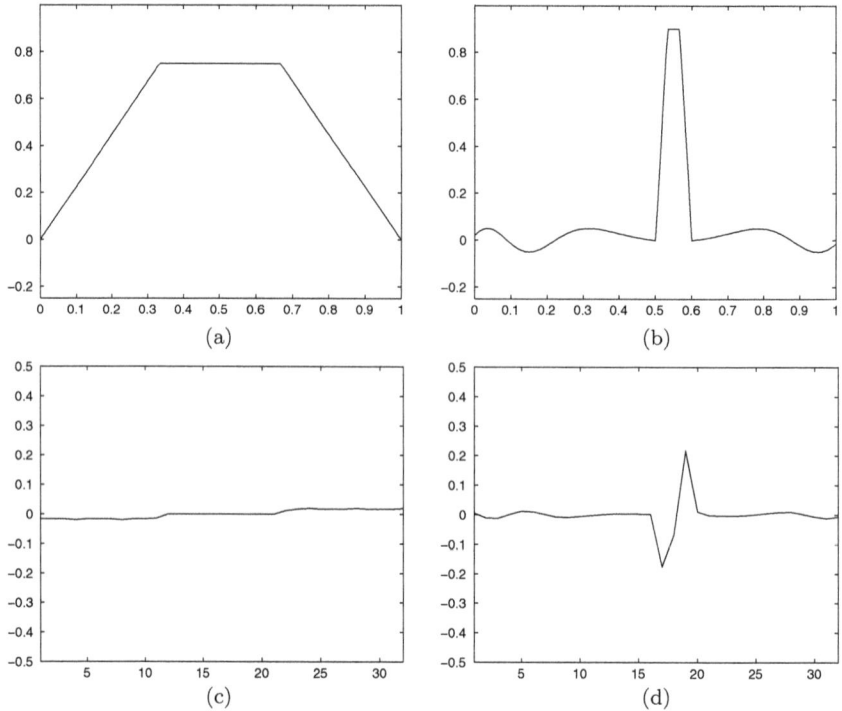

Fig. 1.9 (a) The original signal $f_1(x)$; (b) another original signal $f_2(x)$; (c) the Haar coefficients corresponding to $f_1(x)$; (d) the Haar coefficients corresponding to $f_2(x)$.

It has been mathematically proved that $\{\psi(2^j x - k)|j, k \in \mathbb{Z}\}$ can constitute an orthogonal basis of the finite energy signal space $L^2(\mathbb{R})$. They can also establish an orthonormal basis of $L^2(\mathbb{R})$ through a normalization:

$$\psi_{j,k}(x) := 2^{j/2}\psi(2^j x - k), \quad (j, k \in \mathbb{Z}). \tag{1.4}$$

Thus, any finite energy signal $f(x)$ can be represented by

$$f(x) = \sum_{j \in \mathbb{Z}} \sum_{k \in \mathbb{Z}} c_{j,k} \psi_{j,k}(x), \tag{1.5}$$

where $c_{j,k} := \int_{\mathbb{R}} f(x)\psi_{j,k}(x)dx$. Consequently, the first condition holds for the Haar wavelet.

Now, we will expand signals $f_1(x)$ and $f_2(x)$ with the Haar family. A part of the coefficients of the Haar wavelet expansion for $N = 5$ is shown in Fig. 1.9.

The original signals $f_1(x)$ (without transient components), and $f_2(x)$ (with transient components) are illustrated in Figs. 1.9(a) and 1.9(b), respectively. By applying the Haar transform to these signals, the Haar coefficients $\{c_{N,k}|k = 0,\ldots,2^N-1\}$ of signals $f_1(x)$ and $f_2(x)$ are obtained and plotted in Figs. 1.9(c) and 1.9(d) respectively. The above coefficients can successfully localize the signals. For the non-transient signal $f_1(x)$, nearly all Haar coefficients are close to zero, which implies that no high frequencies are included in the original signal. By contrast, for the transient signal $f_2(x)$, only a few coefficients are affected and produce a vibration in the frequency domain. Two peaks of the vibration in Fig. 1.9(d) correspond to just the positions of the transient components in signal $f_2(x)$ shown in Fig. 1.9(b). Meanwhile, each coefficient is just determined by the local action of the signal. In this way, the Haar family can be employed to analyze the localization of signals in the time domain. Consequently, the second condition also holds for the Haar wavelet.

Next, we turn to the verification of the third condition, i.e., we will check the question: Can we apply the Haar wavelet to approximate signals with small error?

In Eq. (1.5), we denote

$$f_0(x) = \sum_{j=-\infty}^{-1} \sum_{k\in\mathbb{Z}} c_{j,k}\psi_{j,k}(x).$$

Similar to the Fourier expansion, by considering the energy, we have

$$\int_{\mathbb{R}}\left|f(x) - \left[f_0 + \sum_{j=0}^{N}\sum_{k\in\mathbb{Z}} c_{j,k}\psi_{j,k}(x)\right]\right|^2 dx$$

$$= \min_{d_0,d_1,\ldots,d_N}\int_{\mathbb{R}}\left|f(x) - \left[f_0 + \sum_{j=0}^{N}\sum_{k\in\mathbb{Z}} d_{j,k}\psi_{j,k}(x)\right]\right|^2 dx \to 0 \quad (N\to\infty),$$

where

$$f_0(x) + \sum_{j=0}^{N}\sum_{k\in\mathbb{Z}} c_{j,k}\psi_{j,k}(x)$$

is an approximation of $f(x)$.

We take signals $f_1(x)$ and $f_2(x)$ as examples again. The approximations of these signals using the Haar family are presented graphically in Fig. 1.10. The original non-transient $f_1(x)$ and transient signals $f_2(x)$ are illustrated in Figs. 1.10(a) and 1.10(b), respectively. The approximation of $f_1(x)$ by the partial sums of the Haar expansion with $N = 4$ (32 terms) and that with $N = 5$ (64 terms) are shown in Figs. 1.10(c) and 1.10(g), respectively. The similar description of $f_2(x)$ by the partial sums of the Haar coefficients

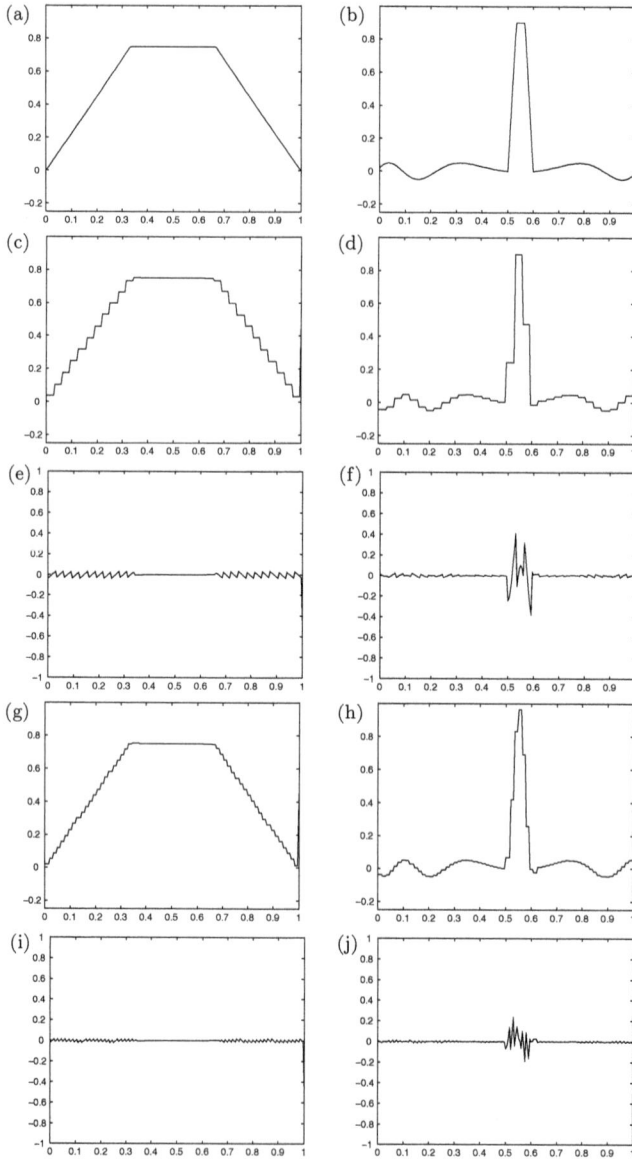

Fig. 1.10  Haar functions approximate signals: The left column (a,c,e,g,i): original non-transient signal $f_1(x)$, the partial sums of the Haar expansion with $N = 4$, 5, and the approximating errors; The right column (b,d,f,h,j): original transient signal $f_2(x)$, the partial sums of the Haar expansion with $N = 4$, 5, and the approximating errors.

with $N = 4$ and that with $N = 5$ are displayed in Figs. 1.10(d) and 1.10(h), respectively. The errors between the original signal $f_1(x)$ and its approximations using the Haar function are presented in Figs. 1.10(e) and 1.10(i), which correspond to $N = 4$ and 5, respectively. The errors between the original signal $f_2(x)$ and its approximations by the Haar coefficients are described graphically in Figs. 1.10(f) and 1.10(j) corresponding to $N = 4$ and 5.

Now, we compare the result using $\sin x$ with that using the Haar function in the case of $N = 4$. As shown in Figs. 1.6(d) and 1.10(h), when the Haar function is employed to approximate a transient signal, the result seems better than that when we use $\sin x$ due to the strong localization of the Haar family. Mathematically, it is said that the Haar function is compactly supported. Unfortunately, it can be found, from Fig. 1.10(g), that when we approximate a non-transient signal with Haar wavelet, the result is not so good as Fourier expansion. In this situation, saw-tooth-wave errors occur due to the discontinuity of the Haar function. In summary, compared to $\sin x$, the Haar wavelet has much better localization and rather worse smoothness. That is why the Haar wavelet did not attract enough attention for a long time. Before various wavelet bases were constructed in the 1980s, it was only a dream to find out a basic wavelet $\psi(x)$, which has both strong localization and smoothness. At the end of the 1980s, wavelets achieved breakthrough development in both theory and application.

It has followed a long and tortuous course from Fourier analysis to wavelet analysis. In fact, as early as 1946, Gabor [1946] has found out that Fourier analysis is lacking the capability to localize signals in the time domain. He applied the Gaussian function as a "window" to improve the Fourier transform. It is referred to as Gabor transform and is applied to analyze the transient signals. Thereafter, Gabor transform led to the general window Fourier transform (or short-time Fourier transform) by the replacement of the Gaussian function with other localized window functions. However, the size of the window in the window Fourier transform is fixed, and it did not improve Fourier transformation thoroughly. Therefore, in this book, we will not discuss it in detail.

Today, wavelet analysis has become an international focus in many research fields and applications. Scientists and engineers all over the world are working on wavelets widely and deeply in mathematics, engineering and many other disciplines.

The fundamental difference between the signal decompositions by wavelet transform and Fourier transform is depicted in Fig. 1.11. A singer

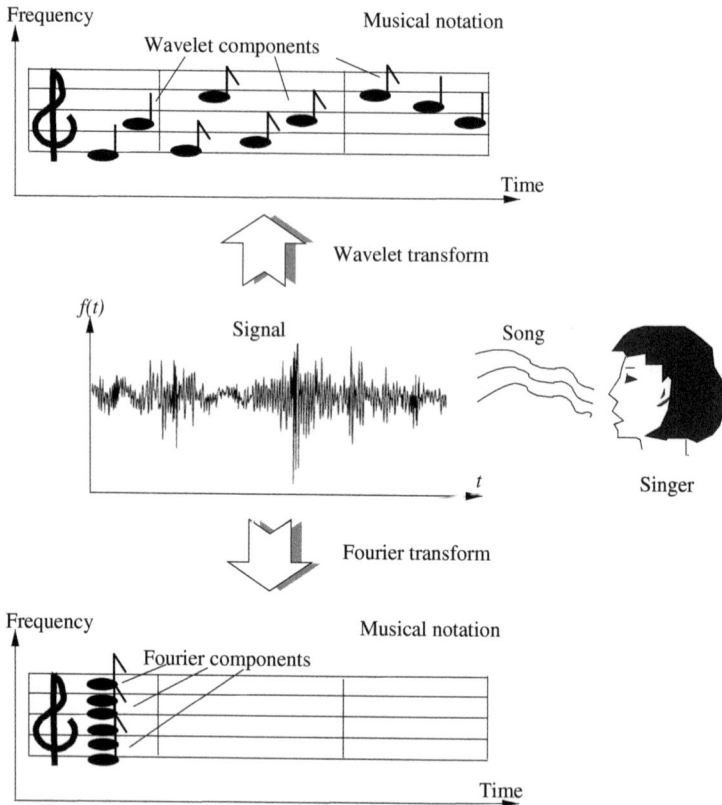

Fig. 1.11   Fundamental difference between the signal decompositions by wavelet transform and Fourier transform: A singer sings a song, which is the signal $f(t)$ of time $t$. It can be decomposed into the musical notes by wavelet transform due to the time–frequency localization. However, it would fail if the Fourier transform is utilized.

sings a song, which is the signal $f(t)$ of time $t$. It can be decomposed into musical notes by wavelet transform due to the property of the time–frequency localization of the wavelet transform. The musical notation can be viewed as depicting a 2D time–frequency space. Frequency (pitch) increases from the bottom of the scale to the top, while time (measured in beats) advances to the right. Each note on the sheet music corresponds to one wavelet component, which appears in the recording of a performance of the song. If we were to analyze a recorded musical performance and write out the corresponding score, we would have a type of wavelet transform. Similarly, a recording of a musician's performance of a song can be viewed as an inverse wavelet transform, since it reconstructs the signal from a time–frequency representation. However, it would fail if we use the Fourier transform to do so.

## 1.2 Brief Review of Pattern Recognition with Wavelet Theory

Wavelet theory is a versatile tool with very rich mathematical content and great applications. It has been employed in many fields and applications, such as signal processing, image analysis, communication systems, biomedical imaging, radar, air acoustics, theoretical mathematics, control system, and endless other areas. A lot of achievements have been made, for instance, many of them have been published [Auslander *et al.*, 1990; Beylkin *et al.*, 1991; Chui, 1992a; Daubechies, 1990; Grossmann and Morlet, 1984; IEEE, 1993; Mallat, 1989b; SPIE, 1994; Daugman, 2003; Kumar and Kumar, 2005; Shankar *et al.*, 2007; You and Tang, 2007; Li, 2008]. However, the research on applying the wavelets to pattern recognition is still too weak; not too many research projects deal with this topic at the present.

In this section, a brief survey of pattern recognition with the wavelet theory is presented.

As for the applications of wavelet theory to pattern recognition, we can consider them to be two hands: (1) system-component-oriented and (2) application-task-oriented. Figure 1.12 displays these two sides. On the left-hand side, four components are enclosed in a recognition system. On the right-hand side, the application-task-oriented ones can be categorized into the following groups:

- Iris pattern recognition [Daugman, 2003],
- Face recognition using wavelet transform [Kouzani and Ong, 2003; Lai *et al.*, 1999; Yang *et al.*, 2003b],
- Hand gestures classification [Kumar and Kumar, 2005; Kumar *et al.*, 2003; Sharnia *et al.*, 2004],
- Classification and clustering [Murtagh and Starck, 1998; Shankar *et al.*, 2007; Tang and Ma, 2000],
- Document analysis with wavelets [Liang *et al.*, 1999; Tang *et al.*, 1996a, 1997a, 1995a, 1997c],
- Analysis and detection of singularities with wavelets [Mallat and Hwang, 1992; Young, 1993; Chen *et al.*, 1995; Chen and Yang, 1995; Deng and

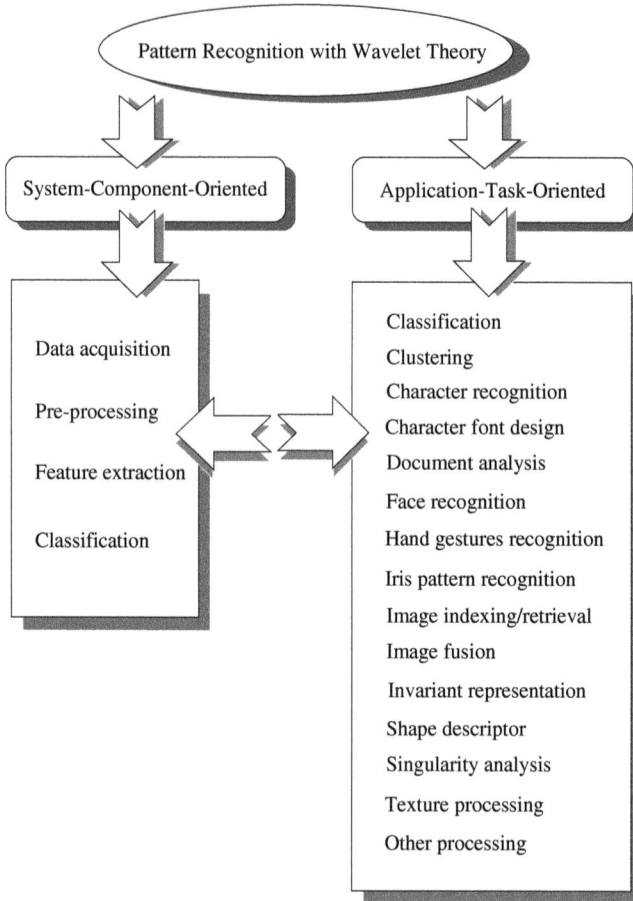

Fig. 1.12    Pattern recognition with wavelet theory.

Lyengar, 1996; Law *et al.*, 1996; Tang *et al.*, 1997c, 1998d, 2000; Thune *et al.*, 1997; Tieng and Boles, 1997a; Tang and You, 2003],

- Wavelet descriptors for shapes of the objects [Chuang and Kuo, 1996; Tang *et al.*, 1998a, 1999; Tieng and Boles, 1997b; Wunsch and Laine, 1995; Yang *et al.*, 2003a; You and Tang, 2007],
- Invariant representation of patterns [Haley and Manjunath, 1999; Shen and Ip, 1999; Tang *et al.*, 1998a; Yang *et al.*, 2003a; Yoon *et al.*, 1998],
- Handwritten and printed character recognition [Lee *et al.*, 1996; Tang *et al.*, 1998a,d, 1999, 1996b; Wunsch and Laine, 1995; Kunte and Samuel, 2007; You and Tang, 2007],

- Texture analysis and classification [de Wouwer *et al.*, 1999a,b; Haley and Manjunath, 1999; Liang and Tjahjadi, 2006; Muneeswaran *et al.*, 2005],
- Image indexing and retrieval [Jain and Merchant, 2004; Ksantini *et al.*, 2006; Kubo *et al.*, 2003; Moghaddam *et al.*, 2005; Special-Issue-Digital-Library, 1996; Smeulders *et al.*, 2000],
- Wavelet-based image fusion [El-Khamy *et al.*, 2006; Li, 2006, 2008],
- Character processing with B-spline wavelet transform [Yang *et al.*, 1998],
- Others [Chambolle *et al.*, 1998; Chen and Yang, 1995; Combettes and Pesquet, 2004; Combettes, 1998; Liao and Tang, 2005; You *et al.*, 2006]

It is clear that the two sides are related to each other. Each group of the right-hand side relates to the component(s) on the left-hand side. For instance, the tasks in the shape descriptor, character recognition, etc. concern all the components of the pattern recognition system; Singularity detector and invariant representation are related with the tasks, which are carried out in the component of the feature extraction. In this section, we discuss the applications of wavelet theory to pattern recognition by means of the right-hand side, i.e., in the application-task-oriented point of view.

## 1.2.1 *Iris Pattern Recognition*

The highest density of biometric degrees of freedom, which is both stable over time and easily measured, is to be found in the complex texture of the iris pattern of the eye [Daugman, 2003]. An example of the human iris pattern is presented in Fig. 1.13(a), which is imaged in near-infrared light at a distance of 30 cm. It can serve as a unique and reliable identifier.

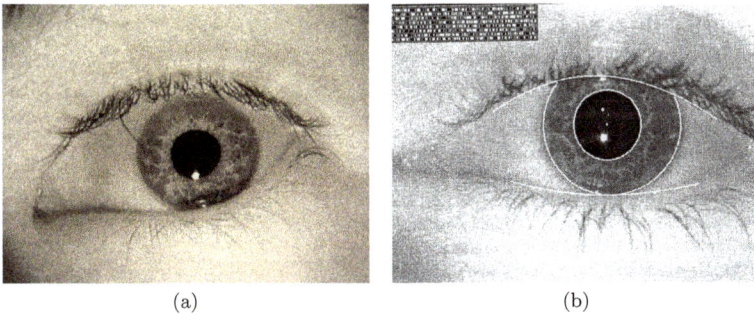

(a)                                    (b)

Fig. 1.13  (a) Example of a human iris pattern; (b) isolation of an iris for encoding and its resulting "IrisCode" [Daugman, 2003].

Daugman [2003] applies complex-valued 2D wavelet to iris pattern recognition. Let $\Psi(x, y)$ be a 2D mother wavelet, we can generate a complete self-similar family of parametrized daughter wavelets $\Psi_{muv\theta}(x, y)$ by

$$\Psi_{muv\theta}(x, y) = 2^{-2m}\Psi(x', y'),$$

$$x' = 2^{-2m}\left[x\cos(\theta) + y\sin(\theta)\right] - u,$$

$$y' = 2^{-2m}\left[-x\sin(\theta) + y\cos(\theta)\right] - v,$$

where the substituted variables $x', y'$ incorporate dilation of the wavelet in size by $2^{-2m}$, translations in position $(u, v)$, and rotations through angle $\theta$.

In the complex-valued 2D wavelet, it is possible to use the real and imaginary parts of their convolution ($*$) with an iris image $f(x, y)$ to extract a description of image structure in terms of local modulus and phase. Let $\Psi_{muv\theta}(x, y)$ be 2D daughter wavelet we used, the amplitude modulation function $A(x, y)$ and phase modulation function $\theta(x, y)$ can be presented, respectively:

$$A(x, y) = \sqrt{(\Re\{\Psi_{muv\theta}(x, y) \star f(x, y)\})^2 + (\Im\{\Psi_{muv\theta}(x, y) \star f(x, y)\})^2}$$

and

$$\theta(x, y) = \tan^{-1}\frac{\Im\{\Psi_{muv\theta}(x, y) \star f(x, y)\}}{\Re\{\Psi_{muv\theta}(x, y) \star f(x, y)\}}.$$

The following operations are accomplished to localize precisely the inner and outer boundaries of the iris and thereafter to detect eyelids if they intrude and exclude them:

$$\max_{r, x_0, y_0}\left|G_\sigma(r) \star \frac{\partial}{\partial r}\oint_{r, x_0, y_0}\frac{f(x, y)}{2\pi r}ds\right|,$$

where contour integration parametrized for size and local coordinates $r$, $x_0$, $y_0$ at a scale of analysis $\sigma$ set by some blurring function $G_\sigma(r)$ is performed over the iris pattern $f(x, y)$. The result of this optimization search is the determination of the circle parameters $r$, $x_0$, $y_0$, which best fit the inner and outer boundaries of the iris. Figure 1.13(b) shows the isolation of an iris for encoding and its resulting "IrisCode" [Daugman, 2003].

## 1.2.2 *Face Recognition Using Wavelet Transform*

**(1) Wavelet-based PCA Approach:** Principal component analysis (PCA) is one of the well-known methods for representing the human face. Sirovich and Kirby [1987] first proposed to use Karhunen–Loeve (K–L) transform to represent human faces. In their method, faces are represented by a linear combination of weighted eigenvector, known as eigenface features. In 1991, Turk and Pentland [1991] developed a face recognition system using PCA (K–L expansion). In 1993, O'Toole *et al.* [1993] demonstrated that whereas the low-dimensional representation is optimal for identifying physical categories of face, such as gender and race, it is not optimal for recognizing a human face. In 1996, Swets and Weng [1996] combined the theories of K–L projection and multi-dimensional discriminant analysis to generate a set of the most discriminating features for recognition.

Very good results are obtained when applying this approach to frontal-view head-and-shoulder images. However, if the face is with different orientations, facial expressions or occluded, the accuracy will be degraded dramatically. Nastar and Ayache [1996] investigated the relationship between variations in facial appearance and their deformation spectrum. They found that facial expressions and small occlusion affect the intensity manifold locally. Under frequency-based representation, only high-frequency spectrum called high-frequency phenomenon is affected. Moreover, changes in pose or scale of a face affect the intensity manifold globally, in which only their low-frequency spectrum, called low-frequency phenomenon is affected. Only a change in the face will affect all frequency components.

The results of Nastar and Ayache [1996] shed light on how to solve the problem in facial appearance variations — use mid-range frequency for recognition. Yuen *et al.* [1998] recently demonstrated that applying PCA method on wavelet transformed (WT) sub-image with mid-range frequency components gives better recognition accuracy than (1) applying PCA on WT sub-image with low frequency components only or (2) applying PCA on the original image that contains all frequency components.

In view of these, Lai *et al.* [1999] combine Fourier transform and wavelet transform for face recognition. Their method is described as follows. First, a compact support and orthogonal wavelet is applied to decompose the face image. The low-frequency subband of the decomposed is selected to represent his mother image. It is an optimal approximate version of the mother image in a lower dimension. Second, Fourier transform (FT) is applied to the low-frequency subband images and represents them.

In Lai *et al.* [1999], Yale and Olivetti databases are selected to evaluate their method. One person from each database is shown in Figs. 1.14 and 1.15. Spectroface with different levels of wavelet decomposition of one face image from the Yale database is presented in Fig. 1.16.

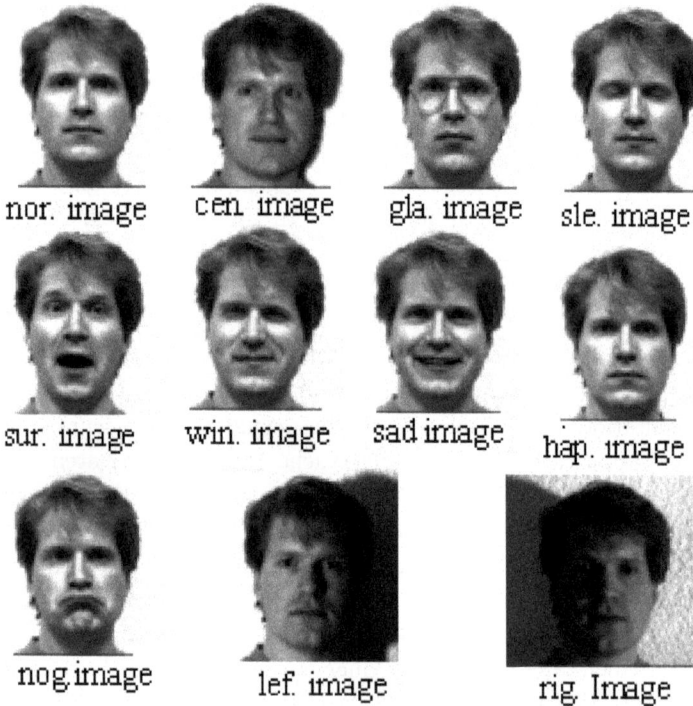

| nor. image | cen. image | gla. image | sle. image |

| sur. image | win. image | sad image | hap. image |

| nog.image | lef. image | rig. Image |

Fig. 1.14   Images of one person in the Yale database.

Following are reported:

- The performance of Spectroface is better than that of Eigenface and Template matching, especially for facial expression and occlusion (by glasses) images.
- The Spectroface approach is invariant to spatial translation. Both Eigenface and Template matching methods are sensitive to the spatial translation.
- The computation complexity of Spectrofaces is dependent on wavelet packet. Expanding $N$ vectors $\{X_n \in R^d : n = 1, 2, \ldots, N\}$ into wavelet packet coefficients results in $O(\text{Nd} \log d)$.

**(2) Nonlinear Wavelet Approximation Method:** For a 2D orthonormal wavelet basis, let $\psi^1$, $\psi^2$ and $\psi^3$ be three 2D wavelets, $\psi^0 := \phi$ be the corresponding scaling function, and

$$\psi_j^{k,e} := 2^{k/2}\psi^e(2^k x - j), \; \forall e \in \Re_k,$$

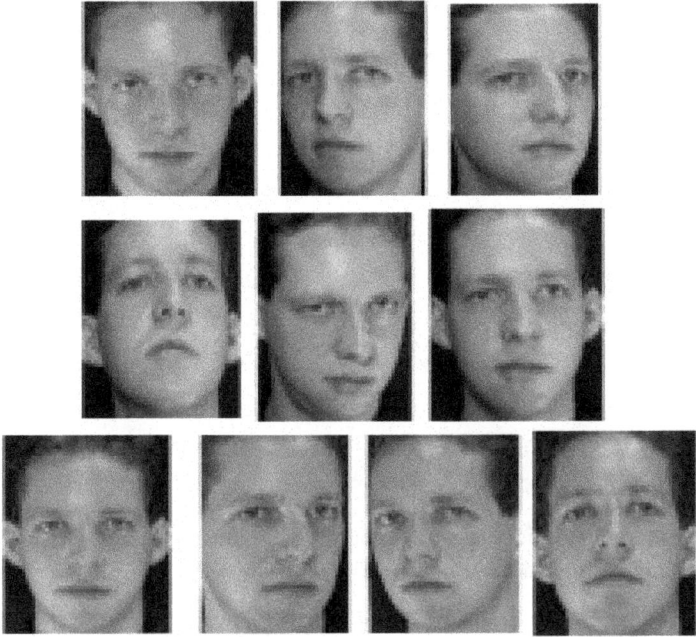

Fig. 1.15   Images of one person in the Olivetti database.

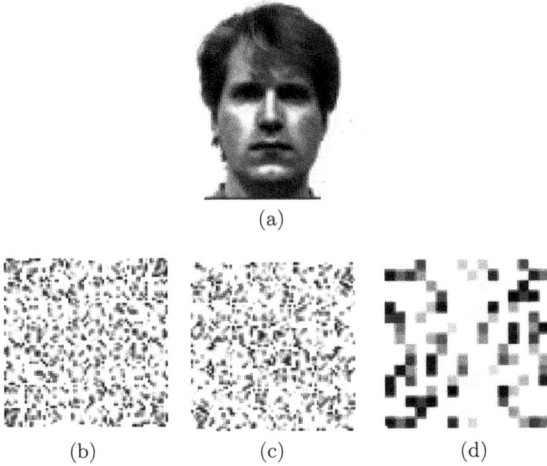

(a)

(b)          (c)          (d)

Fig. 1.16   (a) The original image; (b) the frequency representation of image (a); (c) Spectroface with one-level wavelet decomposition; (d) Spectroface with three-level wavelet decomposition.

where

$$\Re_k := \begin{cases} \{1, 2, 3\}, & k > 0, \\ \{0, 1, 2, 3\}, & k = 0. \end{cases}$$

Thus, any $f \in L^2(I)$ can be written by

$$f = \sum_{k \geq 0} \sum_{j} \sum_{e \in \Re_k} c_j^{k,e} \psi_j^{k,e}.$$

Let $X = span\{\psi_j^{k,e} : e \in \Re_k, \ 0 \leq k < K, \ j \in Z^2\}$, then the approximation element

$$\tilde{f} = \sum_{0 \leq k < K} \sum_{j} \sum_{e \in \Re_k} c_j^{k,e} \psi_j^{k,e}$$

consists of

$$N = 3(2^{K-1} \cdot 2^{K-1} + 2^{K-2} \cdot 2^{K-2} + \cdots + 2^1 \cdot 2^1 + 2^0 \cdot 2^0) = 2^{2K}$$

terms in the sum. It is the unique best approximation element of $f$ in $X$, and the linear approximation is presented by

$$\|f - \tilde{f}\|_{L_2(I)} \leq N^{-\alpha/2} \|f\|_{W^\alpha(L^2(I))}.$$

For any sequence $\{c_j^{k,e}\}$, let $\aleph(\{c_j^{k,e}\})$ be the number of non-zero items. We can denote

$$\sum_N := \left\{ \sum_{k \geq 0} \sum_{j} \sum_{e \in \Re_k} c_j^{k,e} \psi_j^{k,e} | \aleph(\{c_j^{k,e}\}) \leq N \right\}.$$

For any

$$f = \sum_{k \geq 0} \sum_{j} \sum_{e \in \Re_k} c_j^{k,e} \psi_j^{k,e},$$

we keep $N$ coefficients with larger amplitudes, eliminating the small ones. Thus, an approximation $\tilde{f}$ can be obtained, and $\tilde{f} \in \sum_N$. Finally, the nonlinear wavelet approximation can be arrived at:

$$\|f - \tilde{f}\|_{L_2(I)} \leq N^{-\alpha/2} \|f\|_{B_\sigma^\alpha(L^\sigma(I))},$$

where $B_\sigma^\alpha(L^\sigma(I))$ is a Besov space, $0 < \alpha < 1$ and $\sigma = 2/(1 + \alpha)$.

Yang *et al.* [2003b] propose a method for face recognition using the nonlinear wavelet approximation. As mentioned by the authors, the linear wavelet approximation cannot reduce the approximation error by adding more coefficients. The nonlinear wavelet approximation has an important advantage over the linear one due to its ability to do so.

**(3) Wavelet Packet Transform for Classification of Facial Images:**
Kouzani and Ong [2003] utilize wavelet packet transform for lighting-effects
classification in facial images. The wavelet packet transform is a generalization of the wavelet transform. In the wavelet transform, only the low-pass
filter is iterated. It is assumed that lower frequencies contain more important information than higher frequencies. This assumption is not true for
many images. The main difference between the wavelet packet transform
and the wavelet transform is that, in the wavelet packets, the basic two-
channel filter bank can be iterated either over the low-pass branch or the
high-pass branch. This provides an arbitrary tree structure with each tree
corresponding to a wavelet packet basis. It can offer a choice of optimal
bases for the representation of specific signals. The basis can be selected
to minimize the number of significantly non-zero coefficients in the transform [Kouzani and Ong, 2003]. Entropy is a suitable function for choice
of the best basis. Shannon's equation for entropy enables searching for the
smallest entropy expansion of a signal. It is presented as follows:

$$\epsilon^2(v; \{H_i, H_j\}) = -\frac{||v_+||^2}{||v||^2} \ln \left( \frac{||v_+||^2}{||v||^2} \right) - \frac{||v_-||^2}{||v||^2} \ln \left( \frac{||v_-||^2}{||v||^2} \right)$$
$$+ ||v_+||^2 \epsilon^2 \left( \frac{v_+}{||v_+||^2}, \{H_i\} \right)$$
$$+ ||v_-||^2 \epsilon^2 \left( \frac{v_-}{||v_-||^2}, \{H_j\} \right),$$

where $H$ is a Hilbert space, $v \in H$, $||v|| = 1$ and $H = \oplus \sum H_i$ is an
orthogonal decomposition of $H$.

The best basis algorithm [Kouzani and Ong, 2003] minimizes the cost
function for the transform coefficients. It takes a complete decomposition
according to the wavelet packet transform. In each node, which corresponds
to a subspace of the image, the cost of the coefficient of the subspace is
calculated and iterated.

### 1.2.3 *Hand Gestures Classification*

**(1) Visual Hand Gestures Classification Using Wavelet Trans-
forms:** Interaction is a common activity in our daily lives. Hand actions
play a very important role in the interaction. To improve human–machine
interaction and for helping disabled people, it is desirable for machines to
extract more information from human hand actions [Kumar *et al.*, 2003].
Some examples of hand gestures are illustrated in Fig. 1.17. The meanings

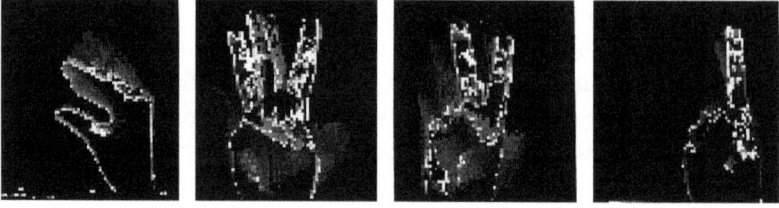

Fig. 1.17    Examples of human hand gestures [Kumar *et al.*, 2003].

of these gestures are "hold", "three", "victory" and "point" from left to right.

Kumar *et al.* [2003] present a novel technique for classifying human hand gestures based on stationary wavelet transform (SWT). It uses the view-based approach for the representation of hand action and artificial neural networks (ANNs) for classification. This approach applies a cumulative image-difference method in the representation of action, which results in the construction of motion history image (MHI). Let $I(x, y, n)$ be an image sequence and let

$$D(x, y, n) = |I(x, y, n) - I(x, y, n - 1)|,$$

where $I(x, y, n)$ is the intensity of each pixel at location $(x, y)$ in the $n$th frame and $D(x, y, n)$ is the difference of consecutive frames representing regions of motion. Binarization of the difference image $D(x, y, n)$ over a threshold $\tau$ is $B(x, y, n)$

$$B(x, y, n) = \begin{cases} 1 & \text{if } D(x, y, n) > \tau, \\ 0 & \text{otherwise.} \end{cases}$$

The motion history image $\text{MHI}(H_N(x, y))$ is

$$\text{MHI}(H_N(x, y)) = \max\left(\bigcup_{n=1}^{N-1} B(x, y, n) \times n\right),$$

where $N$ represents the duration of the time window used to capture the motion.

These MHIs are decomposed into four sub-images $(f_{ll},\ f_{lh},\ f_{hl},\ f_{hh})$ using stationary wavelet transform. The average image $(f_{ll})$ is fed as the global image descriptor to the ANNs for classification. The schematic diagram of the approach is illustrated in Fig. 1.18.

**(2) Hand Gesture Classification by Wavelet Transforms and Moment-Based Features:** Kumar and Kumar [2005] propose a novel

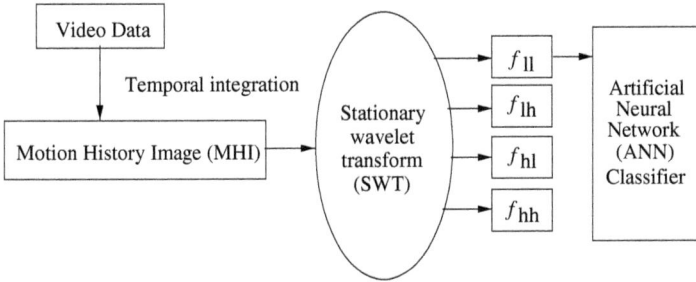

Fig. 1.18   The schematic diagram of the approach [Kumar *et al.*, 2003].

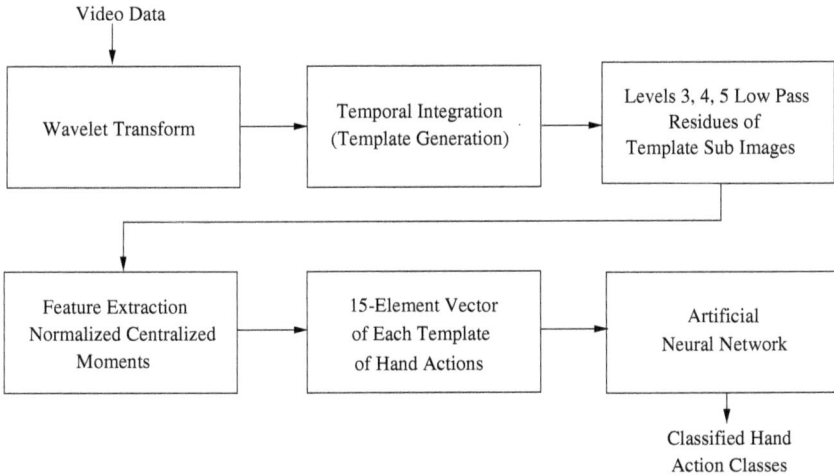

Fig. 1.19   The schematic diagram of the wavelet-moment approach [Kumar and Kumar, 2005].

technique for classifying human hand gestures based on SWT and geometric-based moments. According to uniqueness theory of moments for a digital image of size $(N, M)$, the $(p + q)$th-order moments $m_{pq}$ are calculated by

$$m_{pq} = \frac{1}{NM} \sum_{x=1}^{N} \sum_{y=1}^{M} f(x, y) x^p y^q,$$

where $p, q = [0, 1, 2, \ldots, n]$.

The schematic diagram of the wavelet-moment approach is illustrated in Fig. 1.19.

**(3) Wavelet Directional Histograms of The Spatio-Temporal Templates of Human Gestures:** Sharnia *et al.* [2004] evaluate the efficacy of directional information of wavelet multiresolution decomposition to enhance histogram-based classification of human gestures. The gestures are represented by spatio-temporal templates. This template collapses the spatial and temporal components of motion into a static grayscale image such that no explicit sequence matching or temporal analysis is required, and it reduces the dimensionality to represent motion. These templates are modified to be invariant to translation and scale. Two-dimensional, three-level dyadic wavelet transforms are applied on the template, resulting in one low-pass sub-image and nine high-pass directional sub-images. Histograms of wavelet coefficients at different scales are used for classification purposes. The experiments demonstrate that while the statistical properties of the template provide high level of classification accuracy, the global detail activity available in high-pass decompositions significantly improve the classification accuracy.

### 1.2.4 *Classification and Clustering*

The classification process can be categorized into two types:

1. **Classification with supervised learning:** In this type of classification, there is a supervisor to teach the recognition system how to classify a known set of patterns, and thereafter, it lets the system go ahead freely to classify other patterns. In this way, *a priori* information is needed to form the basis of the learning [Bow, 1992].
2. **Classification with non-supervised learning:** This classification process is not dependent on *a priori* information. Clustering is the nonsupervised classification, which involves the process of generating classes without any *a priori* knowledge about the patterns; neither can proper training pattern sets be obtained.

As for the supervised classification with the wavelet theory, there is Chapter 14 in this book to discuss it. For the non-supervised classification, i.e., the clustering, this section gives an example [Murtagh and Starck, 1998], where the wavelet theory is applied to clustering.

**(1) Classifier Design Based on Orthogonal Wavelet Series:** Tang and Ma [2000] discuss the supervised classification, it providing an in-depth examination of the classifier design problem. First, an overview of the fundamentals in pattern classifier design is presented. In so doing, the emphasis is on *minimum average-loss classifier design* and *minimum error-probability classifier design*. Next, the use of orthogonal wavelet series in classifier design is specifically described and discussed. It addresses the issue of how to derive a probability density estimate based on orthogonal wavelet series. From the multiresolution theory in wavelet analysis, it is know that

$$L^2(R) = \overline{\bigcup_m V_m}.$$

Let $p_m(x)$ denote the orthogonal project of $p(x)$ in space $V_m$. Thus, it can be found that

$$(L^2) \lim_{m \to \infty} p_m(x) = p(x), \quad p_m(x) = \sum_{n=-\infty}^{+\infty} a_{mn} 2^{m/2} \phi(2^m x - n).$$

The minimum mean square error estimator of $p_m(x)$ can be written as

$$\hat{p}_m(x) = \sum_{n=-\infty}^{+\infty} \left[ \frac{1}{N} \sum_{i=1}^{N} 2^{m/2} \phi(2^m X_i - n) \right] 2^{m/2} \phi(2^m x - n).$$

A theorem is proved by Tang and Ma [2000], which indicates that when scaling function $\phi(x)$ and unknown density function $p(x)$ satisfy certain specific properties, orthogonal wavelet series density estimator $\hat{p}_m(x)$ will converge to $p(x)$.

Let scaling function $\phi(x) \in S_r$, and for a certain $\lambda \geq 1$, it satisfies property $Z_\lambda$. Let $X$ be a continuous bounded density function random variable, and $X_1, X_2, \ldots, X_N$ be $N$ independent identically distributed samples of $X$. Thus, if

$$p(x) \in H^\alpha, \quad \alpha > \lambda + \frac{1}{2}, m \approx lgN/(2\lambda + 1)lg2,$$

then

$$E|\hat{p}_m(x) - p(x)|^2 \leq O(2^{-2m\lambda}).$$

From the discussions by Tang and Ma [2000], we can note that the orthogonal wavelet series estimator differs from the kernel estimator and the traditional orthogonal series density estimator. Its basic idea shares some similarities with that of the traditional orthogonal series density estimator. However, it also satisfies several key properties of kernel estimator and exhibits some additional features. Generally speaking, the orthogonal wavelet series density estimator represents a new non-parametric way of estimating density functions, which has great potential for practical applications. For instance, in pattern classifier design, sometimes, the probability density function, $p(x)$, of a certain feature vector may not be available. In such a case, we can readily replace $p(x)$ with $\hat{p}_m(x)$ using the above-described orthogonal wavelet series density estimator and thus effectively design the classifiers.

**(2) Neuro-Wavelet Classifier for Multispectral Remote Sensing Images:** A neuro-wavelet supervised classifier is proposed by Shankar *et al.* [2007] for land cover classification of multispectral remote sensing images. Features extracted from the original pixel information using wavelet transform (WT) are fed as input to a feed-forward multi-layer neural network (MLP). The WT basically provides the spatial and spectral features of a pixel along with its neighbors and these features are used for improved classification. For testing the performance of the proposed method, two IRS-1A satellite images and one SPOT satellite image are used. Results are compared with those of the original spectral feature-based classifiers and found to be consistently better. The simulation study revealed that Biorthogonal 3.3 (Bior3.3) wavelet in combination with MLP performed better compared to all other wavelets. Results are evaluated visually and quantitatively with two measurements, $\beta$ index of homogeneity and Davies–Bouldin (DB) index for compactness and separability of classes. Shankar *et al.* [2007] suggested a modified $\beta$ index in accessing the percentage of accuracy ($PA_\beta$) of the classified images also.

**(3) Pattern Clustering Based on Noise Modeling in Wavelet Space:** Point pattern clustering has constituted one of the major strands in Cluster analysis. Murtagh and Starck [1998] describe an effective approach to object or feature detection in point patterns via noise modeling. Two advantages arrive at in this work: (1) a multiscale approach with wavelet transform is computationally very efficient and (2) a direct treatment of noise and clutter, which leads to improved cluster detection. In this approach, the noise modeling is based on a Poisson process, and a non-pyramidal (or redundant) wavelet transform is applied.

Given a planar point pattern, a 2D image is created by the following:

- producing the tuple $(x, y, 1)$ when a point at $(x, y)$ with value one;
- projecting onto a plane by using a regular discrete grid (image) and assigning the contribution of points to the image pixels by an interpolation function used by a wavelet transform referred to as *à trous* algorithm with a cubic B-spline;
- Employing the *à trous* algorithm to the resulting image: The significant structures are extracted at each wavelet decomposition level, according to the noise model for the original image, $(x, y, 1)$.

A detailed description of the non-pyramidal (or redundant) wavelet transform can be found in the work of Shensa [1992]. A summary of the "*à trous*" wavelet transform is presented in the following:

**Step 1** Initialize $i$ to 0, starting with an image $c_i(k)$, i.e. $c_0(k)$, which is the input image. The index $k$ ranges over all pixels in the image.

**Step 2** Increment $i$ and thereafter carry out a discrete convolution of the image with a filter $h$ to obtain $c_{i-1}(k)$. The distance between a central pixel and adjacent ones is $2^{i-1}$. Note that the filter $h$ is based on a cubic B-spline (5×5 filter).

**Step 3** Obtain the discrete wavelet transform, $w_i(k) = c_{i-1}(k) - c_i(k)$.

**Step 4** Return to step 2 if $i$ is less than the number of resolution levels wanted (let $p$ be the number of resolution levels).

As a result, the following set, representing the wavelet transform of the image, is produced:

$$W = \{w_0, w_1, \ldots, w_p, c_p\},$$

where $c_p$ denotes a residual. Therefore, the following additive decomposition can be applied to the input image, $c_0(k)$:

$$c_0(k) = c_p + \sum_{i=1}^{p} w_i(k).$$

Figure 1.20 shows a point pattern set, which is the simulated Gaussian cluster with 300 and 259 points, and background Poisson noise with 300 points. Figure 1.21 presents the corresponding wavelet transform. Wavelet scales 1–6 are shown in sequence, left to right, starting at the upper left corner.

Images and sets of point patterns generally contain noise. Thus, the wavelet coefficients are noisy too. In most applications, it is necessary to

Fig. 1.20    An example of the point pattern [Murtagh and Starck, 1998].

know if a wavelet coefficient is produced from the signal (i.e., it is significant) or due to the noise. Murtagh and Starck [1998] develop a statistical significance test to treat this problem. It defines that

$$P = Prob(|w_j| < \tau),$$

where $\tau$ denotes the detection threshold, which is defined for each scale. Given an estimation threshold, $\epsilon$, if $P < \epsilon$, the wavelet coefficient value cannot be due to the noise along, and a significant wavelet coefficient can be detected.

The multiresolution support can be obtained by detecting the significant coefficients at each scale level. The multiresolution support is defined as follows [Starck *et al.*, 1995]:

$$M(j, x, y) = \begin{cases} 1 & \text{if } w_j(x,y) \text{ is significant,} \\ 0 & \text{if } w_j(x,y) \text{ is not significant.} \end{cases}$$

The algorithm to create the multiresolution support is presented as follows:

**Step 1** Compute the wavelet transform of the image.

**Step 2** Estimate the noise standard deviation at each scale and thereafter deduce the statistically significant coefficients at each scale level.

**Step 3** Booleanize each scale, which can lead to the multiresolution support.

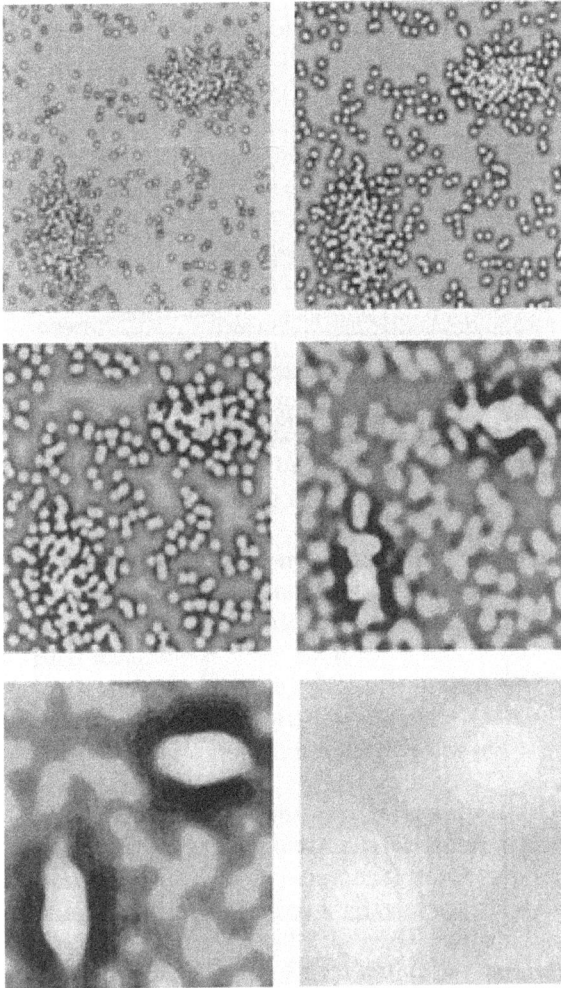

Fig. 1.21 Wavelet transform of the point pattern with scales of 1–6 [Murtagh and Starck, 1998].

**Step 4** Modify using *a priori* knowledge if desired.

In order to visualize the multiresolution support, we can create an image $S$ defined by

$$S(x,y) = \sum_{j=1}^{p} 2^j M(j,x,y). \qquad (1.6)$$

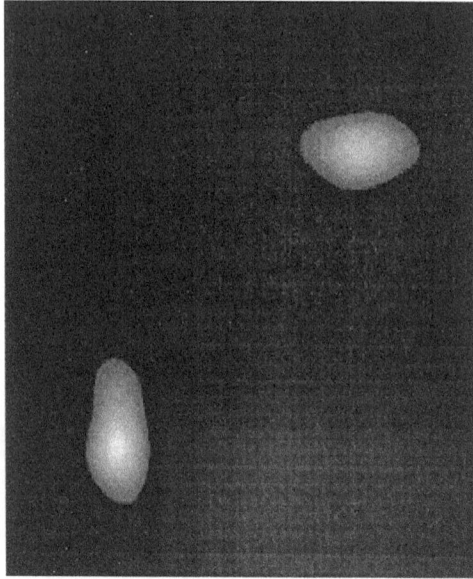

Fig. 1.22    The multiresolution support image at the fifth scale level of the wavelet transform of the point pattern shown in Fig. 1.20 [Murtagh and Starck, 1998].

Figure 1.22 is related to the multiresolution support image of the fifth scale image in Fig. 1.21. Murtagh and Starck [1998] use this method to carry out three examples: (1) excellent recovery of Gaussian clusters, (2) diffusion of rectangular cluster, and (3) diffusion of rectangle and fuzzier Gaussian clusters. The results show that the computational complexity is $O(1)$.

### 1.2.5 *Document Analysis with Wavelets*

Document processing is one of the most active branches in the area of pattern recognition. Documents contain knowledge. Precisely, they are the medium for transferring knowledge. In fact, much knowledge is acquired from documents, such as technical reports, government files, newspapers, books, journals, magazines, letters, bank cheques, to name a few. The acquisition of knowledge from such documents by an information system can involve an extensive amount of hand-crafting. Such hand-crafting is time-consuming and can severely limit the application of information systems. Actually, it is a bottleneck of information systems. Thus, automatic knowledge acquisition from documents has become an important subject. Since the 1960s, much research on document processing has been done

[Tang *et al.*, 1994]. Recently, the wavelet theory has been employed in this research [Liang *et al.*, 1999; Tang *et al.*, 1995a, 1996a, 1997a,c].

In this book, Chapter 11 provides a detailed presentation of wavelet-based document processing. In this section, another achievement is presented in the following:

**(1) Form-Document Analysis by Reference Line Detection with 2D Wavelet Transform:** The major characteristics of forms are analyzed by Tang *et al.* [1997a]:

- In general, a form consists of straight lines, which are oriented mostly in horizontal and vertical directions. These lines are referred to as *reference lines*.
- The reference lines are pre-printed to guide the users to complete the form.
- The information that should be entered into the computer and processed is usually the filled data.
- In order to indicate the filling position, the reference lines can be used and the filled information usually appears either above, beneath, or beside these reference lines. Thus, in form processing, the reference lines have to be detected first, then we can find the useful information from a form based on them and thereafter enter into the computers.

Tang *et al.* [1997a] present a novel wavelet-based method. In this method, 2D MRA, wavelet decomposition algorithm and compactly supported orthonormal wavelets are used to transform a document image into several sub-images. Based on these sub-images, the reference lines of a complex-background document can be extracted, and knowledge about the geometric structure of the document can be acquired. Particularly, this approach appears to be more efficient in processing form documents with multigray-level background.

A document image can be transformed into four sub-images by applying the Mallat algorithm: (1) LL sub-image, (2) LH sub-image, (3) HL sub-image, and (4) HH sub-image. We are interested in the LH and HL sub-images. The LH sub-image is achieved from a filter which allows lower frequency components to reach across along the horizontal direction as well as the higher frequencies along the vertical direction. This gives an "enhancing" effect on the vertical and "smoothing" effect on the horizontal. As a result, only horizontal lines remain in the LH sub-image. The situation of

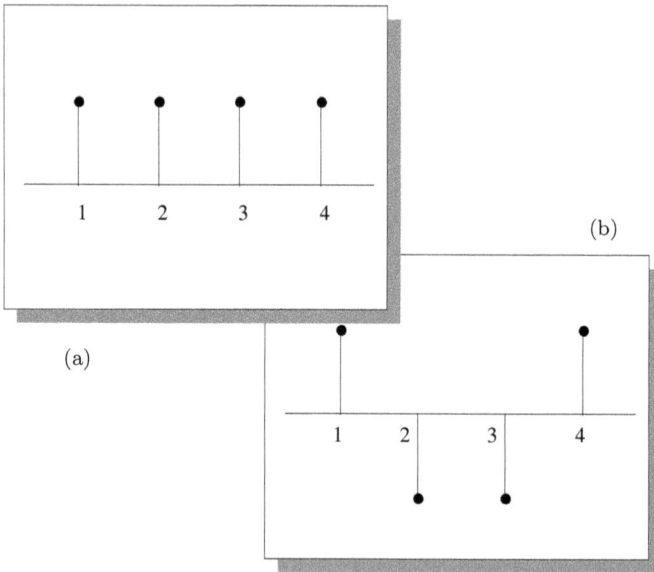

Fig. 1.23    The basic functions of the multiresolution Hadamard representation (MHR): (a) the low-pass filter $h(\cdot)$; (b) the high-pass filter $g(\cdot)$ [Liang *et al.*, 1999].

the HL sub-image is opposite to that of the LH one. In this way, the horizontal direction of the filter opens for the higher frequencies and the vertical direction opens for lower frequency components. This gives an "enhancing" effect on the horizontal and "smoothing" effect on the vertical. Thus, only vertical lines remain in the HL sub-image.

**(2) Multiresolution Hadamard Representation and its Application to Document Image Analysis:** A novel class of wavelet transform referred to as the multiresolution Hadamard representation (MHR) is proposed by Liang *et al.* [1999] for document image analysis.

The MHR is a 2D dyadic wavelet representation which employs two Hadamard coefficients $[1, 1, 1, 1]$ and $[1, -1, -1, 1]$, as shown in Fig. 1.23. These coefficients are further normalized with respect to $l^1$ norm [Mallat, 1989c].

The 2D dyadic wavelet representation is produced by applying the 1D filters $h(\cdot)$ and $g(\cdot)$ to the 2D image in both horizontal and vertical directions. This representation comprises four channels, namely, low-passed $L$, horizontal $H$, vertical $V$ and diagonal $D$, at each level of transform. These channels

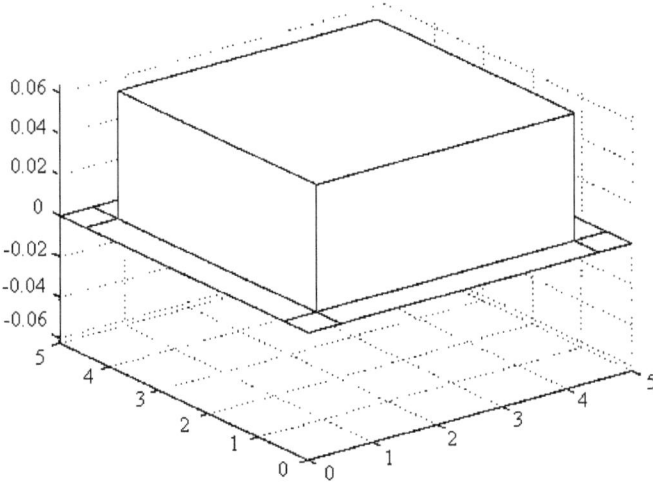

Fig. 1.24  The L channels of the 2D filters $h(\cdot)g(\cdot)$ [Liang *et al.*, 1999].

can be defined by the following iterative formulas:

$$L_{u,v,j+1} = \int_x \int_y L_{u,v,j}h(x-2u)h(y-2v),$$

$$H_{u,v,j+1} = \int_x \int_y L_{u,v,j}h(x-2u)g(y-2v),$$

$$V_{u,v,j+1} = \int_x \int_y L_{u,v,j}g(x-2u)h(y-2v),$$

$$D_{u,v,j+1} = \int_x \int_y L_{u,v,j}g(x-2u)g(y-2v),$$

where $x$ and $u$ are horizontal coordinates and $y$ and $v$ are vertical ones. Note that when $j=0$, $L_{u,v,j}$ denotes the original image.

The $L_{(j+1)}$ channel is obtained by the convolution of $L_j$ with the 2D filter $h(x)h(y)$ shown in Fig. 1.24. The $H_{(j+1)}(V_{(j+1)})$ channel is produced by the convolution of $L_j$ with the 2D filter $h(x)g(y)(h(y)g(x))$. Since the shape of $h(x)g(y)(h(y)g(x))$ is similar to a horizontal (vertical) bar, this 2D filter serves as a detector for horizontal (vertical) bars on $L_j$, which can be graphically illustrated in Fig. 1.25. The $D_{(j+1)}$ channel is obtained by the convolution of $L_j$ with the 2D filter $g(x)g(y)$. It is displayed graphically in Fig. 1.26.

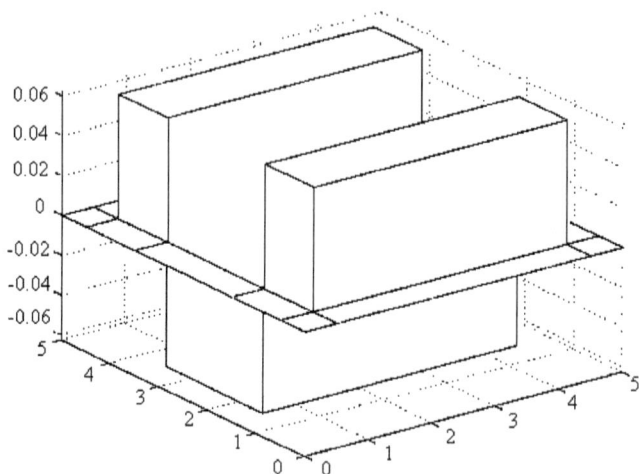

Fig. 1.25   The H and V channels of the 2D filters $h(\cdot)g(\cdot)$ [Liang *et al.*, 1999].

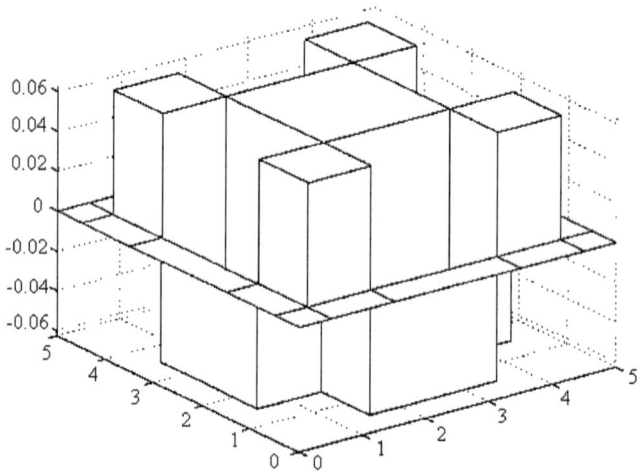

Fig. 1.26   The D channels of the 2D filters $h(\cdot)g(\cdot)$ [Liang *et al.*, 1999].

In Liang *et al.* [1999], the MHR is applied to document image analysis including the following processes:

- the exaction of half-tone picture,
- segmentation of document image into text blocks,
- determination of character scales for each text block.

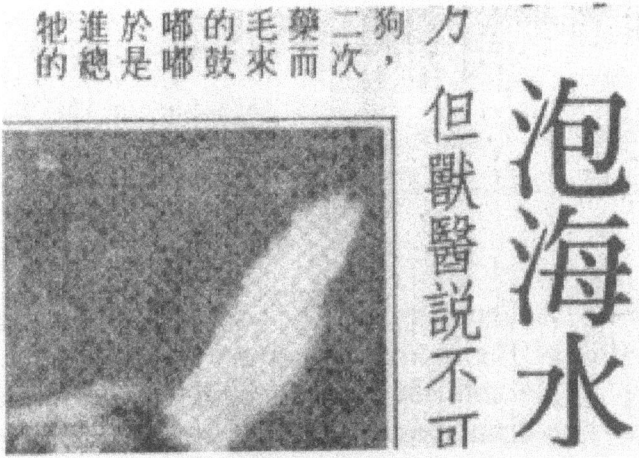

Fig. 1.27   A portion of Chinese newspaper [Liang *et al.*, 1999].

The transformed values of the vertical strokes of characters are very positive in some of the $V$ channels, while the horizontal strokes of characters react strongly in the $H$ channels. The diagonal strokes of characters, on the other hand, react in both $H$ and $V$ channels. The pictures in newspapers are generally produced by half-tones, where the tone of the pictures is produced by small block dots with varying densities. While characters show up their strengths in the $H$ and $V$ channels, a half-tone picture reacts as a significant regular pattern in the $D$ channel. This pattern is produced when the half-tone pictures are filtered by $(g(x)g(y))$.

In Liang *et al.* [1999], the MHR is used to treat a portion of Chinese newspaper, as shown in Fig. 1.27. It contains Chinese characters of three different sizes. The MHR developed by Liang *et al.* [1999] picks up three text blocks, with different scales from the original image, which are presented in Fig. 1.28. The multiresolution Hadamard representation of Fig. 1.27 is illustrated in Fig. 1.29, where the scale is $j = 1$.

A detailed description can be found in the work at Liang *et al.* [1999].

## 1.2.6   *Analysis and Detection of Singularities with Wavelets*

A significant application of wavelet theory is the analysis of singularities. Many methods based on wavelet theory have been developed to analyze the

Fig. 1.28 Three text blocks with different scales [Liang *et al.*, 1999].

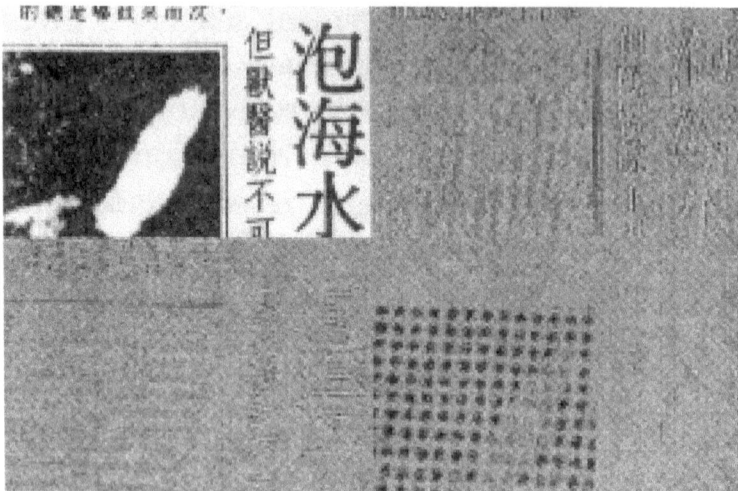

Fig. 1.29 The multiresolution Hadamard representation of Fig. 1.27 at scale $j = 1$ [Liang *et al.*, 1999].

properties of the singularities and detect them from various signals/images, and some examples can be found in the Chen *et al.* [1995], Chen and Yang [1995], Chuang and Kuo [1996], Deng and Lyengar [1996], IEEE [1993], Law *et al.* [1996], SPIE [1994], Tang *et al.* [1997c], Tang *et al.* [1998d], Thune *et al.* [1997], Tieng and Boles [1997a], Young [1993]. The edge is one class of singularities, which commonly appear in both the 1D signals and 2D images. The subject of wavelet transform is a remarkable mathematical tool to analyze the singularities including the edges and, further, to detect them effectively. A significant study related to this research topic has been done by Mallat, Hwang and Zhong, and published in the Special Issue on Wavelet Transforms and Multiresolution Signal Analysis of the *IEEE Transactions on Information Theory* [Mallat and Hwang, 1992] and *IEEE Trans. on Pattern Analysis and Machine Intelligence* [Mallat and Zhong, 1992].

Corners show the special type of edges, which are very attractive features for many applications in pattern recognition and computer vision. In Chen *et al.* [1995], a new gray-level corner detection algorithm based on the wavelet transform is presented. The wavelet transform is used because the evolution across scales of its magnitudes and orientations can be used to characterize localized signals like edges including the corners. Most conventional corner detectors detect corners based on the edge detection information. However, these edge detectors perform poorly at corners, adversely affecting their overall performance. To overcome this drawback, Chen *et al.* [1995] first proposes a new edge detector based on the ratio of the interscale wavelet transform modulus. This edge detector can correctly detect edges at the corner positions, making accurate corner detection possible. To reduce the number of points required to be processed, it applies the non-minima suppression scheme to the edge image and extracts the minima image. Based on the orientation variance, these non-corner edge points are eliminated. In order to locate the corner points, it proposes a new corner indicator based on the scale-invariant property of the corner orientations. By examining the corner indicator, the corner points can be located accurately, as shown by experiments with the algorithm. In addition, since wavelet transform possesses the smoothing effect inherently, this algorithm is insensitive to noise contamination as well.

**(1) Edge Detection with Local Maximal Modulus of Wavelet Transform:** A significant study related to this research topic was carried out by Mallat and Hwang [1992] and Mallat and Zhong [1992]. Many important contributions are made in their papers. They have proven that the maxima of the wavelet transform modulus can detect the locations of the irregular structures. Further, a numerical procedure to calculate their Lipshitz exponents is provided. It also numerically shows that 1D and 2D signals can be reconstructed, with a good approximation, from the local maxima of their wavelet transform modulus. The algorithm of the edge detection with local maximal modulus of the wavelet transform is presented in the following.

---

**Algorithm 1.1.** Given an input digital signal, $\{f(k,l)|k = 0, 1, \ldots, K; l = 0, 1, \ldots, L\}$:

**Step 1** Calculate the modulo of its wavelet transform

$$\{M_s f(k,l)|k = 0, 1, \ldots, K; l = 0, 1, \ldots, L\}$$

as well as the codes

$$\{CodeA_s f(k,l)|k = 0, 1, \ldots, K; l = 0, 1, \ldots, L\}$$

along the gradient directions.

**Step 2** Take a threshold $T > 0$, for $k = 0, 1, \ldots, K; l = 0, 1, \ldots, L$, if

(1) $|M_s f(k,l)| \geq T$,
(2) $|M_s f(k,l)|$ reaches its local maximum along the gradient direction represented by $CodeA_s f(k,l)$,

then $(k,l)$ is an edge pixel.

---

**(2) Detection of Step-Structure Edges by Scale-Independent Algorithm and MASW Wavelet Transform:** The local maxima modulus of the wavelet transform can provide enough information for analyzing the singularities and can detect all singularities. However, it may not identify different structures of singularities [Mallat and Hwang, 1992; Mallat and Zhong, 1992].

In Tang *et al.* [1998c], an important property is proved that the modulus of wavelet transform at each point of the step edge is a non-zero constant which is independent of both the gradient direction and the scale of the wavelet transform. Thus, a novel algorithm called scale-independent algorithm is developed.

---

**Algorithm 1.2.** Given 2D signal, $f(x, y)$:

**Step 1:** Take different scales $s_1, \ldots, s_J$, and calculate $W_{s_j} f(x, y)$, $(1 \leq j \leq J)$ based on

$$W_s^1 f(n, m) = \sum_{k,l} f(n - 1 - k, m - 1 - l)\psi_{k,l}^{s,1},$$

$$W_s^2 f(n, m) = \sum_{k,l} f(n - 1 - k, m - 1 - l)\psi_{k,l}^{s,2}.$$

**Step 2:** Select peak-threshold $T$, such that

$$|\nabla W_{s_j} f(x, y)| \geq T.$$

**Step 3:** Select proportional threshold $R$, such that

$$\frac{1}{R} \leq \frac{|\nabla W_{s_j} f(x, y)|}{|\nabla W_{s_l} f(x, y)|} \leq R, \quad (1 \leq j \leq J).$$

---

This method possesses an important property, i.e., the wavelet transform of a step-structure edge is scale-independent. It can improve the method proposed by Mallat and Hwang [1992] and Mallat and Zhong [1992], where the modulus-angle-separated-wavelet (MASW) is used. The precise definition of the MASW can be found in the work of Tang *et al.* [1998c,d]. After applying the scale-independent algorithm to the images, only the contour of the aircraft is extracted, while all other edges including drawing lines and text are eliminated.

In this book, Chapter 5 provides a detailed example of this application.

## 1.2.7 *Wavelet Descriptors for Shapes of the Objects*

The shape of a pattern is one of the most important features in pattern recognition. The description of such a shape plays a key role in shape analysis. Many research projects [Chuang and Kuo, 1996; Hsieh *et al.*, 1995; Tieng and Boles, 1997b; Wunsch and Laine, 1995] present some shape descriptors, which can represent digitized patterns. These descriptors are derived from the wavelet transform of the contours of a pattern and particularly well suited for the recognition of 2D objects, such as handprinted characters. Five examples are presented in the following:

**(1) Wavelet Descriptor of Planar Curves: Theory and Applications:** By using the wavelet transform, Chuang and Kuo [1996] develop a hierarchical planar curve descriptor that decomposes a curve into components of different scales so that the coarsest scale components carry the global approximation information while the finer scale components contain the local detailed information. It shows that the wavelet descriptor has many desirable properties, such as multiresolution representation, invariance, uniqueness, stability, and spatial localization. A deformable wavelet descriptor is also proposed by interpreting the wavelet coefficients as random variables. The applications of the wavelet descriptor to character recognition and model-based contour extraction from low SNR images are examined. Numerical experiments are performed to illustrate the performance of the wavelet descriptor.

**(2) Wavelet Descriptors for Multiresolution Recognition of Handprinted Characters:** Wunsch and Laine [1995] present a novel set of shape descriptors that represents a digitized pattern in a concise way and that is particularly well suited for the recognition of handprinted characters. The descriptor set is derived from the wavelet transform of a pattern's contour. The approach is closely related to feature extraction methods by Fourier series expansion. The motivation to use an orthonormal wavelet basis rather than the Fourier basis is that wavelet coefficients provide localized frequency information and that wavelets allow us to decompose a function into a multiresolution hierarchy of localized frequency bands. This paper describes a character recognition system that relies upon wavelet descriptors to simultaneously analyze the character shape at multiple levels of resolution. The system was trained and tested on a large database of more than 6,000 samples of handprinted alphanumeric characters. The results show that wavelet descriptors are an efficient representation that can provide for reliable recognition in problems with large input variability.

**(3) Wavelet-Based Shape Form Shading:** Hsieh *et al.* [1995] propose a wavelet-based approach for solving the shape from shading (SFS) problem. The proposed method takes advantage of the nature of wavelet theory, which can be applied to efficiently and accurately represent "things", to develop a faster algorithm for reconstructing better surfaces. To derive the algorithm, the formulation of Horn and Brooks (1989) which combines several constraints into an objective function, is adopted. In order to improve the robustness of the algorithm, two new constraints are introduced into the objective function to strengthen the relation between an estimated surface and its counterpart in the original image. Thus, solving the SFS problem becomes a constrained optimization process. Instead of solving the problem directly by using the Euler equation or numerical techniques, the objective function is first converted into the wavelet format. Due to this format, the set of differential operators of different orders, which is involved in the whole process, can be approximated with connection coefficients of Daubechies bases. In each iteration of the optimization process, an appropriate stem size, which can result in maximum decrease of the objective function, is determined. After finding correct iterative schemes, the solution of the SFS problem can finally be decided. Compared with conventional algorithms, the proposed scheme is a great improvement in the accuracy as well as the convergence speed of the SFS problem. Experimental results, using both synthetic and real images, prove that the proposed method is indeed better than traditional methods.

**(4) Representation of 2D Pattern by 1D Wavelet Sub-patterns:** Tang *et al.* [1998a] present an approach to represent a 2D shape by several 1D wavelet sub-patterns. In this way, first, a 2D pattern is converted into a 1D curve by the dimensionality reduction [Tang *et al.*, 1991]. Thereafter, according to the wavelet orthonormal decomposition, the 1D curve can be decomposed orthogonally into several high-frequency sub-curves and low-frequency ones using the wavelet transform.

**(5) Wavelet Descriptors for Multiresolution Recognition of Hand-printed Characters:** Wunsch and Laine [1995] presents a novel set of shape descriptors that represents a pattern in a concise way and that is particularly well suited for the recognition of handprinted characters. The descriptors are derived from a contour of a pattern using wavelet transform. This method is closely related to the feature extraction by Fourier series expansion. The motivation to use an orthonormal wavelet basis rather than the Fourier one is that wavelet coefficients provide localized frequency information and that wavelets allow us to decompose a function into a multiresolution hierarchy of localized frequency bands. This paper describes a character recognition system where the wavelet descriptors are used to analyze character shape at multiple levels of resolution. A large set of samples of handprinted alphanumeric characters are used to train and test this system. The results show that wavelet descriptors can efficiently represent the shapes of the objects and can provide for reliable recognition with large input variability.

### 1.2.8  *Invariant Representation of Patterns*

The invariance characteristic of size, orientation and translation is a significant subject in pattern recognition. With such capability, a pattern recognition system can find many applications in character recognition, computer vision, document processing, and many others. A lot of researchers paid attention to this subject, many methods have been developed. Recently, new methods in accordance with the wavelet-based approach have been made [Haley and Manjunath, 1999; Shen and Ip, 1999; Tieng and Boles, 1997b; Yoon *et al.*, 1998]. In this section, we introduce three papers, where the wavelet approach has been used [Shen and Ip, 1999; Tang *et al.*, 1998a; Yoon *et al.*, 1998].

**(1) Extraction of Rotation-Invariant Feature by Ring-Projection-Wavelet-Fractal Method:** In our previous work [Tang *et al.*, 1998a], a novel approach to extract features with the rotation-invariant property in pattern recognition is presented, which utilizes ring-projection-wavelet-fractal signatures (RPWFS). In particular, this approach reduces the dimensionality of a two-dimensional pattern by way of a ring-projection method and thereafter performs Daubechies' wavelet transform on the derived 1D pattern to generate a set of wavelet-transformed sub-patterns, namely, curves that are non-self-intersecting. Further from the resulting non-self-intersecting curves, the divider dimensions are readily computed. These divider dimensions constitute a new feature vector for the original 2D pattern defined over the curves' fractal dimensions.

An overall description of this approach is illustrated by a diagram shown in Fig. 1.30. A detailed description of this method can be found in Chapter 9.

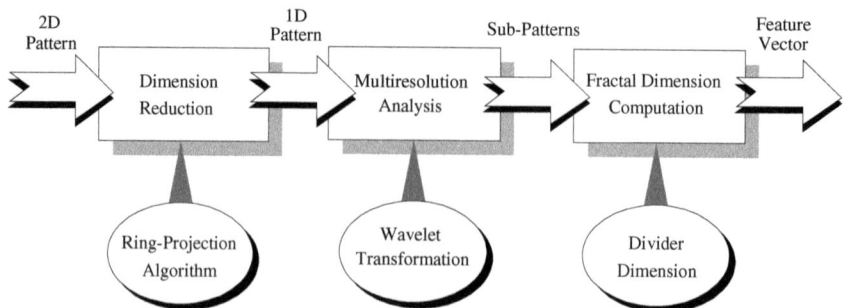

Fig. 1.30   Diagram of ring-projection-wavelet-fractal method.

**(2) Wavelet Rotation-Invariant Shape Descriptors for Recognition of 2D Pattern:** In a pattern recognition system, typically, a set of numerical features are extracted from an image. The selection of discriminative features is a crucial step in the system. The use of moment invariants as features for the identification of 2D shapes has received much attention. Shen and Ip [1999] investigate a set of wavelet rotation invariant moments presented for capturing global and local information from the objects of interest, together with a discriminative feature extraction method, for the classification of seemingly similar 2D objects with subtle differences.

To achieve rotation invariant moment, typically, a generalized expression is used, which can be written by

$$F_{pq} = \iint f(r, \theta) g_p(r) e^{jq\theta} r \, dr \, d\theta, \tag{1.7}$$

where $F_{pq}$ denotes the $pq$-order moment and $g_p(r)$ stands for a function of radial variable $r$. In addition, $p$ and $q$ are integers. The following have been proved.

- The value of $||F_{pq}||$ is rotation invariant, where $||F_{pq}|| = \sqrt{F_{pq} \cdot F_{pq}^*}$, and the symbol $*$ denotes the conjugate of the complex number.
- The combined moments, such as $F_{p_i q} \cdot F_{p_j q}^*$ are also rotation-invariant.

To facilitate the analysis, an expression of 1D sequence is substituted for that of the 2D image. Thus, Eq. (1.7) becomes

$$F_{pq} = \int S_q(r) \cdot g_p(r) r \, dr, \tag{1.8}$$

where $S_q(r) = f(r, \theta) e^{jq\theta} d\theta$. Shen and Ip [1999] treat $\{g_p(r)\}$ in Eq. (1.8) as wavelet basis functions:

$$\psi^{a,b}(r) = \frac{1}{\sqrt{a}} \left( \frac{r - b}{a} \right).$$

This indicates that the basis functions $\{g_p(r)\}$ are replaced by wavelet basis functions $\{\psi^{a,b}(r)\}$. The mother wavelet used in this work is a cubic B-spline in Gaussian approximation form (Fig. 1.31):

$$\psi(r) = \frac{4a^{n-1}}{\sqrt{2\pi(n+1)}} \sigma_w \cos(2\pi f_0(2r - 1))$$

$$\times \exp\left( -\frac{(2r - 1)^2}{2\sigma_w^2(n+1)} \right),$$

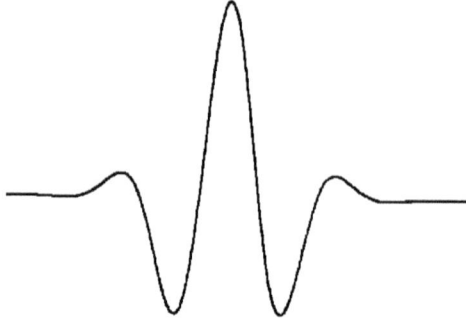

Fig. 1.31   The cubic B-spline mother wavelet [Shen and Ip, 1999].

where $n = 3$, $a = 0.697066$, $f_0 = 0.409177$ and $\sigma_w^2 = 0.561145$. The details of this mother wavelet can be fund in Unser *et al.* [1992]. Let $a = 0.5^m$, $m = 0, 1, 2, 3$, and $b = 0.5n0.5^m$, $n = 0, 1, \ldots, 2^{m+1}$. Thereafter, the wavelet defined along a radial axis in any orientation is denoted by

$$\psi_{m,n}(r) = 2^{m^2}\psi(2^m r - 0.5n).$$

Consequently, a set of wavelet moment invariants for classifying objects can be defined as follows:

$$\|F_{m,n,q}^{\text{wavelet}}\| = \left\|\int S_q(r) \cdot \psi_{m,n}(r)rdr\right\|, \tag{1.9}$$

where $\psi_{m,n}(r)$ replaces $g_g(r)$ in Eq. (1.8). The parameters are as follows: $m = 0, 1, 2, 3$, $n = 0, 1, \ldots, 2^{m+1}$, and $q = 0, 1, 2, 3$. It can be found that $\|F_{m,n,q}^{\text{wavelet}}\|$ is actually a wavelet transform of $S_q(r)r$. $\|F_{m,n,q}^{\text{wavelet}}\|$ can also be considered to be the first moment of $S_q(r)$ at the $m$th scale level with shift index $n$. It has been proved that the wavelet moment invariants, $\|F_{m,n,q}^{\text{wavelet}}\|$, are invariant to the rotation of an object.

Shen and Ip [1999], are the wavelet moment invariants along with a minimum-distance classifier and achieve a high classification rate for four different sets of patterns. For instance, 100% recognition rate is obtained for a test set consisting of 26 upper-cased English letters with different scales and orientations.

**(3) Scale-Invariant Object Recognition Based on the Multiresolution Approximation:** Boundary-based pattern matching is one of the most useful approaches in pattern recognition. Yoon *et al.* [1998] present a multiresolution approximation approach to obtaining boundary representation for object recognition. The summary of this approach is presented in the following.

It establishes the theory of the continuous multiresolution approximation (CMA) and implements a fast algorithm for the continuous wavelet transform (CWT). The CMT is a vehicle for scale-invariant matching or coarse-to-fine matching. In addition, the fast algorithm for the CWT enables us to quickly compute a representation. Yoon *et al.* [1998] propose good representations for boundary-based object matching. The representations are obtained using the CMA. These representations allow us to recognize objects in the presence of noise, occlusion, scale variation, rotation, and translation. It tests various types of objects such as tools, guns, and maps, with occlusion and scale variations. It also tests those objects by adding various types of noise. The test results show that the proposed representations are reliable and consistent.

Object matching in the presence of noise, occlusion, and scale variations is considered the most difficult problem in the area of boundary-based object matching. Yoon *et al.* [1998] call this scale-invariant matching. To solve this problem, the scaling effect of an object by using the CWT is modeled. The model enables us to use the CWT for scale-invariant matching. Further, a scale-invariant representation is proposed to handle the scale-invariant matching.

First, the modeling scaling effect of an object by using the wavelet transform is introduced in the following.

When an analog signal is converted to a digital one, the analog signal is pre-filtered by a lowpass filter to reduce aliasing and it is then sampled as shown in Fig. 1.32. An object in a picture may have different scales as the distance of the camera from the target object changes. By means of the wavelet transform, the scaling effect of an object can be modeled, which can lead to generating a scale-invariant representation for the scaling effect. The mathematical description is shown in Fig. 1.32:

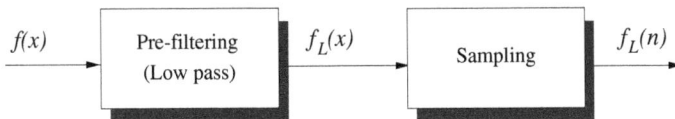

Fig. 1.32  Digitization of analog signal as pre-filtering and sampling [Yoon *et al.*, 1998].

Let $h(x)$ be the low-pass filter, thus the output, $f_L(x)$, of the filtering of an original signal $f(x)$ is

$$f_L(x) = f(x) \cdot h(x) = \int f(\alpha) h(\alpha - x) d\alpha.$$

The filtered output of the scaled signal is

$$f_L(sx) = \frac{1}{s} f(sx) \cdot h(x) = \int \frac{1}{s} f(s\alpha) h(\alpha - x) d\alpha. \tag{1.10}$$

Seting $\alpha' = s\alpha$, Eq. (1.10) becomes

$$f_L(sx) = \int \frac{1}{s} f(\alpha') \cdot h\left(\frac{\alpha'}{s} - x\right) \frac{d\alpha'}{s} = f(x) \cdot \frac{1}{s} h\left(\frac{x}{s}\right). \tag{1.11}$$

Equation (1.11) is the same formulation as the wavelet transform at the scale $s$, and the filter $h(x)$ is considered to be a continuous scaling function. This formulation implies that the scaling effect of an object in the process of digitization of an analog input image can be represented by the wavelet transform. This idea is the motivation for using the wavelet transform for a scale-invariant representation.

Second, the scale-invariant representation is established by means of the CMA. The CMA has two important features to receive significant representations for pattern recognition (1) the CMA provides approximations of objects at various scales, hence scale-invariant representations can be constructed by using the approximations; (2) the representations can be efficiently computed by using the fast algorithm.

The approximation, $\{R_s(k), Q_s(k)\}$ of a boundary at the scale $2^s$ is mathematically described as follows:

$$R_s(k) = [r(t) \cdot \beta_s^3(-t)]_{t=pk}, \tag{1.12}$$

$$Q_s(k) = [q(t) \cdot \beta_s^3(-t)]_{t=pk}, \tag{1.13}$$

where $p$ denotes the sampling period in the CWT. Applying Eq. 1.13 to an original boundary $\{r(t), q(t)\}$ produces a pair of functions for an approximation of a boundary $\{R_s(k), Q_s(k)\}$ at each scale. The example of an approximation is illustrated in Fig. 1.33. An original boundary is shown in Fig. 1.33(a), and an approximation of the original boundary at one-half of the scale is presented in Fig. 1.33(b). The curvature function of the gun is displayed in Fig. 1.33(c). The curvature function is invariant under scaling, rotation, and translation of a curve. Moreover, zero crossings of the curvature function are important features for shape analysis or recognition. Therefore, we can use the zero crossing of the curvature functions of the approximations using the CWT to construct the proposed representations.

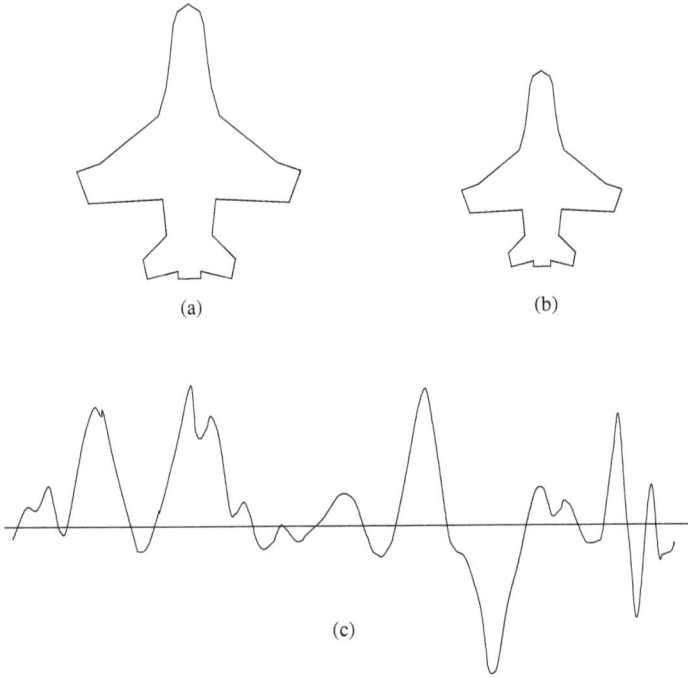

Fig. 1.33  The example of an approximation of a gun boundary.

The detailed procedure to receive the scale-invariant representation can be presented by the following algorithm:

**Step 1 Interpolation:** Interpolate an original boundary with the linear interpolation filter $h_1(n)$ given by

$$M(j, x, y) = \begin{cases} 1 - \frac{|n|}{L} & |n| < L, \\ 0 & \text{otherwise,} \end{cases}$$

where $L$ is an interpolation length between the original samples.

**Step 2 Initialization:** Apply the dilated cubic B-spline filter for initialization to obtain low-resolution boundaries at scales of $0.5 < s < 1$. The low-resolution boundaries of the interpolated signal are given by

$$R_s(k) = [r(t) \cdot 2^s \beta^3 (-2^s t)]_{t=2^{-s}k}.$$

**Step 3 Decomposition:** Obtain the low-resolution boundaries by use of the DWT:

$$R_s(k) = [R_{s-1} \cdot \beta_2^3(k)]_{\downarrow 2},$$

where the discrete filter $\beta_2^3(k)$ is a binomial one.

**Step 4 Construction of scale-invariant representation:** Compute the curvature functions and find zero crossings of the curvature functions over the desired scales. Moreover, to construct the scale-invariant representation by making zero crossing on the $t-k$ plane.

Experiments have been conducted by Yoon *et al.* [1998]. Several images, such as guns, tools, maps, etc. are used. Some objects are taken from a camera, and some, for example, guns are from the X-rays. Maps are taken from a commercial image database. The results have shown the proposed representations are reliable and consistent.

### 1.2.9 *Handwritten and Printed Character Recognition*

Character recognition including the identification of handwritten and printed characters is a major branch in the field of pattern recognition. Quite a number of articles, which deal with this branch, have been published. However, only a few have used wavelets [Lee *et al.*, 1996; Tang *et al.*, 1996b, 1998d,a; Wunsch and Laine, 1995]. In this section, two publications [Lee *et al.*, 1996; Wunsch and Laine, 1995] are briefly introduced. In addition, Chapter 12 also provides a detailed description of character recognition with the wavelet theory.

**(1) Wavelet Descriptors for Recognition of Handprinted Characters:** Wunsch and Laine [1995] describe a character recognition system that relies upon wavelet descriptors to simultaneously analyze character shapes at multiple levels of resolution. The system was trained and tested on a large database of more than 6,000 samples of handprinted alphanumeric characters. The results show that wavelet descriptors are an efficient representation that can provide for reliable recognition in problems with large input variability.

**(2) Extracting Multiresolution Features in Recognition of Handwritten Numerals with 2D Haar Wavelet:** The well-known Haar wavelet is adequate for local detection of line segments and global detection of line structures with fast computation. Lee *et al.* [1996] develop a method based on the Haar wavelet. It enables us to have an invariant interpretation of the character image at different resolutions and presents a multiresolution analysis in the form of coefficient matrices. The details of character image at different resolutions generally characterize different physical structures of the character, and the coefficients obtained from wavelet transform are very useful in recognizing unconstrained handwritten numerals. Therefore, Lee *et al.* [1996] use wavelet transform with a set of Haar wavelets for multiresolution feature extraction in handwriting recognition.

In this way, we take

$$1 = \cos^2 \frac{\omega}{2} + \sin^2 \frac{\omega}{2}.$$

Let us write

$$|H(\omega)|^2 = \cos^2 \frac{\omega}{2} = \frac{1 + \cos \omega}{2}$$

$$= \frac{1}{4} \left[ 1 + 2\cos\omega + \cos^2\omega + \sin^2\omega \right]$$

$$= \frac{1}{4} \left[ (1 + \cos\omega)^2 + \sin^2\omega \right]$$

$$= \left| \frac{1 + \cos\omega - i\sin\omega}{2} \right|^2 = \left| \frac{1 + e^{-i\omega}}{2} \right|^2,$$

so that

$$|H(\omega + \pi)|^2 = \sin^2 \frac{\omega}{2} = \frac{1 - \cos\omega}{2} = \left| \frac{1 - e^{-i\omega}}{2} \right|^2.$$

Thus, we can take

$$H(\omega) = \frac{1 + e^{-i\omega}}{2} = \sum_{k=0}^{1} \frac{1}{\sqrt{2}} h_k e^{-i\omega k},$$

where

$$h_0 = \frac{1}{\sqrt{2}}, \quad h_1 = \frac{1}{\sqrt{2}}.$$

The scaling function $\varphi(x)$ and wavelet function $\psi(x)$ can be represented by

$$\varphi(x) = \begin{cases} 1 & \text{if } 0 < x < 1, \\ 0 & \text{otherwise,} \end{cases} \tag{1.14}$$

and

$$\psi(x) = c_1\varphi(2x) - c_0\varphi(2x - 1)$$

$$= \varphi(2x) - \varphi(2x - 1)$$

$$= \begin{cases} 1 & \text{if } 0 < x < \frac{1}{2}, \\ 1 & \text{if } \frac{1}{2} \le x < 1, \\ 0 & \text{otherwise.} \end{cases} \tag{1.15}$$

An image of the handwritten character can be decomposed into its wavelet coefficients by using Mallat's pyramid algorithm. By using Haar wavelets, an image $F$ is decomposed as follows:

$$F = \begin{bmatrix} & & \\ & w\ x & \\ & y\ z & \end{bmatrix} \Rightarrow \begin{bmatrix} p\ q & \\ r\ a & b \\ c & d \end{bmatrix} \Rightarrow \begin{bmatrix} s\ t & \\ u\ v & b \\ c & d \end{bmatrix},$$

$$\begin{aligned} a &= 1/4(w + x + y + z), & b &= 1/4(w - x + y - z), \\ c &= 1/4(w + x - y - z), & d &= 1/4(w - x - y + z), \\ s &= 1/4(p + q + r + a), & t &= 1/4(p - q + r - a), \\ u &= 1/4(p + q - r - a), & v &= 1/4(p - q - r + a). \end{aligned} \tag{1.16}$$

In Eq. (1.16), the image $\{w, x, y, z\}$ is decomposed into image $\{a\}$, $\{b\}$, $\{c\}$, and $\{d\}$ at resolution $2^{-1}$. The image $\{a\}$ corresponds to the lowest frequencies $(D_1)$, $\{b\}$ gives the vertical high frequencies $(D_2)$, $\{c\}$ the horizontal high frequencies $(D_3)$, and $\{d\}$ the high frequencies in horizontal and vertical directions $(D_4)$. Likewise, the image $\{p, q, r, a\}$ at resolution $2^{-1}$ is decomposed into image $\{s\}$, $\{t\}$, $\{u\}$, and $\{v\}$ at resolution $2^{-2}$. This decomposition can be archived by convolving the $2 \times 2$ image array with the Haar masks as follows:

$$D_k^j = \frac{1}{4}I^{j+1} \otimes H_k \quad k = 1, 2, 3, 4 \tag{1.17}$$

where $\otimes$ is the convolution operator, $D_k^j$ are decomposed images at resolution $2^j$, $I^{j+1}$ is $2 \times 2$ image at resolution $2^{j+1}$, and $H_k$ denote Haar masks defined in Fig. 1.34.

In this application, the decomposed results at resolution $2^{-1}$ and $2^{-2}$ are used as multiresolution features. Figure 1.35 shows the process of multiresolution feature extraction. Figure 1.35(a) gives an input image of the handwritten number "8", and its digitized image is shown in

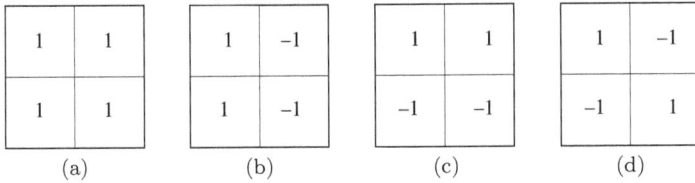

Fig. 1.34  Haar masks: (a) lowest frequencies; (b) vertical high frequencies; (c) horizontal high frequencies; (d) high frequencies in horizontal and vertical directions.

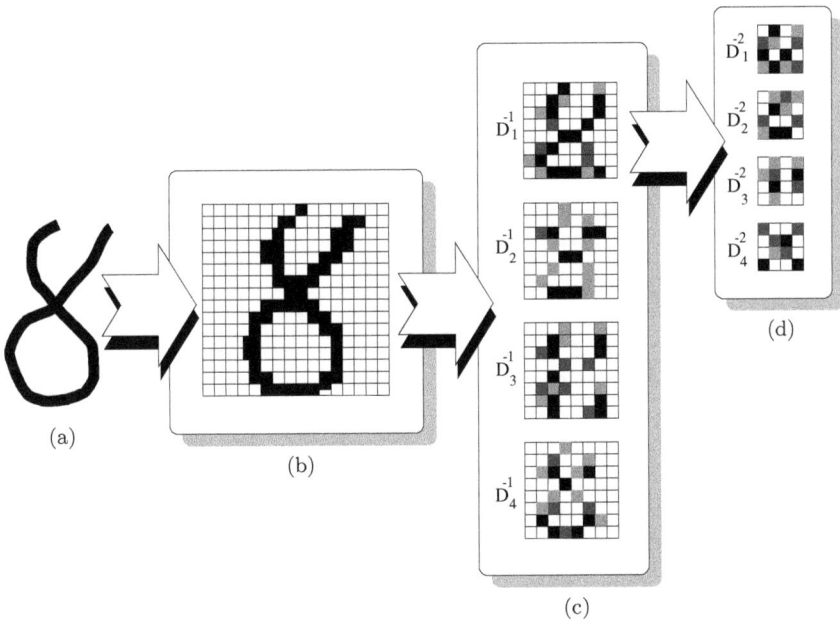

Fig. 1.35  Overview of multiresolution feature extraction.

Fig. 1.35(b). The decomposed feature vector at resolution $2^{-1}$ is illustrated in Fig. 1.35(c), while the decomposed feature vector at resolution $2^{-2}$ is shown in Fig. 1.35(d).

These features are used to recognize the unconstrained handwritten numerals from the database of Concordia University of Canada (Fig. 1.36), Electro-Technical Laboratory of Japan (Fig. 1.37), and Electronics and Telecommunications Research Institute of Korea (Fig. 1.38). The error rates are 3.20%, 0.83%, and 0.75%, respectively. These results have shown that the proposed scheme is very robust in terms of various writing styles

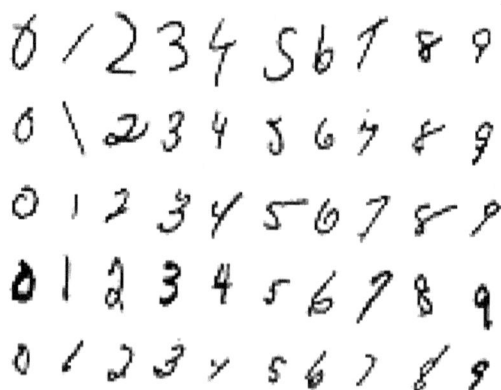

Fig. 1.36    Handwritten numeral database of Concordia University of Canada.

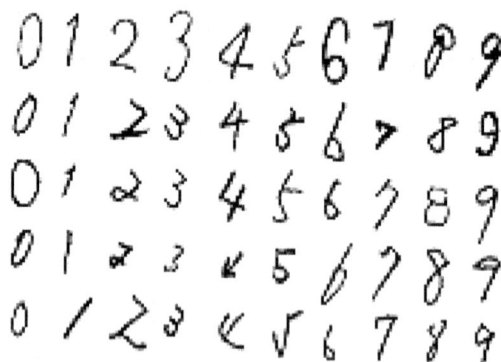

Fig. 1.37    Handwritten numeral database of Electro-Technical Laboratory of Japan.

and sizes. A detailed description of this application can be found in the work of Lee *et al.* [1996].

**(3) Wavelet Descriptors for Recognition of Printed Kannada Text in Indian Languages:** An OCR system for Indian languages, especially for Kannada, a popular South Indian language, is developed by Kunte and Samuel [2007]. Some examples of the Kannada language is presented in Fig. 1.39.

One-dimensional discrete wavelet transform (DWT) is used for feature extraction. Daubechies wavelet from the family of orthonormal wavelets is considered. A DWT when applied to a sequence of coordinates from the character contour returns a set of approximation coefficients and a set

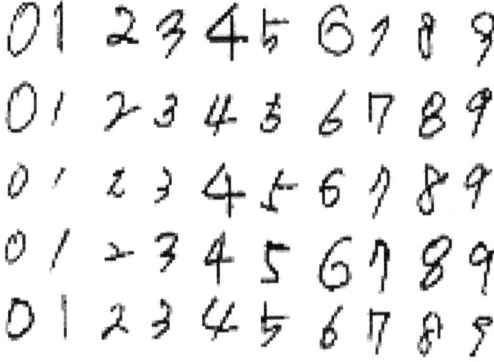

Fig. 1.38 Handwritten numeral database of Electronics and Telecommunications Research Institute of Korea.

Fig. 1.39 Some examples of Kannada text in Indian languages [Kunte and Samuel, 2007].

of detailed coefficients. The approximation coefficients correspond to the basic shape of the contour (low-frequency components) and the detailed coefficients correspond to the details of the contour (high-frequency components), which reflect the contour direction, curvature, etc. Neural classifiers are effectively used for the classification of characters based on wavelet features. The system methodology can be extended for the recognition of other south Indian languages, especially for Telugu.

## 1.2.10 *Texture Analysis and Classification*

Texture is a specific kind of pattern, and the texture analysis is one of the most important techniques used in the analysis and classification of images where repetition or quasi-repetition of fundamental elements occurs. So far, there is no precise definition of texture. Three principal approaches are used in texture analysis; statistical, spectral and structural. In this section, several methods based on the wavelet theory [Haley and Manjunath, 1999; de Wouwer *et al.*, 1999a,b] are introduced.

**(1) Wavelet Correlation Signatures for Color Texture Characterization:** In the last decade, multiscale techniques for gray-level texture analysis have been intensively used. de Wouwer *et al.* [1999a] aim to extend these techniques to color images. It introduces wavelet energy-correlation signatures and derives the transformation of these signatures upon linear color space transformations. Experiments are conducted on a set of 30 natural-colored texture images in which color and gray-level texture classification performances are compared. It is demonstrated that the wavelet correlation features contain more information than the intensity or the energy features of each color plane separately. The influence of image representation in color space is evaluated.

**(2) Rotation-Invariant Texture Classification Using a Complete Space-Frequency Model:** A method of rotation-invariant texture classification based on a complete space–frequency model is introduced by Haley and Manjunath [1999]. A polar, analytic form of a 2D Gabor wavelet is developed, and a multiresolution family of these wavelets is used to compute information-conserving micro-features. From these micro-features, a micro-model, which characterizes spatially localized amplitude, frequency, and directional behavior of the texture, is formed. The essential characteristics of a texture sample, and its macro-features, are derived from the estimated selected parameters of the micro-model. Classification of texture samples is based on the macro-model derived from a rotation-invariant subset of macro features. In experiments, comparatively high correct classification rates were obtained using large sample sets.

**(3) Statistical Texture Characterization from Discrete Wavelet Representation:** Texture analysis plays an important role in many tasks of image processing, pattern recognition, robot vision, computer vision, etc. de Wouwer *et al.* [1999b] conjecture that the texture can be characterized by the statistics of the wavelet detail coefficients and therefore introduces two feature sets:

1. the wavelet histogram signatures, which capture all first-order statistics using a model-based approach;

2. the wavelet co-occurrence signatures, which reflect the coefficients' second-order statistics.

The (average) best results are obtained by combining both feature sets. The introduced features are very promising for many image processing tasks, such as texture recognition, segmentation, and indexing image databases.

In this section, we would like to briefly introduce the basic idea of de Wouwer *et al.* [1999b]. In this work, the authors combine the statistical and multiscale views on texture. They conjecture that texture can be completely characterized from the statistical properties of its multiscale representation. The first-order statistical information is derived from the detail image histogram. The detail histograms of natural textured images can be modeled by a family of exponential functions. Introducing the parameters of this model as texture features completely describes the wavelet coefficients' first-order statistics. Further, improvement in texture description is obtained from the coefficients' second-order statistics, which can be described using the detailed image co-occurrence matrices. The most complete description is obtained by combining both first- and second-order statistical information.

The 2D discrete wavelet transform is applied in this work, which is a separable filterbank:

$$L_n(b_i, b_j) = [H_x * [H_y * L_{n-1}]_{\downarrow 2,1}]_{\downarrow 1,2}(b_i, b_j), \qquad (1.18)$$

$$D_{n1}(b_i, b_j) = [H_x * [G_y * L_{n-1}]_{\downarrow 2,1}]_{\downarrow 1,2}(b_i, b_j), \qquad (1.19)$$

$$D_{n2}(b_i, b_j) = [G_x * [H_y * L_{n-1}]_{\downarrow 2,1}]_{\downarrow 1,2}(b_i, b_j), \qquad (1.20)$$

$$D_{n3}(b_i, b_j) = [G_x * [G_y * L_{n-1}]_{\downarrow 2,1}]_{\downarrow 1,2}(b_i, b_j), \qquad (1.21)$$

where $*$ denotes the convolution operator, $\downarrow 2, 1 (\downarrow 1, 2)$ subsampling along the rows (columns), and $L_0 = I(\vec{x})$ is the original image. $H$ and $G$ are low- and bandpass filters, respectively.

The histogram of the wavelet detail coefficients are noted as $h_{ni}(u)$; thus, $h_{ni}(u)du$ is the probability that a wavelet coefficient $D_{ni}(\vec{b})$ has a value between $u$ and $u + du$. The detail histograms of natural textured images can be modeled by a family of exponential [Mallat, 1989b]:

$$h(u) = Ke^{-(|u|/\alpha)^\beta}. \qquad (1.22)$$

In this model, $\alpha$ and $\beta$ are wavelet histogram signatures, which are easily interpreted as specific, independent characteristics of the detail histogram.

They contain all first-order information present in the detail histogram. Equation (1.22) can be employed for many texture analysis tasks.

When the features based on the first-order statistics do not suffice, the second-order statistics can improve the texture discrimination. The wavelet co-occurrence signatures can reflect the coefficients' second-order statistics. The element $(j, k)$ of the co-occurrence matrix $C_{ni}^{\delta\theta}$ is defined as the joint probability that a wavelet coefficient $\tilde{D}_{ni} = j$ co-occurs with a coefficient $\tilde{D}_{ni} = k$ on a distance $\delta$ in direction $\theta$. Formulas for eight common co-occurrence features are provided by de Wouwer *et al.* [1999b]. These features extracted from the detail images are referred to as the wavelet co-occurrence signatures.

A database consisting of 30 real-world 512×512 images from different natural scenes is used for the experiments, as presented in Fig. 1.40.

Fig. 1.40   Selection of images from the VisTex database used by de Wouwer *et al.*, [1999b].

This database is available at http://www.white.media.mit.edu/vismod/ imagery/VisionTexture, which is provided by MIT Media Lab. Each image region is transformed to an overcomplete wavelet representation of depth four using a biorthogonal spline wavelet of order two Unser *et al.* [1993]. Four different feature sets are generated:

- 12 wavelet energy signatures,
- 24 wavelet histogram signatures,
- 96 wavelet co-occurrence signatures,
- all features from 2 and 3.

The results of the experiments can be found in the work of de Wouwer *et al.* [1999b].

**(4) Adaptive Scale Fixing for Multiscale Texture Segmentation:** Liang and Tjahjadi [2006] address two challenging issues in unsupervised multiscale texture segmentation: determining adequate spatial and feature resolutions for different regions of the image and utilizing information across different scales/resolutions. The center of a homogeneous texture is analyzed using coarse spatial resolution, and its border is detected using fine spatial resolution so as to locate the boundary accurately. The extraction of texture features is achieved via a multiresolution pyramid. The feature values are integrated across scales/resolutions adaptively. The number of textures is determined automatically using the variance ratio criterion. Experimental results on synthetic and real images demonstrate the improvement in performance of the proposed multiscale scheme over single-scale approaches.

Adaptive scale fixing is a method which determines the optimum $\triangle w_s$ at a site $s$ for segmentation. It also facilitates the utilization of information across resolutions. The aim is to maximize the precisions of both texture estimation and boundary localization, overcoming the limitation of single-resolution approaches. This method generates a multiscale segmentation map, as shown in Fig. 1.41. Multiscale representation of a two-texture image is presented in Fig. 1.41(a), where two homogeneous textures are depicted as grey and white. The central regions of homogeneous textures are represented using large windows, while the border regions of homogeneous textures are represented using small windows. This representation guarantees that only one texture exists in each window, thus reducing erroneous texture characterization. An image with textures of metal and straw is illustrated in Fig. 1.41(b).

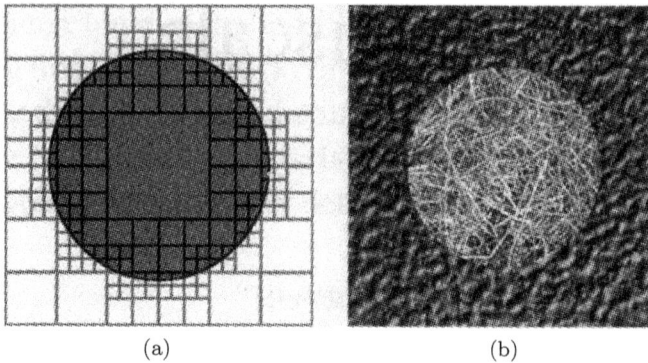

Fig. 1.41   Adaptive scale fixing [Liang and Tjahjadi, 2006].

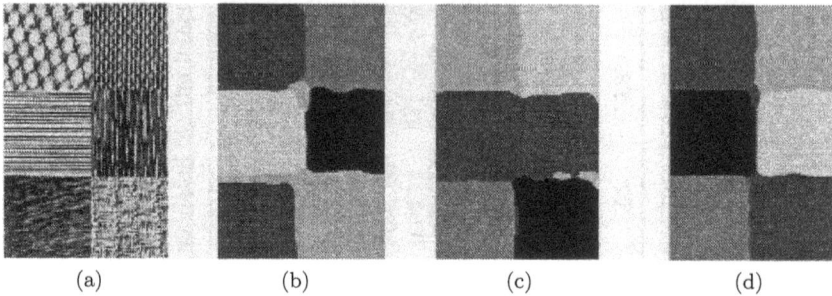

Fig. 1.42   Results of segmentation: (a) input image containing various number of textures; (b) segmentation result by Gabor wavelet; (c) segmentation result by moments; (d) segmentation result by the combination method [Muneeswaran *et al.*, 2005].

**(5) Combination of Gabor Wavelet Transforms and Moments for Texture Segmentation:** Muneeswaran *et al.* [2005] propose a combination approach, as shown in Fig. 1.42. It tries to incorporate into itself the better of the two methodologies, namely, Gabor wavelet and general moments. This method involves forming the combinational feature vector for a texture image from the extracted features by both approaches. This combinational approach has been evolved since both measures tend to look for features of a texture image in different views. The following equation is used to construct the feature vector:

$$F_c = \partial(F_M \oplus F_G),$$

where $F_c$ is the feature vector representing a pixel, constructed by the combinational approach, $F_M$ is the feature vector obtained for the same pixel by applying moments and $F_G$ is the feature vector using Gabor wavelet. $\partial$ is a stabilizing function which is defined as follows:

$$\partial(X) = [\triangle x_i, \forall i \in \{1, \ldots, m\}],$$

where $X$ is a vector and $m$ is the vector length and the weight value $\triangle$ is calculated by

$$\triangle = \left[ \sum_{i=1}^{m} x_i^2 \right]^{-1/2}.$$

## 1.2.11 *Image Indexing and Retrieval*

Digital image libraries and other multimedia databases have been dramatically expanded in recent years. Storage and retrieval of images in such libraries become a real demand in industrial, medical, and other applications. A solution for this problem is content-based image indexing and retrieval (CBIR), in which some features are extracted from every picture and stored as an index vector. Thereafter, the index is compared in the retrieval phase to find some similar pictures to the query image [Special-Issue-Digital-Library, 1996; Smeulders *et al.*, 2000]. Two major approaches can be used in CBIR: spatial domain method and transform domain method.

**(1) Wavelet Correlogram:** An approach called *wavelet correlogram* is proposed by Moghaddam *et al.* [2005], which can take advantage of both spatial and transform domain information. There are three steps in this way. (i) Wavelet coefficients are computed to decompose space–frequency information of the image; (ii) A quantization step is then utilized before computing directional autocorrelograms of the wavelet coefficients; (iii) Finally, index vectors are constructed using these wavelet correlograms. Figure 1.43 shows the diagram of the wavelet correlogram indexing method.

The wavelet autocorrelogram computing equations are

$$\Gamma^{m,n}(i,i,k) = \{(p1,p2)|p1,p2 \in W_{l_i}^{m,n}, |p1 - p2|_n = k\},$$

$$\alpha^{m,n}(i,k) = \frac{|\Gamma^{m,n}(i,i,k)|}{2hl_i(W^{m,n})},$$

where $|\Gamma^{m,n}(i,i,k)|$ represents the size of the wavelet autocorrelogram for each pair of $(m,n)$ and $hl_i(W^{m,n})$ is the total number of pixels of level $l_i$.

For evaluation of the proposed approach, some query images were selected randomly from a 1000-image subset of the COREL database,

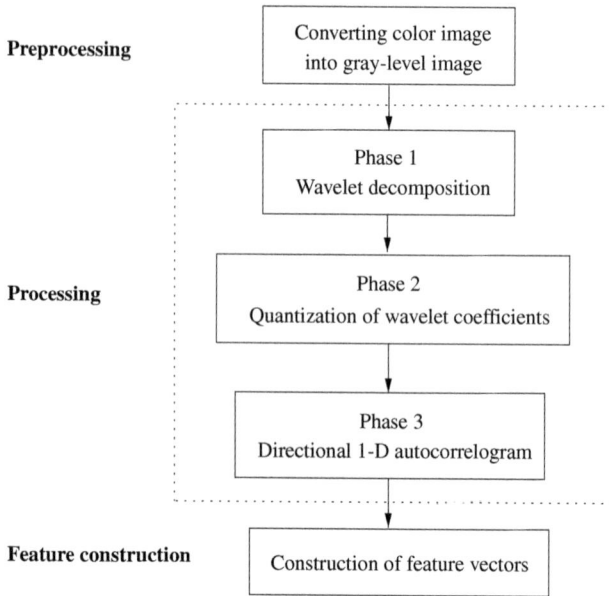

Fig. 1.43   Diagram of the wavelet correlogram indexing method [Moghaddam *et al.*, 2005].

http://wang.ist.psu.edu/docs/related. The experimental results indicate that a total average of 71% matched retrieved images is achievable using the wavelet correlogram indexing retrieval algorithm [Moghaddam *et al.*, 2005].

**(2) Region Separation and Multiresolution Analysis:** A simple and fast querying method for content-based image retrieval is developed by Ksantini *et al.* [2006]. Using the multispectral gradient, a color image is split into two disjoint parts that are the homogeneous color regions and the edge regions.

(1) The homogeneous regions are represented by the traditional color histograms:

$$h_k^h(c) = \sum_{i=0}^{M-1} \sum_{j=0}^{N-1} \delta(I_k(i,j) - c)\chi_{[0,\eta]}(\lambda_{\max}(i,j)),$$

where $\lambda_{\max}$ stands for the largest eigenvalue and $\chi_{[0,\eta]}$ is the characteristic function, for each $c \in \{0, \ldots, 255\}$ and $k = a, b$.

(2) The edge regions are represented by the multispectral gradient module mean histograms:

$$h_k^e h(c) = \sum_{i=0}^{M-1} \sum_{j=0}^{N-1} \delta(I_k(i,j) - c)\chi_{]\eta,+\infty[}(\lambda_{\max}(i,j))\lambda_{\max}(i,j).$$

In order to measure the similarity degree between two color images both quickly and effectively, Ksantini *et al.* use a 1D pseudo-metric, which makes use of the 1D wavelet decomposition and compression of the extracted histograms. This querying method is invariant to the query color image object translations and color intensities.

Apart from the above methods, a content-based image retrieval technique is presented by Kubo *et al.* [2003], who use wavelet-based shift and brightness-invariant edge features. Jain and Merchant [2004] propose a method using wavelet-based multiresolution histogram for fast image retrieval.

## 1.2.12 *Wavelet-Based Image Fusion*

Image fusion is one of the most important techniques to enhance image information. The fused image is more suitable for image processing. Image fusion is a process by which information from different observation images are incorporated into a single image. The importance of image fusion lies in the fact that each observation image contains complementary information. When this complementary information is integrated with that of another observation, an image with the maximum amount of information is obtained.

Different approaches have been adopted for multi-sensor or multiple observation image fusion from the simple image averaging approach to the wavelet transform image fusion approach. A super-resolution linear minimum mean square error (LMMSE) algorithm for image fusion is proposed by El-Khamy *et al.* [2006]. The schematic diagram of the proposed super-resolution LMMSE algorithm can be found in Fig. 1.44.

$$f_h = R_{f_n} \begin{bmatrix} D_1 H_1 M_1 \\ \cdot \\ \cdot \\ \cdot \\ D_p H_p M_p \end{bmatrix}^t$$

$$g_1 \rightarrow \boxed{\substack{\text{Image} \\ \text{Alignment}}} \rightarrow \boxed{\substack{\text{Multi-Channel} \\ \text{LMMSE} \\ \text{Restoration}}} \rightarrow \boxed{\substack{\text{Wavelet-Based} \\ \text{Image Fusion}}} \xrightarrow{y} \boxed{\substack{\text{LMMSE} \\ \text{Interpolation}}} \xrightarrow{f_h}$$

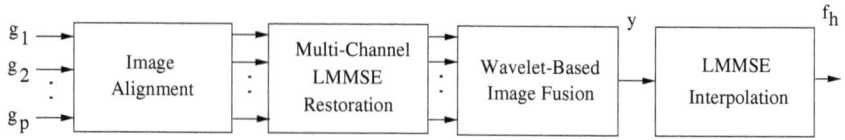

Fig. 1.44    The schematic diagram of the proposed super-resolution algorithm [El-Khamy *et al.*, 2006].

$$\left[ \begin{bmatrix} D_1 H_1 M_1 \\ \cdot \\ \cdot \\ \cdot \\ D_p H_p M_p \end{bmatrix} R_{f_n} \begin{bmatrix} D_1 H_1 M_1 \\ \cdot \\ \cdot \\ \cdot \\ D_p H_p M_p \end{bmatrix}^t + R_n \right]^{-1}$$

$$\begin{bmatrix} g_1 \\ \cdot \\ \cdot \\ \cdot \\ g_p \end{bmatrix},$$

where $g_i$, $i = 1, 2, \ldots, p$ stand for the observed images, $R_{f_n}$ and $R_n$ are the high resolution images and noise correlation matrices respectively, $D_i$, $i = 1, 2, \ldots, p$ is the uniform downsampling matrix, $H_i$, $i = 1, 2, \ldots, p$ is the blur matrix, and $M_i$, $i = 1, 2, \ldots, p$ is the registration shift matrix.

El-Khamy *et al.* [2006] adopts the wavelet transform image fusion approach to integrate the data from the multiple outputs of the LMMSE image restoration step. This is due to the fact that the multiple outputs of the LMMSE restoration step are correctly registered and aligned. The registration of the multiple inputs to the wavelet fusion step is a very important prerequisite to the success of the fusion step. In the application of the simple wavelet image fusion scheme, the wavelet packet decomposition is calculated for each observation to obtain the multiresolution levels of the images to be fused. In the transform domain, the coefficients in all resolution levels whose absolute values are larger are chosen between the available observations. This rule is known as the maximum frequency rule. Using this method, fusion takes place in all resolution levels and dominant features at each scale are preserved. Another alternative to the maximum frequency rule is the area-based selection rule. This rule is called the local variance or the standard deviation rule. The local variance of the wavelet coefficients is calculated as a measure of local activity levels associated with each wavelet coefficient. If the measures of activity of the wavelet coefficients in each of the two images to be fused are close together, the average of the two wavelet coefficients is taken; otherwise, the coefficient with maximum absolute is chosen. Generally, the process of wavelet packet image fusion can be summarized in the following steps: (1) The available images are first registered (this is achieved for the outputs of the LMMSE restoration step). (2) The wavelet packet decomposition of the observations is calculated using a suitable basis function and decomposition level. (3) A suitable fusion rule is used to select the wavelet coefficients from the source observations. (4) A decision map is created according to the fusion rule. (5) A wavelet packet reconstruction is performed on the combinations created coefficients.

Novel pixel-level image fusion schemes are presented by Li [2006] based on multiscale decomposition. In this way, the wavelet coefficients of image are chosen according to the image fusion operators and different fusion rules. Experiments of multispectral image and high-resolution panchromatic images are given. It shows that the wavelet-based human visual system method can achieve better fusion performance than others.

An approach to multisensor remote sensing image fusion using stationary wavelet transform is proposed by Li [2008]. In this paper, the author investigates the effects of orthogonal/biorthogonal filters and decomposition depth on using stationary wavelet analysis for fusion. Spectral discrepancy and spatial distortion are used as quality measures. Empirical results lead to some recommendations on the wavelet filter parameters for use in remote sensing image fusion applications.

### 1.2.13 *Others*

Apart from the above sections, there are many other applications of wavelet theory to pattern recognition. For instance, Chambolle *et al.* [1998] examines the relationship between wavelet-based image processing algorithms and variational problems. Algorithms are derived as exact or approximate minimizers of variational problems; in particular, it shows that wavelet shrinkage can be considered the exact minimizer of the following problem: Given an image $F$ defined on a square $I$, minimize over all $g$ in the Besov space $B_1^1(L_1(I))$ the functional $\|F - g\|_{L_2(I)}^2 + \lambda \|g\|_{B_1^1(L_1(I))}$. It uses the theory of nonlinear wavelet image compression in $L_2(I)$ to derive accurate error bounds for noise removal through wavelet shrinkage applied to images corrupted with i.i,d., mean zero, Gaussian noise. A new signal-to-noise ratio (SNR), which claim more accurately reflects the visual perception of noise in images, arises in this derivation. It presents extensive computations that support the hypothesis that near-optimal shrinkage parameters can be derived if one knows (or can estimate) only two parameters about an image $F$: the largest $\alpha$ for which $F \in B_q^\alpha(L_q(I))$, $1/q = \alpha/2 + 1/2$, and the norm $\|F\|_{B_q^\alpha(L_q(I))}$. Both theoretical and experimental results indicate that this choice of shrinkage parameters yields uniformly better results than Donoho and Johnstone's VisuShrink procedure.

A standard wavelet multiresolution analysis can be defined via a sequence of projectors onto a monotone sequence of closed-vector subspaces possessing certain properties. Combettes [1998] proposes a nonlinear extension of this framework in which the vector subspaces are replaced by convex subsets. These sets are chosen so as to provide a recursive, monotone approximation scheme that allows for various signal and image features to be investigated. Several classes of convex multiresolution analysis are discussed and numerical applications to signal and image processing problems are demonstrated.

Image restoration problems can naturally be cast as constrained convex programming problems in which the constraints arise from *a priori* information and the observation of signals physically related to the image to be recovered. Combettes and Pesquet [2004], place the focus on the construction of constraints based on wavelet representations. Using a mix of statistical and convex-analytical tools, Combettes and Pesquet propose a general framework to construct wavelet-based constraints. The resulting optimization problem is then solved with a block-iterative parallel algorithm which offers great flexibility in terms of implementation. Numerical results illustrate an application of the proposed framework.

A novel approach for thinning character using modulus minima of wavelet transform is developed by You *et al.* [2006]. A method for signal denoising using wavelets and block hidden Markov model is presented by Liao and Tang [2005].

# PART 2

# Chapter 2

# Continuous Wavelet Transforms

Similar to the Fourier transform, the wavelet theory consists of two parts: the wavelet transform and the wavelet basis. In this chapter, we focus on the former and organize it into three sections: (1) the general theory of wavelet transform, (2) the filtering properties of the wavelet transforms, and (3) the characterization of Lipschitz regularity of signals by wavelet transforms.

## 2.1 General Theory of Continuous Wavelet Transforms

In Fourier transform,

$$\mathcal{F}f(\xi) = \int_{\mathbb{R}} f(\xi - x)\overline{e^{i\xi x}}dx,$$

if we replace $e^{i\xi x}$, the dilation of the basic wave function $e^{ix}$, with $\psi(\frac{t-b}{a})$, the translation and dilation of the basic wavelet $\psi(x)$, then the transform is referred to as a continuous wavelet transform (also called an integral wavelet transform), which is defined as follows.

**Definition 2.1.** A function $\psi \in L^2(\mathbb{R})$ is called an admissible wavelet or a basic wavelet if it satisfies the following "admissibility" condition:

$$C_\psi := \int_{\mathbb{R}} \frac{|\hat{\psi}(\xi)|^2}{|\xi|}d\xi < \infty. \tag{2.1}$$

The continuous (or integrable) wavelet transform with kernel $\psi$ is defined by

$$(W_\psi f)(a, b) := |a|^{-\frac{1}{2}} \int_{\mathbb{R}} f(t)\overline{\psi\left(\frac{t-b}{a}\right)}dt, \quad f \in L^2(\mathbb{R}), \tag{2.2}$$

where $a, b \in \mathbb{R}$ and $a \neq 0$ are the dilation parameter and the translation parameter, respectively.

Before showing the functions of the admissibility condition (2.1), we would like to discuss some basic properties of the continuous wavelet transform at first. According to the admissibility condition (2.1), if $\psi \in L^1(\mathbb{R})$, it can be mathematically inferred that

$$\hat{\psi}(0) = 0$$

and

$$\hat{\psi}(\xi) \to 0 \quad (|\xi| \to \infty).$$

This indicates that the function $\psi$ is a bandpass filter.

The characteristics of the localized components of a signal $f(t)$ can be described by the continuous wavelet transform.

- Due to the damp of $\psi(x)$ at infinity, the localized characteristic of $f$ near $x = b$ is described by (2.2). It will be clearer if we extremely assume that $\psi(x)$ will always be zero out of $[-1, 1]$. Then, for all $x \notin [b - |a|, b + |a|]$, we have

$$\psi\left(\frac{t - b}{a}\right) = 0.$$

Therefore,

$$(W_\psi f)(a, b) = |a|^{-\frac{1}{2}} \int_{\mathbb{R}} f(t) \overline{\psi\left(\frac{t - b}{a}\right)} dt$$

$$= |a|^{-\frac{1}{2}} \int_{b-|a|}^{b+|a|} f(t) \overline{\psi\left(\frac{t - b}{a}\right)} dt.$$

This means that $(W_\psi f)(a, b)$ is completely determined by the behaviors of $f$ in $[b - |a|, b + |a|]$ with the center $b$ (Fig. 2.1). It is said that $(W_\psi f)(a, b)$ describes only the localized characteristic of $f$ in $[b - |a|, b + |a|]$. The smaller the $a$, the better the localized characteristic of $f$, which can also be found in Fig. 2.1.

- On the other hand, by the basic properties of the Fourier transform, we have

$$(W_\psi f)(a, b) = \frac{|a|^{-\frac{1}{2}}}{2\pi} \int_{\mathbb{R}} \hat{f}(\xi) \left(\psi\left(\frac{\cdot - b}{a}\right)\right)^{\wedge}(\xi) d\xi$$

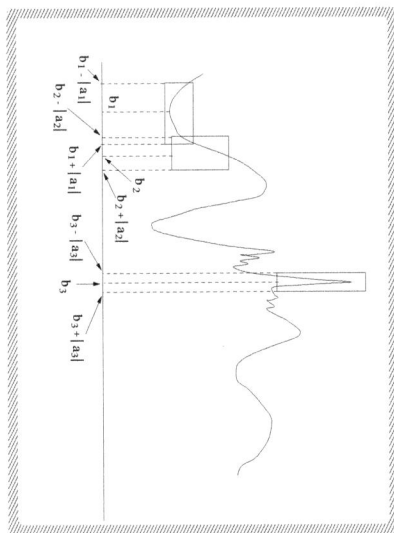

Fig. 2.1    Time–frequency localization.

$$= \frac{|a|^{-\frac{1}{2}}}{2\pi} \int_{\mathbb{R}} \hat{f}(\xi)|a|e^{-ib\xi}\hat{\psi}(a\xi)d\xi$$

$$= \frac{|a|^{\frac{1}{2}}}{2\pi} \int_{\mathbb{R}} \hat{f}(\xi)e^{-ib\xi}\hat{\psi}(a\xi)d\xi.$$

There is only a phase difference between $e^{-ib\xi}\hat{\psi}(a\xi)$ and $\hat{\psi}(a\xi)$. The bandpass property of $\psi$ ensures that the energy of $\hat{\psi}(a\xi)$ concentrates on two bands, while the bandwidth depends on the scale parameter $a$ with positive ratio. Hence, the localization of $\hat{f}$ is illustrated by $(W_\psi f)(a,b)$, and $\hat{f}$ gets worse localization with a smaller $a$.

$\psi$ can be viewed as a basic window function. Through this window, we cannot observe the integrated $f$ or $\hat{f}$, whereas the local performances of $f$ and $\hat{f}$ are very clear. Moving the translation parameter $b$, each part of $f$ and $\hat{f}$ can be traveled. Meanwhile, by adjusting $a$, we can observe $f$ and $\hat{f}$ with different localization. The latter is the so-called focus property of the wavelet.

For the convenience of our discussion, under the meaning of statistics, we define the center of basic window $\psi$ as

$$x_\psi^* := \frac{1}{\|\psi\|_2^2} \int_{\mathbb{R}} t|\psi(t)|^2 dt. \tag{2.3}$$

Moreover, we define the radius of the window, which is the statistic width of the energy of $\psi$ to its center, as

$$\Delta_\psi := \frac{1}{\|\psi\|_2} \left[ \int_{\mathbb{R}} (t - x^*)^2 |\psi(t)|^2 dt \right]^{1/2}, \qquad (2.4)$$

where $\|\psi\|_2$ is the norm of $\psi$, i.e.,

$$\|\psi\|_2 := \left( \int_{\mathbb{R}} |\psi(x)|^2 dx \right)^{1/2}.$$

The window of $\psi$ (also called the time window of $\psi$ in general) is

$$[x_\psi^* - \Delta_\psi, x_\psi^* + \Delta_\psi].$$

In the same way, the window of $\hat{\psi}$ (also called the frequency window of $\psi$) is

$$[x_{\hat{\psi}}^* - \Delta_{\hat{\psi}}, x_{\hat{\psi}}^* + \Delta_{\hat{\psi}}].$$

The following square region

$$[x_\psi^* - \Delta_\psi, x_\psi^* + \Delta_\psi] \times [x_{\hat{\psi}}^* - \Delta_{\hat{\psi}}, x_{\hat{\psi}}^* + \Delta_{\hat{\psi}}]$$

is referred to the time–frequency window of $\psi$. For $a, b \in \mathbb{R}, a \neq 0$, a set of wavelets can be generated by the basic wavelet $\psi$ as follows:

$$\psi_{a,b}(t) := |a|^{-\frac{1}{2}} \psi\left(\frac{t - b}{a}\right). \qquad (2.5)$$

Thus, the time–frequency window of $\psi_{a,b}$ is

$$[b + ax_\psi^* - |a|\Delta_\psi, b + ax_\psi^* + |a|\Delta_\psi] \times \left[ \frac{x_{\hat{\psi}}^*}{a} - \frac{1}{|a|}\Delta_{\hat{\psi}}, \frac{x_{\hat{\psi}}^*}{a} + \frac{1}{|a|}\Delta_{\hat{\psi}} \right]. \qquad (2.6)$$

The procedure of reasoning is as follows:

$$\begin{aligned}
x_{\psi_{a,b}}^* &= \frac{1}{\|\psi\|_2^2} \int_{\mathbb{R}} t|a|^{-1} \left| \psi\left(\frac{t - b}{a}\right) \right|^2 dt \\
&= \frac{|a|^{-1}}{\|\psi\|_2^2} \int_{\mathbb{R}} (ax + b)|\psi(x)|^2 |a| dx
\end{aligned}$$

$$= \frac{1}{\|\psi\|_2^2} \left[ a \int_{\mathbb{R}} x|\psi(x)|^2 dx + b\|\psi\|_2^2 \right]$$

$$= ax_\psi^* + b;$$

$$\Delta_{\psi_{a,b}} = \frac{1}{\|\psi\|_2} \left[ \int_{\mathbb{R}} (t - ax_\psi^* - b)^2 |a|^{-1} \left| \psi\left(\frac{t-b}{a}\right) \right|^2 dt \right]^{1/2}$$

$$= \frac{1}{\|\psi\|_2} \left[ \int_{\mathbb{R}} (ax - ax_\psi^*)^2 |\psi(x)|^2 |a|^{-1} |a| dx \right]^{1/2}$$

$$= \frac{|a|}{\|\psi\|_2} \left[ \int_{\mathbb{R}} (x - x_\psi^*)^2 |\psi(x)|^2 dx \right]^{1/2}$$

$$= |a|\Delta_\psi;$$

$$\hat{\psi}_{a,b}(\xi) = \int_{\mathbb{R}} |a|^{-\frac{1}{2}} \psi\left(\frac{t-b}{a}\right) e^{-i\xi t} dt = |a|^{\frac{1}{2}} e^{-ib\xi} \hat{\psi}(a\xi);$$

$$\|\hat{\psi}_{a,b}\|_2^2 = |a| \int_{\mathbb{R}} |\hat{\psi}(a\xi)|^2 d\xi = \|\hat{\psi}\|_2^2;$$

$$x_{\hat{\psi}_{a,b}}^* = \frac{1}{\|\hat{\psi}_{a,b}\|_2^2} \int_{\mathbb{R}} \xi|a||\hat{\psi}(a\xi)|^2 d\xi$$

$$= \frac{1}{a} \frac{1}{\|\hat{\psi}_{a,b}\|_2^2} \int_{\mathbb{R}} \xi|\hat{\psi}(\xi)|^2 d\xi$$

$$= \frac{1}{a} x_{\hat{\psi}}^*;$$

$$\Delta_{\hat{\psi}_{a,b}} = \frac{1}{\|\hat{\psi}\|_2} \left[ \int_{\mathbb{R}} \left(\xi - \frac{1}{|a|} x_{\hat{\psi}}^*\right)^2 |a||\hat{\psi}(a\xi)|^2 d\xi \right]^{1/2}$$

$$= \frac{1}{|a|} \frac{1}{\|\hat{\psi}\|_2} \left[ \int_{\mathbb{R}} (|a|\xi - x_{\hat{\psi}}^*)^2 |\hat{\psi}(a\xi)|^2 |a| d\xi \right]^{1/2}$$

$$= \frac{1}{|a|} \frac{1}{\|\hat{\psi}\|_2} \left[ \int_{\mathbb{R}} (\xi - x_{\hat{\psi}}^*)^2 |\hat{\psi}(\xi)|^2 d\xi \right]^{1/2}$$

$$= \frac{1}{|a|} \Delta_{\hat{\psi}}.$$

From (2.6), we can easily find out, when $|a|$ changes to be smaller, the time window of $\psi$ becomes more narrow, whereas the frequency window becomes wider. The area of its time–frequency window is a constant, which is irrelevant to $a$ and $b$.

$$(2\Delta_{\psi_{a,b}})(2\Delta_{\hat{\psi}_{a,b}}) = 4\Delta_\psi\Delta_{\hat{\psi}}.$$

From the above discussion, we find an intrinsic fact: For a basic wavelet $\psi$, it is impossible to achieve perfect localization in both the time domain and the frequency domain simultaneously. Is there any $\psi$ that could make the area of the time–frequency to be small enough? Unfortunately, the following theorem, the famous Heisenberg uncertainty principle, gave a negative answer to this question.

**Theorem 2.1.** *Let $\psi \in L^2(\mathbb{R})$ satisfy $x\psi(x)$ and $\xi\hat{\psi}(\xi) \in L^2(\mathbb{R})$. Then,*

$$\Delta_\psi\Delta_{\hat{\psi}} \geq \frac{1}{2}.$$

*Furthermore, the equality in the above equation holds if and only if*

$$\psi(x) = ce^{iax}g_\alpha(x-b),$$

*where $c \neq 0, \alpha > 0$, $a, b \in \mathbb{R}$ and $g_\alpha(x)$ is the Gaussian function defined by*

$$g_\alpha(x) := \frac{1}{2\sqrt{\pi\alpha}}e^{-\frac{x^2}{4\alpha}}. \tag{2.7}$$

**Proof.** We assume that the window centers of $\psi$ and $\hat{\psi}$ are 0 without losing generality. Based on the basic knowledge of Fourier analysis, we have

$$(\Delta_\psi\Delta_{\hat{\psi}})^2 = \frac{\left(\int_\mathbb{R} t^2|\psi(t)|^2dt\right)\left(\int_\mathbb{R} \xi^2|\hat{\psi}(\xi)|^2d\xi\right)}{\|\psi\|_2^2\|\hat{\psi}\|_2^2}$$

$$= \frac{\left(\int_\mathbb{R} t^2|\psi(t)|^2dt\right)\left(\int_\mathbb{R} |\hat{\psi}'(\xi)|^2d\xi\right)}{\|\psi\|_2^2\|\hat{\psi}\|_2^2}$$

$$= \frac{\|x\psi(x)\|_2^2\|\hat{\psi}'(x)\|_2^2}{\|\psi\|_2^2\|\hat{\psi}\|_2^2}$$

$$= \frac{\|x\psi(x)\|_2^2\ 2\pi\|\psi'(x)\|_2^2}{\|\psi\|_2^2\ 2\pi\|\psi\|_2^2}$$

$$\geq \frac{\|x\psi(x)\psi'(x)\|_2^2}{\|\psi\|_2^4}$$

$$\geq \frac{1}{\|\psi\|_2^4}\left|\text{Re}\int_\mathbb{R} x\psi(x)\overline{\psi'(x)}dx\right|^2$$

$$= \frac{1}{\|\psi\|_2^4} \left| \frac{1}{2} \int_{\mathbb{R}} x \frac{d}{dx} |\psi(x)|^2 dx \right|^2$$

$$= \frac{1}{\|\psi\|_2^4} \left( \frac{1}{2} \int_{\mathbb{R}} |\psi(x)|^2 dx \right)^2$$

$$= \frac{1}{4}.$$

Thus, we have

$$\Delta_\psi \Delta_{\hat\psi} \geq \frac{1}{2}.$$

Further, by the conditions, which preserve the equal sign in the Holder inequality, we know that the equalities in the above reasoning will be tenable if and only if there is a constant $\alpha$ such that

$$\begin{cases} -\mathrm{Re}(x\psi(x)\overline{\psi'(x)}) = |x\psi(x)\overline{\psi'(x)}|, \\ |x\psi(x)| = 2\alpha|\psi'(x)|, \\ \|\psi\|_2 \neq 0. \end{cases}$$

First, by the second equality, we have

$$x\psi(x) = 2\alpha\psi'(x)e^{i\theta(x)},$$

where $\theta(x)$ is a real-valued function. Second, by the first equality, we have

$$-x\psi(x)\overline{\psi'(x)} \geq 0.$$

Then,

$$-2\alpha|\psi'(x)|^2 e^{i\theta(x)} \geq 0.$$

It infers $e^{i\theta(x)} = -1$. Thus, we have

$$x\psi(x) = -2\alpha\psi'(x).$$

By resolving this ordinary differential equation, we can obtain

$$\psi(x) = ce^{-\frac{x^2}{4\alpha}}.$$

Finally, by the third equality, we know that $c \neq 0$.

It should be mentioned, at the beginning of our proof, we have assumed that the window centers of $\psi$ and $\hat\psi$ are 0. Otherwise, consider

$$\tilde\psi(x) := e^{-iax}\psi(x+b),$$

where $a := x_{\hat{\psi}}^*$, $b := x_{\psi}^*$. The centers of the time window and the frequency window of $\tilde{\psi}(x)$ are 0. It means that the above reasoning is available to $\tilde{\psi}(x)$. It is easy to know that

$$\Delta_\psi = \Delta_{\tilde{\psi}}, \quad \Delta_{\hat{\psi}} = \Delta_{\hat{\tilde{\psi}}},$$

Hence, $\Delta_\psi \Delta_{\hat{\psi}} \geq \frac{1}{2}$. The equality holds if and only if

$$\tilde{\psi}(x) = ce^{-\frac{x^2}{4\alpha}},$$

i.e.,

$$\psi(x) = ce^{i(x-b)a}e^{-\frac{(x-b)^2}{4\alpha}},$$

where $c \neq 0$. Replacing $\frac{ce^{iab}}{2\sqrt{2\pi\alpha}}$ with $c$, we have

$$\psi(x) = ce^{iax}g_\alpha(x-b),$$

where $c \neq 0, \alpha > 0$, $a, b \in \mathbb{R}$ and $g_\alpha(x)$ is the Gaussian function defined by (2.7). This establishes the theorem.                                    □

An important fact is supported by this theorem: No matter what wavelet $\psi$ we choose, it cannot achieve perfect localization in both the time domain and the frequency domain simultaneously. The time window and the frequency window are just like the length and width of a rectangle with a fixed area. When the time window is narrow, the frequency window must be wide. By the contrary, when the time window is wide, the frequency window must be narrow (see Fig. 2.2). In fact, the narrow time window and the wide frequency window are good for the high frequency part of a signal, whereas the wide time window and the narrow frequency window are good for the low frequency part of a signal. Therefore, this property of wavelet just meets the demand of signal analysis. It can be applied for self-adaptive signal analysis.

The area of the time–frequency window of Gaussian is the smallest, which is $4\Delta_\psi \Delta_{\hat{\psi}} = 2$. Therefore, for the localized time–frequency analysis, Gaussian function is the best. It has been a traditional tool for signal analysis.

Like the same as Fourier transform, the wavelet transform is invertible. The inverse wavelet transform can be viewed as the reconstruction of the original signal. First of all, we need to prove the following theorem, which specifies that wavelet transform keeps the law of energy conservation.

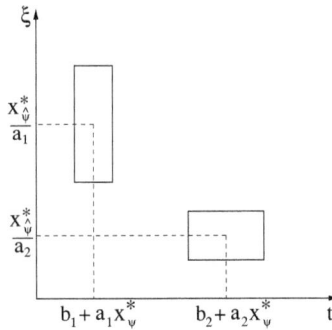

Fig. 2.2 Time–frequency windows, $a_1 < a_2$.

**Theorem 2.2.** *Let $\psi$ be a basic wavelet. Then,*

$$\int_{\mathbb{R}} \int_{\mathbb{R}} W_\psi f(a, b)\overline{W_\psi g(a, b)}\frac{dadb}{a^2} = C_\psi \langle f, \, g \rangle$$

*holds for any $f, \, g \in L^2(\mathbb{R})$, where $C_\psi$ is a constant defined by (2.1), $\langle f, \, g \rangle$ is the inner product of $f$ and $g$. Particularly, for $g = f$, we have*

$$\int_{\mathbb{R}} \int_{\mathbb{R}} |W_\psi f(a, b)|^2 \frac{dadb}{a^2} = C_\psi \|f\|_2^2.$$

**Proof.** Obviously, the second equality is a specific instance of the first equality at $g = f$. Thus, only the first equality needs to be proved. It is easy to know that

$$\left(\psi\left(\frac{t-b}{a}\right)\right)\hat{}(\xi) = |a|e^{-ib\xi}\hat{\psi}(a\xi).$$

Denote that

$$\begin{cases} F(x) := \hat{f}(x)\overline{\hat{\psi}(ax)}, \\ G(x) := \hat{g}(x)\overline{\hat{\psi}(ax)}. \end{cases}$$

Then,

$$(W_\psi f)(a, b) = \left\langle f(x), |a|^{-\frac{1}{2}}\psi\left(\frac{x-b}{a}\right) \right\rangle$$

$$= \frac{1}{2\pi}\langle \hat{f}(\xi), |a|^{\frac{1}{2}}e^{-b\xi}\hat{\psi}(a\xi)\rangle$$

$$= \frac{1}{2\pi}|a|^{\frac{1}{2}} \int_{\mathbb{R}} \hat{f}(\xi)\overline{\hat{\psi}(a\xi)}e^{b\xi}d\xi$$

$$= \frac{1}{2\pi}|a|^{\frac{1}{2}} \int_{\mathbb{R}} F(\xi)e^{b\xi}d\xi$$

$$= \frac{1}{2\pi}|a|^{\frac{1}{2}}\hat{F}(-b).$$

In the same way,

$$(W_\psi g)(a,b) = \frac{1}{2\pi}|a|^{\frac{1}{2}}\hat{G}(-b).$$

Therefore,

$$\int_{\mathbb{R}} (W_\psi f)(a,b)\overline{(W_\psi g)(a,b)}db$$

$$= \left(\frac{1}{2\pi}\right)^2 |a| \int_{\mathbb{R}} \hat{F}(-b)\overline{\hat{G}(-b)}db$$

$$= \left(\frac{1}{2\pi}\right)^2 |a| \int_{\mathbb{R}} \hat{F}(b)\overline{\hat{G}(b)}db$$

$$= \frac{1}{2\pi}|a| \int_{\mathbb{R}} F(x)\overline{G(x)}dx.$$

Furthermore,

$$\int_{\mathbb{R}} \int_{\mathbb{R}} (W_\psi f(a,b)\overline{W_\psi g(a,b)}\frac{dadb}{a^2}$$

$$= \int_{\mathbb{R}} \frac{1}{2\pi}|a| \int_{\mathbb{R}} F(x)\overline{G(x)}dx\frac{da}{a^2}$$

$$= \frac{1}{2\pi} \int_{\mathbb{R}} \int_{\mathbb{R}} \frac{|\psi(ax)|^2}{|a|}\hat{f}(x)\overline{\hat{g}(x)}dxda$$

$$= \frac{1}{2\pi} \int_{\mathbb{R}} \left(\int_{\mathbb{R}} \frac{|\psi(ax)|^2}{|ax|}d(ax)\right)\hat{f}(x)\overline{\hat{g}(x)}dx$$

$$= \frac{1}{2\pi} \left(\int_{\mathbb{R}} \frac{|\psi(\xi)|^2}{|\xi|}d\xi\right) \int_{\mathbb{R}} \hat{f}(x)\overline{\hat{g}(x)}dx$$

$$= \frac{1}{2\pi}C_\psi\langle\hat{f},\hat{g}\rangle$$

$$= C_\psi\langle f,g\rangle.$$

This completes our proof.     $\square$

According to the above theorem, the inverse wavelet transform, which is also called the reconstruction formula, can be formally inferred. We consider

$$\overline{W_\psi g(a,b)} = \left\langle |a|^{-\frac{1}{2}}\psi\left(\frac{x-b}{a}\right), \, g\right\rangle.$$

If we give up the exactness in mathematics temporarily, we have

$$C_\psi\langle f,g\rangle = \int_{\mathbb{R}}\int_{\mathbb{R}} W_\psi f(a,b)\left\langle |a|^{-\frac{1}{2}}\psi\left(\frac{x-b}{a}\right), \, g\right\rangle \frac{dadb}{a^2}$$

$$= \left\langle \int_{\mathbb{R}}\int_{\mathbb{R}} W_\psi f(a,b)|a|^{-\frac{1}{2}}\psi\left(\frac{x-b}{a}\right)\frac{dadb}{a^2}, \, g\right\rangle.$$

Since $g$ can be any function in $L^2(\mathbb{R})$, we can write

$$f(x) = C_\psi^{-1}\int_{\mathbb{R}}\int_{\mathbb{R}} W_\psi f(a,b)|a|^{-\frac{1}{2}}\psi\left(\frac{x-b}{a}\right)\frac{dadb}{a^2}. \tag{2.8}$$

This is the formal inverse wavelet transform, by which we can reconstruct the original signal from the wavelet transform $W_\psi f(a,b)$. The above reasoning is not exact in mathematics. The exact inverse formula is studied as follows.

**Theorem 2.3.** *Let $\psi$ be a basic wavelet. Then, for any $f \in L^2(\mathbb{R})$, the inverse wavelet transform (2.8) holds in the sense of $L^2(\mathbb{R})$-norm:*

$$\left\| f(x) - C_\psi^{-1}\int_{|a|\geq A} da \int_{|b|\leq B}(W_\psi f)(a,b)|a|^{-\frac{1}{2}}\psi\left(\frac{x-b}{a}\right)\frac{dadb}{a^2}\right\|_2 \to 0,$$

*as $A \to 0$, $B \to \infty$, where $C_\psi$ is a constant defined by (2.1).*

**Proof.** For any $A, B > 0$, we prove the following equality at first:

$$\left\langle \iint_{|a|\geq A,|b|\leq B}(W_\psi f)(a,b)|a|^{-\frac{1}{2}}\psi\left(\frac{x-b}{a}\right)\frac{dadb}{a^2}, \, g(x)\right\rangle$$

$$= \iint_{|a|\geq A,|b|\leq B}(W_\psi f)(a,b)\overline{(W_\psi g)(a,b)}\frac{dadb}{a^2}.$$

In fact,

$$\int_{\mathbb{R}} \iint_{|a|\geq A,|b|\leq B} \left| (W_\psi f)(a,b)|a|^{-\frac{1}{2}} \psi\left(\frac{x-b}{a}\right) g(x) \right| \frac{dadb}{a^2} dx$$

$$\leq \iint_{|a|\geq A,|b|\leq B} |(W_\psi f)(a,b)||a|^{-\frac{1}{2}} \left(\int_{\mathbb{R}} |\psi\left(\frac{x-b}{a}\right)|^2 dx\right)^{\frac{1}{2}} \|g\|_2 \frac{dadb}{a^2} dx$$

$$= \|\psi\|_2 \|g\|_2 \iint_{|a|\geq A,|b|\leq B} |(W_\psi f)(a,b)| \frac{dadb}{a^2}$$

$$\leq \|\psi\|_2 \|g\|_2 \left( \iint_{|a|\geq A,|b|\leq B} |(W_\psi f)(a,b)|^2 \frac{dadb}{a^2} \right)^{1/2}$$

$$\left( \iint_{|a|\geq A,|b|\leq B} \frac{dadb}{a^2} \right)^{1/2}$$

$$= \|\psi\|_2 \|g\|_2 C_\psi^{1/2} \|f\|_2 \left( 4B \int_A^\infty \frac{da}{a^2} \right)^{1/2}$$

$$= \|\psi\|_2 \|g\|_2 \|f\|_2 \left( \frac{4B}{A} C_\psi \right)^{1/2} < \infty,$$

where $C_\psi$ is the constant defined by (2.1). By the Fubini theorem in real-analysis [Rudin, 1974], we have

$$\left\langle \iint_{|a|\geq A,|b|\leq B} (W_\psi f)(a,b)|a|^{-\frac{1}{2}} \psi\left(\frac{x-b}{a}\right) \frac{dadb}{a^2}, \ g(x) \right\rangle$$

$$= \int_{\mathbb{R}} \iint_{|a|\geq A,|b|\leq B} (W_\psi f)(a,b)|a|^{-\frac{1}{2}} \psi\left(\frac{x-b}{a}\right) \overline{g(x)} \frac{dadb}{a^2} dx$$

$$= \iint_{|a|\geq A,|b|\leq B} (W_\psi f)(a,b)|a|^{-\frac{1}{2}} \int_{\mathbb{R}} \psi\left(\frac{x-b}{a}\right) \overline{g(x)} dx \frac{dadb}{a^2}$$

$$= \iint_{|a|\geq A,|b|\leq B} (W_\psi f)(a,b)\overline{(W_\psi g)(a,b)} \frac{dadb}{a^2}.$$

This is the equality that we intend to prove. Therefore,

$$\left\| f(x) - C_\psi^{-1} \int_{|a|\geq A} da \int_{|b|\leq B} (W_\psi f)(a,b)|a|^{-\frac{1}{2}} \psi\left(\frac{x-b}{a}\right) \frac{dadb}{a^2} \right\|_2$$

$$= C_\psi^{-1} \sup_{\|g\|_2=1} |C_\psi \langle f, \ g\rangle$$

$$-\left\langle \iint_{|a|\geq A,|b|\leq B}(W_\psi f)(a,b)|a|^{-\frac{1}{2}}\psi\left(\frac{x-b}{a}\right)\frac{dadb}{a^2},\ g(x)\right\rangle\Bigg|$$

$$= C_\psi^{-1}\sup_{\|g\|_2=1}\left|C_\psi\langle f,\ g\rangle-\iint_{|a|\geq A,|b|\leq B}(W_\psi f)(a,b)\overline{(W_\psi g)(a,b)}\frac{dadb}{a^2}\right|$$

$$= C_\psi^{-1}\sup_{\|g\|_2=1}\left|\iint_{|a|<A\ or\ |b|>B}(W_\psi f)(a,b)\overline{(W_\psi g)(a,b)}\frac{dadb}{a^2}\right|$$

$$\leq C_\psi^{-1}\sup_{\|g\|_2=1}\left(\iint_{\mathbb{R}^2}|(W_\psi g)(a,b)|^2\frac{dadb}{a^2}\right)^{1/2}$$

$$\left(\iint_{|a|<A\ or\ |b|>B}|(W_\psi f)(a,b)|^2\frac{dadb}{a^2}\right)^{1/2}$$

$$= C_\psi^{-1}\sup_{\|g\|_2=1}C_\psi^{1/2}\|g\|_2\left(\iint_{|a|<A\ or\ |b|>B}|(W_\psi f)(a,b)|^2\frac{dadb}{a^2}\right)^{1/2}$$

$$= C_\psi^{-1/2}\left(\iint_{|a|<A\ or\ |b|>B}|(W_\psi f)(a,b)|^2\frac{dadb}{a^2}\right)^{1/2}$$

$$\to 0,\quad (A\to 0,\ B\to\infty).$$

The proof of the theorem is complete. $\qquad\qquad\square$

*Note*: The result of Theorem 2.3 is not so clear as formula (2.8). It is important to determine whether formula (2.8) is tenable.

By the result of Theorem 2.3, the following conclusion can be mathematically inferred: As $A\to 0$, $B\to\infty$,

$$f_{A,B}(x):=C_\psi^{-1}\int_{|a|\geq A}da\int_{|b|\leq B}(W_\psi f)(a,b)|a|^{-\frac{1}{2}}\psi\left(\frac{x-b}{a}\right)\frac{dadb}{a^2}$$

converges to $f(x)$ in measure [Rudin, 1974]. According to Riesz theorem [Rudin, 1974], there are two sequences $A_k\to 0+$ and $B_k\to+\infty$, such that $f_{A_k,B_k}(x)\to f(x),\ (k\to\infty),\ a.e.\ x\in\mathbb{R}.$, i.e.,

$$f(x)=\lim_{k\to\infty}C_\psi^{-1}\int_{|a|\geq A_k}da\int_{|b|\leq B_k}(W_\psi f)(a,b)|a|^{-\frac{1}{2}}\psi\left(\frac{x-b}{a}\right)\frac{dadb}{a^2}.$$

If the right integral of (2.8) exists, (2.8) holds *a.e.* $x\in\mathbb{R}$. Based on this conclusion, we have the following theorem.

**Theorem 2.4.** *Let $\psi$ be a basic wavelet. Then for any $f \in L^2(\mathbb{R})$, (2.8) holds a.e. $x \in \mathbb{R}$ if*

$$\int_{\mathbb{R}} \int_{\mathbb{R}} \left| (W_\psi f)(a,b)|a|^{-\frac{1}{2}} \psi\left(\frac{x-b}{a}\right) \right| \frac{dadb}{a^2} < \infty, \qquad (2.9)$$

*where $C_\psi$ is a constant, which can be defined by (2.1).*

The following theorem gives a sufficient condition such that (2.9) holds.

**Theorem 2.5.** *Let $\psi \in L^1(\mathbb{R})$ be a basic wavelet. For any $f \in L^2(\mathbb{R})$, if there exists a non-negative measurable function $F(a)$, such that*

$$|W_\psi f(a,b)| \le F(a), \quad and \quad \int_{\mathbb{R}} \frac{1}{|a|^{3/2}} F(a) da < \infty,$$

*then, (2.9) holds. Consequently, the inverse wavelet transform (2.8) holds a.e. $x \in \mathbb{R}$. Furthermore, if $\psi(x)$ is continuous, the right part of (2.8) is a continuous function on $\mathbb{R}$.*

**Proof.** It is known that

$$\int_{\mathbb{R}} \int_{\mathbb{R}} \left| (W_\psi f)(a,b)|a|^{-\frac{1}{2}} \psi\left(\frac{x-b}{a}\right) \right| \frac{dadb}{a^2}$$

$$\le \int_{\mathbb{R}} \int_{\mathbb{R}} F(a) \left| \psi\left(\frac{x-b}{a}\right) \right| \frac{1}{|a|^{5/2}} dadb$$

$$= \int_{\mathbb{R}} F(a) \frac{1}{|a|^{5/2}} \left( \int_{\mathbb{R}} \left| \psi\left(\frac{x-b}{a}\right) \right| db \right) da$$

$$= \|\psi\|_1 \int_{\mathbb{R}} \frac{1}{|a|^{3/2}} F(a) da$$

$$< \infty.$$

Thus, (2.9) holds. If $\psi(x)$ is continuous, we denote

$$w_x(a,b) := (W_\psi f)(a, x-b)|a|^{-\frac{1}{2}} \psi\left(\frac{b}{a}\right) \frac{1}{a^2}.$$

Then,

$$|w_x(a,b)| \le F(a) \left| \psi\left(\frac{b}{a}\right) \right| \frac{1}{|a|^{5/2}} \in L^1(\mathbb{R}^2).$$

$\forall x_n$, $x \in \mathbb{R}$, $x_n \to x$ $(n \to \infty)$, by the Lebesgue dominated convergence theorem [Rudin, 1974], we have

$$\lim_{n \to \infty} \int_{\mathbb{R}} \int_{\mathbb{R}} w_{x_n}(a, b) da db = \int_{\mathbb{R}} \int_{\mathbb{R}} w_x(a, b) da db.$$

Note that

$$\lim_{n \to \infty} \int_{\mathbb{R}} \int_{\mathbb{R}} (W_\psi f)(a, b) |a|^{-\frac{1}{2}} \psi \left( \frac{x_n - b}{a} \right) \frac{da db}{a^2}$$

$$= \lim_{n \to \infty} \int_{\mathbb{R}} \int_{\mathbb{R}} w_{x_n}(a, b) da db$$

$$= \int_{\mathbb{R}} \int_{\mathbb{R}} w_x(a, b) da db,$$

therefore, the right part of (2.8), $\int_{\mathbb{R}} \int_{\mathbb{R}} w_x(a, b) da db$, is a continuous function on $\mathbb{R}$. Our proof is complete. $\qquad\square$

*Note*: Let $\psi$ be a basic wavelet, $\alpha > 0$ and $f \in L^2(\mathbb{R})$. If there exists constants $C$, $\delta > 0$, such that

$$|W_\psi f(a, b)| \leq C|a|^{\alpha + \frac{1}{2}}, \quad (\forall b \in \mathbb{R}),$$

for $|a| < \delta$, the conditions of the previous theorem are satisfied. In fact, it will be clear if we let

$$F(a) := \begin{cases} C|a|^{\alpha + \frac{1}{2}}, & |a| < \delta; \\ \|f\|_2 \|\psi\|_2, & |a| \geq \delta. \end{cases}$$

The necessity of the admissibility condition (2.1) is very important. It ensures the existence of the inverse wavelet transform. As a conclusion, it is feasible to apply the wavelet transform to signal analysis, image processing, and pattern recognition. It is easy to see that this condition is rather weak and is almost equivalent to $\hat{\psi}(0) = 0$ or

$$\int_{\mathbb{R}} \psi(x) dx = 0. \tag{2.10}$$

Actually, if $\psi(x) \in L^2(\mathbb{R})$ and there exists $\alpha > 0$, such that $(1 + |x|)^\alpha \psi(x) \in L^1(\mathbb{R})$, then (2.1) must be equivalent to (2.10). This conclusion is proved as follows.

By $(1 + |x|)^\alpha \psi(x) \in L^1(\mathbb{R})$, we know that $\psi(x) \in L^1(\mathbb{R})$. It means that $\hat{\psi}(\xi)$ is a continuous function in $\mathbb{R}$. If (2.1) is tenable, then $\hat{\psi}(0) = 0$.

It indicates that (2.10) holds. Contrarily, if (2.10) holds, we assume $\alpha \leq 1$ without losing generality, then

$$
|\hat{\psi}(\xi)| = \left| \int_{\mathbb{R}} \psi(x) e^{-i\xi x} dx \right|
$$

$$
= \left| \int_{\mathbb{R}} \psi(x)[e^{-i\xi x} - 1] dx \right|
$$

$$
\leq \int_{\mathbb{R}} \left| \psi(x) 2 \sin \frac{\xi x}{2} \right| dx
$$

$$
\leq |\xi|^\alpha \int_{|\xi x| \leq 1} |x^\alpha \psi(x)| dx + |\xi|^\alpha \int_{|\xi x| > 1} |x^\alpha \psi(x)| dx
$$

$$
= |\xi|^\alpha \int_{\mathbb{R}} |x^\alpha \psi(x)| dx
$$

$$
= C|\xi|^\alpha,
$$

where $C := \int_{\mathbb{R}} |x^\alpha \psi(x)| dx$. Therefore,

$$
C_\psi := \int_{\mathbb{R}} \frac{|\hat{\psi}(\xi)|^2}{|\xi|} d\xi \leq C^2 \int_{|\xi \leq 1} |\xi|^{2\alpha - 1} d\xi + \int_{|\xi| > 1} |\hat{\psi}(\xi)|^2 d\xi < \infty.
$$

It specifies that (2.1) holds. This finished our proof. $\qquad\square$

In general, the wavelets applied to practice have good damping and satisfy the condition $(1 + |x|)^\alpha \psi(x) \in L^1(\mathbb{R})$ $(\alpha > 0)$. As a result, in engineering, the wavelet is often defined as the function in $L^2(\mathbb{R})$ which satisfies the condition (2.10). This definition is not exact in mathematics, however, it is harmless in the application of wavelet. The condition (2.10) is an objective specification of the vibration of $\psi$. The damping and the vibration are two basic characteristics of basic wavelets.

In signal analysis, only the positive frequency needs to be considered, i.e., $a > 0$. With this premise, we have another theorem.

**Theorem 2.6.** *Let $\psi$ be a basic wavelet satisfying*

$$
\int_0^\infty \frac{|\hat{\psi}(\xi)|^2}{\xi} d\xi = \int_0^\infty \frac{|\hat{\psi}(-\xi)|^2}{\xi} d\xi = \frac{1}{2} C_\psi < \infty.
$$

*Then, for any $f$, $g \in L^2(\mathbb{R})$, we have*

$$\int_0^\infty \int_\mathbb{R} W_\psi f(a,b) \overline{W_\psi g(a,b)} \frac{dadb}{a^2} = \frac{1}{2} C_\psi \langle f, g \rangle.$$

*In particular,*

$$\int_0^\infty \int_\mathbb{R} |W_\psi f(a,b)|^2 \frac{dadb}{a^2} = \frac{1}{2} C_\psi \|f\|_2^2.$$

The proof of this theorem is similar to those of Theorems 2.2 and 2.3.

## 2.2 The Continuous Wavelet Transform as a Filter

Let $\psi$ be a basic wavelet and we denote

$$\tilde{\psi}(x) := \overline{\psi(-x)}. \tag{2.11}$$

We define scale wavelet transform as follows:

$$W_s f(x) := W_s^{\tilde{\psi}} f(x) := (f * \tilde{\psi})(x). \tag{2.12}$$

By the definition of the wavelet transform, for $a > 0$, we have

$$\begin{aligned}
(W_\psi f)(a,b) &:= \int_R f(t) a^{-\frac{1}{2}} \overline{\psi\left(\frac{t-b}{a}\right)} dt \\
&= \int_R f(t) a^{-\frac{1}{2}} \tilde{\psi}\left(\frac{b-t}{a}\right) dt \\
&= a^{\frac{1}{2}} (f * \tilde{\psi}_a)(b),
\end{aligned}$$

where $\tilde{\psi}_a(x) := \frac{1}{a}\tilde{\psi}(\frac{x}{a})$. It specifies that the wavelet transform is actually a convolution, which is also called a filter in engineering.

The design of filters is very important in the filter theory. There are two kinds of digital filters: the finite impulse response (FIR) and the infinite impulse response (IIR). The former is easy to be realized and has good time localization. Therefore, in the above-scale wavelet transforms, the basic wavelet $\psi$, which is the kernel of the transform, should be compactly supported. Although the Heisenberg uncertainty principle has specified that the area of the time–frequency domain cannot be arbitrarily small, the localization in the time–frequency domain still can be perfect if $\psi$ and its Fourier transform $\hat{\psi}$ have compact support simultaneously. Unfortunately, from the following theorem, we can find that this condition cannot be satisfied.

**Theorem 2.7.** *If $f \in L^2(\mathbb{R})$ is a non-zero function, $f$ and its Fourier transform $\hat{f}$ cannot be compactly supported simultaneously.*

**Proof.** If $\hat{f}$ is compactly supported, supp $\hat{f} \subset [-B, B]$, then

$$f(z) := \frac{1}{2\pi} \int_{-B}^{B} \hat{f}(\xi) e^{i\xi z} d\xi$$

is an analytic function on complex plane $\mathbb{C}$. Due to the zero isolation of non-zero analytic functions, $f$ cannot be compactly supported. This finishes our proof. □

We have known that the Fourier transform $\hat{\psi}$ of a compactly supported basic wavelet $\psi$ cannot be compactly supported. Now, we wish its damping property would be good enough. It is essentially equivalent to the fact that the smoothness of $\psi$ would be good enough. The following theorem ensures this fact.

**Theorem 2.8.** *Let $m$ be a non-negative integer and $H^m(\mathbb{R})$ be the Sobolev space of order $m$ defined by*

$$H^m(\mathbb{R}) := \{f \mid f, f', \ldots, f^{(m)} \in L^2(\mathbb{R})\}.$$

*Then, $f \in H^m(\mathbb{R})$ if and only if*

$$\int_{\mathbb{R}} (1 + |x|^m)|\hat{f}(x)| dx < \infty. \tag{2.13}$$

**Proof.** This is a fundamental fact in the theory of Sobolev spaces. The detailed proof can be found in Gilbarg and Trudinger [1977]. □

According to the definition of $H^m(\mathbb{R})$, $f \in H^m(\mathbb{R})$ indicates that $f$ has certainly differentiableness and smoothness (or regularity). However, (2.13) indicates that $\hat{f}$ has damping of order $m$. Therefore, when we choose a basic wavelet $\psi$ as a filter function, in order to ensure that $\psi$ is good for localized analysis in the time frequency, we must consider both its damping (or compact support) and its regularity.

Another very important property of filters is the linear phase. With the linear phase or the generalized linear phase, a filter can avoid distortion.

The linear phase is mathematically defined as follows.

**Definition 2.2.** $f \in L^2(\mathbb{R})$ is said to have a linear phase if its Fourier transform satisfies

$$\hat{f}(\xi) = \epsilon|\hat{f}(\xi)|e^{-ia\xi},$$

where $\epsilon = 1$ or $-1$ and $a$ is a real constant. $f$ is said to have generalized linear phase if there exists a real function $F(\xi)$ and two real constants $a$ and $b$, such that

$$\hat{f}(\xi) = F(\xi)e^{-i(a\xi+b)}.$$

Obviously, if $f$ has a linear phase, it also has a generalized linear phase. The generalized linear phase will degenerate to the linear phase if and only if $e^{-ib} = \pm 1$ and the real function $F(\xi)$ keeps its sign, i.e., identically positive or identically negative.

According to the above mathematical definition, the linear phase or the generalized linear phase of $\psi$ is an attribute in the Fourier transform domain. In the following, we will demonstrate that the generalized linear phase is equivalent to a symmetry of $\psi$ itself.

**Theorem 2.9.** *$f \in L^2(\mathbb{R})$ has a generalized phase if and only if $f$ is skew-symmetric at $a \in \mathbb{R}$, i.e., there exists a constant $b \in \mathbb{R}$ such that*

$$e^{ib}f(a+x) = \overline{e^{ib}f(a-x)}, \quad x \in \mathbb{R}.$$

*In particular, for a real function $f$, it has a generalized linear phase if and only if it is symmetric or antisymmetric at $a \in \mathbb{R}$. More precisely,*

$$f(a+x) = f(a-x), \quad x \in \mathbb{R},$$

*or*

$$f(a+x) = -f(a-x), \quad x \in \mathbb{R}.$$

**Proof.** It is easy to see that $f$ has generalized phase if and only if two constants $a$ and $b \in \mathbb{R}$ exist, such that

$$\hat{f}(\xi)e^{i(a\xi+b)} = F(\xi)$$

is a real function, i.e.,

$$\hat{f}(\xi)e^{i(a\xi+b)} = \overline{\hat{f}(\xi)e^{i(a\xi+b)}}, \quad (\xi \in \mathbb{R}),$$

which is equivalent to

$$\int_{\mathbb{R}} \hat{f}(\xi)e^{i(a\xi+b)}e^{i\xi x}d\xi = \overline{\int_{\mathbb{R}} \hat{f}(\xi)e^{i(a\xi+b)}e^{-i\xi x}d\xi}, \quad (x \in \mathbb{R}).$$

i.e.,

$$e^{ib} f(a+x) = \overline{e^{ib} f(a-x)}, \quad (x \in \mathbb{R}).$$

If $f$ is a real function, $e^{i2b}$ must be a real function. It means $e^{i2b} = 1$ or $-1$. Thus,

$$f(a+x) = f(a-x), \quad x \in \mathbb{R},$$

or

$$f(a+x) = -f(a-x), \quad x \in \mathbb{R}.$$

It specifies that $f$ is symmetric or antisymmetric on $a$. Our proof is complete. $\qquad\square$

Summarily, as a filter, in order to be good for the localized analysis in the time domain, the basic wavelet $\psi$ should have good damping or compact support. On the other hand, $\psi$ should have good regularity or smoothness to obtain a good property for the localized analysis in the frequency domain. At last, $\psi$ should be skew-symmetric to avoid the distortion.

## 2.3 Characterization of Lipschitz Regularity of Signal by Wavelet

In mathematics, a signal is actually a function. A stationary signal always corresponds to a smooth function, while a transient one refers to a singularity. In fact, the concepts of the stability and singularity of the signal are ambiguous without using the tool of mathematics. In this section, we introduce an accurate description of the regularity of the signal mathematically employing the exponent of Lipschitz.

**Definition 2.3.** Let $\alpha$ satisfy $0 \le \alpha \le 1$. A function $f$ is called uniformly Lipschitz $\alpha$ over the interval $(a, b)$ if there exists a positive constant $K$, such that

$$|f(x_1) - f(x_2)| \le K|x_1 - x_2|^\alpha, \quad \forall x_1, x_2 \in (a, b),$$

and we denote $f \in C^\alpha(a, b)$. The constant $\alpha$ is called the Lipschitz exponent.

Function $f$ to be uniformly Lipschitz $\alpha$ over $(a, b)$ is also equivalent to the following definition.

**Definition 2.4.** Let $\alpha$ satisfy $0 \le \alpha \le 1$. A function $f$ is called uniformly Lipschitz $\alpha$ over the interval $(a, b)$ if

$$K := \sup_{x_1, x_2 \in (a,b), x_1 \neq x_2} \frac{|f(x_1) - f(x_2)|}{|x_1 - x_2|} < \infty. \tag{2.14}$$

It shows that the variation of $f$ depends on $|x_1 - x_2|^\alpha$ over $(a, b)$. Function $f$ has neither the "fracture" points over $(a, b)$, nor sharp transient points. The shape of $f$ depends on $\alpha$, and the sharpness of the shape can be decided by the constant $K$ in (5.25).

In the above definition, the Lipschitz $\alpha$ is confined within $\alpha \le 1$. Otherwise, when $\alpha > 1$, the above definition loses its meaning because in this case, the function will be too smooth. Since the definition of the Lipschitz regularity with $\alpha \le 1$ does not deal with the differentiation, we can extend the definition of the Lipschitz exponent to $\alpha > 1$.

**Definition 2.5.** Let $\alpha > 1$ and $n$ be the largest number which is less than $\alpha$. A function $f$ is uniformly Lipschitz $\alpha$ over $(a, b)$ if $f$ is $n$th differentiable and $f^{(n)} \in C^{\alpha-n}(a, b)$. It is symbolized by $f \in C^\alpha(a, b)$.

The concept of uniform Lipschitz $\alpha$ can be extended to $\alpha < 0$ and the definition can be modified as in the following definition:

**Definition 2.6.** Let $\alpha$ be $-1 \le \alpha < 0$. If the primitive function of $f$

$$F(x) := \int_a^x f(t)dt$$

is uniformly Lipschitz $(\alpha+1)$ over $(a, b)$, $f$ is said to be uniformly Lipschitz $\alpha$ over $(a, b)$ and denoted by $f \in C^\alpha(a, b)$.

In the following, we will prove an important property that is the Lipschitz regularity which closely correlates with the decay property of wavelet transform in scale.

**Theorem 2.10.** *Let $\alpha > 0$ and $n$ be the largest integer which is less than $\alpha$, $\psi \in L^2(\mathbb{R})$ satisfy $(1 + |x|)^\alpha \psi(x) \in L^1(\mathbb{R})$ and have vanishing moments of order $n$, i.e.,*

$$\int_\mathbb{R} t^k \psi(t)dt = 0, \quad (k = 0, 1, \ldots, n).$$

Then, there exists a constant $C > 0$, such that $\forall f \in L^2(\mathbb{R}) \cap C^\alpha(\mathbb{R})$, the following holds:

$$|(W_\psi f(a,b)| \le C|a|^{\alpha+\frac{1}{2}}, \quad (\forall a, b \in \mathbb{R}).$$

**Proof.** $\forall a, b \in \mathbb{R}$, we have

$$|(W_\psi f(a,b)| = \left| \int_{\mathbb{R}} f(t) \frac{1}{|a|^{1/2}} \psi\left(\frac{t-x}{a}\right) dt \right|$$

$$= \left| \int_{\mathbb{R}} \left[ f(t) - f(x) - f'(x)(t-x) - \cdots - \frac{f^{(n)}(x)}{n!}(t-x)^n \right] \right.$$
$$\left. \cdot \frac{1}{|a|^{1/2}} \psi\left(\frac{t-x}{a}\right) dt \right|$$

$$= \left| \int_{\mathbb{R}} \left[ \frac{f^{(n)}(\xi)}{n!}(t-x)^n - \frac{f^{(n)}(x)}{n!}(t-x)^n \right] \right.$$
$$\left. \cdot \frac{1}{|a|^{1/2}} \psi\left(\frac{t-x}{a}\right) dt \right| \quad (\xi \text{ is between } t \text{ and } x)$$

$$\le C \int_{\mathbb{R}} \frac{|t-x|^{\alpha-n}}{n!} |t-x|^n \frac{1}{|a|^{1/2}} \left| \psi\left(\frac{t-x}{a}\right) \right| dt$$

$$= \frac{C}{n!} |a|^{\alpha+\frac{1}{2}} \int_{\mathbb{R}} |t|^\alpha |\psi(t)| dt$$

$$\le C|a|^{\alpha+\frac{1}{2}},$$

where $C$ denotes a positive constant. The proof is complete. $\qquad\square$

**Theorem 2.11.** Let $0 < \alpha < 1$ and $\psi \in L^2(\mathbb{R})$ have vanishing moment of order 0 and satisfy

$$(1 + |x|)^\alpha \psi(x) \in L^1(\mathbb{R}), \quad |\psi'(x)| = O\left(\frac{1}{1+|x|^\sigma}\right), \quad (\exists \sigma > 1).$$

Then,

$$f \in C^\alpha(\mathbb{R}) \Longleftrightarrow \exists C > 0: \ |W_\psi f(a,b)| \le C|a|^{\alpha+\frac{1}{2}}, \quad (\forall a, b \in \mathbb{R}).$$

**Proof.** Here, we give only the proof of the sufficiency part. According to the note in Theorem 2.5, we have $f(x) \in C(\mathbb{R})$ and the following inverse wavelet transform holds:

$$f(x) = C_\psi^{-1} \int_{\mathbb{R}} \int_{\mathbb{R}} (W_\psi f)(a, b) \psi_{a,b}(x) \frac{dadb}{a^2}, \quad (\forall x \in \mathbb{R}).$$

Therefore,

$$C_\psi |f(x) - f(y)|$$

$$\leq \int_{\mathbb{R}} \int_{\mathbb{R}} |(W_\psi f)(a, b)| \frac{1}{|a|^{1/2}} \left| \psi\left(\frac{x-b}{a}\right) - \psi\left(\frac{y-b}{a}\right) \right| \frac{dadb}{a^2}$$

$$\leq C \int_{|a| \leq \delta} \int_{\mathbb{R}} |a|^{\alpha + \frac{1}{2}} \frac{1}{|a|^{1/2}} \left( \left| \psi\left(\frac{x-b}{a}\right) \right| + \left| \psi\left(\frac{y-b}{a}\right) \right| \right) \frac{dadb}{a^2}$$

$$+ C \int_{|a| > \delta} \int_{\mathbb{R}} |a|^{\alpha + \frac{1}{2}} \frac{1}{|a|^{1/2}} |\psi'(\xi)| \frac{|x-y|}{|a|} \frac{dadb}{a^2}$$

$$(\text{where } \xi \text{ is between } \frac{x-b}{a} \text{ and } \frac{y-b}{a})$$

$$\leq C \int_{|a| \leq \delta} |a|^{\alpha - 1} da \|\psi\|_1 + C \left( \int_{|a| > \delta} \int_{\mathbb{R}} |a|^{\alpha 3} \frac{1}{1 + |\xi|^\sigma} dadb \right) |x - y|.$$

Since $\xi$ is between $\frac{x-b}{a}$ and $\frac{y-b}{a}$, we have that (See the note followed by this proof)

$$\left| \frac{x-b}{a} \right| \leq |\xi| + \left| \frac{x-b}{a} - \frac{y-b}{a} \right| = |\xi| + \left| \frac{x-y}{a} \right|.$$

In particular, setting $\delta = |x - y|$, for $|a| > \delta$, we can deduce that

$$\left| \frac{x-b}{a} \right| \leq |\xi| + \frac{\delta}{|a|} \leq |\xi| + 1.$$

Therefore, we have

$$1 + \left| \frac{x-b}{a} \right|^\sigma \leq 1 + (|\xi| + 1)^\sigma \leq (1 + 2^\sigma)(|\xi|^\sigma + 1).$$

Subsequently, we arrive at

$$\frac{1}{1 + |\xi|^\sigma} \leq (1 + 2^\sigma) \frac{1}{1 + \left| \frac{x-b}{a} \right|^\sigma},$$

and hence, for $\delta = |x - y|$,

$$C_\psi |f(x) - f(y)|$$

$$\leq C \int_{|a| \leq \delta} |a|^{\alpha - 1} da \|\psi\|_1 + C|x - y|$$

$$\left( \int_{|a| > \delta} \int_{\mathbb{R}} |a|^{\alpha - 3} (1 + 2^\sigma) \frac{1}{1 + \left| \frac{x-b}{a} \right|^\sigma} da db \right)$$

$$= C \frac{1}{\alpha} \delta^\alpha + C|x - y| \int_{|a| > \delta} |a|^{\alpha - 2} da \int_{\mathbb{R}} \frac{db}{1 + |b|^\sigma}$$

$$\leq C \frac{1}{\alpha} \delta^\alpha + C|x - y| \frac{1}{1 - \alpha} \delta^{\alpha - 1}$$

$$\leq C \left( \frac{1}{\alpha} + \frac{1}{1 - \alpha} \right) |x - y|^\alpha.$$

This means $f \in C^\alpha(\mathbb{R})$. The proof is complete. $\qquad\square$

*Note:* In the proof, we have used the following obvious results; If $\xi$, $a$, and $b$ are three real numbers, $\xi$ is between $a$ and $b$, then $|a| \leq |\xi| + |a - b|$. In fact, (1) if $a$, $b$ have different signs, then $|a| \leq |a - b| \leq |\xi| + |a - b|$; (2) if $a$ and $b$ have same sign, then $|a| \leq |\xi| \leq |b|$ or $|b| \leq |\xi| \leq |a|$. In the first case, $|a| \leq |\xi| \leq |\xi| + |a - b|$ already holds; in the second case, we have $|a| \leq |b| + |a - b| \leq |\xi| + |a - b|$. These are the results we wanted.

    The above theorem can be extended to the case of $\alpha > 1$ as follows. We omit the proof here.

**Theorem 2.12.** *Let $n$ be a non-negative integer and $\alpha \in \mathbb{R}$ satisfy $n < \alpha < n + 1$. Suppose $\psi \in L^2(\mathbb{R})$ has vanishing moments of order $n$ and satisfies:*

$$(1 + |x|)^\alpha \psi(x), \psi, \psi', \ldots, \psi^{(n)} \in L^1(\mathbb{R}),$$

$$|\psi^{(n+1)}(x)| = O \left( \frac{1}{1 + |x|^\sigma} \right), \quad (\exists \sigma > 1).$$

*Then, $\forall f \in L^1(\mathbb{R})$, the following holds:*

$$f \in C^\alpha(\mathbb{R}) \iff \exists C > 0 : |W_\psi f(a, b)| \leq C|a|^{\alpha + \frac{1}{2}}, \quad (\forall a, b \in \mathbb{R}).$$

## 2.4 Some Examples of Basic Wavelets

There are a great number of basic wavelets. Generally speaking, the derivative of a compact support function, which is continuous differentiable, is a basic wavelet. Several examples of basic wavelets and their basic properties are given in this section.

- **Gaussian wavelets:**

$$\psi_\alpha(x) = -\frac{x}{4\alpha\sqrt{\pi\alpha}} e^{-\frac{x^2}{4\alpha}}.$$

It is the derivative of the Gaussian function

$$g_\alpha(x) = \frac{1}{2\sqrt{\pi\alpha}} e^{-\frac{x^2}{4\alpha}},$$

i.e., $\psi_\alpha(x) = g'_\alpha(x)$. Its Fourier transform is

$$\hat{\psi}_\alpha(\xi) = i\xi e^{-\alpha\xi^2}.$$

The time–frequency window of $\psi_\alpha$ is

$$[-\sqrt{3\alpha}, \ \sqrt{3\alpha}] \times \left[-\sqrt{\frac{3}{4\alpha}}, \ \sqrt{\frac{3}{4\alpha}}\right].$$

Figure 2.3 graphically shows the Gaussian wavelet with $\alpha = \frac{1}{2}$ and its Fourier transform. Both of them have the same shape. The difference between $\psi_\alpha(x)$ and its Fourier transform is only a constant.

- **Mexico hat-like wavelet:** When $\alpha = \frac{1}{2}$, the second derivative of Gaussian function $g_{\frac{1}{2}}$ is a wavelet, which is referred to as Mexico hat-like

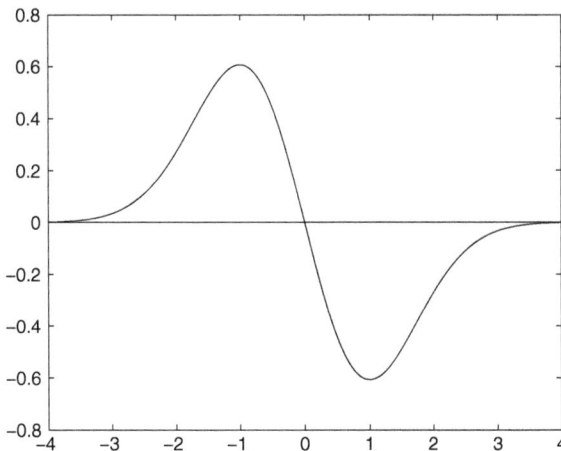

Fig. 2.3   Gaussian wavelet $f(x) = xe^{-\frac{x^2}{2}}$ and its Fourier transform.

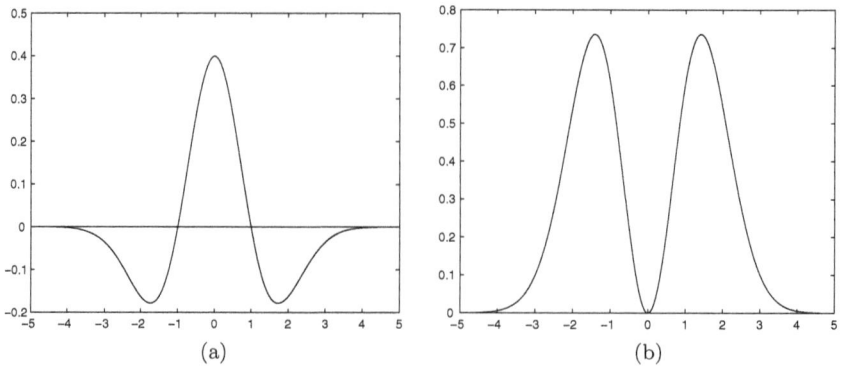

Fig. 2.4   (a) Mexico-hat wavelet $\psi_M$ and (b) its Fourier transform $\hat{\psi}_M$.

wavelet:

$$\psi_M(x) := -g''_{\frac{1}{2}}(x) = \frac{1}{\sqrt{2\pi}}(1 - x^2)e^{-\frac{1}{2}x^2}.$$

Its Fourier transform is

$$\hat{\psi}_M(\xi) = \xi^2 e^{-\frac{1}{2}\xi^2}.$$

The time–frequency window is

$$\left[-\frac{\sqrt{42}}{6}, \frac{\sqrt{42}}{6}\right] \times \left[-\sqrt{\frac{8}{3\sqrt{\pi}}}, \sqrt{\frac{8}{3\sqrt{\pi}}}\right].$$

Mexico hat-like wavelet and its Fourier transform are shown in Fig. 2.4.

- **Spline wavelet:** For $m \geq 1$, the B-spline of order $m$ is defined by

$$N_m(x) = \underbrace{(N_1 * \cdots N_1)}_{m}(x) = (N_{m-1} * N_1)(x) = \int_0^1 N_{m-1}(x - t)dt,$$

$$(2.15)$$

where $N_1(x)$ is a characteristic function of $[0, 1)$ and can be written as

$$N_1(x) := \begin{cases} 1 & x \in [0, 1); \\ 0 & \text{otherwise.} \end{cases}$$

Its Fourier transform is

$$\hat{N}_1(\xi) = \frac{2}{\xi}e^{-i\xi/2}\sin\frac{\xi}{2}.$$

We can easily prove that

$$N_m(x) = \frac{1}{(m-1)!} \sum_{k=0}^{m} (-1)^k \binom{m}{k} (x-k)_+^{m-1}.$$

$N_m(x)$ has compact support $[0, m]$. For $m \geq 2$, the derivative of $N_m(x)$ is a basic wavelet, which is called the spline wavelet of order $m - 1$ and can be represented as

$$\psi_{m-1}(x) := N'_m(x) = \frac{1}{(m-1)!} \sum_{k=0}^{m} (-1)^k \binom{m}{k} (x-k)_+^{m-2}.$$

Its Fourier transform is

$$\hat{\psi}_{m-1}(\xi) = (N'_m)\hat{}(\xi) = i\xi \hat{N}_m(\xi) = i\xi(\hat{N}_1(\xi))^m$$

$$= i\xi \left( \frac{2}{\xi} e^{-i\xi/2} \sin \frac{\xi}{2} \right)^m.$$

In particular, we consider the following two cases, namely, $m = 2$ and $m = 4$:

**(1)** When $m = 2$, the first spline wavelet is

$$\psi_1(x) := \begin{cases} 1 & x \in (0,1), \\ -1 & x \in (1,2), \\ 0 & x \in \mathbb{R}\backslash[0,1]. \end{cases}$$

By taking dilation to the 1D spline wavelet, Haar wavelet $h(x) := \psi_1(2x)$ can be achieved:

$$h(x) := \begin{cases} 1 & x \in (0, \frac{1}{2}), \\ -1 & x \in (\frac{1}{2}, 1), \\ 0 & x \in \mathbb{R}\backslash[0,1]. \end{cases}$$

Its Fourier transform is

$$\hat{h}(\xi) = \frac{1}{2} \hat{\psi}_1 \left( \frac{\xi}{2} \right) = \frac{1}{4} i\xi e^{-i\xi/2} \left( \frac{4}{\xi} \sin \frac{\xi}{4} \right)^2.$$

The time–frequency window is

$$\left[ \frac{1}{2} - \frac{1}{2\sqrt{3}}, \frac{1}{2} + \frac{1}{2\sqrt{3}} \right] \times (-\infty, \infty).$$

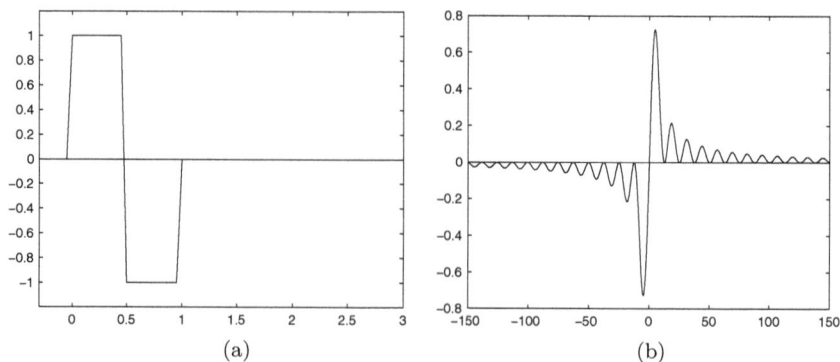

Fig. 2.5   (a) Haar wavelet $h(x)$ and (b) its Fourier transform $-ie^{i\xi/2}\hat{h}(\xi) = \frac{4}{\xi}(\sin\frac{\xi}{4})^2$.

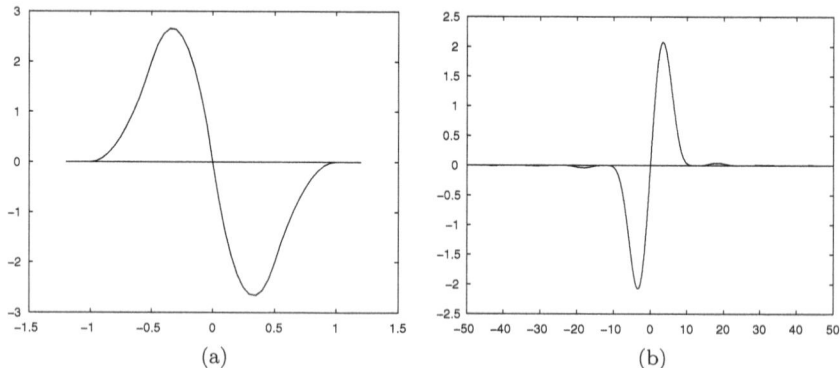

Fig. 2.6   (a) Quadratic spline wavelet $\tilde{\psi}_3$ and (b) its Fourier transform $-i\hat{\tilde{\psi}}_3$.

It is clear that the width of its frequency window is infinite. Haar wavelet is good for the localized analysis in the time domain but in the frequency domain. Haar wavelet and its Fourier transform are graphically displayed in Fig. 2.5.

**(2)** When $m = 4$, the quadratic spline wavelet is

$$\psi_3(x) := \frac{1}{6}\sum_{k=0}^{4}(-1)^k \binom{4}{k}(x-k)_+^2.$$

$\tilde{\psi}_3(x) := 4\psi_3(2x+2)$, the quadratic spline wavelet, is applied widely in signal processing. It is an odd function and has compact support

$[-1, 1]$. In $[0, \infty)$, it can be represented as

$$\tilde{\psi}_3(x) := \begin{cases} 8(3x^2 - 2x) & x \in [0, \frac{1}{2}), \\ -8(x-1)^2 & x \in [\frac{1}{2}, 1), \\ 0 & x \geq 1. \end{cases}$$

Its Fourier transform can be written as

$$\hat{\tilde{\psi}}_3(\xi) = i\xi \left( \frac{4}{\xi} \sin \frac{\xi}{4} \right)^4.$$

The time–frequency window is

$$\left[ -\frac{1}{\sqrt{7}}, \frac{1}{\sqrt{7}} \right] \times [-4, 4].$$

Quadratic spline wavelet wavelet and its Fourier transform are graphically displayed in Fig. 2.6. Note that width 4 of the frequency window was approximately computed with Matlab system; however, a little error may exist.

# Chapter 3

# Multiresolution Analysis and Wavelet Bases

## 3.1 Multiresolution Analysis

### 3.1.1 *Basic Concept of Multiresolution Analysis (MRA)*

It was very difficult to construct a wavelet basis of $L^2(\mathbb{R})$ early on in the history of wavelet development. From the point of view of function analysis, it is also a non-trivial task to find such a function $\psi$ with good regularity and localization to ensure that $\{\psi_{j,k}(x) := 2^{j/2}\psi(2^j x - k)\}_{j \in \mathbb{Z},\ k \in \mathbb{Z}}$ to be an orthonormal basis of $L^2(\mathbb{R})$. For a long time, people doubted the existence of this function. Fortunately, in the 1980s, such functions were found, and a standard scheme to construct wavelet bases was set up as well. It has been proved theoretically that almost all useful wavelet bases can be constructed using the standard scheme: multiresolution analysis (MRA). MRA was first published in 1989 by Mallat [1989a] and Meyer [1990]. Since that time, MRA has become a very important tool in signal processing, image processing, pattern recognition, and other related fields. In this section, we introduce the intuitive meaning of wavelet and MRA. Our purpose is to find a function $\psi$ to ensure that $\{\psi_{j,k}(x) := 2^{j/2}\psi(2^j x - k)\}_{j \in \mathbb{Z},\ k \in \mathbb{Z}}$ is an orthonormal base of $L^2(\mathbb{R})$. Two variables are embedded in them, namely, the translation factor $k \in \mathbb{Z}$ and the dilation factor $j \in \mathbb{Z}$. For $\{\psi_{j,k} | k \in \mathbb{Z}\}$, if $j$ is fixed, there will be a fixed bandwidth:

$$|(\psi_{j,k})\hat{\ }(\xi)| = 2^{-j/2}\left|\hat{\psi}\left(\frac{\xi}{2^j}\right)\right|.$$

Thus, the decomposition of $L^2(\mathbb{R})$ in frequency is as follows:

$$W_j := \overline{span\{\psi_{j,k} | k \in \mathbb{Z}\}},$$

where the $\{\psi_{j,k}\}_{k \in \mathbb{Z}}$ denotes the orthonormal bases of $W_j$. We can write

$$L^2(\mathbb{R}) = \cdots \oplus W_{j-1} \oplus W_j \oplus W_{j+1} \oplus \cdots$$

in which, from left to right, the frequency of $W_j$ changes from low to high. We denote that

$$V_j = \cdots \oplus W_{j-1} \oplus W_j, \quad (j \in \mathbb{Z}),$$

where $V_j$ refers to the function set with lower frequency and satisfies

- $\cdots \subset V_{-1} \subset V_0 \subset V_1 \subset \cdots$,
- $\bigcap_{j \in \mathbb{Z}} V_j = \{0\}$, $\overline{\bigcup_{j \in \mathbb{Z}} V_j} = L^2(\mathbb{R})$,
- $f(x) \in V_j \iff f(2x) \in V_{j+1}$.

We can also write $W_j = V_{j+1} \ominus V_j$. Therefore, $\{W_j\}_{j \in \mathbb{Z}}$ can also be represented by $\{V_j\}_{j \in \mathbb{Z}}$. According to the above definitions, $L^2(\mathbb{R})$ can be divided into the conjoint "concentric annulus" or the expanding "concentric balls" $\{W_j\}_{j \in \mathbb{Z}}$ by $\{V_j\}_{j \in \mathbb{Z}}$. Now, we consider such a question as follows: Is there a function $\phi$ with low frequency such that $\{\phi(\cdot - k)\}_{k \in \mathbb{Z}}$ is an orthonormal or a Riesz basis of $V_0$? The answer is affirmative. In this way, we can find $\psi$ from

$$\psi \in V_1 \ominus V_0.$$

By $\psi \in V_0 \subset V_1$, we obtain the two-scale equation as follows:

$$\phi(x) = 2 \sum_{k \in \mathbb{Z}} h_k \phi(2x - k), \quad \exists \{h_k\} \in l^2(\mathbb{Z}).$$

This is a famous framework referred to as the MRA, by which a wavelet can be constructed. In the following, we show how to construct wavelet bases and biorthonormal bases by using MRA. We give the formal mathematical definitions of MRA.

**Definition 3.1.** Let $H$ be a Hilbert space. Then a sequence in $H$, $\{e_j\}_{j=1}^{\infty}$ is said to be a Riesz basis of $H$, if the following conditions are satisfied:

(1) $\overline{span}\{e_j\}_{j=1}^{\infty} = H$, i.e., $\forall x \in H$ and $\forall \varepsilon > 0$, there exists $\sum_{j=1}^{n} c_j e_j$, such that $\|x - \sum_{j=1}^{n} c_j e_j\| < \varepsilon$.
(2) Constants $A$ and $B$ exist, such that

$$A \sum_{j=1}^{\infty} |c_j|^2 \leq \left\| \sum_{j=1}^{\infty} c_j e_j \right\|^2 \leq B \sum_{j=1}^{\infty} |c_j|^2, \quad \forall \{c_j\}_{j=1}^{\infty} \in l^2.$$

$A$ and $B$ are called the lower and upper bounds of the Riesz basis, respectively.

Particularly, if $A = B = 1$, $\{e_j\}_{j=1}^{\infty}$ is said to be an orthonormal basis.

**Definition 3.2.** A closed subspace $\{V_j\}_{-\infty}^{\infty}$ of $L^2(\mathbb{R})$ is said to be a MRA, if

(1) $\cdots \subset V_{-1} \subset V_0 \subset V_1 \subset \cdots$,
(2) $\bigcap_{j\in\mathbb{Z}} V_j = \{0\}$, $\overline{\bigcup_{j\in\mathbb{Z}} V_j} = L^2(\mathbb{R})$,
(3) $f(x) \in V_j \iff f(2x) \in V_{j+1}$,
(4) $\exists \phi \in V_0$, such that $\{\phi(\cdot - k)\}_{k\in\mathbb{Z}}$ is a Riesz basis of space $V_0$. $\phi$ is called the scaling function of the MRA.

Especially, if the basis of (4) is orthonormal, than the MRA is then called orthonormal MRA. Furthermore, if $\{\phi(\cdot - k)\}_{k\in\mathbb{Z}}$ is the Riesz basis of $V_0$ with both upper bound B and low bound A, we call that $\phi$ generates an MRA with upper bound B and low bound A.

Note that, certainly, $V_j$ is translation-invariant with $2^{-j}\mathbb{Z}$. Meanwhile, $\{\phi_{j,k}\}_{k\in\mathbb{Z}}$ is a Riesz basis of $V_j$ (the upper and lower bounds are constants for any $j \in \mathbb{Z}$), where $\phi_{j,k}(x) := 2^{j/2}\phi(2^j x - k)$.

In the MRA, the key point is how the scale function $\phi$ can be constructed. Since $\phi \in V_0 \subset V_1$, and $\{\phi(2 \cdot -k)|k \in \mathbb{Z}\}$ is the Riesz basis of $V_1$, hence, $\{h_k\} \in l^2(\mathbb{Z})$ exists, such that

$$\phi(x) = 2\sum_{k\in\mathbb{Z}} h_k\phi(2x - k). \tag{3.1}$$

Applying Fourier transform to both sides of the above equation, we can find that it is equivalent to the following fact: There exists $m_0(\xi) = \sum_{k\in\mathbb{Z}} h_k e^{-ik\xi} \in L^2(\mathbb{T})$, such that

$$\hat{\phi}(\xi) = m_0\left(\frac{\xi}{2}\right)\hat{\phi}\left(\frac{\xi}{2}\right). \tag{3.2}$$

Therefore, we can obtain

$$\hat{\phi}(\xi) = \left(\prod_{j=1}^{n} m_0\left(\frac{\xi}{2^j}\right)\right)\hat{\phi}\left(\frac{\xi}{2^n}\right), \quad (\forall n \in \mathbb{N}).$$

It means that if $\hat{\phi}(\xi)$ is continuous at $\xi = 0$ and $\phi(0) \neq 0$, we can infer $\prod_{j=1}^{\infty} m_0(2^{-j}\xi)$ is convergent, and

$$\hat{\phi}(\xi) = \prod_{j=1}^{\infty} m_0\left(\frac{\xi}{2^j}\right). \tag{3.3}$$

The construction of $\phi$ is then equivalent to looking for the function $m_0$ with $2\pi\mathbb{Z}$-period. The $\{h_k\} \in l^2(\mathbb{Z})$ can be regarded as the mask of the two-scale equation, and the $m_0(\xi) = \sum_{k \in \mathbb{Z}} h_k e^{-ik\xi}$ can be considered to be the filter function.

As in the above discussion, most wavelet bases can be constructed by MRA, and the key point of the MRA is how to construct $\phi$. Moreover, $\phi$ can be constructed by solving the two-scale equation (3.1) or (3.2). Therefore, one of the most important problems in MRA is

- the existence of the solution for the two-scale equation (3.1) or (3.2).

From (3.2), we can obtain $\hat{\phi}$ which is represented by (3.3). However, it is only the formal solution: in fact, it is very difficult to find the analytical expression of $\phi$ for the following reasons: First, we do not know whether (3.3) is convergent. Second, we do not know whether $\hat{\phi}$ belongs to $L^2(\mathbb{R})$. Actually, the analytical expressions do not exist for most of the scaling functions, which are solutions of (3.1). A treatment to handle these problems is to study the convergence of the partial product $\prod_{j=1}^{n} m_0(2^{-j}\xi)$ and, thereafter, to check if it converges in $L^2(\mathbb{R})$.

Although the two-scale equation has a solution $\phi$ in $L^2(\mathbb{R})$, it cannot guarantee the generation of MRA from $\phi$. The main reason is that $\{\phi(\cdot - k) \mid k \in \mathbb{Z}\}$ must be a Riesz basis of $V_0$. Consequently, the second important problem is

- whether the solutions of two-scale equation (3.1) can generate an MRA.

Proving that $\{\phi(\cdot - k) \mid k \in \mathbb{Z}\}$ is a Riesz basis of $V_0$ is equivalent to the proof that the $[\hat{\phi}, \hat{\phi}]$ has positive upper and lower bounds.

The orthonormality and biorthonormality are very important in multiresolution analysis and wavelet theory. Thus, the third important problem is

- the orthonormality and biorthonormality of the solutions of two-scale equations.

The orthonormality and biorthonormality can strictly be defined as follows:

**Definition 3.3.** Suppose that $\tilde{\phi}$, $\phi \in L^2(\mathbb{R})$.
(1) If

$$\langle \tilde{\phi}(\cdot - k), \phi(\cdot) \rangle = \delta_{0,k} \quad (\forall k \in \mathbb{Z}),$$

then $\{\tilde{\phi}, \phi\}$ is said to be biorthonormal; furthermore, if $\{\phi, \phi\}$ is biorthonormal, we can say $\phi$ is orthonormal.

(2) If $\{\tilde{\phi}, \phi\}$ is biorthonormal, and

$$\tilde{V}_j := \{\tilde{\phi}_{j,k} | k \in \mathbb{Z}\}, \quad V_j := \{\phi_{j,k} | k \in \mathbb{Z}\}, \quad (j \in \mathbb{Z}),$$

form MRAs in $L^2(\mathbb{R})$ with scale functions $\tilde{\phi}$ and $\phi$, respectively, we say that $\{\tilde{\phi}, \phi\}$ generate a pair of biorthonormal MRAs; particularly, if $\{\phi, \phi\}$ forms a pair of orthonormal MRAs, $\phi$ is said to generate an orthonormal MRA.

### 3.1.2  *The Solution of Two-Scale Equation*

• **The General Solution of Two-Scale Equation**: As mentioned above, the formal solution of the two-scale equation is

$$\prod_{j=1}^{\infty} m_0(\xi/2^j).$$

The following is a sufficient condition for its convergence:

**Theorem 3.1.** *Let $m_0(\xi) \in C(\mathbb{T})$ and satisfy*

$$\exists \varepsilon > 0, \quad \text{when } |\xi| \text{ is small enough: } m_0(\xi) = 1 + O(|\xi|^{\varepsilon}).$$

*Then, $\prod_{j=1}^{\infty} m_0(\xi/2^j)$ is uniformly convergent on any compact subset of $\mathbb{R}$; consequently, a continuous function can be defined in $C(\mathbb{R})$. Furthermore, there are positive constants $C, \tau,$ and $\delta$ such that the following holds:*

$$\left| \prod_{j=1}^{n} m_0(2^{-j}\xi) \right| \leq \begin{cases} \exp(\tau|\xi|^{\varepsilon}), & \text{when } |\xi| \leq \delta; \\ C|\xi|^{\sigma}, & \text{when } |\xi| > \delta, \end{cases} \quad (\forall n \in \mathbb{N} \cup \{+\infty\}),$$

*where $\sigma := \log_2 \|m_0\|_{C(\mathbb{T})}$.*

**Proof.** According to the condition, there must exist positive constants $M$ and $\delta$, such that

$$|m_0(\xi) - 1| \leq M|\xi|^{\varepsilon} < 1, \quad \forall |\xi| \leq \delta.$$

For each compact subset $K$ of $\mathbb{R}$, we choose $J \in \mathbb{N}$ satisfying

$$|2^{-J}\xi| < \delta, \quad \forall \xi \in K.$$

Then, $\forall J_2 \geq J$, we have

$$\ln\left(\prod_{j=J}^{J_2} m_0(2^{-j}\xi)\right) = \sum_{j=J}^{J_2} \ln m_0(2^{-j}\xi)$$

$$= \sum_{j=J}^{J_2} \ln\left[1 + (m_0(2^{-j}\xi) - 1)\right]$$

$$\leq \sum_{j=J}^{J_2} [m_0(2^{-j}\xi) - 1]$$

$$\leq M \sum_{j=J}^{J_2} |2^{-j}\xi|^\varepsilon$$

$$\leq M \frac{2^{-\varepsilon(J-1)}}{2^\varepsilon - 1} |\xi|^\varepsilon. \quad (*)$$

Therefore, when $J$ is large enough, we can write

$$\ln\left(\prod_{j=J}^{J_2} m_0(2^{-j}\xi)\right) = O(2^{-J\varepsilon}) \to 0, \quad (\forall \xi \in K,\ J \to \infty).$$

Consequently, $\prod_{j=1}^\infty m_0(\xi/2^j)$ is uniformly convergent on any compact subset of $\mathbb{R}$. Since $m_0 \in C(\mathbb{T})$, it defines a continuous function.

For the inequality, we only focus on $n \in \mathbb{N}$. It is obvious, for $n = \infty$. $\forall \xi \in \mathbb{R}$ with $\xi \neq 0$, let $J_1 \in \mathbb{Z}$ satisfy

$$\frac{1}{2}\delta < |2^{-J_1}\xi| \leq \delta.$$

Now, we denote $J := \max(1, J_1)$, then $\forall J_2 \geq J$, according to the formula $(*)$, we have

$$\prod_{j=J}^{J_2} m_0(2^{-j}\xi) \leq \exp\left(M \frac{2^{-\varepsilon(J-1)}}{2^\varepsilon - 1} |\xi|^\varepsilon\right).$$

If $0 < |\xi| \leq \delta$, then $J_1 \leq 0$, $J = 1$, and $\forall n \in \mathbb{N}$, we denote $J_2 = n$ and obtain

$$\left|\prod_{j=1}^m m_0(2^{-j}\xi)\right| = \prod_{j=J}^m m_0(2^{-j}\xi) \leq \exp\left(\frac{M}{2^\varepsilon - 1}|\xi|^\varepsilon\right) = \exp\left(\tau|\xi|^\varepsilon\right),$$

where $\tau := \frac{M}{2^\varepsilon - 1} > 0$.

When $\xi = 0$, because $m_0(0) = 1$, it satisfies the inequality.

If $|\xi| > \delta$, then $J_1 \geq 1$, $J = J_1$, therefore, when $n \geq J_1$, we can deduce that

$$\left| \prod_{j=1}^{m} m_0(2^{-j}\xi) \right| = \left| \prod_{j=1}^{J_1-1} m_0(2^{-j}\xi) \right| \left| \prod_{j=J}^{m} m_0(2^{-j}\xi) \right|$$

$$\leq \left( \|m_0\|_{C(\mathbb{T})} \right)^{J_1-1} \exp \left( M \frac{2^{-\varepsilon(J-1)}}{2^{\varepsilon} - 1} |\xi|^{\varepsilon} \right)$$

$$= 2^{(J_1-1)\log_2 \|m_0\|_{C(\mathbb{T})}} \exp \left( M \frac{2^{\varepsilon}}{2^{\varepsilon} - 1} |2^{-J_1}\xi|^{\varepsilon} \right)$$

$$\leq \left( \frac{|\xi|}{\delta} \right)^{\sigma} \exp \left( M \frac{2^{\varepsilon}}{2^{\varepsilon} - 1} \delta^{\varepsilon} \right)$$

$$= C|\xi|^{\sigma},$$

where

$$\sigma := \log_2 \|m_0\|_{C(\mathbb{T})}, \quad C := \delta^{-\sigma} \exp \left( M \frac{2^{\varepsilon}}{2^{\varepsilon} - 1} \delta^{\varepsilon} \right).$$

For $n < J_1$, by $\|m_0\|_{C(\mathbb{T})} \geq |m_0(0)| = 1$, we have

$$\left| \prod_{j=1}^{m} m_0(2^{-j}\xi) \right| \leq \left( \|m_0\|_{C(\mathbb{T})} \right)^{m} \leq \left( \|m_0\|_{C(\mathbb{T})} \right)^{J_1-1} \leq \left( \frac{|\xi|}{\delta} \right)^{\sigma} \leq C|\xi|^{\sigma}.$$

This concludes our proof. $\qquad\qquad\qquad\qquad\qquad\qquad\qquad\qquad$ □

*Note*: Especially, if $m_0$ is a trigonometric polynomial and $m_0(0) = 1$, then the conditions of this theorem are satisfied.

We hope that $\prod_{j=1}^{\infty} m_0(\xi/2^j) \in L^2(\mathbb{R})$ so that the inverse Fourier transform of $\phi$ can be also in $L^2(\mathbb{R})$. This means that $\phi$ is a solution of Eq. (3.1). Unfortunately, under the conditions of (3.1), $\prod_{j=1}^{\infty} m_0(\xi/2^j)$ may not be in $L^2(\mathbb{R})$, which can be shown by a simple example: Let $m_0(\xi) \equiv 1$, then $\prod_{j=1}^{\infty} m_0(\xi/2^j) \equiv 1 \notin L^2(\mathbb{R})$. We can handle this problem in two ways: First, we can enhance the conditions for $m_0(\xi)$, so as to ensure $\prod_{j=1}^{\infty} m_0(\xi/2^j) \in L^2(\mathbb{R})$. Second, we can discuss the solution of Eq. (3.1) in a larger space over $L^2(\mathbb{R})$. We now focus on the second way. In the conditions of (3.1), $\prod_{j=1}^{\infty} m_0(\xi/2^j)$ is continuous and increases at most polynomially fast at infinite. It belongs to $\varphi'$, the space of all the slowly increasing generalized functions, where the operation of Fourier transform is very convenient. We discuss the solution of two-scale equation in $\varphi'$.

As we know, the space $\varphi'$ consists of all the continuous linear functionals on $\varphi$. $\varphi$ is the space of all the fast decreasing $C^\infty$ functions, whose definition is as follows:

$$\varphi := \left\{ \varphi \in C^\infty(\mathbb{R}) \;\middle|\; \sup_{x \in \mathbb{R}} (1 + |x|^2)^{\frac{k}{2}} \left| \partial^\alpha \varphi(x) \right| \leq M_{k,\alpha} < \infty \; (k, |\alpha| = 0, 1, \dots) \right\}.$$

By equipping countable semi-norms as follows,

$$\|\varphi\|_m := \sup_{\substack{|\alpha| \leq m \\ x \in \mathbb{R}}} (1 + |x|^2)^{\frac{m}{2}} |\partial^\alpha \varphi(x)|, \quad (m = 0, 1, \dots),$$

$\varphi$ becomes a countable-norm space, i.e., a $B_0^*$ space [Zhang, 1986].

For $1 \leq p \leq \infty$, we define

$$PL^p := PL^p(\mathbb{R}) := \{ f \mid \exists m \in \mathbb{N} : (1 + |x|)^{-m} f(x) \in L^p(\mathbb{R}) \}. \tag{3.4}$$

Let $f \in PL^1$, and we define

$$\langle f, \, g \rangle := \int_{\mathbb{R}} f(x)g(x)dx \quad (\forall g \in \varphi).$$

It is obvious that

$$PL^\infty \subset PL^p \subset PL^1 \subset \varphi' \quad (1 \leq p \leq \infty).$$

Now, we prove that the following two equations are equivalent to each other:

$$\phi = 2 \sum_{k \in \mathbb{Z}} h_k \phi(2 \cdot -k) \quad \text{(in } \varphi', \text{ the right term is convergent)}, \tag{3.5}$$

$$\hat{\phi} = m_0 \left( \frac{\cdot}{2} \right) \hat{\phi} \left( \frac{\cdot}{2} \right) \quad \text{(in } \varphi'). \tag{3.6}$$

**Theorem 3.2.** *Let* $m_0(\xi) = \sum_{k \in \mathbb{Z}} h_k e^{-ik\xi} \in C(\mathbb{T})$, *and* $\phi \in \varphi'$. *If one of the following two conditions is satisfied,*

(1) $\hat{\phi} \in PL^1$ *and* $\sum_{k \in \mathbb{Z}} |h_k| < \infty$,
(2) $\sum_{k \in \mathbb{Z}} |h_k| |k|^n < \infty \; (\forall n \in \mathbb{N})$,

*then,* $\phi$ *satisfies (3.5) if and only if* $\phi$ *satisfies (3.6).*

**Proof.** $\forall \phi \in \varphi'$, $g \in \varphi$, we denote

$$I_n := \lim_{n \to \infty} \left\langle \left( \sum_{|k| \leq n} h_k e^{-ik\frac{\cdot}{2}} \right) \hat{\phi}\left(\frac{\cdot}{2}\right) - m_0\left(\frac{\cdot}{2}\right) \hat{\phi}\left(\frac{\cdot}{2}\right), g \right\rangle.$$

The satisfaction of the condition (1) produces

$$|I_n| = \left| \int_{\mathbb{R}} \hat{\phi}\left(\frac{\xi}{2}\right) \left( \sum_{|k|>n} h_k e^{-ik\frac{\xi}{2}} \right) g(\xi) d\xi \right|$$

$$\leq \left( \sum_{|k|>n} |h_k| \right) \int_{\mathbb{R}} \left| \hat{\phi}\left(\frac{\xi}{2}\right) g(\xi) \right| d\xi$$

$$\to 0 \ (n \to \infty).$$

The satisfaction of the condition (2) yields

$$g_n(\xi) := \left( \sum_{|k|>n} h_k e^{-ik\frac{\xi}{2}} \right) g(\xi) \in \varphi.$$

Consequently, we have, $\forall m \in \mathbb{Z}^+$,

$$\|g_n\|_m = \sup_{\xi \in \mathbb{R}, |\alpha| \leq m} |(1 + |\xi|^2)^{\frac{m}{2}} D^\alpha g_n(\xi)|$$

$$= \sup_{\xi \in \mathbb{R}, |\alpha| \leq m} \left| (1 + |\xi|^2)^{\frac{m}{2}} \sum_{\beta \leq \alpha} \binom{\alpha}{\beta} \right.$$

$$\times \left. \left( \sum_{|k|>n} h_k \left(-i\frac{k}{2}\right)^\beta e^{-ik\frac{\xi}{2}} \right) D^{\alpha-\beta} g(\xi) \right|$$

$$\leq \sup_{\xi \in \mathbb{R}, |\alpha| \leq m} \left| (1 + |\xi|^2)^{\frac{m}{2}} \sum_{\beta \leq \alpha} \binom{\alpha}{\beta} \left( \sum_{|k|>n} |h_k||k|^{|\beta|} \right) |D^{\alpha-\beta} g(\xi)| \right|$$

$$\leq C \left( \sum_{|k|>n} |h_k||k|^m \right) \sup_{\xi \in \mathbb{R}, |\alpha| \leq m} |(1 + |\xi|^2)^{\frac{m}{2}} D^\alpha g(\xi)|$$

$$= C \left( \sum_{|k|>n} |h_k||k|^m \right) \|g\|_m$$

$$\to 0 \ (n \to \infty),$$

where $C$ stands for a constant which depends only on $m$. Thus, we can obtain $g_n \to 0$ (in $\varphi$ as $n \to 0$). Therefore, we have

$$|I_n| = \left| \left\langle \hat{\phi}\left(\frac{\cdot}{2}\right), \left(\sum_{|k|>n} h_k e^{-ik\frac{\cdot}{2}}\right) g \right\rangle \right| = \left| \left\langle \hat{\phi}\left(\frac{\cdot}{2}\right), g_n \right\rangle \right| \to 0 \; (n \to 0).$$

It implies that

$$\phi \text{ satisfies (3.5)} \iff \hat{\phi} = 2 \sum_{k\in\mathbb{Z}} h_k (\phi(2\cdot -k))\check{}$$

$$\iff \hat{\phi} = \lim_{n\to\infty} \sum_{|k|\leq n} h_k e^{-ik\frac{\cdot}{2}} \hat{\phi}\left(\frac{\cdot}{2}\right)$$

$$\iff \langle \hat{\phi}, g \rangle = \lim_{n\to\infty} \left\langle \left(\sum_{|k|\leq n} h_k e^{-ik\frac{\cdot}{2}}\right) \hat{\phi}\left(\frac{\cdot}{2}\right), g \right\rangle \; (\forall g \in \varphi)$$

$$\iff \left\langle \hat{\phi} - m_0\left(\frac{\cdot}{2}\right) \hat{\phi}(\frac{\cdot}{2}), g \right\rangle$$

$$= \lim_{n\to\infty} \left\langle \left[\sum_{|k|\leq n} h_k e^{-ik\frac{\cdot}{2}} - m_0\left(\frac{\cdot}{2}\right)\right] \hat{\phi}\left(\frac{\cdot}{2}\right), g \right\rangle$$

$$(\forall g \in \varphi)$$

$$\iff \left\langle \hat{\phi} - m_0\left(\frac{\cdot}{2}\right) \hat{\phi}\left(\frac{\cdot}{2}\right), g \right\rangle = \lim_{n\to\infty} I_n = 0 \; (\forall g \in \varphi)$$

$$\iff \phi \text{ satisfy (3.6).}$$

This establishes our proof. $\qquad\square$

As for the solution of two-scale equation in $\varphi'$, we have the following results:

**Theorem 3.3.** *Let* $m_0(\xi) = \sum_{k\in\mathbb{Z}} h_k e^{-ik\xi} \in C(\mathbb{T})$, *such that* $\prod_{j=1}^{\infty} m_0(\frac{\xi}{2^j})$ *converges pointwise to* $M(\xi) \in PL^1$. *Then,* $\phi := \check{M} \in \varphi'$ *is a solution of* (3.6). *Furthermore, if* $M(\xi)$ *is continuous at* $\xi = 0$ *and* $M(0) = 1$, *then* $\phi$ *is the unique solution of* (3.6) *in* $\varphi'_0$ $(\varphi'_0 \subset \varphi)$ *which is defined by*

$$\varphi'_0 := \{\phi \in \varphi' | \hat{\phi} \in PL^1, \hat{\phi}(\xi) \text{ is continuous at } \xi = 0, \hat{\phi}(0) = 1\}. \quad (3.7)$$

**Proof.** It is clear that

$$M(\xi) = m_0\left(\frac{\xi}{2}\right) M\left(\frac{\xi}{2}\right), \quad a.e. \ \xi \in \mathbb{R}.$$

Therefore, as a generalized function in $\varphi'$, we have $M(\cdot) = m_0(\frac{\cdot}{2})M(\frac{\cdot}{2})$, that is, $\phi := \check{M} \in \varphi'$ satisfies

$$\hat{\phi}(\cdot) = m_0\left(\frac{\cdot}{2}\right) \hat{\phi}\left(\frac{\cdot}{2}\right) \quad \text{in } \varphi'.$$

Consequently, $\phi$ is the solution of (3.6).

If $M(\xi)$ is continuous at $\xi = 0$, and $M(0) = 1$, it is obvious that $\phi \in \varphi'_0$, i.e., $\phi := \check{M}$ is the solution of (3.6) in $\varphi'_0$.

Now, we prove the unicity of the solution. Suppose $\phi \in \varphi'_0$ is a solution of (3.6), hence, as a generalized function of $\varphi'$, it satisfies

$$\hat{\phi} = m_0\left(\frac{\cdot}{2}\right) \hat{\phi}\left(\frac{\cdot}{2}\right) \quad (\text{in } \varphi').$$

As $\hat{\phi} \in PL^1$, the above result is equivalent to that of ordinary function; thus, we obtain

$$\hat{\phi}(\xi) = m_0\left(\frac{\xi}{2}\right) \hat{\phi}\left(\frac{\xi}{2}\right) \quad a.e. \ \xi \in \mathbb{R}.$$

Therefore,

$$\hat{\phi}(\xi) = m_0\left(\frac{\xi}{2}\right) \hat{\phi}\left(\frac{\xi}{2}\right) = \cdots = \left(\prod_{j=1}^{n} m_0\left(\frac{\xi}{2^j}\right)\right) \hat{\phi}\left(\frac{\xi}{2^n}\right) \quad a.e. \ \xi \in \mathbb{R}.$$

Let $n \to \infty$, we have

$$\hat{\phi}(\xi) = \prod_{j=1}^{\infty} m_0\left(\frac{\xi}{2^j}\right) = M(\xi) \quad a.e. \ \xi \in \mathbb{R}.$$

As a conclusion, $\phi = \check{M}$ holds. The proof is complete. $\qquad\square$

**Corollary 3.1.** *Let* $m_0(\xi) = \sum_{k \in \mathbb{Z}} h_k e^{-ik\xi} \in C(\mathbb{T})$, *such that*

$$\sum_{k \in \mathbb{Z}} |h_k| < \infty$$

*and*

$$\exists \varepsilon > 0, \ when \ |\xi| \ is \ small \ enough, \quad m_0(\xi) = 1 + O(|\xi|^\varepsilon).$$

*Then, (3.5) and (3.6) have the same solution in* $\varphi'_0$, *which is just the inverse Fourier transform of* $\prod_{j=1}^{\infty} m_0(\frac{\xi}{2^j})$.

**Proof.** According to Theorem 3.2, it is known that the two-scale equations (3.5) and (3.6) have the same solution in $\varphi'$. By Theorems 3.1 and 3.3, a unique solution exists in $\varphi'_0$, which is just the inverse Fourier transform of $\prod_{j=1}^{\infty} m_0(\frac{\xi}{2^j})$. $\qquad\square$

This result is enough for wavelet analysis in $L^2(\mathbb{R})$, although the solution of two-scale equation is in $\varphi'_0$. Solving the unique solution of two-scale equation in $L^2(\mathbb{R})$ is much more difficult than that in $\varphi'$. Thus, we discussed the case of $\varphi'$ first. Meanwhile, from the above analysis, we found that the unicity of the solution in $\varphi'_0$ implicates the unicity in $L^2(\mathbb{R})$. Consequently, our task is changed to investigate the conditions which can guarantee that the solution in $\varphi'_0$ belongs to $L^2(\mathbb{R})$.

**The Cascade Algorithm to Solve Two-Scale Equation**

It is assumed that $m_0(\xi) = \sum_{k\in\mathbb{Z}} h_k e^{-ik\xi} \in C(\mathbb{R})$; we take $\eta_0 \in L^\infty(\mathbb{R}) \cap L^1(\mathbb{R})$, such that $\hat{\eta}_0(0) = 1$. Hence, the general scheme of the Cascade algorithm is

$$\eta_n(x) := 2\sum_{k\in\mathbb{Z}} h_k \eta_{n-1}(2x - k), \quad (n = 1, 2, \ldots). \tag{3.8}$$

In order to guarantee the convergence of the series in the Cascade algorithm, we suppose that $\sum_{j=1}^{\infty} |h_k| < \infty$. It is easy to see that the right-hand side of the Cascade algorithm is absolutely convergent, $\eta_n \in L^\infty(\mathbb{R}) \cap L^1(\mathbb{R})$ $(n = 1, 2, \ldots)$, and

$$\|\eta_n\|_\infty \le \left(2\sum_{k\in\mathbb{Z}} |h_k|\right) \|\eta_{n-1}\|_\infty, \quad \|\eta_n\|_1 \le \left(\sum_{k\in\mathbb{Z}} |h_k|\right) \|\eta_{n-1}\|_1,$$

for $n = 1, 2, \ldots$. Since

$$\hat{\eta}_n(\xi) = \left(\prod_{j=1}^{n} m_0\left(\frac{\xi}{2^j}\right)\right) \hat{\eta}_0\left(\frac{\xi}{2^n}\right),$$

we have, under the condition of Theorem 3.1, that

$$\hat{\eta}_n(\xi) \to \prod_{j=1}^{\infty} m_0\left(\frac{\xi}{2^j}\right) \quad a.e.\ \xi \in \mathbb{R},$$

thus, positive constants $C$ and $\sigma$ exist, such that

$$|\hat{\eta}_n(\xi)| \le C(1 + |\xi|)^\sigma, \quad (\forall \xi \in \mathbb{R}).$$

Therefore, $\forall g \in \varphi$, using Lebesgue's dominated convergence theorem, we conclude that

$$\langle \hat{\eta}_n, g \rangle = \int_{\mathbb{R}} \hat{\eta}_n(\xi) g(\xi) d\xi \to 0, \quad (n \to \infty),$$

that is,

$$\hat{\eta}_n \to \left( \prod_{j=1}^{\infty} m_0 \left( \frac{\xi}{2^j} \right) \right) \quad \text{in } \varphi'.$$

Therefore,

$$\eta_n \to \left( \prod_{j=1}^{\infty} m_0 \left( \frac{\xi}{2^j} \right) \right)^{\check{}} \quad \text{in } \varphi'.$$

Consequently, the limit of $\{\eta_n\}$ in $\varphi'_0$ is just the unique solution of (3.5) in $\varphi'$. In summary, we have the following theorem:

**Theorem 3.4.** *Let* $m_0(\xi) = \sum_{k \in \mathbb{Z}} h_k e^{-ik\xi} \in C(\mathbb{T})$ *satisfy* $\sum_{k \in \mathbb{Z}} |h_k| < \infty$, *and*

$$\exists \varepsilon > 0, \quad \text{when } |\xi| \text{ is small enough, } m_0(\xi) = 1 + O(|\xi|^{\varepsilon}).$$

*Let* $\eta_0 \in L^{\infty}(\mathbb{R})$ *be a function with compact support* $[A, B]$ *and satisfy* $\hat{\eta}_0(0) = 1$. *Then the function sequence* $\{\eta_n\}$ *defined by Cascade algorithm* (3.8) *is convergent in* $\varphi'$. *The limit is the unique solution of* (3.5) *in* $\varphi'_0$.

If $\{h_k\}_{k \in \mathbb{Z}}$ has finite length, that is, the non-zero items in $\{h_k\}_{k \in \mathbb{Z}}$ are finite, it can be proved that the solution of (3.5) has compact support in $\varphi'_0$. The support of $\phi \in \varphi'$ is defined as

$$\text{supp} \phi := \{x \in \mathbb{R} | \forall \text{ open set } O_x \ni x, \exists g \in C_c^{\infty}(O_x) : \langle \phi, g \rangle \neq 0\}, \quad (3.9)$$

where $C_c^{\infty}(O_x)$ denotes the space of all the $C^{\infty}$ functions which is compactly supported in $O_x$.

Obviously, the support defined in the above is a closed set in $\mathbb{R}$. When $\phi$ degenerates to an ordinary locally integrable function, the definition of its support is the same as that of the ordinary meaning. The definition of the support of an ordinary function is defined by

$$\text{supp} \phi := \{x \in \mathbb{R}^d | \forall O_x \ni x, \text{ always has : } |O_x \cap \{\phi(x) \neq 0\}| > 0\}, \quad (3.10)$$

where $O_x$ denotes an open set containing $x$ and $|O_x \cap \{\phi(x) \neq 0\}|$ denotes the Lebesgue measure of the set $O_x \cap \{\phi(x) \neq 0\}$. In order to prove that the unique solution of Eq. (3.5) in $\varphi_0'$ is compactly supported when $\{h_k\}_{k\in\mathbb{Z}}$ has finite length, we first consider the following lemma:

**Lemma 3.1.** *Let $\{h_k\}_{k\in\mathbb{Z}}$ be a complex sequence supported in $[N, M]$ in the meaning that $h_k = 0$ if $k \notin [N, M]$. Let $\eta_0 \in L^\infty(\mathbb{R})$ be compactly supported on $[A, B]$. Then, for the function sequence $\{\eta_n\}$ which is given by Cascade algorithm (3.8), we have*

$$\mathrm{supp}\,\eta_n \subset [A^{(n)}, B^{(n)}], \quad (n = 0, 1, \ldots),$$

*where $A^{(n)}$, $B^{(n)}$ are defined by*

$$\begin{cases} A^{(0)} = A \\ B^{(0)} = B \end{cases} \quad \begin{cases} A^{(n)} = \dfrac{1}{2}(N + A^{(n-1)}) \\ B_i^{(n)} = \dfrac{1}{2}(M + B^{(n-1)}) \end{cases}, \quad (n = 1, 2, \ldots)$$

*and satisfy the following:*

$\{A^{(n)}\}_{n=1}^\infty$ *is a monotonous sequence between $N$ and $A$, and $A^{(n)} \to N$ $(n \to \infty)$.*

$\{B^{(n)}\}_{n=1}^\infty$ *is a monotonous sequence between $M$ and $B$, and $B^{(n)} \to M$ $(n \to \infty)$.*

**Proof.** We use induction for $n$ to prove $\mathrm{supp}\,\eta_n \subset [A^{(n)}, B^{(n)}]$. Obviously, for $n = 0$, our conclusion is established. Suppose it is right for $n - 1$. As for $\mathrm{supp}\,\eta_n$, let $\eta_n(x) \neq 0$, then $\exists k \in [N, M]$, such that $2x - k \in \mathrm{supp}\,\eta_{n-1} \subset [A^{(n-1)}, B^{(n-1)}]$. This means $A^{(n-1)} \leq 2x - k \leq B^{(n-1)}$ or $\frac{1}{2}(A^{(n-1)} + k) \leq x \leq \frac{1}{2}(B^{(n-1)} + k)$. Therefore, $\frac{1}{2}(A^{(n-1)} + N) \leq x \leq \frac{1}{2}(B^{(n-1)} + M)$. So far, we proved that

$$\mathrm{supp}\,\eta_n \subset \left[\frac{(A^{n-1}) + N}{2}, \frac{B^{(n-1)} + M}{2}\right] = [A^{(n)}, B^{(n)}].$$

By induction, the result holds for $n \in \mathbb{Z}_+$.

We now discuss the monotonicity of $\{A^{(n)}\}_{n=0}^\infty$ in two cases:

If $A^{(0)} = A \geq N$, then, by $A^{(n)} = \frac{1}{2}(N + A^{(n-1)})$, it is easy to prove that $A^{(n)} \geq N$, so $A^{(n)} \leq A^{(n-1)} \leq \cdots \leq A^{(0)}$. That is, $\{A^{(n)}\}_{n=1}^\infty$ is a monotonous sequence valued between $N$ and $A$.

If $A^{(0)} = A \leq N$, the same conclusion can be proved in the same way.

Letting $n \to \infty$ on both sides of $A^{(n)} = \frac{1}{2}(N + A^{(n-1)})$, we obtain that $A^{(n)} \to N$.

We can give the proof in the same way for $B^{(n)}$.

The proof of the lemma is complete. $\qquad\square$

**Theorem 3.5.** *Let* $\{h_k\}_{k \in \mathbb{Z}}$ *be a complex sequence which is supported in* $[N, M]$, $\sum_{k \in \mathbb{Z}} h_k = 1$. *Then, the unique solution* $\phi$ *of Eq. (3.5) in* $\varphi_0'$ *is compactly supported on* $[N, M]$.

**Proof.** Let $\eta_0 \in L^\infty(\mathbb{R})$ satisfy supp$\eta_0 \subset [A, B]$, and $\eta_0(0) = 1$. Then, for the function sequence $\{\eta_n\}$ derivated by the cascade algorithm, we have $\eta_n \to \phi$ (in $\varphi'$), where $\phi$ is the unique solution of (3.5)in $\varphi_0'$, and

$$\text{supp}\eta_n \subset [A^{(n)}, B^{(n)}] \quad (n = 0, 1, \ldots),$$

where $A^{(n)}$, $B^{(n)}$ satisfy

$$A^{(n)} \to N, \quad B^{(n)} \to M, \quad (n \to \infty).$$

$\forall x \notin [N, M]$, there exists a neighborhood $O_x$ of $x$, such that the distance between $[N, M]$ and $O_x$ is positive. Then, when $n$ is large enough, we have

$$([A^{(n)}, B^{(n)}]) \bigcap O_x = \text{empty set}.$$

For any $g \in C_c^\infty(O_x)$, we have

$$\langle \phi, g \rangle = \lim_{n \to \infty} \langle \eta_n, g \rangle = 0.$$

Therefore, $x \notin \text{supp}\phi$. Now that $x$ is an arbitrary point of $\mathbb{R} \backslash [N, M]$, we have

$$\mathbb{R} \backslash [N, M] \subset \mathbb{R} \backslash \text{supp}\phi,$$

i.e., supp$\phi \subset [N, M]$.
    This ends the proof. □

## 3.2 The Construction of MRAs

To facilitate the discussion, we assume the $2\pi\mathbb{Z}$-periodic function, $m_0(\xi)$, is continuous on $\mathbb{R}$. We study how MRAs can be constructed from such $m_0(\xi)$ that satisfies some basic conditions as follows:

**Definition 3.4.** Suppose $m_0(\xi)$ is a $2\pi\mathbb{Z}$-periodic measurable function on $\mathbb{R}$:

- $m_0$ is said to satisfy the basic condition I, if $m_0(\xi) \in C(\mathbb{T})$, $m_0(0) = 1$ and $m_0(\pi) = 0$.
- $m_0$ is said to satisfy the basic condition II, if $\prod_{j=1}^\infty m_0(\xi/2^j)$ *a.e.* converges to a non-zero function, which is denoted by $\hat{\phi}$ and is called the corresponding limit function (where $\hat{\phi}$ denotes a function instead of the Fourier transform of a function temporarily):

To study the sufficient conditions such that the solution of the two-scale equation belongs to $L^2(\mathbb{R})$, the transition operators are introduced first.

Let $\tilde{m}_0$, $m_0$ be $2\pi\mathbb{Z}$-periodic functions on $\mathbb{R}$; we define

$$Tf(\xi) := \left|m_0\left(\frac{\xi}{2}\right)\right|^2 f\left(\frac{\xi}{2}\right) + \left|m_0\left(\frac{\xi}{2}+\pi\right)\right|^2 f\left(\frac{\xi}{2}+\pi\right), \quad (3.11)$$

$$\tilde{T}f(\xi) := \left|\tilde{m}_0\left(\frac{\xi}{2}\right)\right|^2 f\left(\frac{\xi}{2}\right) + \left|\tilde{m}_0\left(\frac{\xi}{2}+\pi\right)\right|^2 f\left(\frac{\xi}{2}+\pi\right), \quad (3.12)$$

$$Sf(\xi) := \tilde{m}_0\left(\frac{\xi}{2}\right) \overline{m}_0\left(\frac{\xi}{2}\right) f\left(\frac{\xi}{2}\right) + \tilde{m}_0\left(\frac{\xi}{2}+\pi\right)$$

$$\times \overline{m}_0\left(\frac{\xi}{2}+\pi\right) f\left(\frac{\xi}{2}+\pi\right), \quad (3.13)$$

$$|S|f(\xi) := \left|\tilde{m}_0\left(\frac{\xi}{2}\right) \overline{m}_0\left(\frac{\xi}{2}\right)\right| f\left(\frac{\xi}{2}\right)$$

$$+ \left|\tilde{m}_0\left(\frac{\xi}{2}+\pi\right) \overline{m}_0\left(\frac{\xi}{2}+\pi\right)\right| f\left(\frac{\xi}{2}+\pi\right). \quad (3.14)$$

The following lemma is important:

**Lemma 3.2.** *Let $m_0 \in C(\mathbb{T})$ satisfy the basic condition II and the corresponding limit function $\hat{\phi} \in C(\mathbb{R})$. Let $X$ be a closed sub-space of $C(\mathbb{T})$ and satisfy both of the following conditions:*

(1) *$T(X) \subset X$ and the spectral radius $r_\sigma(T|_X)$ of $T|_X$, which is the restriction of the transition operator $T$ on $X$, satisfies $r_\sigma(T|_X) < 1$.*
(2) *There exists a non-negative function $f \in X$ which has at most one zero point $\xi = 0$ in $[-\pi, \pi]$.*

*Then $\hat{\phi} \in L^2(\mathbb{R})$ and $\|\hat{\phi}_n - \hat{\phi}\|_2 \to 0$ $(n \to \infty)$, where $\hat{\phi}_n$ is defined by*

$$\hat{\phi}_n(\xi) := \prod_{j=1}^{n} m_0(2^{-j}\xi)\chi_{2^n[-\pi,\pi]}(\xi). \quad (3.15)$$

**Proof.** It can easily be deduced that $\hat{\phi}(0) = 1$. Let $\delta : 0 < \delta < \pi$ satisfy

$$\frac{1}{2} \leq |\hat{\phi}(\xi)| \leq 2, \quad \forall \xi \in [-\delta, \delta].$$

Since $f$ is continuous and positive on $[-\pi, \pi]\backslash(-\frac{1}{2}\delta, \frac{1}{2}\delta)$, a constant $C > 0$ must exist, such that

$$C^{-1} \leq f(\xi) \leq C, \quad \forall \xi \in [-\pi, \pi]\backslash\left(-\frac{1}{2}\delta, \frac{1}{2}\delta\right).$$

We denote

$$E_n := [-2^n \delta, 2^n \delta], \quad (n = 0, 1, \ldots).$$

Then $2^{-n}\xi \in E_0 = [-\delta, \delta]$ for any $\xi \in E_n$. Therefore,

$$|\hat{\phi}(\xi)| = |\hat{\phi}_n(\xi)\hat{\phi}(2^n\xi)| \le |\hat{\phi}_n(\xi)| \sup_{\xi \in E_0} |\hat{\phi}(\xi)| \le 2|\hat{\phi}_n(\xi)|.$$

Hence, we have

$$\int_{E_n \setminus E_{n-1}} |\hat{\phi}(\xi)|^2 d\xi \le 4 \int_{E_n \setminus E_{n-1}} |\hat{\phi}_n(\xi)|^2 d\xi$$

$$\le 4 \int_{[-2^n\pi, 2^n\pi] \setminus E_{n-1}} |\hat{\phi}_n(\xi)|^2 d\xi$$

$$\le 4C \int_{[-2^n\pi, 2^n\pi] \setminus E_{n-1}} |\hat{\phi}_n(\xi)|^2 f(2^{-n}\xi) d\xi$$

$$\le 4C \int_{\mathbb{R}} |\hat{\phi}_n(\xi)|^2 f(2^{-n}\xi) d\xi$$

$$= 4C \int_{\mathbb{T}} (T^n f)(\xi) d\xi$$

$$= 4C \int_{\mathbb{T}} ((T|_X)^n f)(\xi) d\xi$$

$$\le 4C(2\pi) \|f\|_{C(\mathbb{T})} \|(T|_X)^n\|.$$

Let $\rho$ satisfy $r_\sigma(T|_X) < \rho < 1$. According to the spectral radius formula, we deduce that $\lim_{n \to \infty} \|(T|_X)^n\|^{1/n} = r_\sigma(T|_X)$. Thus, there exists $N > 0$ such that $\|(T|_X)^n\| \le \rho^n$ for $n > N$, which concludes that

$$\int_{E_n \setminus E_{n-1}} |\hat{\phi}(\xi)|^2 d\xi \le 4 \int_{[-2^n\pi, 2^n\pi] \setminus E_{n-1}} |\hat{\phi}_n(\xi)|^2 d\xi \le 4C(2\pi) \|f\|_{C(\mathbb{T})} \rho^n.$$

Consequently,

$$\int_{\mathbb{R}} |\hat{\phi}(\xi)|^2 d\xi = \int_{E_N} |\hat{\phi}(\xi)|^2 d\xi + \sum_{n=N+1}^{\infty} \int_{E_n \setminus E_{n-1}} |\hat{\phi}(\xi)|^2 d\xi$$

$$\le (2^{N+1}\delta) \|\hat{\phi}\|_{C(E_N)} + 4C(2\pi) \|f\|_{C(\mathbb{T})} \sum_{n=N+1}^{\infty} \rho^n$$

$$< \infty.$$

$\hat{\phi} \in L^2(\mathbb{R})$ is proved.

To prove $\|\hat{\phi}_n - \hat{\phi}\|_2 \to 0$ $(n \to \infty)$, we deduce that

$$\|\hat{\phi}_n\|_2^2 = \int_{\mathbb{R}} |\hat{\phi}_n(\xi)|^2 d\xi = \int_{[-2^n\pi, 2^n\pi]} |\hat{\phi}_n(\xi)|^2 d\xi$$

$$= \int_{[-2^n\pi, 2^n\pi]\backslash E_{n-1}} |\hat{\phi}_n(\xi)|^2 d\xi + \int_{E_{n-1}} |\hat{\phi}_n(\xi)|^2 d\xi.$$

Since $2^{-n}\xi \in [-\frac{1}{2}\delta, \frac{1}{2}\delta] \subset [-\delta, \delta]$ for any $\xi \in E_{n-1}$, we have

$$|\hat{\phi}(\xi)| = |\hat{\phi}_n(\xi)\hat{\phi}(2^{-n}\xi)| \geq \frac{1}{2}|\hat{\phi}_n(\xi)|.$$

Hence,

$$|\hat{\phi}_n(\xi)|^2 \chi_{E_{n-1}}(\xi) \leq 4|\hat{\phi}(\xi)|^2, \quad \forall \xi \in \mathbb{R}.$$

Now that $\hat{\phi}_n(\xi)\chi_{E_{n-1}}(\xi) \to \hat{\phi}(\xi)$ a.e. $\xi \in \mathbb{R}$, the Lebesgue dominated convergence theorem deduces that

$$\int_{E_{n-1}} |\hat{\phi}_n(\xi)|^2 d\xi = \int_{\mathbb{R}} |\hat{\phi}_n(\xi)|^2 \chi_{E_{n-1}}(\xi) d\xi \to \int_{\mathbb{R}} |\hat{\phi}(\xi)|^2 d\xi = \|\hat{\phi}\|_2^2,$$

which together with the following result

$$\int_{[-2^n\pi, 2^n\pi]\backslash E_{n-1}} |\hat{\phi}_n(\xi)|^2 d\xi \leq C(2\pi)\|f\|_{C(\mathbb{T})}\rho^n \to 0 \quad (n \to 0)$$

concludes that $\|\hat{\phi}_n\|_2 \to \|\hat{\phi}\|_2$ $(n \to \infty)$. Note that $\hat{\phi}_n(\xi) \to \hat{\phi}(\xi)$ a.e. $\xi \in \mathbb{R}$ $(n \to \infty)$; we obtain $\|\hat{\phi}_n - \hat{\phi}\|_2 \to 0$ $(n \to \infty)$. Thus, the proof of the lemma is complete. $\qquad\square$

Particularly, if $m_0$ satisfies the basic conditions I and II, we set

$$X = \dot{C}(\mathbb{T}) := \{f \in C(\mathbb{T})|f(0) = 0\} \quad \text{or} \quad X = \dot{\mathcal{P}}_N := \{f \in \mathcal{P}_N|f(0) = 0\}$$

corresponding to $m_0 \in C(\mathbb{T})$ or $m_0 \in \mathcal{P}_N^+$, respectively, where

$$\mathcal{P}_N := \left\{ \sum_{n=-N}^{N} c_n e^{-in\xi} | \forall \text{ sequence } \{c_n\}_{n\in\mathbb{Z}} \right\},$$

$$\mathcal{P}_N^+ := \left\{ \sum_{n=0}^{N} c_n e^{-in\xi} | \forall \text{ sequence } \{c_n\}_{n\in\mathbb{Z}} \right\}.$$

Then we have $T(X) \subset X$. According to the above lemma, we obtain the following theorem:

**Theorem 3.6.** *Suppose $m_0$ meets the basic conditions I and II and its limit function satisfies $\hat{\phi} \in C(\mathbb{R})$. If the spectral radius $r_\sigma(T|_{\dot{C}(\mathbb{T})})$ of $T|_X$, the restriction of $T$ defined by (3.11), satisfies $r_\sigma(T|_{\dot{C}(\mathbb{T})}) < 1$, then $\hat{\phi} \in L^2(\mathbb{R})$ and $\|\hat{\phi}_n - \hat{\phi}\|_2 \to 0$ $(n \to \infty)$, where $\hat{\phi}_n$ is defined by (3.15).*

**Proof.** According to the above lemma, the theorem holds obviously if we set $X = \dot{C}(\mathbb{T})$ and $f(\xi) := (1 - \cos\xi)^2$. □

Now, we give a theorem for the case that $m_0$ is a trigonometric polynomial. The readers can refer to the work of Long [1995] for the proof.

**Theorem 3.7.** *Let* $m_0 \in \mathcal{P}_N^+$ ($N \neq 0$), $m_0(0) = 1$, $m_0(\pi) = 0$, *and* $\hat{\phi}$ *be its limit function. We denote*

$$\dot{\mathcal{P}}_N := \{f \in \mathcal{P}_N | f(0) = 0\}.$$

*Then the following three conditions are equivalent to each other:*

(1) $\lambda = 1$ *is the simple eigenvalue of the restriction* $T|_{\mathcal{P}_N}$ *of* $T$ *on* $\mathcal{P}_N$, *and each of its other eigenvalues* $\lambda$ *satisfies* $|\lambda| < 1$, *where* $T$ *is defined by* (3.11).

(2) *Each eigenvalue* $\lambda$ *of the restriction* $T|_{\dot{\mathcal{P}}_N}$ *of* $T$ *on* $\dot{\mathcal{P}}_N$ *satisfies* $|\lambda| < 1$.

(3) $\hat{\phi} \in L^2(\mathbb{R})$ *and* $\|\hat{\phi}_n - \hat{\phi}\|_2 \to 0$ ($n \to \infty$), *where* $\hat{\phi}_n$ *is defined by* (3.15).

*Furthermore, each of them implies the following results:*

(i) $\Phi := [\hat{\phi}, \hat{\phi}] \in \mathcal{P}_N$ *and* $\Phi$ *is the eigenvector of* $T|_{\mathcal{P}_N}$ *corresponding to the simple eigenvalue* $\lambda = 1$.

(ii) *There exist constants* $0 < \varepsilon < 2$ *and* $C > 0$ *such that*

$$\sum_{k \in \mathbb{Z}} |\hat{\phi}(\xi + 2k\pi)|^{2-\varepsilon} \leq C, \quad |\hat{\phi}(\xi)| \leq C(1 + |\xi|)^{-\varepsilon}, \quad (\forall \xi \in \mathbb{R}),$$

*where*

$$\Phi(\xi) := [\hat{\phi}, \hat{\phi}] := \sum_{k \in \mathbb{Z}} |\hat{\phi}(\xi + 2\pi k)|^2. \tag{3.16}$$

We consider whether $\phi$ generates an orthonormal MRA or a biorthogonal MRA. The estimation of the lower and upper bounds of $\Phi = [\hat{\phi}, \hat{\phi}]$ is key to this question. In mathematics, it is difficult since $\Phi$ is defined by an infinite sum and not easily be expressed analytically in general. The above theorem, however, tells us that under certain conditions, $\Phi$ is just the eigenvector of $T|_{\mathcal{P}_N}$ corresponding to the simple eigenvalue $\lambda = 1$. The latter can be solved with a computer.

A sufficient condition such that $\hat{\phi} \in L^2(\mathbb{R})$ and $\|\hat{\phi}_n - \hat{\phi}\|_2 \to 0$ ($n \to \infty$) is given as follows. Its proof is omitted here and can be found in some references, such as the works of Daubechies [1992] and Long [1995].

**Theorem 3.8.** *Suppose* $m_0$ *can be written as*

$$m_0(\xi) = \left(\frac{1 + e^{-i\xi}}{2}\right)^L M_0(\xi),$$

*where* $L \in \mathbb{Z}_+$ *and* $M_0 \in C(\mathbb{T})$ *satisfies the following:*

(1) $M_0(\pi) \neq 0$.
(2) There exists a constant $0 < \delta < 1$ satisfying $M_0(\xi) = 1 + O(|\xi|^\delta)$ in some neighborhood of $\xi = 0$.
(3) There exists $k \in \mathbb{N}$ such that

$$B_k := \max_{\xi \in \mathbb{R}} |M_0(\xi) \cdots M_0(2^{k-1}\xi)|^{1/k} < 2^{L-\frac{1}{2}}.$$

Then, $\hat{\phi} \in L^2(\mathbb{R})$, $\|\hat{\phi}_n - \hat{\phi}\|_2 \to 0$ $(n \to \infty)$, and a constant $C > 0$ exists, such that

$$|\hat{\phi}(\xi)| \leq \frac{C}{(1+|\xi|)^{\frac{1}{2}+\varepsilon}}, \quad (\forall \xi \in \mathbb{R}),$$

where $\varepsilon := L - \frac{1}{2} - \log B_k > 0$ and $\hat{\phi}_n$ is defined by (3.15).

The following is a sufficient and necessary condition, which can guarantee that $\phi$ generates an MRA.

**Theorem 3.9.** *Let $m_0$ satisfy the basic conditions I and II and $\hat{\phi} \in C(\mathbb{R})$ be its limit function. Then $\phi$ generates MRA if and only if both of the following two conditions hold:*

(1) $\hat{\phi} \in L^2(\mathbb{R})$ *and* $\|\hat{\phi}_n - \hat{\phi}\|_2 \to 0$ $(n \to \infty)$, *where $\hat{\phi}_n$ is defined by (3.15).*
(2) *Two positive constants $A$ and $B$ exist, such that $A \leq (T^n 1)(\xi) \leq B$ a.e. $\xi \in \mathbb{T}$ for any $n \in \mathbb{N}$, where $T$ is the transition operator defined by (3.11).*

**Proof.** To prove the necessity, we assume that $\phi$ generates an MRA of $L^2(\mathbb{R})$. Two constants $A$ and $B$ exist, such that $\Phi$ defined by (3.16) satisfies

$$A \leq \Phi(\xi) \leq B \quad a.e. \ \xi \in \mathbb{T},$$

which concludes (1).

Now, we prove (2). Since $T^n$ is a positive linear operator for any $n \in \mathbb{N}$, we deduce that $T^n f(\xi) \leq T^n g(\xi)$ a.e. $\xi$ for any $f(\xi) \leq g(\xi)$ a.e. $\xi$. Using $T^n \Phi = \Phi$, we have

$$A \leq \Phi(\xi) = T^n \Phi(\xi) \leq (T^n B)(\xi) = B(T^n 1)(\xi), \quad a.e. \ \xi \in \mathbb{T},$$
$$B \geq \Phi(\xi) = T^n \Phi(\xi) \geq (T^n A)(\xi) = A(T^n 1)(\xi), \quad a.e. \ \xi \in \mathbb{T}.$$

Therefore, $\frac{A}{B} \leq (T^n 1)(\xi) \leq \frac{B}{A}$ a.e. $\xi \in \mathbb{T}$. The proof of (2) is complete.

To prove the sufficiency, we suppose (1) and (2) hold. For any function $g$ which is $2\pi\mathbb{Z}$-periodic, bounded, non-negative, and measurable, the following holds:

$$\int_{\mathbb{R}} |\hat{\phi}_n(\xi)|^2 g(\xi) d\xi = \int_{\mathbb{T}} (T^n 1)(\xi) g(\xi) d\xi.$$

Hence,

$$A \int_T g(\xi) d\xi \leq \int_{\mathbb{R}} |\hat{\phi}_n(\xi)|^2 g(\xi) d\xi \leq B \int_T g(\xi) d\xi.$$

Let $n \to \infty$, we have

$$A \int_T g(\xi) d\xi \leq \int_{\mathbb{R}} |\hat{\phi}(\xi)|^2 g(\xi) d\xi \leq B \int_T g(\xi) d\xi,$$

i.e.,

$$A \int_T g(\xi) d\xi \leq \int_{\mathbb{T}} \Phi(\xi) g(\xi) d\xi \leq B \int_T g(\xi) d\xi.$$

Thus, $A \leq \Phi(\xi) \leq B$ a.e. $\xi \in \mathbb{T}$ which implies consequently that $\{\phi(\cdot - k) | k \in \mathbb{Z}\}$ constitute a Riesz basis of $V_0 := \overline{span}\{\phi(\cdot - k) | k \in \mathbb{Z}\}$. By the fact of $\hat{\phi}(0) = m_0(0) = 1$, we deduce that $\overline{\cup_{j \in \mathbb{Z}} V_j} = L^2(\mathbb{R})$ for $V_j := \{f(2^j \cdot) | f \in V_0\}$ $(j \in \mathbb{Z})$. Therefore, $\phi$ generates an MRA of $L^2(\mathbb{R})$. This ends the proof of the theorem. $\qquad\square$

For the case that $m_0$ is a trigonometric polynomial, we further have the following result:

**Corollary 3.2.** *Let* $m_0 \in \mathcal{P}_N^+$ *(where* $N \in \mathbb{Z}_+$, $N \neq 0$*). Then* $\hat{\phi}(\xi) := \prod_{j=1}^{\infty} m_0(\xi/2^j)$ *converges a.e.* $\xi \in \mathbb{R}$ *belongs to* $L^2(\mathbb{R})$, *and* $\phi$ *generates an MRA, if and only if the following three conditions hold:*

(1) $m_0(0) = 1$ *and* $m_0(\pi) = 0$.
(2) $\lambda = 1$ *is the simple eigenvalue of* $T|_{\mathcal{P}_N}$ *which is the restriction of* $T$ *defined by (3.11) on* $\mathcal{P}_N$, *and each of its other eigenvalues* $\lambda$ *satisfies* $|\lambda| < 1$.
(3) *The eigenvector* $g \in \mathcal{P}_N$ *of* $T|_{\mathcal{P}_N}$ *corresponding to the eigenvalue 1 has positive lower bound if* $g(0) = 1$, *that is, there is a constant* $C > 0$ *such that* $g(\xi) \geq C$ $(\forall \xi \in \mathbb{T})$.

**Proof.** To prove the necessity, we choose $\xi_0 \in \mathbb{R}$ such that $\prod_{j=1}^{\infty} m_0(\xi_0/2^j)$ converges. Then $\lim_{j \to \infty} m_0(\xi_0/2^j) = 1$, i.e., $m_0(0) = 1$.

Since $\hat{\phi} \in L^2(\mathbb{R})$, and $m_0 \in \mathcal{P}_N^+$ implies that $\phi$ is compactly supported, we conclude that $\Phi := [\hat{\phi}, \hat{\phi}]$ is a trigonometric polynomial. According to the fact that $\phi$ generates an MRA, it can be shown easily that $\Phi$ has positive upper and lower bounds on $\mathbb{T}$. Therefore, (1) holds.

By Theorems 3.7 and 3.9, it can be concluded that (2) holds and $\Phi = [\hat{\phi}, \hat{\phi}]$ is the eigenvector of $T|_{\mathcal{P}_N}$ corresponding to the eigenvalue 1. Since 1 is the simple eigenvalue of $T|_{\mathcal{P}_N}$, we deduce that $g \in \mathcal{P}_N$ is just some constant times of $\Phi$. Now, $g(0) = \Phi(0) = 1$, we further have $g \equiv \Phi$. Hence, $g$ has positive lower bound and (3) is proved.

To prove the sufficiency, we deduce that $\hat{\phi} \in L^2(\mathbb{R})$, $\Phi \in \mathcal{P}_N$, and $\Phi$ has positive lower bound according to Theorem 3.7 and condition (3). On the other hand, $\Phi \in \mathcal{P}_N$ implies that $\Phi$ has positive upper bound. Hence, $\Phi$ has positive upper and lower bounds, which together with the equality $\hat{\phi}(0) = 1$ concludes that $\phi$ generates an MRA. The proof is complete. $\square$

This corollary is more convenient to be applied in practice. It can be easily implemented with a computer.

### 3.2.1  *The Biorthonormal MRA*

The purpose of this section is to construct such $\tilde{\phi}$, $\phi \in L^2(\mathbb{R})$ that generate a pair of biorthonormal MRA. The first task is to construct the biorthonormal functions $\{\tilde{\phi}, \phi\}$, or equivalently, to find out the conditions for masks $\tilde{m}_0$ and $m_0$, such that following two-scale relations:

$$\hat{\phi}(\xi) = m_0\left(\frac{\xi}{2}\right)\hat{\phi}\left(\frac{\xi}{2}\right), \tag{3.17}$$

$$\hat{\tilde{\phi}}(\xi) = \tilde{m}_0\left(\frac{\xi}{2}\right)\hat{\tilde{\phi}}\left(\frac{\xi}{2}\right). \tag{3.18}$$

Generally, it is very difficult to find out the sufficient and necessary conditions. To avoid such difficulty, we first study some basic necessary conditions for $\{\tilde{m}_0, m_0\}$, Thereafter, by enhancing the conditions until we meet the sufficient ones.

Let $\{\tilde{\phi}, \phi\}$ generate a pair of biorthonormal MRA. There exists a constant $C > 0$ such that

$$C^{-1} \leq \tilde{\Phi}(\xi), \quad \Phi(\xi) \leq C \quad a.e. \ \xi \in \mathbb{R},$$

where $\Phi(\xi)$ is defined by (3.16), and

$$\tilde{\Phi}(\xi) := [\hat{\tilde{\phi}}, \hat{\tilde{\phi}}] := \sum_{k \in \mathbb{Z}} |\hat{\tilde{\phi}}(\xi + 2k\pi)|^2. \tag{3.19}$$

Enhance the above inequality by letting
$$C^{-1} \leq \tilde{\Phi}(\xi), \quad \Phi(\xi) \leq C \quad (\forall \xi \in \mathbb{R}).$$
Then it can be deduced that
$$m_0(0) = 1, \quad m_0(\pi) = 0.$$

In order to define $\tilde{\phi}, \phi$ from $\tilde{m}_0, m_0$ based on the two-scale relation, we assume that $\prod_{j=1}^{\infty} \tilde{m}_0(\xi/2^j)$ and $\prod_{j=1}^{\infty} m_0(\xi/2^j)$ are pointwise convergents. This assumption is necessary if $\hat{\tilde{\phi}}(\xi)$ and $\hat{\phi}(\xi)$ are expected to be continuous at $\xi = 0$ and are not zero functions.

According to the biorthonormal property, we have (see Theorem 3.11)
$$\tilde{m}_0(\xi)\overline{m_0(\xi)} + \tilde{m}_0(\xi+\pi)\overline{m_0(\xi+\pi)} = 1 \quad a.e. \ \xi \in \mathbb{R}.$$

To facilitate the discussion, we assume that $2\pi\mathbb{Z}$-period functions $\tilde{m}_0(\xi)$ and $m_0(\xi)$ are continuous and satisfy the following basic conditions. We study such sufficient conditions $\{\tilde{m}_0(\xi), m_0(\xi)\}$ satisfy that $\{\tilde{\phi}(\xi), \phi(\xi)\}$ generate a pair of biorthonomals.

**Definition 3.5.** Let $\{\tilde{m}_0(\xi), m_0(\xi)\}$ be a pair of $2\pi\mathbb{Z}$-period measurable functions on $\mathbb{R}$. The following three conditions are called the basic conditions:

- **Basic condition I:** $\tilde{m}_0(\xi), m_0(\xi) \in C(\mathbb{T})$, $\tilde{m}_0(0) = m_0(0) = 1$, $\tilde{m}_0(\pi) = m_0(\pi) = 0$.
- **Basic condition II:** $\prod_{j=1}^{\infty} \tilde{m}_0(\xi/2^j)$ and $\prod_{j=1}^{\infty} m_0(\xi/2^j)$ are pointwise convergent to non-zero functions $\hat{\tilde{\phi}}$ and $\hat{\phi}$, respectively, which are called the corresponding limit functions. (Here $\hat{\tilde{\phi}}, \hat{\phi}$ denote two functions instead of the Fourier transforms of functions temporarily.)
- **Basic condition III:** $\tilde{m}_0(\xi)\overline{m_0(\xi)} + \tilde{m}_0(\xi+\pi)\overline{m_0(\xi+\pi)} = 1$, $a.e. \ \xi \in \mathbb{R}$.

$\{\tilde{m}_0(\xi), m_0(\xi)\}$ is called to be basic if it satisfies the above three conditions.

Particularly, if $\tilde{m}_0 = m_0$, the corresponding basic conditions are defined as follows:

**Definition 3.6.** Let $m_0(\xi)$ be a $2\pi\mathbb{Z}$-period measurable function on $\mathbb{R}$. The following three conditions are called the basic conditions:

- **Basic condition I:** $m_0(\xi) \in C(\mathbb{T})$ and $m_0(0) = 1, m_0(\pi) = 0$.
- **Basic condition II:** $\prod_{j=1}^{\infty} m_0(\xi/2^j)$ is pointwise convergent to a non-zero function $\hat{\phi}$, which is called the corresponding limit function. (Here $\hat{\phi}$ denotes a function instead of the Fourier transform of a function temporarily.)
- **Basic condition III:** $|m_0(\xi)|^2 + |m_0(\xi+\pi)|^2 = 1$, $a.e. \ \xi \in \mathbb{R}$.

$m_0(\xi)$ is called to be basic if it satisfies the above three conditions.

*Note 1:* If $m_0(\xi)$ is basic and its limit function $\hat{\phi} \in L^2(\mathbb{R}) \cap C(\mathbb{R})$, then $\hat{\phi}(2\pi\alpha) = \delta_{0,\alpha} \ (\forall \alpha \in \mathbb{Z})$.

*Note 2:* Differing from the case of function pair in Definition 3.5, the conditions $m_0(0) = 1$ and $m_0(\pi) = 0$ in the basic condition I of Definition 3.6 are implied in the basic conditions II and III of Definition 3.6. Furthermore, $\hat{\phi} \in L^2(\mathbb{R})$ can also be concluded by them. The details are included in the following theorem:

**Theorem 3.10.** *Assume that $\{\tilde{m}_0, m_0\} \subset C(\mathbb{R})$ satisfies the basic condition II, and $\hat{\tilde{\phi}}, \hat{\phi}$ are the corresponding limit functions, then we have the following:*

(1) *If there is a constant $B > 0$, such that operator $|S|$, which is defined by (3.14), satisfies $|S|^n 1(\xi) \leq B$ (a.e. $\xi \in \mathbb{T}$, for all $n \in N$), then $\hat{\tilde{\phi}}\hat{\phi} \in L^1(\mathbb{R})$.*
(2) *If $m_0$ satisfies the basic condition III, then $\hat{\phi} \in L^2(\mathbb{R})$, $m_0(0) = 1$, and $m_0(\pi) = 0$.*

**Proof.** We prove (1) first. Similar to (3.15), we denote

$$\hat{\tilde{\phi}}_n(\xi) := \prod_{j=1}^{n} \tilde{m}_0(2^{-j}\xi)\chi_{2^n[-\pi,\pi]}(\xi), \tag{3.20}$$

then it is easy to see that

$$\int_{\mathbb{R}} |\hat{\tilde{\phi}}_n(x)\hat{\phi}_n(x)|dx = \int_{\mathbb{R}}(|S|^n 1)(\xi)d\xi \leq B(2\pi) \quad (\forall n \in \mathbb{N}).$$

Since

$$\hat{\tilde{\phi}}_n(\xi) \to \hat{\tilde{\phi}}(\xi) \quad \hat{\phi}_n(\xi) \to \hat{\phi}(\xi) \quad a.e. \ \xi \in \mathbb{R}.$$

By Fatou lemma, letting $n \to \infty$, we have

$$\int_{\mathbb{R}} |\hat{\tilde{\phi}}(\xi)\hat{\phi}(\xi)|d\xi \leq B2\pi.$$

This establishes $\hat{\tilde{\phi}}\hat{\phi} \in L^1(\mathbb{R})$.

To prove (2), let $\hat{\tilde{\phi}} = \hat{\phi}$. It is clear that the corresponding operator $|S|$ satisfies $|S|^n 1(\xi) \equiv 1$ (*a.e.* $\xi \in \mathbb{T}$, $\forall n \in \mathbb{N}$), therefore, $\hat{\phi}\hat{\tilde{\phi}} \in L^1(\mathbb{R})$ or, equivalently, $\hat{\phi} \in L^2(\mathbb{R})$. Since

$$\hat{\phi}_{n+1}(\xi) = m_0(\xi/2^{n+1})\hat{\phi}_n(\xi/2), \quad (\forall \xi \in 2^n[-\pi, \pi)),$$

by letting $n \to \infty$, we have

$$\hat{\phi}(\xi) = m_0(0)\hat{\phi}(\xi) \quad a.e. \ \xi \in \mathbb{R}.$$

It can be observed easily that $\hat{\phi}(0) = 1$, therefore $m_0(0) = 1$. Furthermore, since $m_0$ satisfies the basic condition III, (2) is proven. $\qquad \square$

**Theorem 3.11.** *Let* $\tilde{\phi}$, $\phi \in L^2(\mathbb{R})$. *Then,*

$$\{\tilde{\phi}, \phi\} \ \text{is biorthonormal} \iff F(\xi) = 1 \quad a.e. \ \xi \in \mathbb{R}$$

$$\implies \exists C > 0 : \ |F|(\xi) \geq C \quad a.e. \ \xi \in \mathbb{R}$$

$$\implies \exists C > 0 : \ \Phi(\xi)\tilde{\Phi}(\xi) \geq C \quad a.e. \ \xi \in \mathbb{T},$$

*where*

$$F(\xi) := \sum_{k \in \mathbb{Z}} \hat{\tilde{\phi}}(\xi + 2k\pi)\overline{\hat{\phi}}(\xi + 2k\pi). \tag{3.21}$$

*Furthermore,*

(1) *if there are* $2\pi\mathbb{Z}$-*periodic measurable functions* $\tilde{m}_0(\xi)$ *and* $m_0(\xi)$ *such that two-scale equations (3.17) and (3.18) hold, then*

$$\{\tilde{\phi}, \phi\} \ \text{is biorthonormal} \implies$$

$$\tilde{m}_0(\xi)\overline{m}_0(\xi) + \tilde{m}_0(\xi + \pi)\overline{m}_0(\xi + \pi) = 1 \quad a.e. \ \xi \in \mathbb{R},$$

(2) *if* $\tilde{\Phi}$, $\Phi \in L^\infty(\mathbb{T})$, *then*

$$\{\tilde{\phi}, \phi\} \ \text{is biorthonormal} \implies \text{There is a constant } C > 0, \ \text{such that}$$

$$C^{-1} \leq \tilde{\Phi}(\xi), \ \Phi(\xi) \leq C \quad a.e. \ \xi \in \mathbb{T}.$$

**Proof.** By

$$\int_{\mathbb{R}} \tilde{\phi}(x - k)\overline{\phi}(x)dx = \left(\frac{1}{2\pi}\right)\int_{\mathbb{R}} \hat{\tilde{\phi}}(\xi)\overline{\hat{\phi}}(\xi)e^{-ik\xi}d\xi$$

$$= \left(\frac{1}{2\pi}\right)\int_{\mathbb{T}} F(\xi)e^{-ik\xi}d\xi \quad (\forall k \in \mathbb{Z}),$$

we have

$$\int_{\mathbb{R}} \tilde{\phi}(x-k)\bar{\phi}(x)dx = \delta_{0,k} \ (\forall k \in \mathbb{Z}) \iff F(\xi) = 1 \quad a.e. \ \xi \in \mathbb{T}.$$

(1) If there are $2\pi\mathbb{Z}$-periodic measurable functions $\tilde{m}_0(\xi)$ and $m_0(\xi)$ such that two-scale equations (3.17) and (3.18) hold, by

$$F(2\xi) = \tilde{m}_0(\xi)\bar{m}_0(\xi)F(\xi) + \tilde{m}_0(\xi+\pi)\bar{m}_0(\xi+\pi)F(\xi+\pi),$$

we conclude that

$$F(\xi) = 1 \quad a.e. \ \xi \in \mathbb{R}$$

implies

$$\tilde{m}_0(\xi)\bar{m}_0(\xi) + \tilde{m}_0(\xi+\pi)\bar{m}_0(\xi+\pi) = 1, \quad a.e. \ \xi \in \mathbb{R}.$$

(2) If $\tilde{\Phi}, \Phi \in L^\infty(\mathbb{T})$, then

$$\{\tilde{\phi}, \phi\} \text{ is biorthonormal}$$

$$\implies \exists C > 0: \ \Phi(\xi)\tilde{\Phi}(\xi) \geq C \quad a.e. \ \xi \in \mathbb{T}$$

$$\implies \exists C > 0: \ \tilde{\Phi}(\xi), \ \Phi(\xi) \geq C \quad a.e. \ \xi \in \mathbb{T}$$

$$\implies \exists C > 0: \ C^{-1} \leq \tilde{\Phi}(\xi), \ \Phi(\xi) \leq C \quad a.e. \ \xi \in \mathbb{T}.$$

This finishes our proof. $\qquad\qquad\qquad\qquad\qquad\qquad\qquad\qquad\qquad\quad\square$

*Note*: According to the above discussion, $\forall \tilde{\phi}, \phi \in L^2(\mathbb{R})$, if $\{\tilde{\phi}, \phi\}$ is biorthonormal and $\tilde{\Phi}, \Phi \in L^\infty(\mathbb{T})$, then $\{\tilde{\phi}(\cdot - k)|k \in \mathbb{Z}\}$ is a Riesz basis of $\tilde{V}_0 := \overline{span}\{\tilde{\phi}(\cdot - k)|k \in \mathbb{Z}\}$ and $\{\phi(\cdot - k)|k \in \mathbb{Z}\}$ is a Riesz basis of $V_0 := \overline{span}\{\phi(\cdot - k)|k \in \mathbb{Z}\}$. Furthermore, if $\tilde{\phi}, \ \phi \in L(\mathbb{R})$, then $\hat{\tilde{\phi}}(0) = \hat{\phi}(0) = 1$. Consequently, biorthonormal MRA $\{\tilde{V}_j\}, \{V_j\}$ can be generated from $\{\tilde{\phi}, \phi\}$, here,

$$\tilde{V}_j := \{f(2^j \cdot)|f \in \tilde{V}_0\}, \quad V_j := \{f(2^j \cdot)|f \in \tilde{V}_0\} \quad (\forall j \in \mathbb{Z}).$$

**Theorem 3.12.** *Let* $\{\tilde{m}_0(\xi), \ m_0(\xi)\}$ *be basic and* $\{\hat{\tilde{\phi}}, \ \hat{\phi}\} \subset L^2(\mathbb{R}) \cap C(\mathbb{R})$ *be their limit functions. Suppose* $\{\hat{\phi}_n\}$ *is defined by* (3.15) *and* $\{\hat{\tilde{\phi}}_n\}$ *is defined by* (3.20). *Then,*

$$\{\tilde{\phi}, \phi\} \text{ is biorthonormal} \iff F(\xi) = 1 \quad a.e. \ \xi \in \mathbb{R}$$

$$\iff \exists C > 0: \ |F|(\xi) \geq C \quad a.e. \ \xi \in \mathbb{R}$$

$$\Longleftrightarrow \ \exists C > 0 : \ \Phi(\xi)\tilde{\Phi}(\xi) \geq C \quad a.e. \ \xi \in \mathbb{T}$$

$$\Longleftrightarrow \ \|\hat{\tilde{\phi}}_n \overline{\hat{\phi}_n} - \hat{\tilde{\phi}}\overline{\hat{\phi}}\|_1 \to 0 \quad (n \to \infty).$$

*Furthermore, if* $\tilde{\Phi}, \Phi \in L^\infty(\mathbb{T})$, *then*

$$\{\tilde{\phi}, \phi\} \quad generate \ a \ pair \ of \ biorthonormal \ MRAs$$

$$\Longleftrightarrow \ F(\xi) = 1 \quad a.e. \ \xi \in \mathbb{R}$$

$$\Longleftrightarrow \ \exists C > 0 : \ |F|(\xi) \geq C \quad a.e. \ \xi \in \mathbb{R}$$

$$\Longleftrightarrow \ \exists C > 0 : \ \Phi(\xi)\tilde{\Phi}(\xi) \geq C \quad a.e. \ \xi \in \mathbb{T}$$

$$\Longleftrightarrow \ \|\hat{\tilde{\phi}}_n \overline{\hat{\phi}_n} - \hat{\tilde{\phi}}\overline{\hat{\phi}}\|_1 \to 0 \quad (n \to \infty).$$

**Proof.** According to Theorem 3.11, we only need to prove that

$$\exists C > 0 : \Phi(\xi)\tilde{\Phi}(\xi) \geq C \ a.e. \ \xi \in \mathbb{T} \Longrightarrow \|\hat{\tilde{\phi}}_n \overline{\hat{\phi}_n} - \hat{\tilde{\phi}}\overline{\hat{\phi}}\|_1 \to 0 \ (n \to \infty)$$
$$(3.22)$$

and

$$\|\hat{\tilde{\phi}}_n \overline{\hat{\phi}_n} - \hat{\tilde{\phi}}\overline{\hat{\phi}}\|_1 \to 0 \ (n \to \infty) \Longrightarrow \{\tilde{\phi}, \phi\} \text{ is biorthonormal.} \quad (3.23)$$

It is known that

$$\int_{\mathbb{R}} \tilde{\phi}(x)\overline{\phi}(x - k)dx = \left(\frac{1}{2\pi}\right) \int_{\mathbb{R}} \hat{\tilde{\phi}}(\xi)\overline{\hat{\phi}}(\xi)e^{ik\xi}d\xi$$

$$= \lim_{n \to \infty} \left(\frac{1}{2\pi}\right) \int_{\mathbb{R}} \hat{\tilde{\phi}}_n(\xi)\overline{\hat{\phi}}_n(\xi)e^{ik\xi}d\xi$$

$$= \delta_{0,k}.$$

Therefore, the second implication is proven. We ignore the proof of the first implication for short [Long, 1995].

When $\tilde{\Phi}, \ \Phi \in L^\infty(\mathbb{T})$, the equivalent equations in the theorem can easily be proved based on the note of Theorem 3.11. □

A key problem of the theory of the wavelet construction is to find out the sufficient conditions for $\{\tilde{m}_0, m_0\}$ to ensure that $\{\tilde{\phi}, \phi\}$ is biorthonormal. Two important results are given in the following:

**Theorem 3.13.** *Let* $m_0$ *satisfy the conditions in Theorem* 3.8, *and*

$$|m_0(\xi)|^2 + |m_0(\xi + \pi)|^2 = 1, \ \forall \xi \in \mathbb{R}.$$

*Then,* $\prod_{j=1}^{\infty} m_0(\xi/2^j)$ *is pointwise convergent to* $\hat{\phi} \in L^2(\mathbb{R}) \cap C(\mathbb{R})$, *and* $\phi$ *can generate an orthonormal MRA.*

**Proof.** By Theorems 3.8 and 3.12, it is clear that the theorem holds.    □

The famous Daubechies wavelet is based on this construction method. The key problem is to ensure that $m_0$ satisfies both Theorem 3.8 and the basic condition Ⅲ. This means to construct the $M_0$ in Theorem 3.8 and ensure that $m_0$ satisfies the basic condition Ⅲ (see Chapter 4 and the work of Daubechies [1992]).

**Theorem 3.14.** *Let $\{\tilde{m}_0, m_0\} \subset \mathcal{P}_N^+$ $(N \neq 0)$ be basic and $\{\tilde{\hat{\phi}}, \hat{\phi}\}$ be their limit functions. Then, the following two statements are equivalent:*

(1) *$\{\tilde{\phi}, \phi\}$ can generate a pair of biorthonormal MRAs.*
(2) *Each eigenvalue $\lambda$ of $\tilde{T}|_{\dot{\mathcal{P}}_N}$ and $T|_{\dot{\mathcal{P}}_N}$, which are the restrictions of the translation operators $\tilde{T}$ and $T$ in $\dot{\mathcal{P}}_N$, satisfies $|\lambda| < 1$.*

**Corollary 3.3.** *Let $\{\tilde{m}_0, m_0\} \subset \mathcal{P}_N$ $(N \neq 0)$ satisfy the basic condition I and $\{\tilde{\hat{\phi}}, \hat{\phi}\}$ be their limit functions. Then, the following two statements are equivalent:*

(1) *$\tilde{\hat{\phi}}, \hat{\phi} \in L^2(\mathbb{R})$, and $\{\tilde{\phi}, \phi\}$ can generate a pair of biorthonormal MRAs.*
(2) *$\tilde{m}_0(\xi)\bar{m}_0(\xi) + \tilde{m}_0(\xi + \pi)\bar{m}_0(\xi + \pi) = 1$ a.e. $\xi \in \mathbb{R}$, and each eigenvalue $\lambda$ of $\tilde{T}|_{\dot{\mathcal{P}}_N}$ and $T|_{\dot{\mathcal{P}}_N}$, which are the restrictions of the translation operators $\tilde{T}$ and $T$ in $\dot{\mathcal{P}}_N$, satisfies $|\lambda| < 1$.*

**Proof.** Obviously, $\{\tilde{m}_0, m_0\}$ satisfies the basic condition Ⅱ.
    "(1) $\Rightarrow$ (2)": By (1) in Theorem 3.11, we obtain

$$\tilde{m}_0(\xi)\bar{m}_0(\xi) + \tilde{m}_0(\xi + \pi)\bar{m}_0(\xi + \pi) = 1 \quad a.e. \; \xi \in \mathbb{R},$$

i.e., $\{\tilde{m}_0, m_0\}$ satisfies the basic condition Ⅲ. Thus, $\{\tilde{m}_0, m_0\}$ is basic. By Theorem 3.14, (2) is proven.
    "(2) $\Rightarrow$ (1)": Since $\{\tilde{m}_0, m_0\}$ is basic, according to Theorem 3.14, (1) is true.    □

### 3.2.2 *Examples of Constructing MRA*

Examples of constructing MRAs based on Theorem 3.3 are discussed in this section.
    First of all, the basic condition I can be equivalently illustrated by masks. That is, we have the following:

**Theorem 3.15.** *Let $m_0(\xi) := \sum_{k \in \mathbb{Z}} h_k e^{-ik\xi}$, where $\{h_k\} \in l^1(\mathbb{Z})$. Then*

$$m_0(0) = 1 \quad and \quad m_0(\pi) = 0 \iff \sum_{k \in \mathbb{Z}} h_{2k} = \sum_{k \in \mathbb{Z}} h_{2k+1} = \frac{1}{2}.$$

**Proof.** For $\mu = 0$ or $1$, we have

$$m_0(\pi\mu) = \sum_{k \in \mathbb{Z}} h_k e^{-i\pi k\mu}$$

$$= \left(\sum_{k \in \mathbb{Z}} h_{2k}\right) + \left(\sum_{k \in \mathbb{Z}} h_{2k+1}\right) e^{-i\pi\mu}$$

or

$$\begin{cases} m_0(0) = \sum_{k \in \mathbb{Z}} h_{2k} + \sum_{k \in \mathbb{Z}} h_{2k+1} \\ m_0(\pi) = \sum_{k \in \mathbb{Z}} h_{2k} - \sum_{k \in \mathbb{Z}} h_{2k+1} \end{cases}.$$

Therefore,

$$\begin{cases} m_0(0) = 1 \\ m_0(\pi) = 0 \end{cases} \iff \begin{cases} \sum_{k \in \mathbb{Z}} h_{2k} + \sum_{k \in \mathbb{Z}} h_{2k+1} = 1 \\ \sum_{k \in \mathbb{Z}} h_{2k} - \sum_{k \in \mathbb{Z}} h_{2k+1} = 0 \end{cases}$$

$$\iff \sum_{k \in \mathbb{Z}} h_{2k} = \sum_{k \in \mathbb{Z}} h_{2k+1} = \frac{1}{2}.$$

This finishes our proof. $\square$

For $m_0 \in \mathcal{P}_N^+$ ($N \neq 0$), by Corollary 3.2, we see that whether $\phi$, which is defined by $m_0$, can generate an MRA, depends on the properties of the eigenvalues and eigenvectors of $T|_{\mathcal{P}_N}$. It is easy to see that $T|_{\mathcal{P}_N}$ can be represented by a $2N+1$-order matrix. Since the eigenvalues and eigenvectors of a matrix can be calculated easily with a computer, the construction of MRAs in this case is always feasible.

Let $m_0(\xi) = \sum_{k \in \mathbb{Z}} h_k e^{-ik\xi} \in \mathcal{P}_N^+$ ($N \neq 0$) and $\{e^{-ik\xi} | k \in \mathbb{Z} \cap [-N, N]\}$ be a set of basis of $\mathcal{P}_N$. Our first step is to get the matrix of $T|_{\mathcal{P}_N}$ based on this basis. We have that

$$(Te^{-il x})(\xi) = \left|m_0\left(\frac{\xi}{2}\right)\right|^2 e^{-il\frac{\xi}{2}} + \left|m_0\left(\frac{\xi}{2}\pi\right)\right|^2 e^{-il\frac{\xi}{2}} e^{-il\pi}$$

$$= \left(\left|m_0\left(\frac{\xi}{2}\right)\right|^2 + \left|m_0\left(\frac{\xi}{2} + \pi\right)\right|^2 e^{-il\pi}\right) e^{-il\frac{\xi}{2}}$$

$$= \sum_{n\in\mathbb{Z}}\sum_{k\in\mathbb{Z}} h_n \bar{h}_k e^{-i(n-k)\frac{\xi}{2}}[1 + e^{-i(n-k+l)\pi}]e^{-il\frac{\xi}{2}}$$

$$= \sum_{n\in\mathbb{Z}}\sum_{k\in\mathbb{Z}} h_n \bar{h}_k e^{-i(n-k+l)\frac{\xi}{2}}[1 + e^{-i(n-k+l)\pi}]$$

$$= \sum_{n\in\mathbb{Z}}\sum_{k\in\mathbb{Z}} h_n \bar{h}_{n+l-k} e^{-ik\frac{\xi}{2}}[1 +^{-ik\pi}]$$

$$= \sum_{k\in\mathbb{Z}}\left(2\sum_{n\in\mathbb{Z}} h_n \bar{h}_{n+l-2k}\right) e^{-ik\xi}.$$

Let $\{r_1,\ldots,r_s\}$ $(s := 2N+1)$ be a permutation of the set $\mathbb{Z}\cap[-N,N]$, and

$$(Te^{-ir_1\xi},\ldots,Te^{-ir_s\xi}) = (e^{-ir_1\xi},\ldots,e^{-ir_s\xi})A,$$

where $A$, the corresponding matrix of $T|_{\mathcal{P}_N}$ based on the basis $\{e^{-ir_1\xi},\ldots,e^{-ir_s\xi}\}$, is a $2N+1$-order matrix:

$$A = \begin{pmatrix} a_{1,1} & a_{1,2} & \cdots & a_{1,s} \\ a_{2,1} & a_{2,2} & \cdots & a_{2,s} \\ \vdots & \vdots & \vdots & \vdots \\ a_{s,1} & a_{s,2} & \cdots & a_{s,s} \end{pmatrix}.$$

The $(k,l)$th element of $A$ is

$$a_{k,l} = 2\sum_{n\in\mathbb{Z}} h_n \bar{h}_{n+r_l-2r_k} \quad (k,l=1,\ldots,s).$$

By Corollary 3.2, we have the following:

**Theorem 3.16.** *Let $m_0(\xi) = \sum_{k\in\mathbb{Z}} h_k e^{-ik\xi} \in \mathcal{P}_N^+, (N\neq 0), \{r_1,\ldots,r_s\}$ be a permutation of $\{-N,-N+1,\ldots,N\}(s := 2N+1)$, and the $(k,l)$-th element of $A$ is*

$$a_{k,l} = 2\sum_{n=0}^{N} h_n \bar{h}_{n+r_l-2r_k} \quad (k,l=1,\ldots,s).$$

*Then, $\hat{\phi}(\xi) := \prod_{j=1}^{\infty} m_0(\xi/2^j)$ is pointwise convergent on $\mathbb{R}$ and $\phi$ generates an MRA of $L^2(\mathbb{R})$, if and only if the following three conditions are satisfied:*

(1) $\sum_{k\in\mathbb{Z}} h_{2k} = \sum_{k\in\mathbb{Z}} h_{2k} = \frac{1}{2}.$
(2) *$1$ is the simple eigenvalue of matrix $A$, and each of the other eigenvalues $\lambda$ satisfies $|\lambda| < 1$.*

(3) *The eigenpolynomial,* $g(\xi) := \sum_{k=1}^{s} b_k e^{-ir_k\xi}$, *has a positive lower bound, where* $(b_1, \ldots, b_s)^T$ *is the eigenvector of matrix A, which corresponds to the eigenvalue 1 and satisfies* $\sum_{k=1}^{s} b_k = 1$.

**Proof.** Due to that $A$ is the corresponding matrix of $T|_{\mathcal{P}_N}$ based on the basis $\{e^{-ir_1\xi}, \ldots, e^{-ir_s\xi}\}$ of $\mathcal{P}_N$, it is clear that (1) and (2) in the theorem are equivalent to (1) and (2) in Corollary 3.2, respectively. We conclude that $g(\xi) := \sum_{k=1}^{s} b_k e^{-ir_k\xi}$ is the eigenvector of $T|_{\mathcal{P}_N}$ corresponding to the eigenvalue 1 and satisfies $g(0) = 1$, if and only if $(b_1, \ldots, b_s)^T$ is the eigenvector of $A$ corresponding to the eigenvalue 1 and satisfies $\sum_{k=1}^{s} b_k = 1$. In fact, it is easy to see that

$$Tg(\xi) = g(\xi)$$

$$\Longleftrightarrow (Te^{-ir_1\xi}, \ldots, Te^{-ir_s\xi}) \begin{pmatrix} b_1 \\ \vdots \\ b_s \end{pmatrix} = (e^{-ir_1\xi}, \ldots, e^{-ir_s\xi}) \begin{pmatrix} b_1 \\ \vdots \\ b_s \end{pmatrix}$$

$$\Longleftrightarrow (e^{-ir_1\xi}, \ldots, e^{-ir_s\xi}) A \begin{pmatrix} b_1 \\ \vdots \\ b_s \end{pmatrix} = (e^{-ir_1\xi}, \ldots, e^{-ir_s\xi}) \begin{pmatrix} b_1 \\ \vdots \\ b_s \end{pmatrix}$$

$$\Longleftrightarrow A \begin{pmatrix} b_1 \\ \vdots \\ b_s \end{pmatrix} = \begin{pmatrix} b_1 \\ \vdots \\ b_s \end{pmatrix}.$$

Obviously, $g(0) = 1 \Longleftrightarrow \sum_{k=1}^{s} b_k = 1$. By Corollary 3.2, our proof is complete. □

For a natural permutation of $\mathbb{Z} \cap [-N, N] = \{-N, -N+1, \ldots, N-1, N\}$,

$$r_1 = -N, \ldots, r_k = -N - 1, \ldots, r_{2N+1} = N.$$

The corresponding matrix $A$ is a $2N+1$-order matrix whose $(k, l)$th elements is

$$a_{k,l} = 2 \sum_{n=0}^{N} h_n \bar{h}_{n+N+1+l-2k}, \quad (k, l = 1, \ldots, 2N + 1),$$

where $h_k = 0$ for $k < 0$ and $k > N$.

To construct MRA, a necessary condition that $\{h_0, \ldots, h_N\}$ should satisfy is

$$\sum_{k \in \mathbb{Z}, 0 \leq 2k \leq N} h_{2k} = \sum_{k \in \mathbb{Z}, 0 \leq 2k+1 \leq N} h_{2k+1} = \frac{1}{2}.$$

Now, for $N = 1, 2$, we discuss the conditions that $\{h_0, \ldots, h_N\}$ should satisfy so that it can generate an MRA. We focus on the three conditions in Theorem 3.16.

(1) For $N = 1$, we have $h_0 = h_1 = \frac{1}{2}$. $A$ is a third-order matrix whose elements are

$$a_{k,l} = 2(h_0 \bar{h}_{2+l-2k} + h_1 \bar{h}_{3+l-2k}) = h_{2+l-2k} + h_{3+l-2k}, \quad (k, l = 1, 2, 3).$$

Thus, $A$ can be represented as

$$A = \begin{pmatrix} \frac{1}{2} & 0 & 0 \\ \frac{1}{2} & 1 & \frac{1}{2} \\ 0 & 0 & \frac{1}{2} \end{pmatrix}.$$

The eigenvalues of $A$ are $1$, $\frac{1}{2}$. The eigenvector corresponding to 1 is $(b_1, b_2, b_3)^t = (0, 1, 0)^t$, and the sum of its components equals 1. Hence, the eigenpolynomial

$$g(\xi) = e^{-ir_2 \xi} = e^{-i0\xi} \equiv 1$$

has a positive lower bound. This tells us that there is only one MRA when $N = 1$. Now, we intend to find out its corresponding scale function. By

$$m_0(\xi) = e^{-i\frac{1}{2}\xi} \cos \frac{\xi}{2},$$

we have

$$\hat{\phi}(\xi) = \lim_{n \to \infty} \prod_{j=1}^{n} m_0 \left( \frac{\xi}{2^j} \right)$$

$$= \lim_{n \to \infty} \prod_{j=1}^{n} \left( e^{-i\frac{\xi}{2^{j+1}}} \cos \frac{\xi}{2^{j+1}} \right)$$

$$= e^{-i\frac{\xi}{2}} \lim_{n \to \infty} \prod_{j=1}^{n} \cos \frac{\xi}{2^{j+1}}$$

$$= e^{-i\frac{\xi}{2}} \lim_{n \to \infty} \frac{\sin \frac{\xi}{2}}{2^n \sin \frac{\xi}{2^{n+1}}}$$

$$= e^{-i\frac{\xi}{2}} \frac{2}{\xi} \sin \frac{\xi}{2}.$$

This is just the Fourier transform of the characteristic function $\chi_{[0,1]}(x)$ of interval $[0,1]$. Therefore,

$$\phi(x) = \chi_{[0,1]}(x) := \begin{cases} 1 & \text{when } x \in [0,1] \\ 0 & \text{otherwise.} \end{cases}$$

As a conclusion, for $N = 1$, there is only one MRA, which is generated by $\chi_{[0,1]}(x)$.

(2) For $N = 2$, we have

$$h_0 + h_2 = \frac{1}{2}, \quad h_1 = \frac{1}{2}.$$

Therefore, $A$ is a fifth-order matrix whose $(k,l)$th elements is

$$a_{k,l} = 2\left(h_0\bar{h}_{3+l-2k} + \frac{1}{2}\bar{h}_{4+l-2k} + h_2\bar{h}_{5+l-2k}\right), \quad (k,l = 1,\ldots,5).$$

Thus,

$$\begin{cases} a_{1,l} = 2h_0\bar{h}_{1+l} \\ a_{2,l} = 2\left(h_0\bar{h}_{l-1} + \frac{1}{2}\bar{h}_l + h_2\bar{h}_{l+1}\right) \\ a_{3,l} = 2\left(h_0\bar{h}_{l-3} + \frac{1}{2}\bar{h}_{l-2} + h_2\bar{h}_{l-1}\right) \\ a_{4,l} = 2\left(h_0\bar{h}_{l-5} + \frac{1}{2}\bar{h}_{l-4} + h_2\bar{h}_{l-3}\right) \\ a_{5,l} = 2h_2\bar{h}_{l-5}. \end{cases}$$

Hence, matrix $A$ can be written as.

$$\begin{pmatrix} h_0 - 2|h_0|^2 & 0 & 0 & 0 & 0 \\ 1 - h_0 - \bar{h}_0 + 4|h_0|^2 & \frac{1}{2} + h_0 - \bar{h}_0 & h_0 - 2|h_0|^2 & 0 & 0 \\ \bar{h}_0 - 2|h_0|^2 & \frac{1}{2} + \bar{h}_0 - h_0 & 1 - h_0 - \bar{h}_0 + 4|h_0|^2 & \frac{1}{2} + h_0 - \bar{h}_0 & h_0 - 2|h_0|^2 \\ 0 & 0 & \bar{h}_0 - 2|h_0|^2 & \frac{1}{2} + \bar{h}_0 - h_0 & 1 - h_0 - \bar{h}_0 + 4|h_0|^2 \\ 0 & 0 & 0 & 0 & \bar{h}_0 - 2|h_0|^2 \end{pmatrix}.$$

The eigenpolynomial of $A$ is

$$|\lambda I - A| = (\lambda - h_0 + 2|h_0|^2)(\lambda - \bar{h}_0 + 2|h_0|^2)$$

$$\cdot \begin{vmatrix} \lambda - \frac{1}{2} - h_0 + \bar{h}_0 & -h_0 + 2|h_0|^2 & 0 \\ -\frac{1}{2} - \bar{h}_0 + h_0 & \lambda - 1 + h_0 + \bar{h}_0 - 4|h_0|^2 & -\frac{1}{2} - h_0 + \bar{h}_0 \\ 0 & -\bar{h}_0 + 2|h_0|^2 & \lambda - \frac{1}{2} - \bar{h}_0 + h_0 \end{vmatrix}$$

$$= (\lambda - h_0 + 2|h_0|^2)(\lambda - \bar{h}_0 + 2|h_0|^2)$$

$$\cdot \begin{vmatrix} \lambda - \frac{1}{2} - h_0 + \bar{h}_0 & -h_0 + 2|h_0|^2 & 0 \\ \lambda - 1 & \lambda - 1 & \lambda - 1 \\ 0 & -\bar{h}_0 + 2|h_0|^2 & \lambda - \frac{1}{2} - \bar{h}_0 + h_0 \end{vmatrix}$$

$$= (\lambda - h_0 + 2|h_0|^2)(\lambda - \bar{h}_0 + 2|h_0|^2)(\lambda - 1)$$

$$\cdot \begin{vmatrix} \lambda - \frac{1}{2} - h_0 + \bar{h}_0 & -h_0 + 2|h_0|^2 & 0 \\ 1 & 1 & 1 \\ 0 & -\bar{h}_0 + 2|h_0|^2 & \lambda - \frac{1}{2} - \bar{h}_0 + h_0 \end{vmatrix}$$

$$= (\lambda - h_0 + 2|h_0|^2)(\lambda - \bar{h}_0 + 2|h_0|^2)(\lambda - 1)$$

$$\cdot \begin{vmatrix} \lambda - \frac{1}{2} - h_0 + \bar{h}_0 & 1 & 0 \\ -h_0 + 2|h_0|^2 & 1 & -\bar{h}_0 + 2|h_0|^2 \\ 0 & 1 & \lambda - \frac{1}{2} - \bar{h}_0 + h_0 \end{vmatrix}$$

$$= -(\lambda - h_0 + 2|h_0|^2)(\lambda - \bar{h}_0 + 2|h_0|^2)(\lambda - 1)$$

$$\cdot \begin{vmatrix} 1 & \lambda - \frac{1}{2} - h_0 + \bar{h}_0 & 0 \\ 1 & -h_0 + 2|h_0|^2 & -\bar{h}_0 + 2|h_0|^2 \\ 1 & 0 & \lambda - \frac{1}{2} - \bar{h}_0 + h_0 \end{vmatrix}$$

$$= -(\lambda - h_0 + 2|h_0|^2)(\lambda - \bar{h}_0 + 2|h_0|^2)(\lambda - 1)$$

$$\cdot \begin{vmatrix} 1 & \lambda - \frac{1}{2} - h_0 + \bar{h}_0 & 0 \\ 0 & -\lambda + \frac{1}{2} + 2|h_0|^2 - \bar{h}_0 & 2|h_0|^2 - \bar{h}_0 \\ 0 & h_0 - 2|h_0|^2 & \lambda - \frac{1}{2} + h_0 - 2|h_0|^2 \end{vmatrix}$$

$$= -(\lambda - h_0 + 2|h_0|^2)(\lambda - \bar{h}_0 + 2|h_0|^2)(\lambda - 1)$$

$$\cdot \begin{vmatrix} -\lambda + \frac{1}{2} + 2|h_0|^2 - \bar{h}_0 & 2|h_0|^2 - \bar{h}_0 \\ h_0 - 2|h_0|^2 & \lambda - \frac{1}{2} + h_0 - 2|h_0|^2 \end{vmatrix}$$

$$= -(\lambda - h_0 + 2|h_0|^2)(\lambda - \bar{h}_0 + 2|h_0|^2)(\lambda - 1)$$

$$\cdot \begin{vmatrix} -\lambda + \frac{1}{2} & 2|h_0|^2 - \bar{h}_0 \\ -\lambda + \frac{1}{2} & \lambda - \frac{1}{2} + h_0 - 2|h_0|^2 \end{vmatrix}$$

$$= (\lambda - h_0 + 2|h_0|^2)(\lambda - \bar{h}_0 + 2|h_0|^2)(\lambda - 1)\left(\lambda - \frac{1}{2}\right)$$

$$\cdot \begin{vmatrix} 1 & 2|h_0|^2 - \bar{h}_0 \\ 1 & \lambda - \frac{1}{2} + h_0 - 2|h_0|^2 \end{vmatrix}$$

$$= (\lambda - h_0 + 2|h_0|^2)(\lambda - \bar{h}_0 + 2|h_0|^2)(\lambda - 1)\left(\lambda - \frac{1}{2}\right)$$

$$\cdot \begin{vmatrix} 1 & 2|h_0|^2 - \bar{h}_0 \\ 0 & \lambda - \frac{1}{2} + h_0 + \bar{h}_0 - 4|h_0|^2 \end{vmatrix}$$

$$= (\lambda - h_0 + 2|h_0|^2)(\lambda - \bar{h}_0 + 2|h_0|^2)(\lambda - 1)\left(\lambda - \frac{1}{2}\right)$$

$$\cdot \left(\lambda - \frac{1}{2} + h_0 + \bar{h}_0 - 4|h_0|^2\right).$$

Therefore, the eigenvalues of $A$ are

$$1, \ \frac{1}{2}, \ h_0 - 2|h_0|^2, \ \bar{h}_0 - 2|h_0|^2, \ \frac{1}{2} + 4|h_0|^2 - h_0 - \bar{h}_0.$$

In order to generate MRA, the moduli of the above eigenvalues, except 1, must be smaller than 1, that is,

$$\begin{cases} |h_0 - 2|h_0|^2| < 1 \\ |\frac{1}{2} + 4|h_0|^2 - h_0 - \bar{h}_0| < 1. \end{cases}$$

We now simplify these conditions as follows. By

$$\frac{1}{2}|h_0|^2 - h_0 - \bar{h}_0 = \frac{1}{2} \ _0 \bar{h}_0 - h_0 - \bar{h}_0 = \frac{1}{4}\left|h_0 - \frac{1}{4}\right|^2 > 0,$$

we get

$$\left|\frac{1}{2} + 4|h_0|^2 - h_0 - \bar{h}_0\right| < 1 \iff \left|h_0 - \frac{1}{4}\right| < \frac{\sqrt{3}}{4}.$$

For $\left|h_0 - \frac{1}{4}\right| < \frac{\sqrt{3}}{4}$, we have

$$\left|h_0 - 2|h_0|^2\right| = \sqrt{|2|h_0|^2 - Re(h_0)|^2 + |Im(h_0)|^2}$$

$$= \sqrt{|2|h_0|^2 - \frac{1}{2}(h_0 + \bar{h}_0)|^2 + |Im(h_0)|^2}$$

$$= \sqrt{\frac{1}{4}\left|4|h_0|^2 - h_0 - \bar{h}_0\right|^2 + |Im(h_0)|^2}$$

$$= \sqrt{\frac{1}{4}\left|4|h_0 - \frac{1}{4}|^2 - \frac{1}{4}\right|^2 + |Im(h_0)|^2}$$

$$< \sqrt{\frac{1}{4}\left(\frac{1}{2}\right)^2 + |Im(h_0)|^2}$$

$$= \frac{1}{2}\sqrt{\frac{1}{4} + 4|Im(h_0)|^2}.$$

And by

$$\left|\frac{1}{2} - h_0 + \bar{h}_0\right| = \left|\frac{1}{2} - 2iIm(h_0)\right| = \sqrt{\frac{1}{4} + 4|Im(h_0)|^2},$$

we have

$$\left|h_0 - 2|h_0|^2\right| < \frac{1}{2}\left|\frac{1}{2} - h_0 + \bar{h}_0\right|$$

$$= \frac{1}{2}\left|\frac{1}{2} - \left(h_0 - \frac{1}{4}\right) + \left(\bar{h}_0 - \frac{1}{2}\right)\right|$$

$$< \frac{1}{2}\left(\frac{1}{2} + \frac{\sqrt{3}}{4} + \frac{\sqrt{3}}{4}\right)$$

$$= \frac{1 + \sqrt{3}}{4} < 1.$$

Hence,

$$\begin{cases} \left|h_0 - 2|h_0|^2\right| < 1 \\ \left|\frac{1}{2} + 4|h_0|^2 - h_0 - \bar{h}_0\right| < 1 \end{cases} \iff \left|h_0 - \frac{1}{4}\right| < \frac{\sqrt{3}}{4}.$$

Now, we suppose $|h_0 - \frac{1}{4}| < \frac{\sqrt{3}}{4}$. To solve the eigenvector corresponding to 1, such that the sum of whose components equals 1, we apply the elementary row transformation to matrix $I - A$ as follows:

$I - A \Longrightarrow$

$$
\begin{pmatrix}
1 - h_0 + 2|h_0|^2 & 0 & 0 \\
h_0 + \bar{h}_0 - 1 - 4|h_0|^2 & \frac{1}{2} - h_0 + \bar{h}_0 & 2|h_0|^2 - h_0 \\
2|h_0|^2 - \bar{h}_0 & h_0 - \bar{h}_0 - \frac{1}{2} & -4|h_0|^2 + h_0 + \bar{h}_0 \\
0 & 0 & 2|h_0|^2 - \bar{h}_0 \\
0 & 0 & 0
\end{pmatrix}
$$

$$
\begin{pmatrix}
0 & 0 \\
0 & 0 \\
\bar{h}_0 - h_0 - \frac{1}{2} & 2|h_0|^2 - h_0 \\
\frac{1}{2} + h_0 - \bar{h}_0 & h_0 + \bar{h}_0 - 1 - 4|h_0|^2 \\
0 & 1 + 2|h_0|^2 - \bar{h}_0
\end{pmatrix}
$$

$$
\Longrightarrow
\begin{pmatrix}
1 - h_0 + 2|h_0|^2 & 0 & 0 \\
h_0 + \bar{h}_0 - 1 - 4|h_0|^2 & \frac{1}{2} - h_0 + \bar{h}_0 & 2|h_0|^2 - h_0 \\
0 & 0 & 0 \\
0 & 0 & 2|h_0|^2 - \bar{h}_0 \\
0 & 0 & 0
\end{pmatrix}
$$

$$
\begin{pmatrix}
0 & 0 \\
0 & 0 \\
0 & 0 \\
\frac{1}{20} + h_0 - \bar{h}_0 & h_0 + \bar{h}_0 - 1 - 4|h_0|^2 \\
0 & 1 + 2|h_0|^2 - \bar{h}_0
\end{pmatrix}
$$

$$
\Longrightarrow
\begin{pmatrix}
1 & 0 & 0 & 0 & 0 \\
0 & \frac{1}{2} - h_0 + \bar{h}_0 & 2|h_0|^2 - h_0 & 0 & 0 \\
0 & 0 & 2|h_0|^2 - \bar{h}_0 & \frac{1}{2} + h_0 - \bar{h}_0 & 0 \\
0 & 0 & 0 & 0 & 1 \\
0 & 0 & 0 & 0 & 0
\end{pmatrix}.
$$

By solving the following linear system:

$$\begin{cases} \begin{pmatrix} 1 & 0 & 0 & 0 & 0 \\ 0 & \frac{1}{2} - h_0 + \bar{h}_0 & 2|h_0|^2 - h_0 & 0 & 0 \\ 0 & 0 & 2|h_0|^2 - \bar{h}_0 & \frac{1}{2} + h_0 - \bar{h}_0 & 0 \\ 0 & 0 & 0 & 0 & 1 \\ 0 & 0 & 0 & 0 & 0 \end{pmatrix} \begin{pmatrix} b_1 \\ b_2 \\ b_3 \\ b_4 \\ b_5 \end{pmatrix} = 0 \\ b_1 + b_2 + b_3 + b_4 + b_5 = 1 \end{cases},$$

we obtain the eigenvector corresponding to the eigenvalue 1 so that the sum of its components equals 1. It is

$$\begin{pmatrix} b_1 \\ b_2 \\ b_3 \\ b_4 \\ b_5 \end{pmatrix} = \frac{8|\frac{1}{2} - h_0 + \bar{h}_0|^2}{3 - 16|h_0 - \frac{1}{4}|^2} \begin{pmatrix} 0 \\ \frac{h_0 - 2|h_0|^2}{\frac{1}{2} - h_0 + \bar{h}_0} \\ 1 \\ \frac{\bar{h}_0 - 2|h_0|^2}{\frac{1}{2} - \bar{h}_0 + h_0} \\ 0 \end{pmatrix}.$$

Thus, the eigenpolynomial is

$$g(\xi) = b_1 e^{i2\xi} + b_2 e^{i\xi} + b_3 + b_4 e^{-i\xi} + b_5 e^{-i2\xi}$$

$$= \frac{8|\frac{1}{2} - h_0 + \bar{h}_0|^2}{3 - 16|h_0 - \frac{1}{4}|^2} \left( \frac{h_0 - 2|h_0|^2}{\frac{1}{2} - h_0 \bar{h}_0} e^{i\xi} + 1 + \frac{\bar{h}_0 - 2|h_0|^2}{\frac{1}{2} - \bar{h}_0 + h_0} e^{-i\xi} \right)$$

$$= \frac{8|\frac{1}{2} - h_0 + \bar{h}_0|^2}{3 - 16|h_0 - \frac{1}{4}|^2} \left( 1 + 2\,\mathrm{Re}\left( \frac{h_0 - 2|h_0|^2}{\frac{1}{2} - h_0 + \bar{h}_0} e^{i\xi} \right) \right)$$

$$\geq \frac{8|\frac{1}{2} - h_0 + \bar{h}_0|^2}{3 - 16|h_0 - \frac{1}{4}|^2} \left( 1 - 2\left| \frac{h_0 - 2|h_0|^2}{\frac{1}{2} - h_0 + \bar{h}_0} \right| \right)$$

$$> 0,$$

which shows that $g(\xi)$ has a positive lower bound.

All the discussion above shows that, for $N = 2$, $m_0(\xi) := h_0 + h_1 e^{-i\xi} + h_2 e^{-i2\xi}$ can generate an MRA based on Theorem 3.16 if and only if

$$\begin{cases} h_0 \neq 0,\ h_0 \neq \frac{1}{2} \quad \text{and} \quad \left| h_0 - \frac{1}{4} \right| < \frac{\sqrt{3}}{4} \\ h_1 = \frac{1}{2} \\ h_2 = \frac{1}{2} - h_0. \end{cases} \tag{3.24}$$

There are infinite banks of $\{h_0, h_1, h_2\}$ which satisfy the above conditions. Generally speaking, it is very difficult to find the corresponding scale function $\phi$. Now, we consider a particular case.

For $h_0 = \frac{1}{4}$, we have

$$
\begin{aligned}
m_0(\xi) &= \frac{1}{4} + \frac{1}{2} e^{-i\xi} + \frac{1}{4} e^{-2i\xi} \\
&= \frac{1}{4}(1 + e^{-i\xi})^2 \\
&= \left( e^{-i\xi/2} \cos\left(\frac{1}{2}\xi\right) \right)^2.
\end{aligned}
$$

Obviously, it is just the square of the filter function $m_0$ for $N = 1$. Therefore,

$$
\hat{\phi}(\xi) = \left( \hat{\chi}_{[0,1)}(\xi) \right)^2 = \left( \chi_{[0,1)} * \chi_{[0,1)} \right)^\hat{}(\xi),
$$

i.e.,

$$
\phi(x) = \chi_{[0,1)} * \chi_{[0,1)}(x) = \begin{cases} x & x \in [0,1) \\ 2 - x & x \in [1,2) \\ 0 & \text{otherwise.} \end{cases}
$$

This is the well-known first-order B-spline function.

## 3.3 The Construction of Biorthonormal Wavelet Bases

MRA, since it was proposed by Mallat and Meyer in the late 1980s, has become the standard scheme for construction of wavelet bases. It has shown that almost all the wavelet bases with usual properties can be constructed from MRAs. In this section, we study how to construct a general (orthonormal or not) wavelet basis $\{\psi_{j,k}\}$ from an MRA.

**Definition 3.7.** $\phi \in L^2(\mathbb{R})$ is said to satisfy the basic smooth condition if there exist positive numbers $\delta < 2$ and $C$ such that

$$
\begin{cases} \displaystyle\sum_{k\in\mathbb{Z}} |\hat{\phi}(\xi + 2\pi k)|^{2-\delta} \le C & \text{a.e. } \xi \in \mathbb{T} \\ \displaystyle\sum_{j=0}^{\infty} |\hat{\phi}(2^j\xi)|^\delta \le C & \text{a.e. } 1 \le |\xi| \le 2. \end{cases}
$$

*Note* 1: The condition in the definition is called the basic smooth condition since the smoothness of $\phi$ corresponds to the decreasing in its Fourier transform.

*Note* 2: It can be proved that $\phi$ defined by $\hat{\phi}(\xi) := \prod_{j=1}^{\infty} m_0(\xi/2^j)$ satisfies the basic smooth condition if $m_0$ satisfies either of the following conditions:

(1) the conditions of Theorem 3.8,
(2) $m_0 \in \mathcal{P}_N^+$ $(N \neq 0)$, $m_0(0) = 1$, $m_0(\pi) = 0$, and any eigenvalue $\lambda$ of $T_{\dot{\mathcal{P}}_N}$, the restriction of transition operator $T$ on $\dot{\mathcal{P}}_N$, satisfies $|\lambda| < 1$.

**Theorem 3.17.** *Suppose* $\tilde{\phi}, \phi \in L^2(\mathbb{R})$, *generate biorthonormal MRA* $\{\tilde{V}_j\}, \{V_j\}$, *satisfy the basic smooth condition and the corresponding filter function* $\tilde{m}_0, m_0 \in L^\infty(\mathbb{T})$ *satisfy*

$$|m_0(\xi)|^2 + |m_0(\xi + \pi)|^2 \neq 0$$

*and*

$$\tilde{m}_0(\xi)m_0(\xi) + \tilde{m}_0(\xi + \pi)m_0(\xi + \pi) = 1, \quad a.e. \ \xi \in \mathbb{R}.$$

*Then, for*

$$m_1(\xi) = a(2\xi)\overline{\tilde{m}}_0(\xi + \pi)e^{-i\xi}, \quad \tilde{m}_1(\xi) = \overline{a^{-1}(2\xi)}\overline{m}_0(\xi + \pi)e^{-i\xi},$$

*where* $a(\xi)$ *denotes a* $2\pi$-*periodic function bounded by positive numbers from both the above and below,* $\{\tilde{\psi}_{j,k}, \psi_{j,k}\}_{j,k \in \mathbb{Z}}$ *which is defined by*

$$\hat{\tilde{\psi}}(\xi) := \tilde{m}_1\left(\frac{\xi}{2}\right)\hat{\tilde{\phi}}\left(\frac{\xi}{2}\right), \quad \hat{\psi}(\xi) := m_1\left(\frac{\xi}{2}\right)\hat{\phi}\left(\frac{\xi}{2}\right), \qquad (3.25)$$

*constitutes a pair of biorthonormal wavelet bases of* $L^2(\mathbb{R})$, *that is, both* $\{\tilde{\psi}_{j,k}\}_{j,k \in \mathbb{Z}}$ *and* $\{\psi_{j,k}\}_{j,k \in \mathbb{Z}}$ *constitute a Riesz basis of* $L^2(\mathbb{R})$ *and they are biorthonormal to each other, i.e.,*

$$\langle \tilde{\psi}_{j,k}, \psi_{j',k'} \rangle = \delta_{j,j'}\delta_{k,k'} \quad (\forall j, j' \in \mathbb{Z}, k, \ k' \in \mathbb{Z}).$$

*Further, the following results hold:*

(1) *The operators defined by*

$$\tilde{P}_j f := \sum_{k \in \mathbb{Z}}\langle f, \phi_{j,k} \rangle \tilde{\phi}_{j,k} \quad P_j f := \sum_{k \in \mathbb{Z}}\langle f, \tilde{\phi}_{j,k} \rangle \phi_{j,k}, \qquad (3.26)$$

$$\tilde{Q}_j f := \sum_{k \in \mathbb{Z}}\langle f, \psi_{j,k} \rangle \tilde{\psi}_{j,k} \quad Q_j f := \sum_{k \in \mathbb{Z}}\langle f, \tilde{\psi}_{j,k} \rangle \psi_{j,k}, \qquad (3.27)$$

*satisfy that,* $\forall j \in \mathbb{Z}$,

$$\begin{cases} \tilde{P}_{j+1} = \tilde{P}_j + \tilde{Q}_j, \quad P_{j+1} = P_j + Q_j, \\ \lim_{j \to -\infty} \|\tilde{P}_j f\|_2 = \lim_{j \to -\infty} \|P_j f\|_2 = 0 \quad \forall f \in L^2(\mathbb{R}), \\ \lim_{j \to +\infty} \|\tilde{P}_j f - f\|_2 = \lim_{j \to +\infty} \|P_j f - f\|_2 = 0 \quad \forall f \in L^2(\mathbb{R}). \end{cases}$$

(2) *If we denote*

$$\tilde{W}_j := \overline{span}\{\tilde{\psi}_{j,k}|k \in \mathbb{Z}\}, \quad W_j := \overline{span}\{\psi_{j,k}|k \in \mathbb{Z}\}, \qquad (3.28)$$

*then, for any* $j \in \mathbb{Z}$, *operators* $\tilde{P}_j$, $P_j$, $\tilde{Q}_j$, $Q_j$ *are the projectors from* $L^2(\mathbb{R})$ *to* $\tilde{V}_j, V_j, \tilde{W}_j$, *and* $W_j$ *respectively satisfying*

$$\begin{cases} V_j \cap \tilde{V}_j^{\perp} = \tilde{V}_j \cap V_j^{\perp} = \{0\} \\ W_j \cap \tilde{W}_j^{\perp} = \tilde{W}_j \cap W_j^{\perp} = \{0\} \end{cases}$$

$$\begin{cases} \tilde{W}_j = \tilde{V}_{j+1} \cap V_j^{\perp} \\ W_j = V_{j+1} \cap \tilde{V}_j^{\perp} \end{cases} \quad \begin{cases} \tilde{V}_{j+1} = \tilde{V}_j \dotplus \tilde{W}_j \\ V_{j+1} = V_j \dotplus W_j, \end{cases}$$

*where* $\dotplus$ *denotes the direct sum.*

We omit the proof of the theorem because it is complicated and refers to some mathematical analysis which is not included in this book. The reader can obtain the details from the works of Daubechies [1992], Long [1995] and other related references.

Sometimes, $V_j \perp W_j$ is useful in practice. We discuss this question here. A lemma is given first.

**Lemma 3.3.** *Let* $m_0$ *be a* $2\pi$-*periodic measurable function on* $\mathbb{R}$ *satisfying* $|m_0(\xi)|^2 + |m_0(\xi + \pi)|^2 \neq 0$ *a.e.* $\xi \in \mathbb{R}$. *Then, a* $2\pi$-*periodic measurable function* $m_1$ *satisfies*

$$m_0(\xi)m_1(\xi) + m_0(\xi + \pi)m_1(\xi + \pi) = 0, \quad a.e. \ \xi \in \mathbb{R},$$

*if and only if a* $2\pi$-*periodic measurable function* $\nu$ *exists such that*

$$m_1(\xi) = e^{-i\xi}\nu(2\xi)m_0(\xi + \pi), \quad a.e. \ \xi \in \mathbb{R}.$$

**Proof.** The part of the sufficiency holds obviously. We need to prove only the necessity, and we assume without losing generality that, $\forall \xi \in \mathbb{R}$,

$$
\begin{cases}
m_0(\xi) = m_0(\xi + 2\pi), \quad m_1(\xi) = m_1(\xi + 2\pi) \\
|m_0(\xi)|^2 + |m_0(\xi + \pi)|^2 \neq 0 \\
m_0(\xi)m_1(\xi) + m_0(\xi + \pi)m_1(\xi + \pi) = 0.
\end{cases}
$$

It is easy to see that

$$
\lambda(\xi) := \begin{cases}
-\dfrac{m_1(\xi + \pi)}{m_0(\xi)} & \text{for } m_0(\xi) \neq 0 \\[2ex]
\dfrac{m_1(\xi)}{m_0(\xi + \pi)} & \text{for } m_0(\xi) = 0,
\end{cases}
$$

is a $2\pi$-periodic measurable function. We claim that

$$
\lambda(\xi) + \lambda(\xi + \pi) = 0, \quad \forall \xi \in \mathbb{T}.
$$

In fact, we have the following:

(1) If $m_0(\xi) = 0$, it is clear that $m_0(\xi + \pi) \neq 0$, therefore,

$$
\lambda(\xi) + \lambda(\xi + \pi) = \frac{m_1(\xi)}{m_0(\xi + \pi)} - \frac{m_1(\xi + 2\pi)}{m_0(\xi + \pi)} = 0.
$$

(2) If $m_0(\xi + \pi) = 0$, it is still easy to deduce that $m_0(\xi) \neq 0$. By the result of (1), we have $\lambda(\xi + \pi) + \lambda(\xi + 2\pi) = 0$, i.e.,

$$
\lambda(\xi) + \lambda(\xi + \pi) = 0.
$$

(3) At last, if $m_0(\xi) \neq 0$ and $m_0(\xi + \pi) \neq 0$, we can obtain

$$
\begin{aligned}
\lambda(\xi) + \lambda(\xi + \pi) &= -\frac{m_1(\xi + \pi)}{m_0(\xi)} - \frac{m_1(\xi)}{m_0(\xi + \pi)} \\
&= -\frac{m_1(\xi)m_0(\xi) + m_1(\xi + \pi)m_0(\xi + \pi)}{m_0(\xi)m_0(\xi + \pi)} \\
&= 0.
\end{aligned}
$$

Our claim is proved.

If $m_0(\xi) = 0$, we have

$$
m_1(\xi) = \lambda(\xi)m_0(\xi + \pi).
$$

If $m_0(\xi) \neq 0$, it is obvious that

$$m_1(\xi + \pi) = -\lambda(\xi)m_0(\xi),$$

which concludes that

$$m_1(\xi)m_0(\xi) = -m_1(\xi + \pi)m_0(\xi + \pi)$$
$$= \lambda(\xi)m_0(\xi)m_0(\xi + \pi).$$

Hence, we also have

$$m_1(\xi) = \lambda(\xi)m_0(\xi + \pi).$$

Obviously, $\lambda(\xi) + \lambda(\xi + \pi) = 0$ implies that $\lambda(\xi)e^{-i\xi}$ is $\pi$-periodic. We denote

$$\nu(\xi) := \lambda\left(\frac{\xi}{2}\right)e^{-i\frac{1}{2}\xi},$$

then $\nu(\xi)$ is a $2\pi$-periodic measurable function and $\lambda(\xi) = \nu(2\xi)e^{i\xi}$. Therefore,

$$m_1(\xi) = e^{i\xi}\nu(2\xi)m_0(\xi + \pi).$$

This ends the proof. $\qquad\qquad\qquad\qquad\qquad\qquad\qquad\qquad\square$

According to the above lemma, the conditions of $V_j \perp W_j$ can be characterized as follows:

**Theorem 3.18.** *Let $m_0$ and $\phi$ be the filter function and scale function of an MRA $\{V_j\}_{j\in\mathbb{Z}}$, respectively, $m_1 \in L^\infty(\mathbb{T})$. We denote*

$$\hat{\psi}(\xi) := m_1\left(\frac{\xi}{2}\right)\hat{\phi}\left(\frac{\xi}{2}\right),$$
$$W_j := \overline{span}\{\psi(2^j \cdot -k)|k \in \mathbb{Z}\}, \quad (\forall j \in \mathbb{Z}).$$

*Then $V_j \perp W_j$ ($\forall j \in \mathbb{Z}$) if and only if there is a $2\pi$-periodic measurable function $\nu(\xi)$ such that*

$$m_1(\xi) = e^{-i\xi}\nu(2\xi)\bar{m}_0(\xi + \pi)\Phi(\xi + \pi),$$

*where $\Phi := [\hat{\phi}, \hat{\phi}]$.*

**Proof.** It is easy to see that

$$V_j \perp W_j \ (\forall j \in \mathbb{Z}) \iff V_0 \perp W_0$$

$$\iff \langle \psi(\cdot - k), \phi(\cdot) \rangle = 0 \quad (\forall k \in \mathbb{Z})$$

$$\iff [\hat{\psi}, \hat{\phi}](\xi) = 0 \quad \text{a.e. } \xi \in \mathbb{T}.$$

Since

$$[\hat{\psi}, \hat{\phi}](\xi) = \sum_{k \in \mathbb{Z}} \hat{\psi}(\xi + 2\pi k)\overline{\hat{\phi}}(\xi + 2\pi k)$$

$$= \sum_{k \in \mathbb{Z}} m_1 \left( \frac{\xi}{2} + \pi k \right) \hat{\phi} \left( \frac{\xi}{2} + \pi k \right) \bar{m}_0 \left( \frac{\xi}{2} + \pi k \right) \overline{\hat{\phi}} \left( \frac{\xi}{2} + \pi k \right)$$

$$= \sum_{k \in \mathbb{Z}} m_1 \left( \frac{\xi}{2} + \pi k \right) \bar{m}_0 \left( \frac{\xi}{2} + \pi k \right) \left| \hat{\phi} \left( \frac{\xi}{2} + \pi k \right) \right|^2$$

$$= \sum_{\alpha \in \mathbb{Z}} \left[ m_1 \left( \frac{\xi}{2} \right) \bar{m}_0 \left( \frac{\xi}{2} \right) \right) \left| \hat{\phi} \left( \frac{\xi}{2} + 2\pi\alpha \right) \right|^2$$

$$+ m_1 \left( \frac{\xi}{2} + \pi \right) \bar{m}_0 \left( \frac{\xi}{2} + \pi \right) \right) \left| \hat{\phi} \left( \frac{\xi}{2} + 2\pi\alpha + \pi \right) \right|^2 \right]$$

$$= m_1 \left( \frac{\xi}{2} \right) \bar{m}_0 \left( \frac{\xi}{2} \right) \Phi \left( \frac{\xi}{2} \right) + m_1 \left( \frac{\xi}{2} + \pi \right)$$

$$\times \bar{m}_0 \left( \frac{\xi}{2} + \pi \right) \Phi \left( \frac{\xi}{2} + \pi \right),$$

we conclude that

$$V_j \perp W_j \ (\forall j \in \mathbb{Z}) \iff m_1(\xi)\bar{m}_0(\xi)\Phi(\xi)$$

$$+ m_1(\xi + \pi)\bar{m}_0(\xi + \pi)\Phi(\xi + \pi)$$

$$= 0 \quad \text{a.e. } \xi \in \mathbb{T}.$$

Set $m_0$ in Lemma 3.3 to be $\bar{m}_0(\xi)\Phi(\xi)$ here and notice that $\phi$ generates an MRA; it can be concluded easily that there exist positive numbers $A$ and $B$ such that

$$A \le |\bar{m}_0(\xi)\Phi(\xi)|^2 + |\bar{m}_0(\xi + \pi)\Phi(\xi + \pi)|^2 \le B, \quad \text{a.e. } \xi \in \mathbb{R}.$$

By Lemma 3.3, our theorem is proved. $\qquad \square$

## 3.4 Mallat Algorithms

Under the conditions of Theorem 3.17, the well-known Mallat algorithm can be deduced easily. This algorithm provides a recursive scheme to calculate the coefficients of the biorthonormal wavelet expansion of a signal from one layer to the next layer. It can be applied to image processing, pattern recognition, and other related subjects effectively.

**Theorem 3.19.** *Suppose the conditions of Theorem 3.17 are satisfied and denote $\tilde{m}_0$, $m_0$, $\tilde{m}_1$, $m_1$ of Theorem 3.17 as follows:*

$$\tilde{m}_0(\xi) = \sum_{k \in \mathbb{Z}} \tilde{h}_k e^{-ik\xi}, \quad m_0(\xi) = \sum_{k \in \mathbb{Z}} h_k e^{-ik\xi},$$

$$\tilde{m}_1(\xi) = \sum_{k \in \mathbb{Z}} \tilde{g}_k e^{-ik\xi}, \quad m_1(\xi) = \sum_{k \in \mathbb{Z}} g_k e^{-ik\xi}.$$

*Then, $\forall f \in L^2(\mathbb{R})$, the following Mallat decomposition algorithm holds:*

$$\begin{cases} \langle f, \phi_{j,k} \rangle = \sqrt{2} \sum_{l \in \mathbb{Z}} h_l \langle f, \phi_{j+1,2k+l} \rangle \\ \langle f, \psi_{\mu,j,k} \rangle = \sqrt{2} \sum_{l \in \mathbb{Z}} g_l \langle f, \phi_{j+1,2k+l} \rangle \end{cases} \quad (\forall j \in \mathbb{Z}, k \in \mathbb{Z}),$$

$$\begin{cases} \langle f, \tilde{\phi}_{j,k} \rangle = \sqrt{2} \sum_{l \in \mathbb{Z}} \tilde{h}_l \langle f, \tilde{\phi}_{j+1,2k+l} \rangle \\ \langle f, \tilde{\psi}_{j,k} \rangle = \sqrt{2} \sum_{l \in} \tilde{g}_l \langle f, \tilde{\phi}_{j+1,2k+l} \rangle \end{cases} \quad (\forall j \in \mathbb{Z}, \ k \in \mathbb{Z}).$$

*And the corresponding reconstruction algorithm holds as follows:*

$$\langle f, \phi_{j+1,k} \rangle = \sqrt{2} \sum_{l \in \mathbb{Z}} \bar{\tilde{h}}_{k-2l} \langle f, \phi_{j,l} \rangle + \sqrt{2} \sum_{l \in \mathbb{Z}} \bar{\tilde{g}}_{k-2l} \langle f, \psi_{j,l} \rangle,$$

$$(\forall j \in \mathbb{Z}, \ k \in \mathbb{Z}),$$

$$\langle f, \tilde{\phi}_{j+1,k} \rangle = \sqrt{2} \sum_{l \in \mathbb{Z}} \bar{h}_{k-2l} \langle f, \tilde{\phi}_{j,l} \rangle + \sqrt{2} \sum_{l \in \mathbb{Z}} \bar{g}_{k-2l} \langle f, \tilde{\psi}_{j,l} \rangle,$$

$$(\forall j \in \mathbb{Z}, \ k \in \mathbb{Z}).$$

**Proof.** Using the two-scale relation and the expression of $m_0$, we have

$$\phi(x) = 2 \sum_{l \in \mathbb{Z}} h_l \phi(2x - l),$$

which is equivalent to

$$\phi_{j,k}(x) = \sum_{l \in \mathbb{Z}} h_l \phi_{j+1,2k+l}(x).$$

Hence,

$$\langle f, \phi_{j,k} \rangle = \sum_{l \in \mathbb{Z}} h_l \langle f, \phi_{j+1,2k+l} \rangle.$$

The other formulae of the decomposition algorithm can be proved similarly. Now, we turn to the proof of the reconstruction algorithm.

By Theorem 3.17, we deduce that, $\forall j \in \mathbb{Z}, l \in \mathbb{Z}$,

$$\tilde{P}_{j+1}\tilde{\phi}_{j+1,l} = \tilde{P}_j \tilde{\phi}_{j+1,l} + \tilde{Q}_j \tilde{\phi}_{j+1,l},$$

i.e.,

$$\begin{aligned}
\tilde{\phi}_{j+1,l} &= \sum_{k \in \mathbb{Z}} \langle \tilde{\phi}_{j+1,l}, \phi_{j,k} \rangle \tilde{\phi}_{j,k} + \sum_{k \in \mathbb{Z}} \langle \tilde{\phi}_{j+1,l}, \psi_{j,k} \rangle \tilde{\psi}_{j,k} \\
&= \sum_{k \in \mathbb{Z}} \langle \tilde{\phi}_{j+1,l}, \sum_{m \in \mathbb{Z}} h_m \phi_{j+1,2k+m} \rangle \tilde{\phi}_{j,k} \\
&\quad + \sum_{k \in \mathbb{Z}} \langle \tilde{\phi}_{j+1,l}, \sum_{m \in \mathbb{Z}} g_{\mu,m} \phi_{j+1,2k+m} \rangle \tilde{\psi}_{j,k} \\
&= \sum_{k \in \mathbb{Z}} \bar{h}_{l-2k} \tilde{\phi}_{j,k} + \sum_{k \in \mathbb{Z}} \bar{g}_{l-2k} \tilde{\psi}_{j,k}.
\end{aligned}$$

Therefore, we have

$$\langle f, \tilde{\phi}_{j+1,l} \rangle = \sum_{k \in \mathbb{Z}} \bar{h}_{l-2k} \langle f, \tilde{\phi}_{j,k} \rangle + \sum_{k \in \mathbb{Z}} \bar{g}_{l-2k} \langle f, \tilde{\psi}_{j,k} \rangle.$$

The other formula of the reconstruction algorithm can be shown similarly. The proof of this theorem is complete. □

*Note*: Mallat decomposition algorithm and reconstruction algorithm is illustrated in Table 3.1.

Mallat algorithm can be applied effectively to image processing, pattern recognition, fast computation of singular integrals, and some other areas. It decomposes a signal from the lower frequency to the higher frequency locally both in time and frequency domains. In practical applications, we usually need the discrete form of Mallat algorithm. Let

$$\begin{cases} s_k^j := 2^{j/2} \langle f, \phi_{j,k} \rangle \\ \tilde{s}_k^j := 2^{j/2} \langle f, \tilde{\phi}_{j,k} \rangle, \end{cases} \qquad \begin{cases} t_k^j := 2^{j/2} \langle f, \psi_{j,k} \rangle \\ \tilde{t}_k^j := 2^{j/2} \langle f, \tilde{\psi}_{j,k} \rangle. \end{cases} \tag{3.29}$$

Table 3.1.  Mallat algorithm.

| Decomposition: |
|---|

$$\cdots \; \longrightarrow \; \langle f, \phi_{j+1,k}\rangle \; \longrightarrow \; \langle f, \phi_{j,k}\rangle \; \longrightarrow \; \langle f, \phi_{j-1,k}\rangle \; \longrightarrow \; \cdots$$
$$\searrow \qquad\qquad \searrow \qquad\qquad \searrow \qquad\qquad \searrow$$
$$\langle f, \psi_{j+1,k}\rangle \qquad \langle f, \psi_{j,k}\rangle \qquad \langle f, \psi_{j-1,k}\rangle \qquad \cdots$$

| Reconstruction: |
|---|

$$\cdots \; \longrightarrow \; \langle f, \phi_{j-1,k}\rangle \; \longrightarrow \; \langle f, \phi_{j,k}\rangle \; \longrightarrow \; \langle f, \phi_{j+1,k}\rangle \; \longrightarrow \; \cdots$$
$$\nearrow \qquad\qquad \nearrow \qquad\qquad \nearrow \qquad\qquad \nearrow$$
$$\cdots \qquad \langle f, \psi_{j-1,k}\rangle \qquad \langle f, \psi_{j,k}\rangle \qquad \langle f, \psi_{j+1,k}\rangle$$

Table 3.2.  Discrete form of Mallat algorithm

| | | | | | | | |
|---|---|---|---|---|---|---|---|
| Decomposition: | $\cdots \; \longrightarrow$ | $s_k^{j+1}$ | $\longrightarrow$ | $s_k^{j}$ | $\longrightarrow$ | $s_k^{j-1}$ | $\longrightarrow \; \cdots$ |
| | | $\searrow$ | | $\searrow$ | | $\searrow$ | |
| | | $t_k^{j+1}$ | | $t_k^{j}$ | | $t_k^{j-1}$ | $\cdots$ |
| Reconstruction: | $\cdots \; \longrightarrow$ | $s_k^{j-1}$ | $\longrightarrow$ | $s_k^{j}$ | $\longrightarrow$ | $s_k^{j+1}$ | $\longrightarrow \; \cdots$ |
| | | $\nearrow$ | | $\nearrow$ | | $\nearrow$ | |
| | $\cdots$ | $t_k^{j-1}$ | | $t_k^{j}$ | | $t_k^{j+1}$ | |

Then, its discrete form can be written as follows: $\forall j \in \mathbb{Z}, k \in \mathbb{Z}$, we have the following:

**Decomposition Algorithm:**

$$
\begin{cases}
s_k^j = \displaystyle\sum_{l\in\mathbb{Z}} h_l s_{2k+l}^{j+1} \\[2mm]
t_k^j = \displaystyle\sum_{l\in\mathbb{Z}} g_l s_{2k+l}^{j+1}
\end{cases}
\qquad
\begin{cases}
\tilde{s}_k^j = \displaystyle\sum_{l\in\mathbb{Z}} \tilde{h}_l \tilde{s}_{2k+l}^{j+1} \\[2mm]
\tilde{t}_k^j = \displaystyle\sum_{l\in\mathbb{Z}} \tilde{g}_l \tilde{s}_{2k+l}^{j+1}.
\end{cases}
\tag{3.30}
$$

**Reconstruction Algorithm:**

$$
\begin{cases}
s_k^{j+1} = 2^d \displaystyle\sum_{l\in\mathbb{Z}} \bar{\tilde{h}}_{k-2l} s_l^{j} + 2^d \displaystyle\sum_{l\in\mathbb{Z}} \bar{\tilde{g}}_{k-2l} t_l^{j} \\[2mm]
\tilde{s}_k^{j+1} = 2^d \displaystyle\sum_{l\in\mathbb{Z}} \bar{h}_{k-2l} \tilde{s}_l^{j} + 2^d \displaystyle\sum_{l\in\mathbb{Z}} \bar{g}_{k-2l} \tilde{s}_l^{j}.
\end{cases}
\tag{3.31}
$$

The discrete Mallat algorithm is depicted in Table 3.2.

The first formula of the decomposition algorithm (3.30) can be explained as follows: An input signal $\{s_k^{j+1}\}_{k\in\mathbb{Z}}$ is first filtered by a filter $\{h_{-k}\}_{k\in\mathbb{Z}}$, which corresponds to a convolution in mathematics, and then the output signal $\{\sum_{l\in\mathbb{Z}} h_l s_{k+l}^{j+1}\}$ is sampled alternately (i.e., only the points with even indices are kept), which is called "downsample" and denoted by "$2\downarrow$". At last, we get $\{s_k^j\}_{k\in\mathbb{Z}}$. The procedure is represented as follows:

$$\{s_k^{j+1}\}_{k\in\mathbb{Z}} \; \longrightarrow \; \boxed{\{h_{-k}\}_{k\in\mathbb{Z}}} \; \longrightarrow \; \boxed{2\downarrow} \; \longrightarrow \; \{s_k^j\}_{k\in\mathbb{Z}}.$$

The other formulae of (3.30) can be illustrated similarly.

The implementation of the first part of the first formula of the recon-struction algorithm (3.31), $x_k^{j+1} := \sum_{l \in \mathbb{Z}} \bar{\bar{h}}_{k-2l} s_l^j$, is just in inverse order. At first, the input signal is upsampled. More precisely, a zero is placed in between every two consecutive terms of the input sequence $\{s_k^j\}_{k \in \mathbb{Z}}$, which is denoted by "2 ↑". Then the upsampled sequence is filtered by $\{\bar{\bar{h}}_k\}_{k \in \mathbb{Z}}$ and at last the output signal $\{x_k^{j+1}\}_{k \in \mathbb{Z}}$ is obtained. The following is an illustration of the procedure:

$$\{s_k^j\}_{k \in \mathbb{Z}} \longrightarrow \boxed{\quad 2 \uparrow \quad} \longrightarrow \boxed{\{\bar{\bar{h}}_k\}_{k \in \mathbb{Z}}} \longrightarrow \{x_k^{j+1}\}_{k \in \mathbb{Z}}.$$

The other parts of (3.31) can be explained similarly.

# Chapter 4

# Some Typical Wavelet Bases

The purpose of this chapter is to give readers more complete information about the wavelet bases commonly used in the areas of signal processing, image processing, and pattern recognition. Some important properties of these wavelet bases are presented. To facilitate the understanding of these properties and further application of these wavelet bases for engineers and scientists, the properties are described in detail while the construction of their mathematical formula is avoided as much as possible. The reason is that in pattern recognition, wavelet transform is only a tool or method which is similar to the Fourier transform in signal analysis.

In practice, we usually apply real wavelet $\psi(t)$, which is a real-valued function. For such a wavelet, it is easy to show that its Fourier transform $\hat{\psi}(\omega)$ satisfies

$$\hat{\psi}(-\omega) = \overline{\hat{\psi}(\omega)},$$

and $\hat{\psi}(\omega)$ is also a real-valued function if and only if $\psi(t)$ is an even function.

In fact, for real-valued function $\psi(t)$, there holds

$$\hat{\psi}(-\omega) = \int_{\mathbb{R}} \psi(t) e^{it\omega} dt$$

$$= \int_{\mathbb{R}} \overline{\psi(t) e^{-it\omega}} dt$$

$$= \overline{\int_{\mathbb{R}} \psi(t) e^{-it\omega} dt}$$

$$= \overline{\hat{\psi}(\omega)}.$$

And since

$$\overline{\hat{\psi}(\omega)} = \overline{\int_{\mathbb{R}} \psi(t)e^{-it\omega}\,dt}$$

$$= \int_{\mathbb{R}} \psi(t)e^{it\omega}\,dt$$

$$= \int_{\mathbb{R}} \psi(-t)e^{-it\omega}\,dt$$

$$= \hat{\psi(-t)}(\omega),$$

$\hat{\psi}(\omega)$ is a real-valued function if and only if

$$\overline{\hat{\psi}(\omega)} = \hat{\psi}(\omega),$$

i.e.,

$$(\psi(-t))\hat{\ }(\omega) = (\psi(t))\hat{\ }(\omega),$$

i.e.,

$$\psi(-t) = \psi(t),$$

which means $\psi(t)$ is an even function.

In general, wavelet $\psi(t)$ is not an even function, which implies that $\hat{\psi}(\omega)$ is an imaginary function. Therefore, we usually indicate the graph of $|\hat{\psi}(\omega)|$ instead. Now that $\hat{\psi}(-\omega) = \overline{\hat{\psi}(\omega)}$; we have that

$$|\hat{\psi}(-\omega)| = |\hat{\psi}(\omega)|,$$

that is, $|\hat{\psi}(\omega)|$ is an even function. Hence we only need to indicate the graph of $|\hat{\psi}(\omega)|$ on positive semi-axis $[0, \infty)$.

Two important constructions of orthonormal wavelet bases from MRAs are listed in the following:

### • Construction of Orthonormal Wavelet Basis from an Orthonormal MRA

**Theorem 4.1.** *Let $\phi$ be the scaling function of an orthonormal MRA with filter function $m_0$. Then $\psi$, which is defined by*

$$\hat{\psi}(\xi) := e^{-i\frac{\xi}{2}} \overline{m_0} \left( \frac{\xi}{2} + \pi \right) \hat{\phi} \left( \frac{\xi}{2} + \pi \right),$$

*generates an orthonormal wavelet basis of $L^2(\mathbb{R})$.*

- **Construction of Orthonormal Wavelet Basis from Non-orthonormal MRA**

**Theorem 4.2.** *Let* $m_0 \in C(\mathbb{T})$ *satisfy the following conditions:*

(1) $\prod_{j=1}^{\infty} m_0(\xi/2^j)$ *converges for almost every* $\xi \in \mathbb{R}$.
(2) $\Phi(\xi) \in C(\mathbb{T})$ *and there exist positive constants* $A$ *and* $B$ *such that* $A \le \Phi(\xi) \le B$ $(\forall \xi \in \mathbb{R})$, *where*

$$\Phi(\xi) := \sum_{k \in \mathbb{Z}} |\hat{\phi}(\xi + 2k\pi)|^2$$

*and*

$$\hat{\phi}(\xi) := \prod_{j=1}^{\infty} m_0(\xi/2^j).$$

*Then,*

$$\hat{\phi}^{\sharp}(\xi) := \frac{\hat{\phi}(\xi)}{\sqrt{\Phi(\xi)}}$$

*is in* $L^2(\mathbb{R})$ *and generates an orthonormal MRA of* $L^2(\mathbb{R})$. *The corresponding scaling function and filter function are, respectively,* $\phi^{\sharp}(\xi)$ *and* $m_0^{\sharp}(\xi)$ *defined by*

$$m_0^{\sharp}(\xi) := m_0(\xi)\sqrt{\frac{\Phi(\xi)}{\Phi(2\xi)}}.$$

**Proof.** It is obvious that $m_0^{\sharp} \in C(\mathbb{T})$ and we can prove that $\Phi(0) = 1$ by Condition (2). It is also easy to see that

$$\prod_{j=1}^{\infty} m_0^{\sharp}(\xi/2^j) = \sqrt{\prod_{j=1}^{\infty} \frac{\Phi(\xi/2^j)}{\Phi(\xi/2^{j-1})}} \prod_{j=1}^{\infty} m_0(\xi/2^j) = \frac{\hat{\phi}(\xi)}{\sqrt{\Phi(\xi)}}.$$

Hence $\phi^{\sharp}$, which is defined by

$$\hat{\phi}^{\sharp}(\xi) := \frac{\hat{\phi}(\xi)}{\sqrt{\Phi(\xi)}} = \prod_{j=1}^{\infty} m_0^{\sharp}(\xi/2^j),$$

is refineable and the filter function is $m_0^{\sharp}$. Since

$$\sum_{k \in \mathbb{Z}} |\hat{\phi}^{\sharp}(\xi + 2k\pi)|^2 = \frac{\sum_{k \in \mathbb{Z}} |\hat{\phi}(\xi + 2k\pi)|^2}{\Phi(\xi)} = 1, \quad (\forall \xi \in \mathbb{R}),$$

$\phi^\sharp$ generates an orthonormal MRA of $L^2(\mathbb{R})$. The corresponding scaling function and filter function are $\phi^\sharp(\xi)$ and $m_0^\sharp(\xi)$, respectively. The proof is complete.                                        □

## 4.1 Orthonormal Wavelet Bases

### 4.1.1 *Haar Wavelet*

The simplest well-known wavelet is the Haar wavelet, which was presented in 1910 by Haar and defined by

$$\psi(t) = \begin{cases} 1 & 0 \le t < \dfrac{1}{2}; \\ -1 & \dfrac{1}{2} \le t < 1; \\ 0 & \text{otherwise.} \end{cases}$$

Its Fourier transform is

$$\hat{\psi}(\xi) == \frac{1}{4} i\xi e^{-i\xi/2} \left( \frac{4}{\xi} \sin \frac{\xi}{4} \right)^2. \tag{4.1}$$

The Haar wavelet has bad decay in the frequency domain, i.e., its localization in the frequency domain is not good. But in the time domain, it is compactly supported on $[0,1]$. It is also orthonormal and antisymmetric. Figure 4.1 shows the Haar wavelet and its frequency spectrum.

The corresponding scale function of Haar wavelet is as follows:

$$\phi(t) = \begin{cases} 1 & 0 \le t \le 1, \\ 0 & \text{otherwise.} \end{cases} \tag{4.2}$$

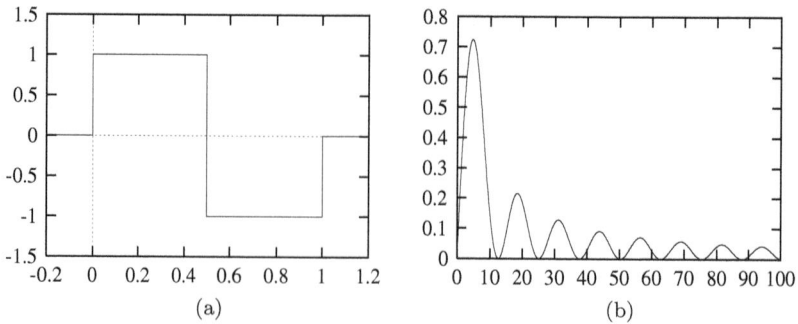

Fig. 4.1   Haar wavelet and its frequency spectrum: (a) Haar wavelet in time domain; (b) frequency spectrum.

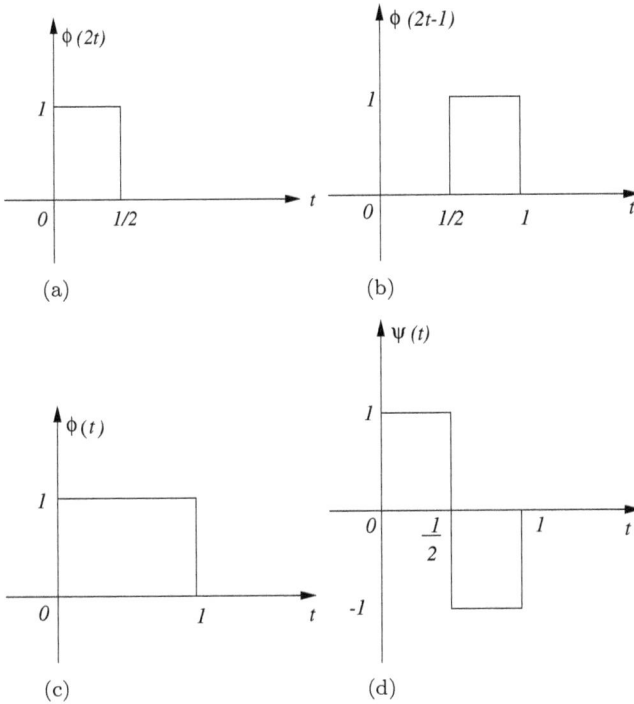

Fig. 4.2 Haar scale function and Haar wavelet: (a) $\phi(2t)$; (b) $\phi(2t-1)$; (c) $\phi(t)$; (d) $\psi(t)$.

It is easy to see that $\phi(t)$ satisfies the following two-scale equation:

$$\phi(t) = \phi(2t) + \phi(2t - 1), \qquad (4.3)$$

and $\psi$ can be generated by $\phi$ according to the following equation:

$$\psi(t) = \phi(2t) - \phi(2t - 1). \qquad (4.4)$$

Equations (4.2)–(4.4) are graphically illustrated in Fig. 4.2.

## 4.1.2 *Littlewood–Paley (LP) Wavelet*

Corresponding to the Haar wavelet, an orthonormal wavelet that is compactly supported in the frequency domain was given by Littlewood and Paley. Its analytical expression in the frequency domain is

$$\hat{\psi}(\xi) = \begin{cases} (2\pi)^{-\frac{1}{2}} & \pi \le |\xi| \le 2\pi, \\ 0 & \text{otherwise.} \end{cases} \tag{4.5}$$

Its inverse Fourier transform is

$$\psi(t) = (\pi t)^{-1}(\sin 2\pi t - \sin \pi t). \tag{4.6}$$

Figure 4.3 shows the LP wavelet and its frequency spectrum. LP wavelet $\psi(t) \in C^\infty$ has order one vanishing moment and is not compactly supported. The slow decay of $\psi(t)$ makes it no localization in time domain, $(\psi(t) \sim| x |^{-1}$ for $t \to \infty)$ but excellent localization in frequency domain since its Fourier transform is compactly supported. Table 4.1 shows the centers and radii of time and frequency windows and the area of the time–frequency window of the LP wavelet.

The corresponding scale function of Littlewood–Paley wavelet is

$$\phi(t) = \frac{\sin \pi t}{\pi t},$$

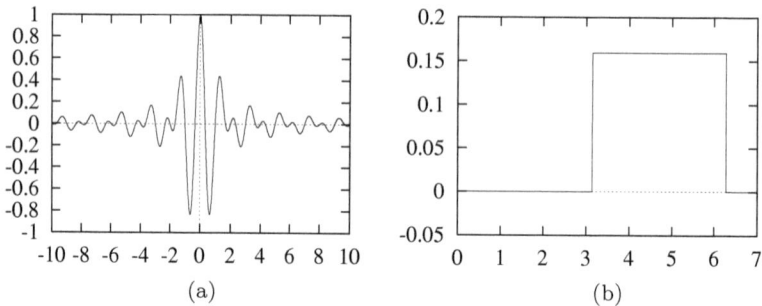

Fig. 4.3   LP wavelet and its frequency spectrum: (a) LP wavelet; (b) frequency spectrum.

Table 4.1.   The centers, radii, and area of time window, frequency window, and time–frequency window of the LP wavelet.

| | Time | Window | Freq. | Window | Time–freq. |
|---|---|---|---|---|---|
| Wavelet | Center | Radius | Center | Radius | area |
| LP | 0 | $\infty$ | $\frac{\pi}{2}\sqrt{3}$ | 12.029 | $\infty$ |

whose Fourier transform is as follows:

$$\hat{\phi}(\xi) = \begin{cases} 1 & -\pi \leq \xi < \pi, \\ 0 & \text{otherwise.} \end{cases}$$

The two-scale equation $\phi(t)$ satisfies

$$\hat{\phi}(\xi) = m_0 \left(\frac{\xi}{2}\right) \hat{\phi} \left(\frac{\xi}{2}\right),$$

where $m_0(\xi)$ is a $2\pi$-periodic function whose definition on $[-\frac{\pi}{2}, \frac{3\pi}{2})$ is as follows:

$$m_0(\xi) = \begin{cases} 1 & \xi \in \left[-\frac{\pi}{2}, \frac{\pi}{2}\right) \\ 0 & \xi \in \left[\frac{\pi}{2}, \frac{3\pi}{2}\right). \end{cases}$$

Since

$$|m_0(\xi)|^2 + |m_0(\xi + \pi)|^2 = 1, \quad \forall \xi \in \mathbb{T},$$

$\phi$ generates an orthonormal MRA, and consequently $\psi$, which is defined by

$$\hat{\psi}(\xi) = e^{-i\frac{\xi}{2}} m_0 \left(\frac{\xi}{2} + \pi\right) \hat{\phi} \left(\frac{\xi}{2}\right),$$

generates an orthonormal wavelet basis.

It is easy to see that $\phi(t)$ happens to be the Shannon sampling function. Figure 4.4 shows it and its frequency spectrum.

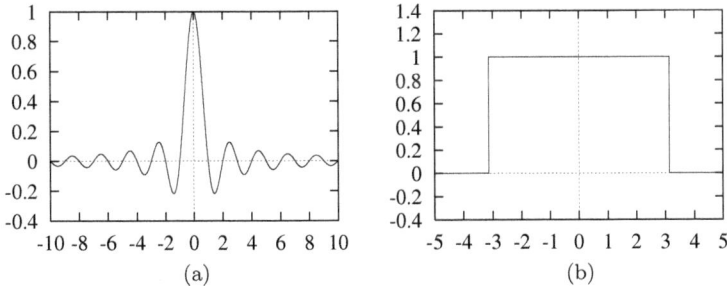

Fig. 4.4 Shannon sampling function and its frequency spectrum: (a) Time domain: (b) frequency domain.

### 4.1.3 *Meyer Wavelet*

Meyer wavelet was constructed by Y. Meyer in 1985. The Fourier transform of the corresponding scale function is defined by

$$
\hat{\phi}(\xi) := \begin{cases} 1 & |\xi| \leq \dfrac{2}{3}\pi \\[2mm] \cos\left[\dfrac{\pi}{2}\nu\left(\dfrac{3}{2\pi}|\xi|-1\right)\right] & \dfrac{2}{3}\pi \leq |\xi| \leq \dfrac{4}{3}\pi \\[2mm] 0 & \text{otherwise,} \end{cases}
$$

where $\nu(\xi)$ is an arbitrary $C^\infty$ function satisfying

$$
\nu(x) = \begin{cases} 0 & x \leq 0 \\ 1 & x \geq 1 \end{cases} \quad \text{and} \quad \nu(x) + \nu(1-x) = 1 \ (\forall x \in [0,1]). \tag{4.7}
$$

$\phi$ generates an orthonormal MRA since we can deduce that

$$
\Phi(\xi) = \sum_{k=-\infty}^{\infty} |\hat{\phi}(\xi+2k\pi)|^2 = 1, \quad (\forall\ \xi \in \mathbb{T}).
$$

It can be proven that the filter function is

$$
m_0(\xi) = \sum_{k=-\infty}^{\infty} \hat{\phi}(2(\xi+2\pi k)),
$$

and Meyer wavelet $\psi$, which is an orthonormal wavelet, is then defined by

$$
\hat{\psi}(\xi) = e^{-i\frac{\xi}{2}} \bar{m}_0\left(\frac{\xi}{2}+\pi\right)\hat{\phi}\left(\frac{\xi}{2}\right).
$$

It can be shown that

$$
\hat{\psi}(\xi) = \begin{cases} e^{-i\frac{\xi}{2}} \sin\left[\dfrac{\pi}{2}\nu\left(\dfrac{3}{2\pi}|\,\xi\,|-1\right)\right] & \dfrac{2\pi}{3} \leq |\,\xi\,| \dfrac{4\pi}{3} \\[2mm] e^{-i\frac{\xi}{2}} \cos\left[\dfrac{\pi}{2}\nu\left(\dfrac{3}{4\pi}|\,\xi\,|-1\right)\right] & \dfrac{4\pi}{3} \leq |\,\xi\,| \dfrac{8\pi}{3} \\[2mm] 0 & \text{otherwise.} \end{cases} \tag{4.8}
$$

Obviously, $\hat{\psi}(\xi)$ is compactly supported and its regularity is the same as that of $\nu$.

Particularly, if we take $\nu(t) = 0$ for $t \in (-\infty, 0) \cup (1, \infty)$ and

$$
\nu(t) = t^4(35 - 84t + 70t^2 - 20t^3), \quad t \in [0,1], \tag{4.9}
$$

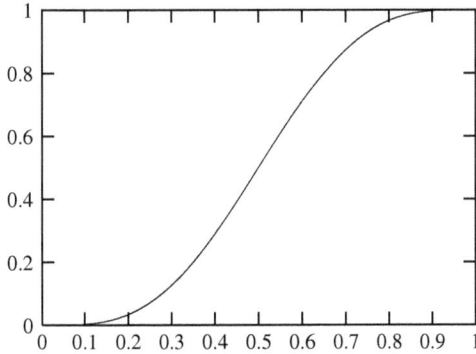

Fig. 4.5　Function $\nu(t)$ as given by (4.7).

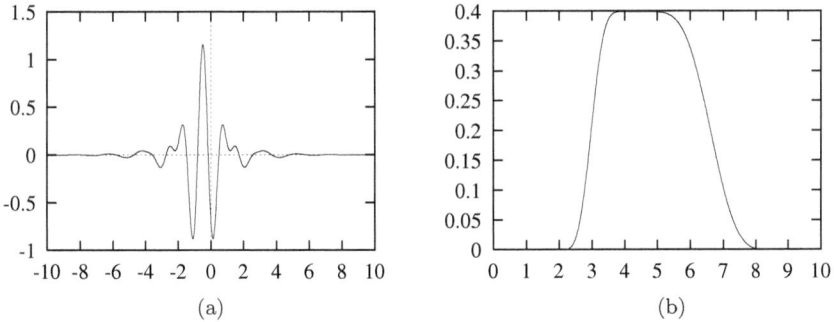

Fig. 4.6　Meyer wavelet and its frequency spectrum: (a) Time domain; (b) frequency domain.

it is easy to see that $\nu(t)$, which is plotted in Fig. 4.5, satisfies (4.7). The corresponding Meyer wavelet, which is the inverse Fourier transform of (4.8), and its frequency property are shown in Fig. 4.6 for $\nu$ defined by (4.9). The regularity of $\psi(t)$ is $C^\infty$. The decay of $\psi(t)$ is faster than any inverse polynomial, i.e., for any natural numbers $N$, there exists $C_N < \infty$ such that

$$|\psi(t)| \le C_N (1 + |t|^2)^{-N}.$$

But it cannot decay exponentially fast as $t \to \infty$. Moreover, Meyer wavelet has linear phase since it is symmetric.

　　The centers, radii, and area of the time window, frequency window, and time–frequency window of this Meyer wavelet are listed in Table 4.2. The order of its vanishing moment is 0.

Table 4.2.   The centers, radii, and area of time window, frequency window, and time–frequency window of the Meyer wavelet.

| | Time | Window | Freq. | Window | Time–freq. |
|---|---|---|---|---|---|
| Wavelet | Center | Radius | Center | Radius | area |
| Meyer | 0.5 | 0.6774 | 2.369 | 1.81349 | 4.9140 |

## 4.1.4 *Battle–Lemaré Spline Wavelet*

The Battle–Lemaré spline wavelet is based on central B-splines, which are polynomial spline functions with equal distance simple knots. For $m \in \mathbb{N}$, the scaling function $\phi$ is defined by

$$\phi(x) := \begin{cases} \underbrace{\chi_{[-\frac{1}{2},\frac{1}{2}]} * \cdots * \chi_{[-\frac{1}{2},\frac{1}{2}]}}_{m}(x), & \text{if } m \text{ is even;} \\ \underbrace{\chi_{[-\frac{1}{2},\frac{1}{2}]} * \cdots * \chi_{[-\frac{1}{2},\frac{1}{2}]}}_{m}(x - \tfrac{1}{2}), & \text{if } m \text{ is odd,} \end{cases} \tag{4.10}$$

where $\chi_{[-\frac{1}{2},\frac{1}{2}]}(x)$ is the characteristic function of $\left[-\frac{1}{2},\frac{1}{2}\right]$ defined by

$$\chi_{[-\frac{1}{2},\frac{1}{2}]}(x) = \begin{cases} 1 & x \in \left[-\frac{1}{2},\frac{1}{2}\right] \\ 0 & \text{otherwise} \end{cases}$$

and "$*$" is the convolution operation. It is easy to see that

$$\phi(x) := \begin{cases} N_m\left(x + \dfrac{1}{2}m\right) & \text{if } m \text{ is even;} \\ N_m\left(x + \dfrac{1}{2}(m - 1)\right) & \text{if } m \text{ is odd,} \end{cases}$$

where $N_m(x)$ is the spline function of order $m$ defined by (2.15). Thus, $\phi(x)$ is a spline function of order $m$ with integer knots. It is a polynomial on interval $[n, n + 1]$ for any $n \in \mathbb{N}$ and $\phi \in C^{m-2}(\mathbb{R})$. The support of $\phi$ is $\left[-\frac{m}{2}, \frac{m}{2}\right]$ if $m$ is even and $\left[-\frac{m}{2} + \frac{1}{2}, \frac{m}{2} + \frac{1}{2}\right]$ if $m$ is odd. A simple calculation yields that

$$\hat{\phi}(\xi) := \begin{cases} \left(\dfrac{2}{\xi} \sin \dfrac{\xi}{2}\right)^m & \text{if } m \text{ is even;} \\ e^{-i\frac{\xi}{2}} \left(\dfrac{2}{\xi} \sin \dfrac{\xi}{2}\right)^m & \text{if } m \text{ is odd.} \end{cases}$$

Hence the filter function is

$$m_0(\xi) = \begin{cases} \left(\cos\frac{\xi}{2}\right)^m & \text{if } m \text{ is even;} \\ e^{-i\frac{\xi}{2}}\left(\cos\frac{\xi}{2}\right)^m & \text{if } m \text{ is odd.} \end{cases}$$

Since

$$\left(\cos\frac{\xi}{2}\right)^m = \left(\frac{e^{i\frac{\xi}{2}} + e^{-i\frac{\xi}{2}}}{2}\right)^m = e^{i\frac{m}{2}\xi}\left(\frac{1 + e^{-i\xi}}{2}\right)^m,$$

we get

$$m_0(\xi) = \begin{cases} e^{i\frac{m}{2}\xi}\left(\frac{1+e^{-i\xi}}{2}\right)^m & \text{if } m \text{ is even;} \\ e^{i\frac{m-1}{2}\xi}\left(\frac{1+e^{-i\xi}}{2}\right)^m & \text{if } m \text{ is odd;} \end{cases}$$

$$= e^{i\left[\frac{m}{2}\right]\xi}\left(\frac{1 + e^{-i\xi}}{2}\right)^m,$$

where $\left[\frac{m}{2}\right]$ is the largest integer, not larger than $\frac{m}{2}$.

It is easy to see that

$$|m_0(\xi)|^2 + |m_0(\xi+\pi)|^2 = \left|\cos\frac{\xi}{2}\right|^{2m} + \left|\sin\frac{\xi}{2}\right|^{2m}.$$

Since

$$\left|\cos\frac{\xi}{2}\right|^{2m} + \left|\sin\frac{\xi}{2}\right|^{2m} \begin{cases} = |\cos\frac{\xi}{2}|^2 + |\sin\frac{\xi}{2}|^2 = 1 & \text{if } m = 1; \\ < |\cos\frac{\xi}{2}|^2 + |\sin\frac{\xi}{2}|^2 = 1 & \text{if } m > 1, \end{cases}$$

we have that $\forall \xi \in [-\pi, \pi]$,

$$\begin{cases} |m_0(\xi)|^2 + |m_0(\xi+\pi)|^2 = 1 & \text{if } m = 1; \\ 0 < |m_0(\xi)|^2 + |m_0(\xi+\pi)|^2 < 1 & \text{if } m > 1. \end{cases}$$

Therefore, if $m = 1$, $\phi$, which is defined by (4.10), i.e.,

$$\phi(x) = \chi_{[-\frac{1}{2},\frac{1}{2}]}\left(x - \frac{1}{2}\right)$$

$$= \begin{cases} 1 & x \in [0, 1] \\ 0 & \text{otherwise} \end{cases}$$

generates an orthonormal MRA. Let

$$\psi(t) = \phi(2t) - \phi(2t - 1),$$

we get the well-known Haar wavelet basis $\{\psi_{j,k}(x)\}$, which was discussed in (4.1.1).

And if $m > 1$, $\phi$, which is defined by (4.10), cannot generate an orthonormal MRA. However, $\forall \xi \in [-\pi, \pi]$, it is easy to see that

$$\Phi(\xi) := \sum_{k=-\infty}^{\infty} |\hat{\phi}(\xi + 2k\pi)|^2$$

$$= \left|2\sin\frac{\xi}{2}\right|^{2m} \sum_{k=-\infty}^{\infty} \frac{1}{|\xi + 2k\pi|^{2m}}$$

$$= \left|2\sin\frac{\xi}{2}\right|^{2m} \left[\frac{1}{|\xi|^{2m}} + \sum_{k=1}^{\infty} \frac{1}{|\xi + 2k\pi|^{2m}} + \sum_{k=1}^{\infty} \frac{1}{|\xi - 2k\pi|^{2m}}\right]$$

and

$$\begin{cases} \sum_{k=1}^{\infty} \frac{1}{|\xi + 2k\pi|^{2m}}, \ \sum_{k=1}^{\infty} \frac{1}{|\xi - 2k\pi|^{2m}} \le \left(\frac{1}{\pi}\right)^{2m} \sum_{k=1}^{\infty} \frac{1}{(2k-1)^{2m}}, \\ \left(\frac{2}{\pi}\right)^{2m} \le \frac{1}{|\xi|^{2m}} \left|2\sin\frac{\xi}{2}\right|^{2m} \le 1, \\ \left|2\sin\frac{\xi}{2}\right|^{2m} \le 2^{2m}, \end{cases}$$

therefore,

$$\left(\frac{2}{\pi}\right)^{2m} \le \Phi(\xi) \le 1 + 2^{2m+1} \left(\frac{1}{\pi}\right)^{2m} \sum_{k=1}^{\infty} \frac{1}{(2k-1)^{2m}}.$$

This inequality implies that $\phi(x)$ can generate a non-orthonormal MRA, i.e., $\{\phi(x - k) | k \in \mathbb{Z}\}$ is a Riesz basis of

$$V_0 := \overline{span\{\phi(x - k) | k \in \mathbb{Z}\}},$$

and $V_j := \{f(2^j x) | f \in V_0\}$ $(j \in \mathbb{Z})$ constitutes an MRA.

To construct an orthonormal MRA, let

$$\hat{\phi}^\sharp(\xi) := \frac{\hat{\phi}(\xi)}{\sqrt{\Phi(\xi)}},$$

and

$$m_0^\sharp(\xi) := m_0(\xi)\sqrt{\frac{\Phi(\xi)}{\Phi(2\xi)}}.$$

Then, by Theorem 4.2, $\phi^\sharp$ generates an orthonormal MRA of $L^2(\mathbb{R})$. The corresponding scaling function and filter function are $\phi^\sharp(\xi)$ and $m_0^\sharp(\xi)$, respectively.

Let

$$\hat{\psi}^\sharp(\xi) := e^{-i\frac{\xi}{2}}\overline{m_0^\sharp\left(\frac{\xi}{2}+\pi\right)}\hat{\phi}^\sharp\left(\frac{\xi}{2}\right).$$

Then, by Theorem 4.1, $\psi^\sharp$ generates an orthonormal wavelet basis of $L^2(\mathbb{R})$, which is the well-known Battle–Lemaré spline wavelet basis or B-spline wavelet basis of order $m-1$.

B-spline wavelet has no analytical expression. Figures 4.7–4.10 show B-spline wavelets of orders 1–4 and their frequency properties, respectively.

The centers, radii, and areas of the time windows, frequency windows, and time–frequency windows of the B-splines wavelets of orders 1–10 are shown in Table 4.3.

According to these figures and the table, we easily see that B-spline wavelets are symmetric, with $-\frac{1}{2}$ as the symmetric center. We also observe that the B-spline wavelet of order 2 has the smallest time–frequency localization property. Also, B-spline wavelets are orthonormal and decay exponentially. Table 4.3 shows the relationship between the order number of B-spline wavelets and their time–frequency localization properties. The time–frequency localization property becomes bad with increasing order number except for the quadratic spline wavelet.

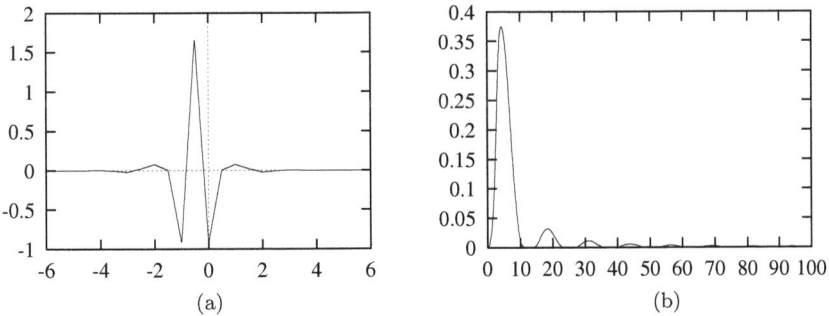

Fig. 4.7   B-spline wavelet of order 1 and its frequency spectrum: (a) Time domain; (b) frequency domain.

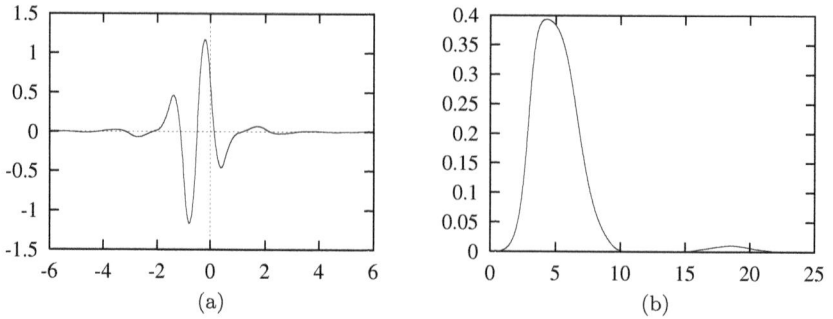

Fig. 4.8   B-spline wavelet of order 2 and its frequency spectrum: (a) Time domain; (b) frequency domain.

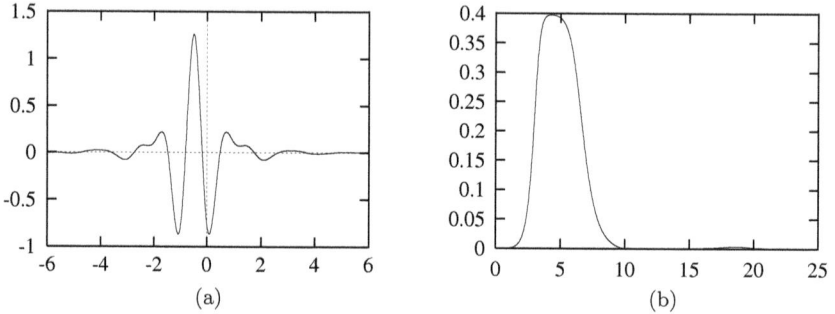

Fig. 4.9   B-spline wavelet of order 3 and its frequency spectrum: (a) Time domain; (b) frequency domain.

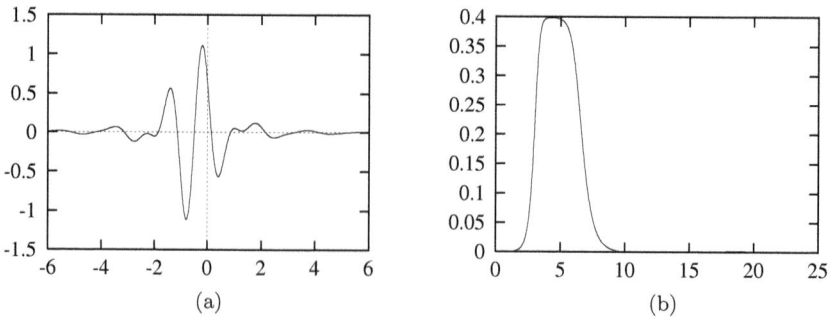

Fig. 4.10   B-spline wavelet of order 4 and its frequency spectrum: (a) Time domain; (b) frequency domain.

Table 4.3. The centers, radii, and area of time window, frequency window, and time–frequency window of B-spline wavelets of orders 1–10.

| B-spline | Time Window Center | Window Radius | Freq. Center | Window Radius | Time–freq. area |
|---|---|---|---|---|---|
| 1 | −0.5 | 0.3941 | 2.5177 | 2.3504 | 3.70514 |
| 2 | −0.5 | 0.4818 | 2.4104 | 1.9093 | 3.67968 |
| 3 | −0.5 | 0.5519 | 2.3857 | 1.8476 | 4.07853 |
| 4 | −0.5 | 0.5626 | 2.3772 | 1.8252 | 4.10739 |
| 5 | −0.5 | 0.6693 | 2.3896 | 1.8127 | 4.85290 |
| 6 | −0.5 | 0.7151 | 2.3607 | 1.8028 | 5.15675 |
| 7 | −0.5 | 0.7502 | 2.3524 | 1.9910 | 5.39881 |
| 8 | −0.5 | 0.7961 | 2.3377 | 1.7980 | 5.72522 |
| 9 | −0.5 | 0.8144 | 2.3087 | 1.7999 | 5.86324 |
| 10 | −0.5 | 0.8234 | 2.3940 | 1.7916 | 5.9008 |

It can be proven that Battle–Lemaré spline wavelet decreases exponentially at infinity. Readers are refered to the work of Daubechies [1992] for details.

## 4.1.5 *Daubechies' Compactly Supported Orthonormal Wavelets*

In the above discussion, except Haar wavelet, all presented orthonormal wavelets, though decay exponentially fast at infinity, are not compactly supported. The compactly supported orthonormal wavelets play an important role in wavelet decomposition of signals and data compression. In practice, the computation with compactly supported orthonormal wavelets need not be truncated and has higher speed and precision.

Daubechies' compactly supported orthonormal wavelets are constructed by letting the filter function as follows according to Theorem 3.8:

$$m_0(\xi) = \left(\frac{1 + e^{-i\xi}}{2}\right)^L M_0(\xi).$$

To ensure the orthogonality of the wavelet basis, $m_0$ is constructed so that

$$|m_0(\xi)|^2 + |m_0(\xi + \pi)|^2 = 1, \quad (\forall \xi \in [-\pi, \pi]). \tag{4.11}$$

It is clear that

$$|m_0(\xi)|^2 = \left|\frac{1 + e^{-i\xi}}{2}\right|^{2L} |M_0(\xi)|^2$$

$$= \left|\cos\frac{\xi}{2}\right|^{2L} |M_0(\xi)|^2$$

$$= \left(1 - \sin^2\frac{\xi}{2}\right)^{L} |M_0(\xi)|^2.$$

On the other hand, to ensure that $\phi$ and $\psi$ are real-valued functions, we let $M_0(\xi)$ be real-valued functions on $e^{-i\xi}$, then

$$|M_0(\xi)|^2 = M_0(\xi)\overline{M_0(\xi)} = M_0(\xi)M_0(-\xi),$$

which concludes that $|M_0(\xi)|^2$ is an even function on $\xi$ or, equivalently, a function on $\cos\xi$ and therefore on $\sin^2\frac{\xi}{2}$ since

$$\cos\xi = 1 - 2\sin^2\frac{\xi}{2}.$$

Let us denote

$$|M_0(\xi)|^2 = P\left(\sin^2\frac{\xi}{2}\right) = P(y), \ y = y(\xi) = \sin^2\frac{\xi}{2}.$$

Then

$$|m_0(\xi)|^2 = \left(1 - \sin^2\frac{\xi}{2}\right)^{L} P(y) = (1 - y)^L P(y).$$

Note that

$$y(\xi + \pi) = \sin^2\frac{\xi + \pi}{2} = \cos^2\frac{\xi}{2} = 1 - \sin^2\frac{\xi}{2} = 1 - y(\xi) = 1 - y,$$

by (4.11), $P(y)$ should satisfy

$$(1 - y)^L P(y) + y^L P(1 - y) = 1. \tag{4.12}$$

The general solution of Eq. (4.12) is as follows [Daubechies, 1992, Chapter 6]:

$$P(y) = P_L(y) + y^L R\left(\frac{1}{2} - y\right),$$

where $R(x)$ is an odd polynomial on $x$ and

$$P_L(y) := \sum_{k=0}^{L-1} \binom{L + k - 1}{k} y^k.$$

Particularly, letting $R(x) \equiv 0$, we have $P(y) = P_L(y)$, therefore

$$|M_0(\xi)|^2 = P\left(\sin^2\frac{\xi}{2}\right) = P_L\left(\sin^2\frac{\xi}{2}\right) = \sum_{k=0}^{L-1}\binom{L+k-1}{k}\sin^{2k}\frac{\xi}{2}.$$

$$(4.13)$$

In general, there cannot be more than one real-valued polynomial $M_0(\xi)$ satisfying (4.13). Riesz lemma [Daubechies, 1992, Chapter 6] ensures the existence of such an $M_0(\xi)$: it is a real-valued polynomial of order $L-1$ on $e^{i\xi}$. Since $|M_0(\xi)|^2 = |M_0(-\xi)|^2$, replacing $\xi$ by $-\xi$ we get another solution, which we denote by

$$M_0(\xi) = \sum_{n=0}^{L-1} b_n e^{-in\xi}, \quad b_n \in \mathbb{R}.$$

Therefore,

$$m_0(\xi) = \left(\frac{1+e^{-i\xi}}{2}\right)^L M_0(\xi) = \left(\frac{1+e^{-i\xi}}{2}\right)^L \sum_{n=0}^{L-1} b_n e^{-in\xi}$$

satisfies (4.11). It is easy to see that $m_0(\xi)$ can be rewritten as follows:

$$m_0(\xi) = \sum_{n=0}^{2L-1} h_n e^{-in\xi}. \tag{4.14}$$

A family of $m_0(\xi)$ is constructed by Daubechies [1988] by choosing $M_0(\xi)$ satisfying (4.13) ($L \geq 2$) so that the corresponding $\psi$ is a compactly supported wavelet with extremal phase and the highest number of vanishing moments compatible with its support width. The corresponding wavelet bases are called Daubechies' wavelet bases. They are orthonormal, compactly supported, and regular wavelet bases. It can be proven that $\psi(t)$ is compactly supported on $[-L+1, L]$ and $\phi(t), \psi(t) \in C^{\mu N}$, where $\mu = \frac{\log 3}{2\log 2} \approx 0.2075$. The filter coefficients $\{h_n\}$ of (4.14) are listed in Table 4.4 for $L = 1$–10.

It is easy to observe that Daubechies' wavelets $\psi$ are asymmetric. It can be shown that any compactly supported, orthonormal, and regular wavelet is always asymmetric. For "least asymmetric", Daubechies chooses another family of $M_0(\xi)$. The corresponding wavelets are called the second type of Daubechies orthonormal wavelets here to differ from those discussed above, the first type. Tables 4.5 and 4.6 list the centers, radii, and areas of time window, frequency window, and time-frequency window for the two Daubechies wavelets. The corresponding filter coefficients $\{h_n\}$

Table 4.4. The filter coefficients $\{h_n\}$ for compactly supported wavelet with extremal phase and the highest number of vanishing moments compatible with their support width.

| $n$ | $h_n\ (L=1)$ |
|---|---|
| 0 | 0.500000000000 |
| 1 | 0.500000000000 |

| $n$ | $h_n\ (L=2)$ |
|---|---|
| 0 | 0.341506350946 |
| 1 | 0.591506350946 |
| 2 | 0.158493649054 |
| 3 | −0.091506350946 |

| $n$ | $h_n\ (L=3)$ |
|---|---|
| 0 | 0.235233603892 |
| 1 | 0.570558457916 |
| 2 | 0.325182500263 |
| 3 | −0.095467207784 |
| 4 | −0.060416104155 |
| 5 | 0.024908749866 |

| $n$ | $h_n\ (L=4)$ |
|---|---|
| 0 | 0.162901714026 |
| 1 | 0.505472857546 |
| 2 | 0.446100069123 |
| 3 | −0.019787513118 |
| 4 | −0.132253583684 |
| 5 | 0.021808150237 |
| 6 | 0.023251800536 |
| 7 | −0.007493494665 |

| $n$ | $h_n\ (L=5)$ |
|---|---|
| 0 | 0.113209491292 |
| 1 | 0.426971771352 |
| 2 | 0.512163472130 |
| 3 | 0.097883480674 |
| 4 | −0.171328357691 |
| 5 | −0.022800565942 |
| 6 | 0.054851329321 |
| 7 | −0.004413400054 |
| 8 | −0.008895935051 |
| 9 | 0.002358713969 |

| $n$ | $h_n\ (L=6)$ |
|---|---|
| 0 | 0.078871216001 |
| 1 | 0.349751907037 |
| 2 | 0.531131879941 |
| 3 | 0.222915661465 |
| 4 | −0.159993299446 |
| 5 | −0.091759032030 |
| 6 | 0.068944046487 |
| 7 | 0.019461604854 |
| 8 | −0.022331874165 |
| 9 | 0.000391625576 |
| 10 | 0.003378031182 |
| 11 | −0.000761766903 |

| $n$ | $h_n\ (L=7)$ |
|---|---|
| 0 | 0.055049715373 |
| 1 | 0.280395641813 |
| 2 | 0.515574245831 |
| 3 | 0.332186241105 |
| 4 | −0.101756911232 |
| 5 | −0.158417505640 |
| 6 | 0.050423232505 |
| 7 | 0.057001722580 |
| 8 | −0.026891226295 |
| 9 | −0.011719970782 |
| 10 | 0.008874896190 |
| 11 | 0.000303757498 |
| 12 | −0.001273952359 |
| 13 | 0.000250113427 |

| $n$ | $h_n\ (L=8)$ |
|---|---|
| 0 | 0.038477811054 |
| 1 | 0.221233623576 |
| 2 | 0.477743075214 |
| 3 | 0.413908266211 |
| 4 | −0.011192867667 |
| 5 | −0.200829316391 |
| 6 | 0.000334097046 |
| 7 | 0.091038178423 |
| 8 | −0.012281950523 |
| 9 | −0.031175103325 |
| 10 | 0.009886079648 |
| 11 | 0.006184422410 |
| 12 | −0.003443859628 |
| 13 | −0.000277002274 |
| 14 | 0.000477614855 |
| 15 | −0.000083068631 |

| $n$ | $h_n\ (L=9)$ |
|---|---|
| 0 | 0.026925174795 |
| 1 | 0.172417151907 |
| 2 | 0.427674532180 |
| 3 | 0.464772857183 |
| 4 | 0.094184774753 |
| 5 | −0.207375880901 |
| 6 | −0.068476774512 |
| 7 | 0.105034171139 |
| 8 | 0.021726337730 |
| 9 | −0.047823632060 |
| 10 | 0.000177446407 |
| 11 | 0.015812082926 |
| 12 | −0.003339810113 |
| 13 | −0.003027480287 |
| 14 | 0.001306483640 |
| 15 | 0.000162907336 |
| 16 | −0.000178164880 |
| 17 | 0.000027822757 |

| $n$ | $h_n\ (L=10)$ |
|---|---|
| 0 | 0.018858578796 |
| 1 | 0.133061091397 |
| 2 | 0.372787535743 |
| 3 | 0.486814055367 |
| 4 | 0.198818870885 |
| 5 | −0.176668100897 |
| 6 | −0.138554939360 |
| 7 | 0.090063724267 |
| 8 | 0.065801493551 |
| 9 | −0.050483285598 |
| 10 | −0.020829624044 |
| 11 | 0.023484907048 |
| 12 | 0.002550218484 |
| 13 | −0.007589501168 |
| 14 | 0.000986662682 |
| 15 | 0.001408843295 |
| 16 | −0.000484973920 |
| 17 | −0.000082354503 |
| 18 | 0.000066177183 |
| 19 | −0.000009379208 |

*Note:* The $\{h_n\}$ are normalized so that $\sum_n h_n = m_0(0) = 1$.

Table 4.5.  The centers, radii, and area of time window, frequency window, and time–frequency window of the first type Daubechies wavelets, whose compactly supported lengths are 3, 5, 7, 9, 11, 13, 15, 17, 19, respectively.

| Daub. I | Time Center | Window Radius | Freq. Center | Window Radius | Time–freq. area |
|---|---|---|---|---|---|
| 1 | 0.4353 | 0.3308 | 1.3559 | 3.0720 | 4.0650 |
| 2 | 0.4793 | 0.3486 | 1.3755 | 3.1644 | 4.4122 |
| 3 | 0.4748 | 0.4243 | 1.2911 | 3.5206 | 4.2774 |
| 4 | 0.4682 | 0.5000 | 1.2502 | 2.2042 | 4.4080 |
| 5 | 0.4597 | 0.5754 | 1.2241 | 2.0580 | 4.7368 |
| 6 | 0.4585 | 0.6511 | 1.2218 | 2.0000 | 5.2092 |
| 7 | 0.4512 | 0.7237 | 1.2121 | 1.9440 | 5.6274 |
| 8 | 0.4446 | 0.7960 | 1.2071 | 1.9330 | 6.1548 |
| 9 | 0.4378 | 0.8666 | 1.2038 | 1.9212 | 6.6542 |

Table 4.6.  The centers, radii, and area of time window, frequency window, and time–frequency window of the second Daubechies wavelets, whose compactly supported lengths are 7, 9, 11, 13, 15, 17, 19, respectively.

| Daub.II | Time Center | Window Radius | Freq. Center | Window Radius | Time–freq. area |
|---|---|---|---|---|---|
| 1 | 0.4077 | 0.3956 | 1.2486 | 2.2000 | 3.4813 |
| 2 | 0.3809 | 0.4519 | 1.1238 | 2.0537 | 3.7124 |
| 3 | 0.3417 | 0.4394 | 1.2202 | 2.0038 | 3.5217 |
| 4 | 0.2727 | 0.4722 | 1.2105 | 1.9462 | 3.6756 |
| 5 | 0.2542 | 0.4786 | 1.2073 | 1.9335 | 3.7022 |
| 6 | 0.2226 | 0.5189 | 1.2068 | 1.9333 | 4.0153 |
| 7 | 0.2110 | 0.5135 | 1.1941 | 1.8898 | 3.9820 |

are listed in Table 4.7. For details, readers are referred to the work of Daubechies [1992]. Figures 4.11–4.19 and 4.20–4.26 show the scaling functions and the wavelet functions of these two types of Daubechies wavelets, as well as the Fourier spectra of the scaling functions and the frequency properties of the wavelet functions, respectively.

## 4.1.6  *Coiflet*

In the applications of wavelet to the fast computation of singular integrals, a high vanishing moment of the wavelet applied is very important. To quicken the computation, Coifman suggested in the spring of 1989 that it might be

Table 4.7.  The filter coefficients $\{h_n\}$ for the "least asymmetric" wavelet basis, the second Daubechies' wavelet basis.

| $n$ | $h_n\ (L=4)$ |
|---|---|
| 0 | $-0.053574450709$ |
| 1 | $-0.020955482562$ |
| 2 | $0.351869534328$ |
| 3 | $0.568329121704$ |
| 4 | $0.210617267102$ |
| 5 | $-0.070158812090$ |
| 6 | $-0.008912350721$ |
| 7 | $0.022785172948$ |

| $n$ | $h_n\ (L=5)$ |
|---|---|
| 0 | $0.019327397978$ |
| 1 | $0.020873432211$ |
| 2 | $-0.027672093058$ |
| 3 | $0.140995348427$ |
| 4 | $0.511526483447$ |
| 5 | $0.448290824190$ |
| 6 | $0.011739461568$ |
| 7 | $-0.123975681307$ |
| 8 | $-0.014921249934$ |
| 9 | $0.013816076479$ |

| $n$ | $h_n\ (L=6)$ |
|---|---|
| 0 | $0.010892350164$ |
| 1 | $0.002468306186$ |
| 2 | $-0.083431607706$ |
| 3 | $-0.034161560794$ |
| 4 | $0.347228986479$ |
| 5 | $0.556946391963$ |
| 6 | $0.238952185667$ |
| 7 | $-0.051362484931$ |
| 8 | $-0.014891875650$ |
| 9 | $0.031625281330$ |
| 10 | $0.001249961047$ |
| 11 | $-0.005515933754$ |

| $n$ | $h_n\ (L=7)$ |
|---|---|
| 0 | $0.001896329267$ |
| 1 | $-0.000740612958$ |
| 2 | $-0.008935215825$ |
| 3 | $0.021577726291$ |
| 4 | $0.048007383968$ |
| 5 | $-0.035039145611$ |
| 6 | $0.012332829745$ |
| 7 | $0.379081300982$ |
| 8 | $0.542891354907$ |
| 9 | $0.204091969863$ |
| 10 | $-0.099028353403$ |
| 11 | $-0.076231935948$ |
| 12 | $0.002835671343$ |
| 13 | $0.007260697381$ |

| $n$ | $h_n\ (L=8)$ |
|---|---|
| 0 | $0.001336396696$ |
| 1 | $-0.000214197150$ |
| 2 | $-0.010572843264$ |
| 3 | $0.002693194377$ |
| 4 | $0.034745232955$ |
| 5 | $-0.019246760632$ |
| 6 | $-0.036731254380$ |
| 7 | $0.257699335187$ |
| 8 | $0.549553315269$ |
| 9 | $0.340372673595$ |
| 10 | $-0.043326807703$ |
| 11 | $-0.101324327643$ |
| 12 | $0.005379305875$ |
| 13 | $0.022411811521$ |
| 14 | $-0.000383345448$ |
| 15 | $-0.002391729256$ |

| $n$ | $h_n\ (L=9)$ |
|---|---|
| 0 | $0.000756243654$ |
| 1 | $-0.000334570754$ |
| 2 | $-0.007257789276$ |
| 3 | $0.006264448121$ |
| 4 | $0.043895625777$ |
| 5 | $-0.012893222965$ |
| 6 | $-0.135446891751$ |
| 7 | $0.024941415480$ |
| 8 | $0.436524203675$ |
| 9 | $0.507629895416$ |
| 10 | $0.168829461801$ |
| 11 | $-0.038586080548$ |
| 12 | $0.000412570465$ |
| 13 | $0.021372216801$ |
| 14 | $-0.008151675613$ |
| 15 | $-0.009384698418$ |
| 16 | $0.000438251270$ |
| 17 | $0.000990596868$ |

| $n$ | $h_n\ (L=10)$ |
|---|---|
| 0 | $0.000544585223$ |
| 1 | $0.000067622510$ |
| 2 | $-0.006110321315$ |
| 3 | $-0.001036181962$ |
| 4 | $0.032475462289$ |
| 5 | $0.008209434713$ |
| 6 | $-0.112779486117$ |
| 7 | $-0.050120107516$ |
| 8 | $0.333535669077$ |
| 9 | $0.544125765250$ |
| 10 | $0.271406505606$ |
| 11 | $-0.025128270046$ |
| 12 | $-0.022620386109$ |
| 13 | $0.035351783775$ |
| 14 | $0.004076408400$ |
| 15 | $-0.014393115963$ |
| 16 | $-0.000568767657$ |
| 17 | $0.003247864187$ |
| 18 | $0.000040330602$ |
| 19 | $-0.000324794948$ |

*Note:* The $\{h_n\}$ are normalized so that $\sum_n h_n = m_0(0) = 1$.

worthwhile to construct orthonormal wavelet bases with vanishing moments not only for $\psi$ but also for $\phi$. That is, we hope to find $\psi$ and $\phi$ such that

$$\int_{\mathbb{R}} x^l \psi(x)\,dx = 0, \quad (l = 0, \ldots, L-1) \tag{4.15}$$

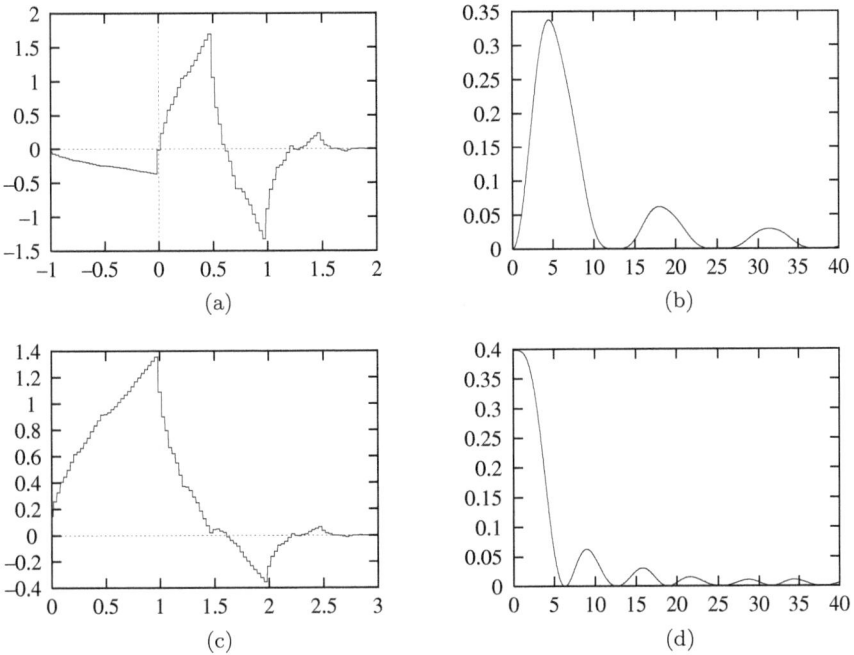

Fig. 4.11  The first type Daubechies orthonormal wavelet and its scaling function $(L = 2)$: (a) $\psi(t)$; (b) the FT of $\psi(t)$; (c) $\phi(t)$; (d) the FT of $\phi(t)$.

and

$$\int_{\mathbb{R}} \phi(x)dx = 0, \quad \int_{\mathbb{R}} x^l \phi(x)dx = 0, \quad (l = 1, \ldots, L-1). \qquad (4.16)$$

Daubechies studied the construction of such wavelets [Daubechies, 1988, 1992], which are called "coiflet" because they were first requested by Coifman and the order is then called the order of coiflet.

To construct compactly supported orthonormal wavelet bases satisfying (4.15) and (4.16), it can be proven that $m_0$ has the form Daubechies [1992]

$$m_0(\xi) = \left(\frac{1 + e^{-i\xi}}{2}\right)^L M_0(\xi)$$

$$= \sum_{n \in \mathbb{Z}} h_n e^{-in\xi},$$

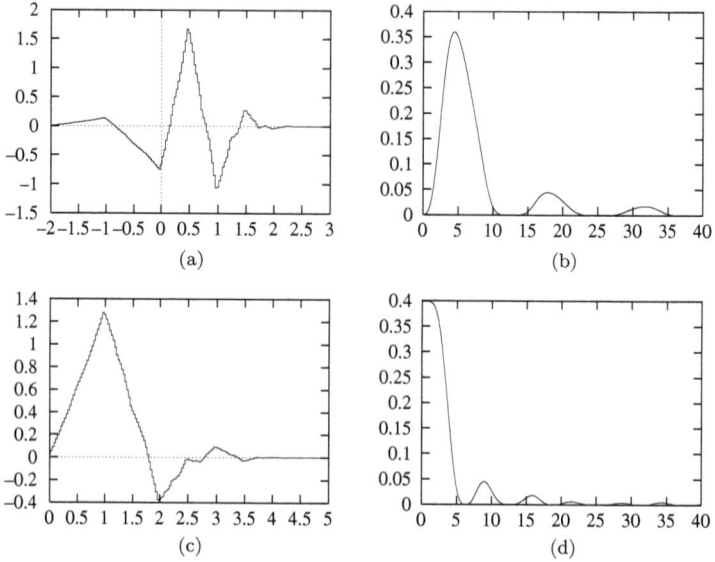

Fig. 4.12  The first type Daubechies orthonormal wavelet and its scaling function $(L = 3)$: (a) $\psi(t)$; (b) the FT of $\psi(t)$; (c) $\phi(t)$; (d) the FT of $\phi(t)$.

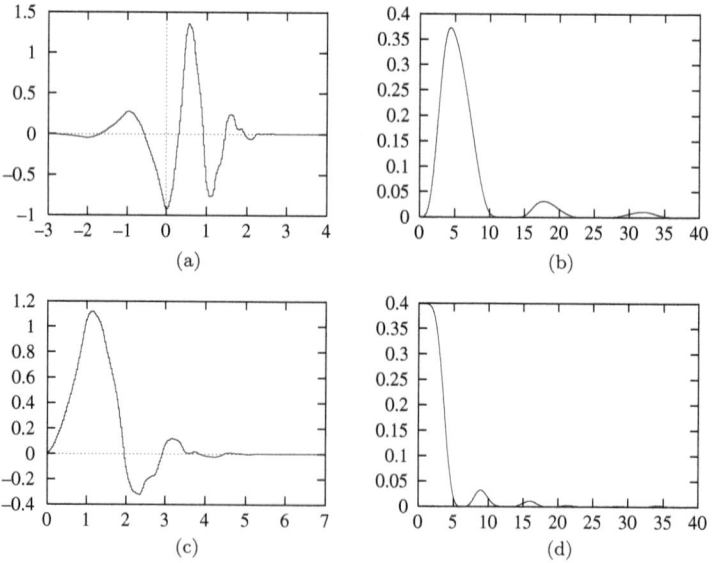

Fig. 4.13  The first type Daubechies orthonormal wavelet and its scaling function $(L = 4)$: (a) $\psi(t)$; (b) the FT of $\psi(t)$; (c) $\phi(t)$; (d) the FT of $\phi(t)$.

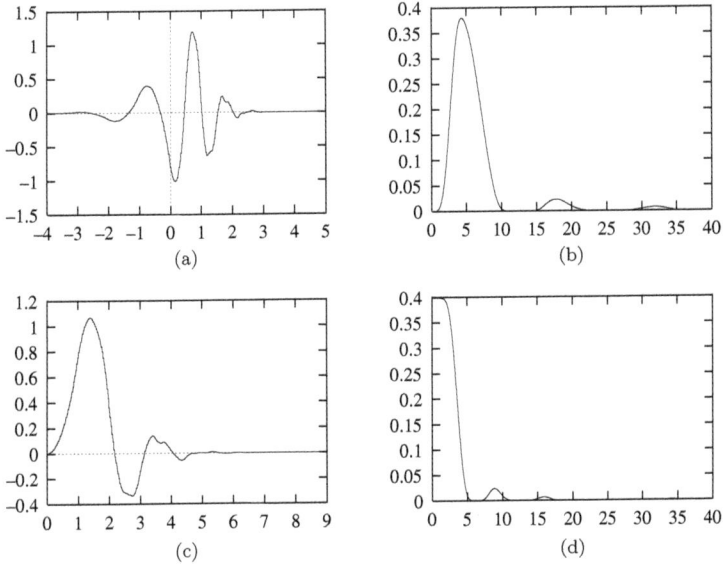

Fig. 4.14 The first type Daubechies orthonormal wavelet and its scaling function ($L = 5$): (a) $\psi(t)$; (b) the FT of $\psi(t)$; (c) $\phi(t)$; (d) the FT of $\phi(t)$.

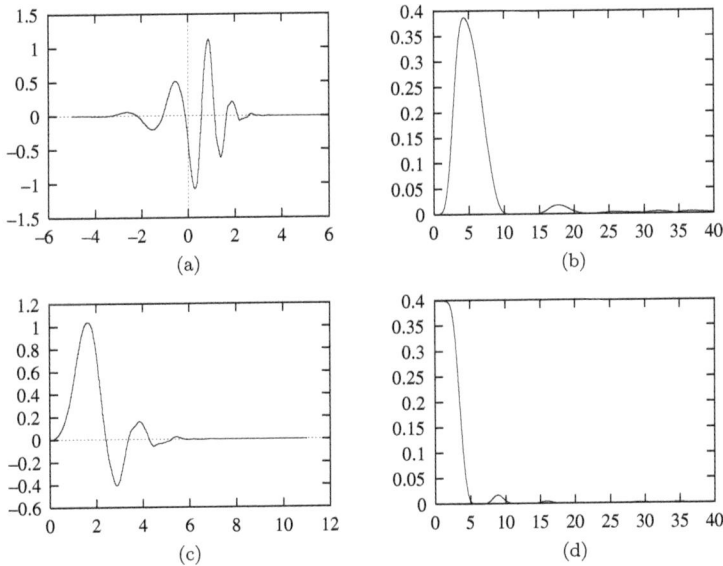

Fig. 4.15 The first type Daubechies orthonormal wavelet and its scaling function ($L = 6$): (a) $\psi(t)$; (b) the FT of $\psi(t)$; (c) $\phi(t)$; (d) the FT of $\phi(t)$.

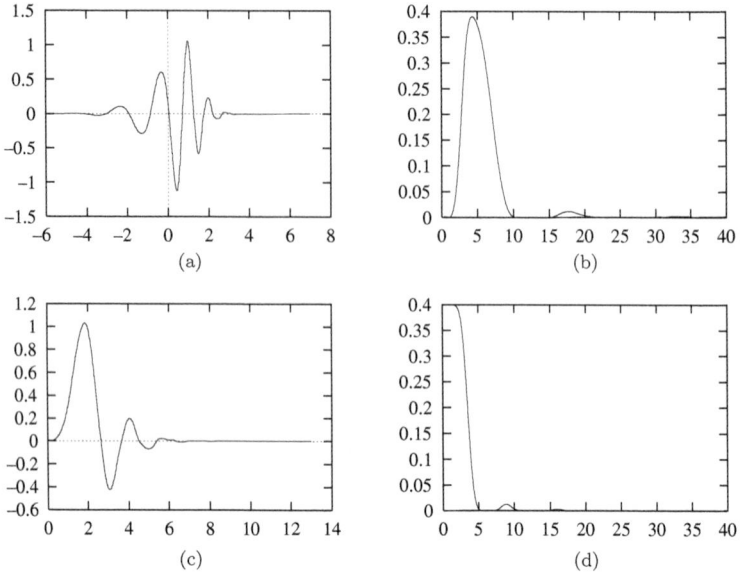

Fig. 4.16  The first type Daubechies orthonormal wavelet and its scaling function $(L = 7)$: (a) $\psi(t)$; (b) the FT of $\psi(t)$; (c) $\phi(t)$; (d) the FT of $\phi(t)$.

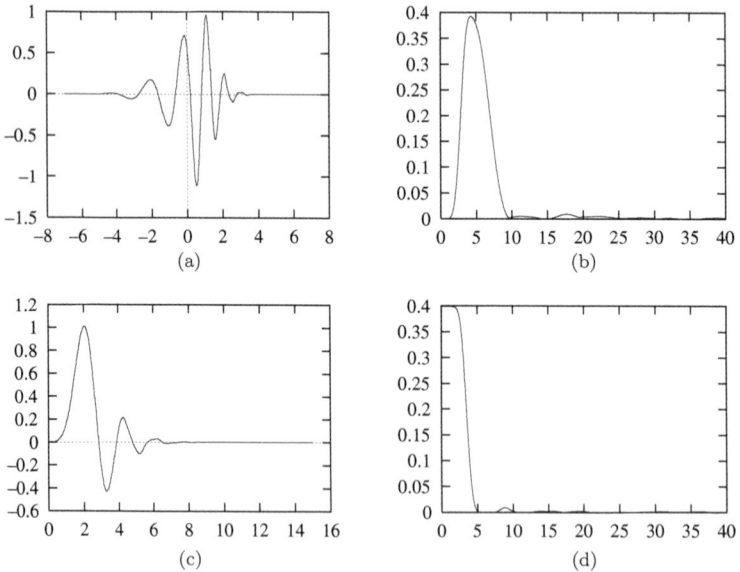

Fig. 4.17  The first type Daubechies orthonormal wavelet and its scaling function $(L = 8)$: (a) $\psi(t)$; (b) the FT of $\psi(t)$; (c) $\phi(t)$; (d) the FT of $\phi(t)$.

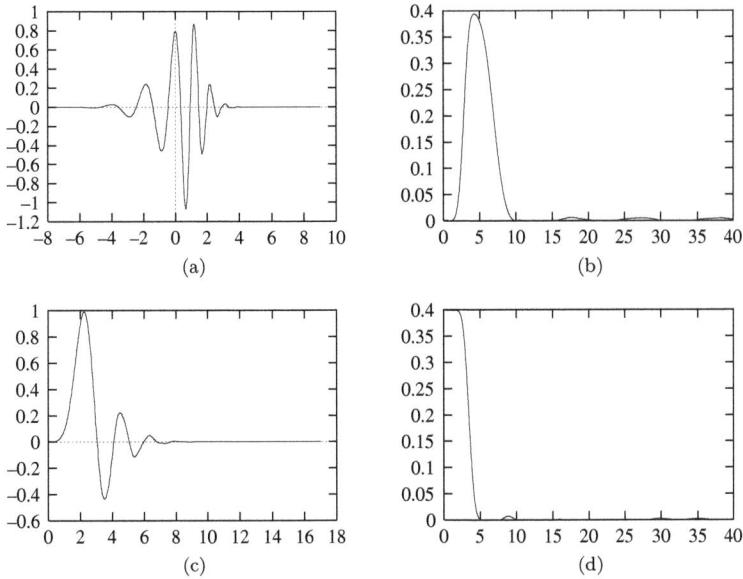

Fig. 4.18 The first type Daubechies orthonormal wavelet and its scaling function ($L = 9$): (a) $\psi(t)$; (b) the FT of $\psi(t)$; (c) $\phi(t)$; (d) the FT of $\phi(t)$.

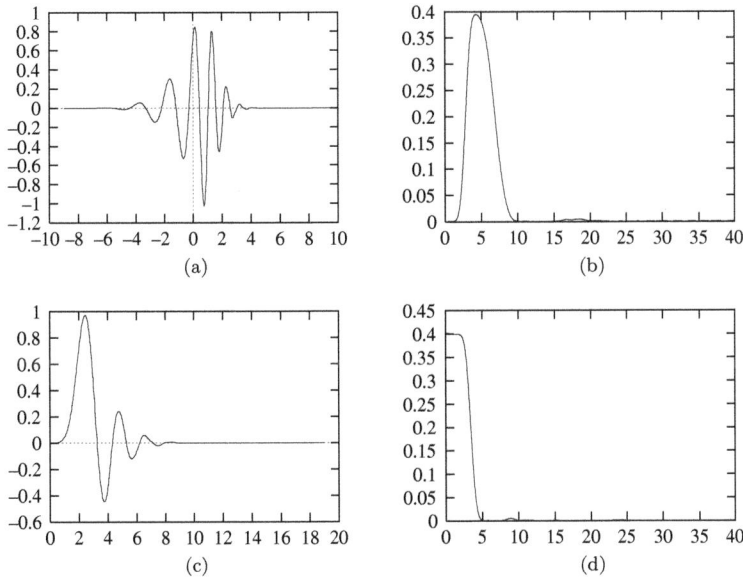

Fig. 4.19 The first type Daubechies orthonormal wavelet and its scaling function ($L = 10$): (a) $\psi(t)$; (b) the FT of $\psi(t)$; (c) $\phi(t)$; (d) the FT of $\phi(t)$.

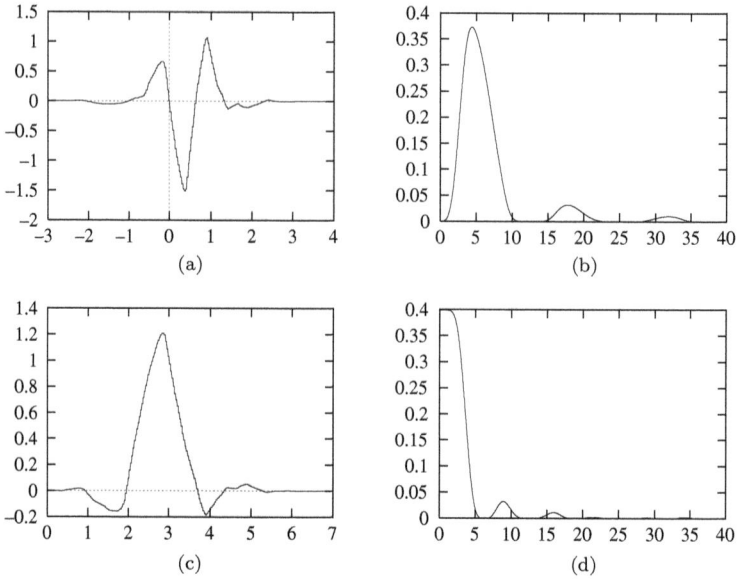

Fig. 4.20  The second type Daubechies orthonormal wavelet and its scaling function ($L = 4$): (a) $\psi(t)$; (b) the FT of $\psi(t)$ ; (c) $\phi(t)$; (d) the FT of $\phi(t)$.

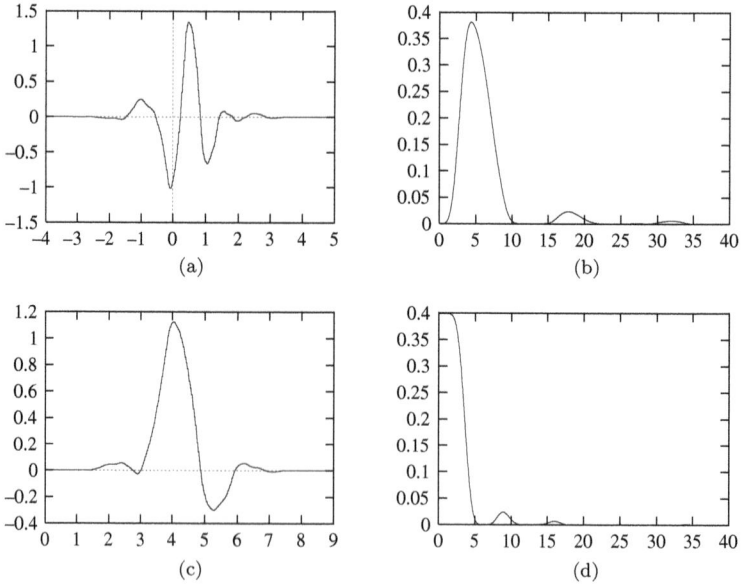

Fig. 4.21  The second type Daubechies orthonormal wavelet and its scaling function ($L = 5$). (a) $\psi(t)$; (b) the FT of $\psi(t)$; (c) $\phi(t)$; (d) the FT of $\phi(t)$.

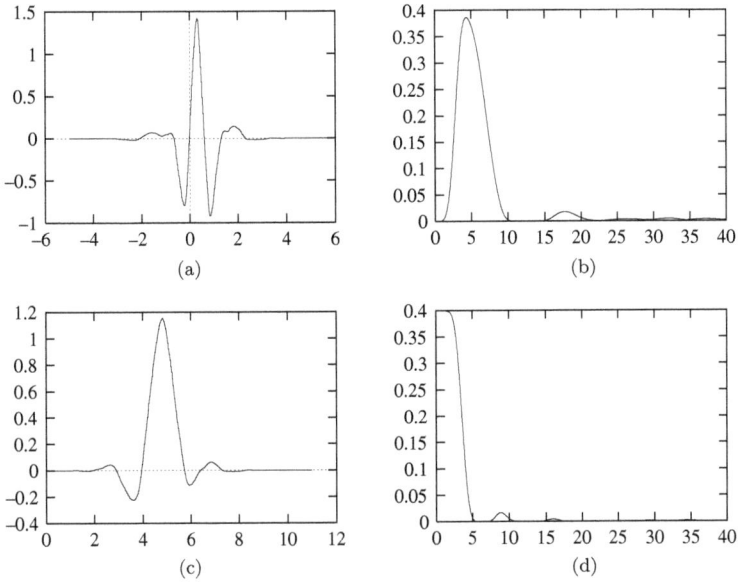

Fig. 4.22 The second type Daubechies orthonormal wavelet and its scaling function ($L = 6$): (a) $\psi(t)$; (b) the FT of $\psi(t)$; (c) $\phi(t)$; (d) the FT of $\phi(t)$.

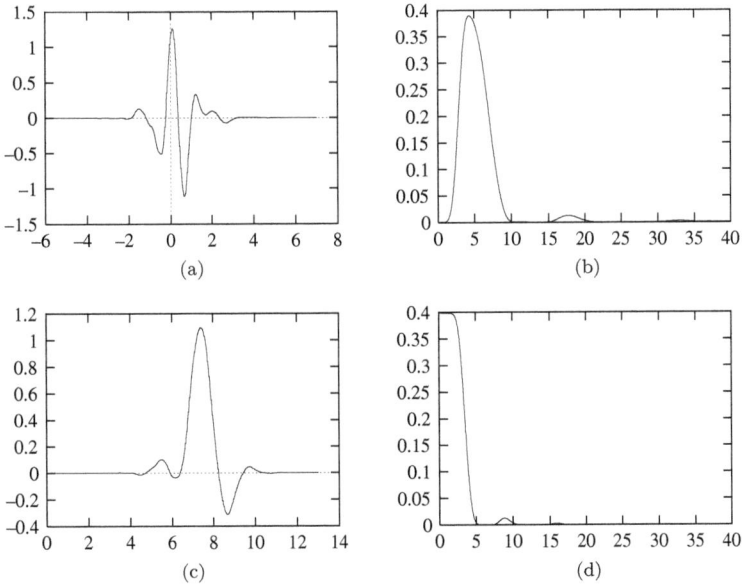

Fig. 4.23 The second type Daubechies orthonormal wavelet and its scaling function ($L = 7$): (a) $\psi(t)$; (b) the FT of $\psi(t)$; (c) $\phi(t)$; (d) the FT of $\phi(t)$.

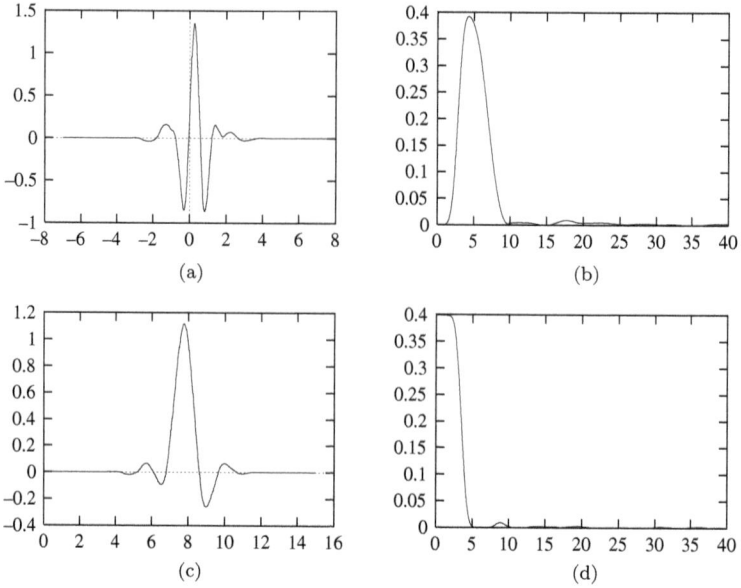

Fig. 4.24 The second type Daubechies orthonormal wavelet and its scaling function ($L = 8$): (a) $\psi(t)$; (b) the FT of $\psi(t)$; (c) $\phi(t)$; (d) the FT of $\phi(t)$.

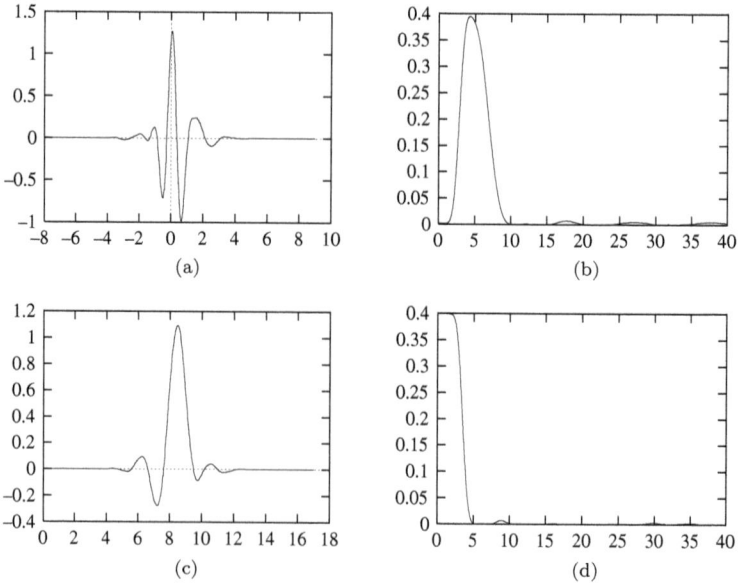

Fig. 4.25 The second type Daubechies orthonormal wavelet and its scaling function ($L = 9$): (a) $\psi(t)$; (b) the FT of $\psi(t)$; (c) $\phi(t)$; (d) the FT of $\phi(t)$.

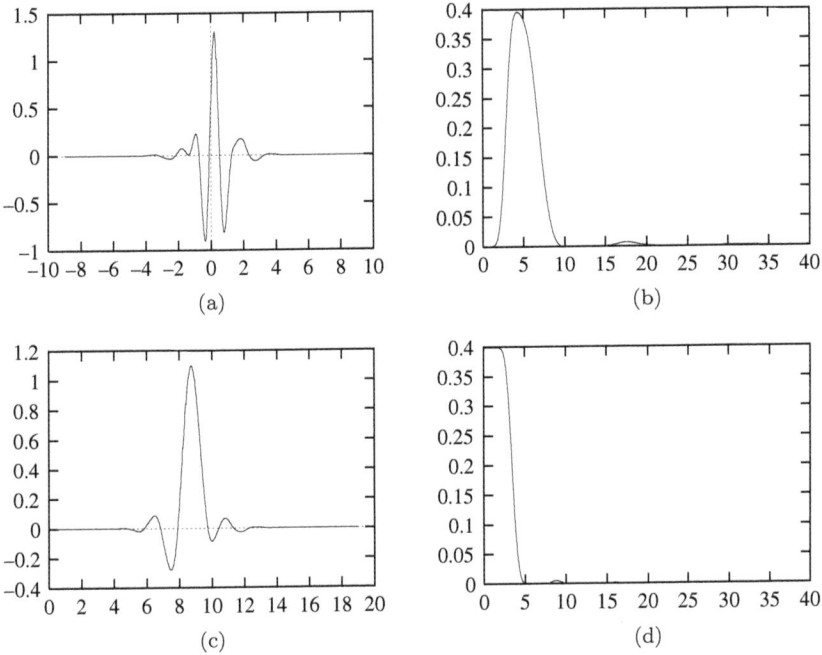

(a)

(b)

(c)

(d)

Fig. 4.26   The second type Daubechies orthonormal wavelet and its scaling function ($L = 10$): (a) $\psi(t)$; (b) the FT of $\psi(t)$; (c) $\phi(t)$; (d) the FT of $\phi(t)$.

where $L = 2K$ is a positive even integer,

$$M_0(\xi) := \sum_{k=0}^{K-1} \binom{K-1+k}{k} \left(\sin^2 \frac{\xi}{2}\right)^k + \left(\sin^2 \frac{\xi}{2}\right)^K f(\xi),$$

and $f(\xi) := \sum_{n=0}^{2K-1} f_n e^{-in\xi}$ is a trigonometric polynomial such that

$$|m_0(\xi)|^2 + |m_0(\xi + \pi)|^2 = 1 \quad (\forall \xi \in \mathbb{R}).$$

Coiflet of order $L$ is a compactly supported orthonormal wavelet basis with support width $3L - 1$. It has vanishing moment of order $L - 1$ and the corresponding scaling function $\phi$ satisfies (4.16). Table 4.8 shows the localization property of coiflets. Figures 4.27–4.31 show the scaling function and its frequency property, as well as the wavelet function and its frequency property. The filter coefficients of the coiflets of orders $L = 2, 4, 6, 8, 10$ are listed in Table 4.9.

Table 4.8.   The centers, radii, and area of time window, frequency window, and time–frequency window of R. Coifman wavelets, whose compactly supported lengths are 5, 11, 17, 23, 29, respectively.

| | Time | Window | Freq. | Window | Time–freq. |
|---|---|---|---|---|---|
| Coifman | Center | Radius | Center | Radius | area |
| 1 | 0.4050 | 0.3808 | 1.3559 | 3.0720 | 4.6972 |
| 2 | 0.3756 | 0.3981 | 1.2458 | 2.1827 | 3.4759 |
| 3 | 0.3135 | 0.4430 | 1.2155 | 1.9945 | 3.5340 |
| 4 | 0.3761 | 0.4740 | 1.2080 | 1.9682 | 3.7314 |
| 5 | 0.3767 | 0.5050 | 1.2005 | 1.9419 | 4.2020 |

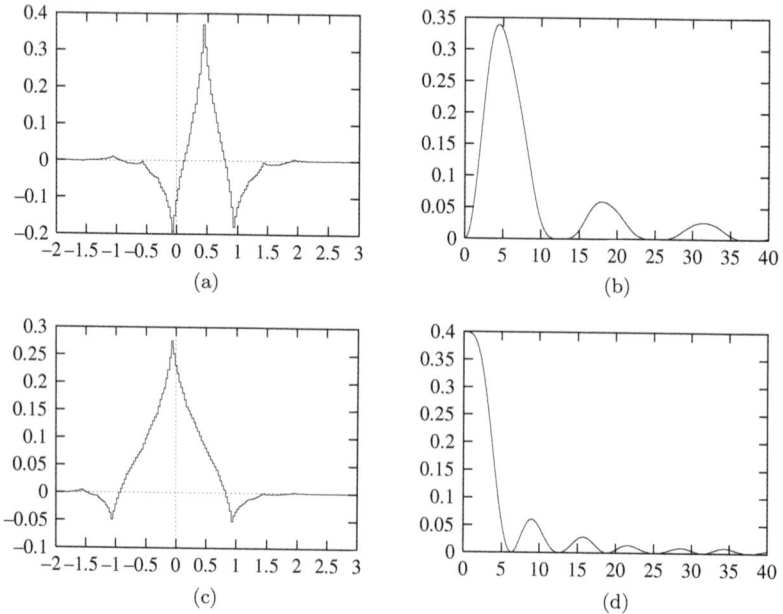

Fig. 4.27   Coiflet of order $L = 2$ and the corresponding scaling function: (a) $\psi(t)$; (b) the FT of $\psi(t)$; (c) $\phi(t)$; (d) the FT of $\phi(t)$.

## 4.2  Non-Orthonormal Wavelet Bases

### 4.2.1  *Cardinal Spline Wavelet*

In this section, we aim at constructing interpolatory spline wavelets based on B-splines, that is, fundamental cardinal splines, and discussing their properties.

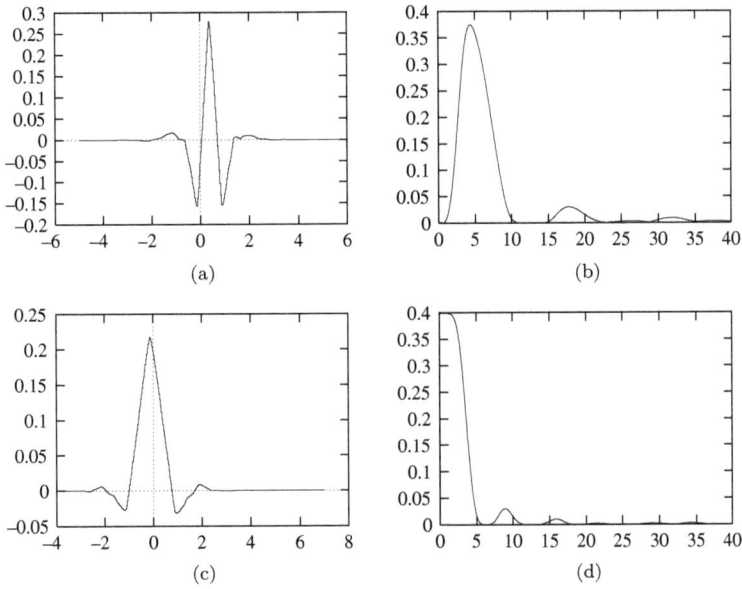

Fig. 4.28   Coiflet of order $L = 4$ and the corresponding scaling function: (a) $\psi(t)$; (b) the FT of $\psi(t)$; (c) $\phi(t)$; (d) the FT of $\phi(t)$.

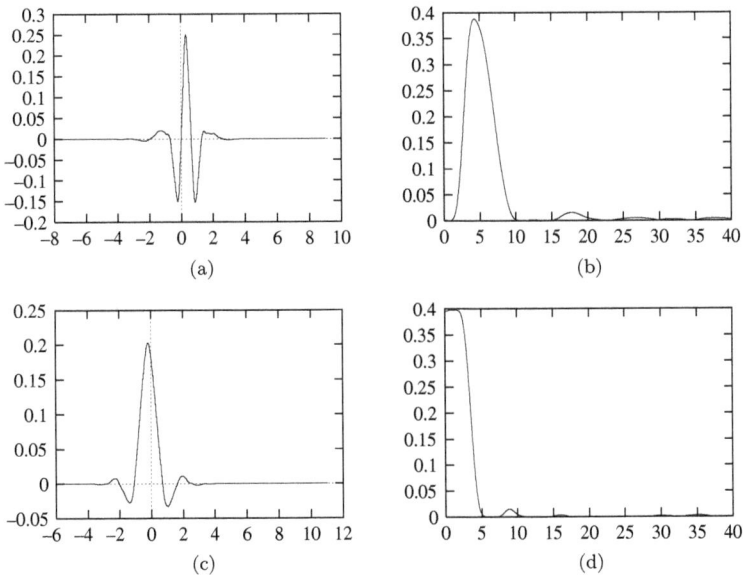

Fig. 4.29   Coiflet of order $L = 6$ and the corresponding scaling function: (a) $\psi(t)$; (b) the FT of $\psi(t)$; (c) $\phi(t)$; (d) the FT of $\phi(t)$.

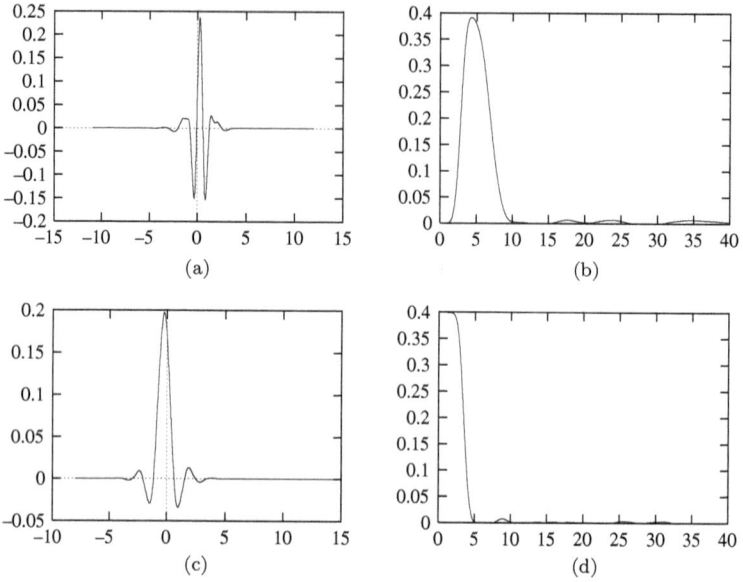

Fig. 4.30  Coiflet of order $L = 8$ and the corresponding scaling function: (a) $\psi(t)$; (b) the FT of $\psi(t)$; (c) $\phi(t)$; (d) the FT of $\phi(t)$.

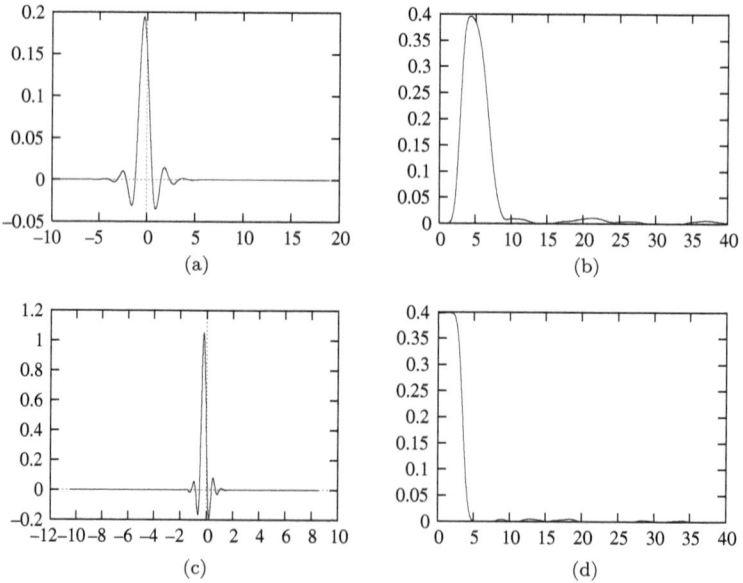

Fig. 4.31  Coiflet of order $L = 10$ and the corresponding scaling function: (a) $\psi(t)$; (b) the FT of $\psi(t)$; (c) $\phi(t)$; (d) the FT of $\phi(t)$.

Table 4.9.  The coefficients $\{h_n\}$ for coiflets of orders $L = 2, 4, 6, 8, 10$.

| $n$ | $h_n\ (L=2)$ |
|---|---|
| $-2$ | $-0.051429728471$ |
| $-1$ | $0.238929728471$ |
| $0$ | $0.602859456942$ |
| $1$ | $0.272140543058$ |
| $2$ | $-0.051429972847$ |
| $3$ | $-0.011070271529$ |

| $n$ | $h_n\ (L=4)$ |
|---|---|
| $-4$ | $0.011587596739$ |
| $-3$ | $-0.029320137980$ |
| $-2$ | $-0.047639590310$ |
| $-1$ | $0.273021046535$ |
| $0$ | $0.574682393857$ |
| $1$ | $0.294867193696$ |
| $2$ | $-0.054085607092$ |
| $3$ | $-0.042026480461$ |
| $4$ | $0.016744410163$ |
| $5$ | $0.003967883613$ |
| $6$ | $-0.001289203356$ |
| $7$ | $-0.000509505399$ |

| $n$ | $h_n\ (L=6)$ |
|---|---|
| $-6$ | $-0.002682418671$ |
| $-5$ | $0.005503126709$ |
| $-4$ | $0.016583560479$ |
| $-3$ | $-0.046507764479$ |
| $-2$ | $-0.043220763560$ |
| $-1$ | $0.286503335274$ |
| $0$ | $0.561285256870$ |
| $1$ | $0.302983571773$ |
| $2$ | $-0.050770140755$ |
| $3$ | $-0.058196250762$ |

| $n$ | (continued $L=2$) |
|---|---|
| $4$ | $0.024434094321$ |
| $5$ | $0.011229240962$ |
| $6$ | $-0.006369601011$ |
| $7$ | $-0.001820458916$ |
| $8$ | $0.000790205101$ |
| $9$ | $0.000329665174$ |
| $10$ | $-0.000050192775$ |
| $11$ | $-0.000024465734$ |

| $n$ | $h_n\ (L=8)$ |
|---|---|
| $-8$ | $0.000630961046$ |
| $-7$ | $-0.001152224852$ |
| $-6$ | $-0.005194524026$ |
| $-5$ | $0.011362459244$ |
| $-4$ | $0.018867235378$ |
| $-3$ | $-0.057464234429$ |
| $-2$ | $-0.039652648517$ |
| $-1$ | $0.293667390895$ |
| $0$ | $0.553126452562$ |
| $1$ | $0.307157326198$ |
| $2$ | $-0.047112738865$ |
| $3$ | $-0.068038127051$ |
| $4$ | $0.027813640153$ |
| $5$ | $0.017735837438$ |
| $6$ | $-0.010756318517$ |
| $7$ | $-0.004001012886$ |
| $8$ | $0.002652665946$ |
| $9$ | $0.000895594529$ |
| $10$ | $-0.000416500571$ |
| $11$ | $-0.000183829769$ |
| $12$ | $0.000044080354$ |
| $13$ | $0.000022082857$ |
| $14$ | $-0.000002304942$ |
| $15$ | $-0.000001262175$ |

| $n$ | $h_n\ (L=10)$ |
|---|---|
| $-10$ | $-0.0001499638$ |
| $-9$ | $0.0002535612$ |
| $-8$ | $0.0015402457$ |
| $-7$ | $-0.0029411108$ |
| $-6$ | $-0.0071637819$ |
| $-5$ | $0.0165520664$ |
| $-4$ | $0.0199178043$ |
| $-3$ | $-0.0649972628$ |
| $-2$ | $-0.0368000736$ |
| $-1$ | $0.2980923235$ |
| $0$ | $0.5475054294$ |
| $1$ | $0.3097068490$ |
| $2$ | $-0.438660508$ |
| $3$ | $-0.0746522389$ |
| $4$ | $0.0291958795$ |
| $5$ | $0.0231107770$ |
| $6$ | $-0.0139736879$ |
| $7$ | $-0.0064800900$ |
| $8$ | $0.0047830014$ |
| $9$ | $0.0017206547$ |
| $10$ | $-0.0011758222$ |
| $11$ | $-0.0004512270$ |
| $12$ | $0.0002137298$ |
| $13$ | $0.0000993776$ |
| $14$ | $-0.0000292321$ |
| $15$ | $-0.0000150720$ |
| $16$ | $0.0000026408$ |
| $17$ | $0.0000014593$ |
| $18$ | $-0.0000001184$ |
| $19$ | $-0.0000000673$ |

*Note*: They are normalized so that $\sum_n h_n = 1$.

For $m \geq 1$, the B-spline of order $m$ $N_m(x)$ is defined by

$$N_m(x) = \underbrace{N_1 * \cdots * N_1}_{m}(x), \qquad (4.17)$$

where

$$N_1(x) = \chi_{[0,1)}(x) = \begin{cases} 1 & x \in [0,1); \\ 0 & \text{otherwise.} \end{cases}$$

Then it can be deduced that

$$\hat{N}_m(\xi) = \left( \frac{2}{\xi} e^{-i\xi/2} \sin \frac{\xi}{2} \right)^m .$$

Therefore, $N_m(x)$ satisfies the following two-scale equation:

$$\hat{N}_m(\xi) = \left(\frac{1 + e^{-i\xi/2}}{2}\right)^m \hat{N}_m(\xi/2).$$

By the expression of $\hat{N}_m(\xi)$, we can conclude that

$$\Phi_m(\xi) := \sum_{k \in \mathbb{Z}} |\hat{N}_m(\xi + 2k\pi)|^2$$

satisfies that

$$\left(\frac{2}{\pi}\right)^{2m} \le \Phi_m(\xi) \le 1 + 2^{2m+1} \left(\frac{2}{\pi}\right)^{2m} \sum_{k=1}^{\infty} \frac{1}{(2k-1)^{2m}} \qquad (4.18)$$

for $m > 1$, and $\Phi_m(\xi) \equiv 1$ for $m = 1$. Hence, $\{N_m(x - k)|k \in \mathbb{Z}\}$ is a Riesz basis of

$$V_0^m := \overline{\text{span}\{N_m(x - k)|k \in \mathbb{Z}\}} \qquad (4.19)$$

and

$$V_j^m := \{f(2^j x)|f(x) \in V_0^m\}, \quad (j \in \mathbb{Z})$$

constitutes an MRA (non-orthonormal except for the case of $m = 1$) of $L^2(\mathbb{R})$.

To estimate the bounds of $\Phi_m(\xi)$ in (4.18) more exactly, we use Poission's summation formula to deduce that [Chui, 1992a, p. 48, 89]

$$\Phi_m(\xi) = \sum_{k \in \mathbb{Z}} \left( \int_{\mathbb{R}} N_m(t + k)\overline{N_m(t)}dt \right) e^{-ik\xi}$$

$$= \sum_{k=-m+1}^{m-1} N_{2m}(m + k)e^{-ik\xi}.$$

Note that $\Phi_m(\xi)$ and $N_m(\xi)$ are always non-positive, we get the upper bound of $\Phi_m(\xi)$, i.e.,

$$\Phi_m(\xi) \le \sum_{k=-m+1}^{m-1} N_{2m}(m + k) = 1.$$

The estimation of the lower bound of $\Phi_m(\xi)$ is more complicated though. Let

$$E_{2m-1}(z) := (2m - 1)!z^{m-1} \sum_{k=-m+1}^{m-1} N_{2m}(m + k)z^k \qquad (4.20)$$

be the Euler–Frobenius polynomials of order $2m - 1$, which is in fact a polynomial on $z$ of degree $2m - 2$. It can be shown that (see the work of

Chui [1992a, Chapter 6] for details) the $2m - 2$ roots, $\lambda_1, \ldots, \lambda_{2m-2}$, of $E_{2m-1}(z)$ are simple, real, and negative, and furthermore, when they are arranged in decreasing order, say,

$$0 > \lambda_1 > \cdot > \lambda_{2m-2},$$

they satisfy

$$\lambda_1 \lambda_{2m-2} = \cdots = \lambda_{m-1} \lambda_m = 1.$$

Hence, we have

$$\Phi_m(\xi) = \sum_{k=-m+1}^{m-1} N_{2m}(m+k) e^{-ik\xi}$$

$$= \frac{1}{(2m-1)!} e^{i(m-1)\xi} E_{2m-1}(e^{-i\xi}).$$

It is easy to see that the coefficient of $z^{2m-2}$ in $E_{2m-1}(z)$ is

$$(2m-1)! N_{2m}(2m-1) = 1.$$

Therefore,

$$E_{2m-1}(z) = \prod_{k=1}^{2m-2} (z - \lambda_k),$$

consequently,

$$\Phi_m(\xi) = \frac{1}{(2m-1)!} e^{i(m-1)\xi} \prod_{k=1}^{2m-2} (e^{-i\xi} - \lambda_k)$$

$$= \frac{1}{(2m-1)!} \prod_{k=1}^{2m-2} |e^{-i\xi} - \lambda_k|$$

$$= \frac{1}{(2m-1)!} \left[ \prod_{k=1}^{m-1} |e^{-i\xi} - \lambda_k| \right] \left[ \prod_{k=m}^{2m-2} |e^{-i\xi} - \lambda_k| \right]$$

$$= \frac{1}{(2m-1)!} \left[ \prod_{k=1}^{m-1} |e^{-i\xi} - \lambda_k| \right] \left[ \prod_{k=1}^{m-1} |e^{-i\xi} - \lambda_{2m-1-k}| \right]$$

$$= \frac{1}{(2m-1)!} \prod_{k=1}^{m-1} \left[ |e^{-i\xi} - \lambda_k| |e^{-i\xi} - \frac{1}{\lambda_k}| \right]$$

$$= \frac{1}{(2m-1)!} \prod_{k=1}^{m-1} \frac{|(e^{-i\xi} - \lambda_k)(\lambda_k e^{-i\xi} - 1)|}{|\lambda_k|}$$

$$= \frac{1}{(2m-1)!} \prod_{k=1}^{m-1} \frac{1 - 2\lambda_k \cos \xi + \lambda_k^2}{|\lambda_k|}$$

$$\geq \frac{1}{(2m-1)!} \prod_{k=1}^{m-1} \frac{1 + 2\lambda_k + \lambda_k^2}{|\lambda_k|}$$

$$= \Phi_m(\pi).$$

Hence, $\Phi_m(\pi)$ and $\Phi_m(0) = 1$ are the lower and upper bounds of $\Phi_m(\xi)$, respectively, and these bounds are the best possible obviously.

The idea behind the construction of Battle–Lemarie wavelet basis is to orthonormalize the non-orthonormal Riesz basis by letting

$$\hat{\phi}^\sharp(\xi) := \frac{\hat{N}_m(\xi)}{\sqrt{\Phi_m(\xi)}}.$$

A drawback of such construction is that $\phi^\sharp(\xi)$ can neither be expressed explicitly nor be supported compactly as $N_m(\xi)$. To overcome this drawback, we keep $V_j^m$ as above and let $W_j^m$ be the orthogonal complement of $V_{j+1}^m$ and $V_j^m$, i.e.,

$$V_{j+1}^m = V_j^m \oplus W_j^m. \tag{4.21}$$

Now our task is to construct a wavelet $\psi$ such that

$$\psi_{j,k}(x) := 2^{j/2}\psi(2^j x - k), \quad (\forall j, k \in \mathbb{Z})$$

constitutes a Riesz basis of $W_j$ for any $j \in \mathbb{Z}$.

Let the $m$th "fundamental cardinal spline function" $L_m(x)$ be defined by

$$L_m(x) := \sum_{k=-\infty}^{\infty} c_k^{(m)} N_m \left(x + \frac{m}{2} - k\right), \tag{4.22}$$

where $\{c_k^{(m)}\}$ is obtained by solving the following bi-infinite system:

$$\sum_{k=-\infty}^{\infty} c_k^{(m)} N_m \left(j + \frac{m}{2} - k\right) = \delta_{j,0}. \tag{4.23}$$

We observe that (4.23) is equivalent to the following equation:

$$\left( \sum_{k=-\infty}^{\infty} c_k^{(m)} z^k \right) \left( \sum_{k=-\infty}^{\infty} N_m \left( \frac{m}{2} + k \right) z^k \right) = 1. \tag{4.24}$$

Since

$$\tilde{N}_m(z) := \sum_{k=-\infty}^{\infty} N_m \left( \frac{m}{2} + k \right) z^k$$

is a Laurent polynomial, which does not vanish on the unit circle $|z| = 1$ (see the work of Chui [1992a, p. 111] for details), there exist $r, R : \ r < 1 < R$ such that

$$\sum_{k=-\infty}^{\infty} c_k^{(m)} z^k = \frac{1}{\tilde{N}_m(z)}$$

is analytic on $r < |z| < R$. Choose $z_0 : 1 < |z_0| < R$, then the convergence of $\sum_{k=-\infty}^{\infty} c_k^{(m)} z_0^k$ implies $\lim_{k \to \infty} c_k^{(m)} z_0^k = 0$. Therefore, $|c_k^{(m)}| = O(|z_0|^{-k})$, $k \to \infty$, i.e., $c_k^{(m)}$ decays to zero exponentially fast as $k \to \infty$. Similarly, we have that $c_k^{(m)}$ decays to zero exponentially fast as $k \to -\infty$. Moreover, it should be pointed out that $\{c_k^{(m)}\}$ is not finite for $m \geq 3$. In fact, it is easy to see that

$$\tilde{N}_m(z) = \frac{1}{z^{[\frac{m-1}{2}]}} \sum_{k=0}^{2[\frac{m-1}{2}]} N_m \left( \frac{m}{2} - \left[ \frac{m-1}{2} \right] + k \right) z^k,$$

where $[x]$ denotes the largest integer not exceeding $x$. If $\{c_k^{(m)}\}$ is finite, by denoting

$$\sum_{k=-\infty}^{\infty} c_k^{(m)} z_0^k = \frac{1}{z^p} (a_0 + a_1 z + \cdots + a_q z^q),$$

where $a_0 \neq 0$ and $p$, $q$ are negative integers and using (4.24), we have

$$\frac{1}{z^p} (a_0 + a_1 z + \cdots + a_q z^q) \frac{1}{z^{[\frac{m-1}{2}]}} \sum_{k=0}^{2[\frac{m-1}{2}]} N_m \left( \frac{m}{2} - \left[ \frac{m-1}{2} \right] + k \right) z^k = 1,$$

i.e.,

$$(a_0 + a_1 z + \cdots + a_q z^q) \sum_{k=0}^{2[\frac{m-1}{2}]} N_m \left( \frac{m}{2} - \left[ \frac{m-1}{2} \right] + k \right) z^k = z^{p + [\frac{m-1}{2}]}. \tag{4.25}$$

On one hand, the constant item of the left side of (4.25) is $a_0 N_m(\frac{m}{2} - [\frac{m-1}{2}])$, which does not vanish; on the other hand, the constant item of the right side of (4.25) is obviously 0 since $[\frac{m-1}{2}] \geq 1$, provided that $m \geq 3$. The contradiction proves that $\{c_k^{(m)}\}$ must be infinite.

It is now easy to see that $L_m$ satisfies the following interpolation property:

$$L_m(j) = \delta_{j,0}.$$

Therefore, for any series $\{f(k)\}$ with at most polynomial growth, if we define an interpolation spline operator as follows:

$$J_m f(x) := \sum_{k=-\infty}^{\infty} f(k) L_m(x - k), \qquad (4.26)$$

then the right side of (4.26) certainly converges at every $x \in \mathbb{R}$ due to the growth limit of $\{f(k)\}$, and $J_m$ satisfies the following interpolation property:

$$(J_m f)(j) = f(j), \quad (\forall j \in \mathbb{Z}).$$

It can be proven that the coefficient sequence $\{c_k^{(m)}\}$ is not finite for each $m \geq 3$ so that the fundamental cardinal spline $L_m(x)$ does not vanish identically outside a compact set, even though it certainly decays to zero exponentially fast as $x \to \infty$ [Chui, 1992a]. The following theorem gives a construction of the so-called fundamental cardinal spline wavelets.

**Theorem 4.3.** *Let $m$ be any positive integer, and define*

$$\psi_{I,m}(x) := L_{2m}^{(m)}(2x - 1),$$

*where $L_{2m}(x)$ is the $(2m)$th-order fundamental cardinal spline. Then*

$$\{\psi_{I,m}(x - k) | k \in \mathbb{Z}\}$$

*constitutes a Riesz basis of $W_0^m$. Consequently,*

$$\{2^{j/2} \psi_{I,m}(2^j x - k) | j, k \in \mathbb{Z}\}$$

*constitutes a (Riesz) wavelet basis of $L^2(\mathbb{R})$.*

For details of the proof, readers are referred to the work of Chui [1992a, pp. 142–145, 178–181] and are mentioned to note that $W_i \perp W_j$ ($\forall i \neq j$).

The wavelet $\psi_{I,m}(x)$ constructed by Theorem 4.3 has an explicit expression. It can be deduced that

$$\psi_{I,m}(x) = \sum_{n=-\infty}^{\infty} q_n N_m(2x - n), \tag{4.27}$$

with

$$q_n := \sum_{l=0}^{m} (-1)^l \binom{m}{l} c_{m+n-1-l}^{(2m)},$$

which decays to zero exponentially fast as $k \to \pm\infty$ as $c_k^{(m)}$ does.

Denote

$$Q(z) := \sum_{n=-\infty}^{\infty} q_n z^n.$$

It is easy to deduce that

$$Q(z) = z^{1-m}(1 - z)^m \sum_{n=-\infty}^{\infty} c_n^{(2m)} z^n.$$

Therefore, to calculate $\{q_n\}$ equals to calculate $\{c_n^{(2m)}\}$. We introduce

$$F_m(z) := \sum_{k=-m+1}^{m-1} N_{2m}(m + k)z^k = \sum_{k=-\infty}^{\infty} N_{2m}(m + k)z^k.$$

Since

$$F_m(z) \sum_{n=-\infty}^{\infty} c_n^{(2m)} z^n = \sum_{k=-\infty}^{\infty} \sum_{n=-\infty}^{\infty} c_n^{(2m)} N_{2m}(m + k)z^{n+k}$$

$$= \sum_{n=-\infty}^{\infty} \sum_{k=-\infty}^{\infty} c_n^{(2m)} N_{2m}(m + k - n)z^k$$

$$= \sum_{k=-\infty}^{\infty} [\sum_{n=-\infty}^{\infty} c_n^{(2m)} N_{2m}(m + k - n)]z^k$$

$$= \sum_{k=-\infty}^{\infty} L_{2m}(k)z^k$$

$$= 1,$$

we have

$$\sum_{n=-\infty}^{\infty} c_n^{(2m)} z^n = \frac{1}{F_m(z)}.$$

Table 4.10.    Modified Euler–Frobenius polynomials.

| $n$ | $\tilde{E}_n(z)$ |
| --- | --- |
| 1 | $1$ |
| 2 | $1 + z$ |
| 3 | $1 + 4z + z^2$ |
| 4 | $1 + 11z + 11z^2 + z^3$ |
| 5 | $1 + 26z + 66z^2 + 26z^3 + z^4$ |
| 6 | $1 + 57z + 302z^2 + 302z^3 + 57z^4 + z^5$ |
| 7 | $1 + 120z + 1191z^2 + 2416z^3 + 1191z^4 + 120z^5 + z^6$ |
| 8 | $1 + 247z + 4293z^2 + 15619z^3 + 15619z^4 + 4293z^5 + 247z^6 + z^7$ |
| 9 | $1 + 502z + 14608z^2 + 88234z^3 + 156190z^4 + 88234z^5 + 14608z^6$ $+ 502z^7 + z^8$ |

Consequently,

$$Q(z) = z^{1-m}(1-z)^m \frac{1}{F_m(z)} = (2m-1)! \frac{(1-z)^m}{E_{2m-1}(z)},$$

where

$$E_{2m-1}(z) := (2m-1)! z^{m-1} F_m(z)$$

$$= (2m-1)! z^{m-1} \sum_{k=-m+1}^{m-1} N_{2m}(m+k) z^k$$

is the Euler–Frobenius polynomial of order $2m - 1$ defined by (4.20). Generally, the modified Euler–Frobenius polynomial of order $n$ is defined by

$$\tilde{E}_n(z) := n! \sum_{k=0}^{n-1} N_{n+1}(k+1) z^k.$$

In particular, Table 4.10 lists nine modified Euler–Frobenius polynomials. Although $\psi_{I,m}(x)$ can be expressed analytically, it is however not supported compactly for $m \geq 2$. The reason is as follows: $\{c_n^{2m}\}$ is not finite for $m \geq 2$, therefore $L_{2m}$ cannot be supported compactly, finally, neither is $\psi_{I,m}(x)$ for $m \geq 2$.

For $m = 1$, by

$$Q(z) = \frac{1-z}{E_1(z)} = 1 - z,$$

we get

$$\psi_{I,1}(x) = N_1(2x) - N_1(2x - 1),$$

which is the well-known Haar wavelet.

For $m = 2$, by

$$Q(z) = 6\frac{(1-z)^2}{E_3(z)} = 6\frac{(1-z)^2}{1+4z+z^2}$$

and

$$\frac{1}{1+4z+z^2}$$

$$= -\frac{1}{2\sqrt{3}}\left(\frac{1}{z+2+\sqrt{3}} - \frac{1}{z+2-\sqrt{3}}\right)$$

$$= \frac{1}{2\sqrt{3}}\left(\frac{1}{z}\cdot\frac{1}{1+(2-\sqrt{3})z^{-1}} - \frac{1}{2+\sqrt{3}}\cdot\frac{1}{1+z(2+\sqrt{3})^{-1}}\right)$$

$$= \frac{1}{2\sqrt{3}}\left[\frac{1}{z}\sum_{k=0}^{\infty}(\sqrt{3}-2)^k z^{-k} + \sum_{k=0}^{\infty}(\sqrt{3}+2)^{-(k+1)}z^k\right]$$

$$= \frac{1}{2\sqrt{3}}\left[\sum_{k=0}^{\infty}(\sqrt{3}-2)^k z^{-(k+1)} + \sum_{k=0}^{\infty}(\sqrt{3}-2)^{k+1}z^k\right]$$

$$= \frac{1}{2\sqrt{3}}\sum_{k=-\infty}^{\infty}(\sqrt{3}-2)^{|k+1|}z^k,$$

we get

$$Q(z) = 6(1-z)^2\frac{1}{2\sqrt{3}}\sum_{k=-\infty}^{\infty}(\sqrt{3}-2)^{|k+1|}z^k$$

$$= \sqrt{3}(1-2z+z^2)\sum_{k=-\infty}^{\infty}(\sqrt{3}-2)^{|k+1|}z^k$$

$$= \sqrt{3}\left[\sum_{k=-\infty}^{\infty}(\sqrt{3}-2)^{|k+1|}z^k - 2\sum_{k=-\infty}^{\infty}(\sqrt{3}-2)^{|k+1|}z^{k+1}\right.$$

$$\left. + \sum_{k=-\infty}^{\infty}(\sqrt{3}-2)^{|k+1|}z^{k+2}\right]$$

$$= \sqrt{3}\left[\sum_{k=-\infty}^{\infty}(\sqrt{3}-2)^{|k+1|}z^k\right.$$

$$-2\sum_{k=-\infty}^{\infty}(\sqrt{3}-2)^{|k|}z^k + \sum_{k=-\infty}^{\infty}(\sqrt{3}-2)^{|k-1|}z^k\Bigg]$$

$$= \sqrt{3}\sum_{k=-\infty}^{\infty}[(\sqrt{3}-2)^{|k+1|} - 2(\sqrt{3}-2)^{|k|} + (\sqrt{3}-2)^{|k-1|}]z^k.$$

It can be deduced that

$$q_0 = \sqrt{3}\,[(\sqrt{3}-2) - 2 + (\sqrt{3}-2)] = 6(1-\sqrt{3}),$$

and for $k \neq 0$,

$$
\begin{aligned}
q_k &= \sqrt{3}[(\sqrt{3}-2)^{|k+1|} - 2(\sqrt{3}-2)^{|k|} + (\sqrt{3}-2)^{|k-1|}]\\
&= \sqrt{3}(\sqrt{3}-2)^{|k|}[(\sqrt{3}-2)^{|k+1|-|k|} - 2 + (\sqrt{3}-2)^{|k-1|-|k|}]\\
&= \sqrt{3}(\sqrt{3}-2)^{|k|}[(\sqrt{3}-2) - 2 + (\sqrt{3}-2)^{-1}]\\
&= \sqrt{3}(\sqrt{3}-2)^{|k|}[(\sqrt{3}-2) - 2 - (\sqrt{3}+2)]\\
&= \sqrt{3}(\sqrt{3}-2)^{|k|}[(\sqrt{3}-2) - 2 - (\sqrt{3}+2)]\\
&= -6\sqrt{3}(\sqrt{3}-2)^{|k|}.
\end{aligned}
$$

Therefore,

$$\psi_{I,2}(x) = -6\sqrt{3}\sum_{k=-\infty}^{\infty}(\sqrt{3}-2)^{|k|}N_2(2x-k).$$

C-spline wavelets $\psi_{I,2}$, $\psi_{I,4}$ of orders 2, 4 and their frequency properties are illustrated in Figs. 4.32 and 4.33, respectively.

The centers, radii, and areas of the time windows, frequency windows, and time–frequency windows of the orders 2–10 C-splines wavelets are given in Table 4.11.

## 4.2.2 *Compactly Supported Spline Wavelet*

As discussed in the previous section, $\psi_{I,m}$ is not supported compactly. To improve this property, another kind of spline wavelet, which is supported compactly, is introduced in this section.

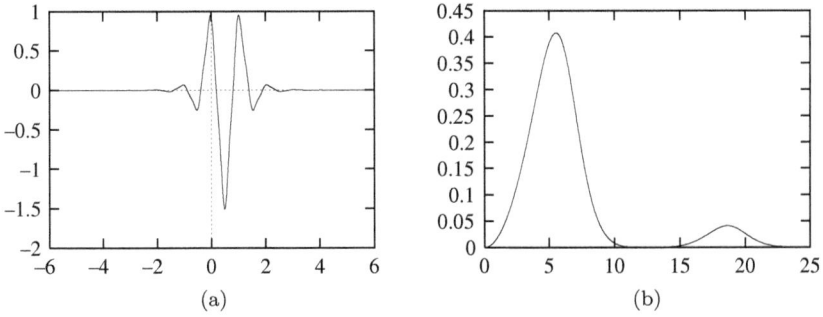

Fig. 4.32   C-spline wavelet $\psi_{I,2}$ of order 2 and its frequency spectrum: (a) Time domain; (b) frequency domain.

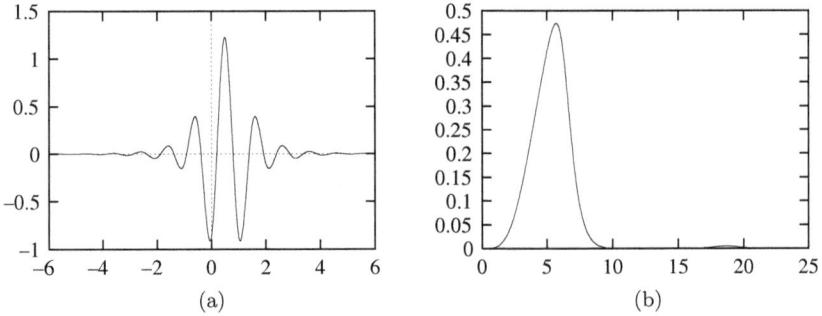

Fig. 4.33   C-spline wavelet $\psi_{I,4}$ of order 4 and its frequency spectrum: (a) Time domain; (b) frequency domain.

Let

$$\psi_m(x) := 2 \sum_{n=-\infty}^{\infty} q_n N_m(2x - n), \tag{4.28}$$

with

$$q_n := \frac{(-1)^n}{2^m} \sum_{l=0}^{m} \binom{m}{l} N_{2m}(n + 1 - l)$$

$$= \begin{cases} \dfrac{(-1)^n}{2^m} \displaystyle\sum_{l=0}^{m} \binom{m}{l} N_{2m}(n + 1 - l), & n = 0, 1, \ldots, 3m - 2, \\ 0, & \text{otherwise.} \end{cases}$$

Table 4.11.    The centers, radii, and area of time window, frequency window, and time–frequency window of C-spline wavelets of orders 2–10.

| C-spline | Time Window | | Freq. Window | | Time–freq. area |
|---|---|---|---|---|---|
| | Center | Radius | Center | Radius | |
| 2 | 0.5 | 0.4156 | 2.7547 | 2.4243 | 4.03025 |
| 3 | 0.5 | 0.4944 | 2.6757 | 2.0549 | 4.06344 |
| 4 | 0.5 | 0.5675 | 2.6767 | 2.0084 | 4.55896 |
| 5 | 0.5 | 0.6352 | 2.6854 | 1.9960 | 5.07147 |
| 6 | 0.5 | 0.6977 | 2.6930 | 1.9903 | 5.55437 |
| 7 | 0.5 | 0.7539 | 2.6969 | 1.9858 | 5.98852 |
| 8 | 0.5 | 0.8048 | 2.6979 | 1.9814 | 6.37891 |
| 9 | 0.5 | 0.8483 | 2.6946 | 1.9757 | 6.70379 |
| 10 | 0.5 | 0.8892 | 2.6885 | 1.9691 | 7.00385 |

Our main result in this section is as follows:

**Theorem 4.4.** *Let $m$ be a positive integer and $N_m(x)$ the $m$th-order cardinal B-spline defined by (4.17). Then $\psi_m(x)$, which is defined by (4.28), is a Riesz basis of $W_0^m$ defined by (4.21) and*

$$\mathrm{supp}\psi_m = [0, 2m - 1].$$

*Consequently,*

$$\{2^{j/2}\psi_m(2^j x - k)|j, k \in \mathbb{Z}\}$$

*constitutes a (Riesz) wavelet basis of $L^2(\mathbb{R})$.*

Furthermore, the dual wavelet $\tilde{\psi}_m$ of $\psi_m$ can be expressed as follows:

$$\tilde{\psi}_m(x) = \frac{(-1)^{m+1}}{2^{m-1}} \sum_{k=-\infty}^{\infty} c_k^{(2m)} \psi_{I,m}(x + m + 1 - k),$$

where $\psi_{I,m}$ is the wavelet defined by (4.27) and $\{c_k^{(2m)}\}$ is determined by bi-infinite system (4.23). On the symmetricity of the wavelets, we can prove the following theorem:

**Theorem 4.5.** *Wavelets $\psi_m, \tilde{\psi}_m$, and $\psi_{I,m}$ are all symmetric for even $m$ and antisymmetric for odd $m$. Consequently, they all have generalized linear phases.*

The details of the proof for the theorems are omitted here; readers are referred to the work of Chui [1992a, pp. 183–184].

Figures 4.34–4.36 show the compactly supported spline wavelets $\psi_m$ of orders 2, 4, 5 and their frequency properties, respectively. The centers, radii,

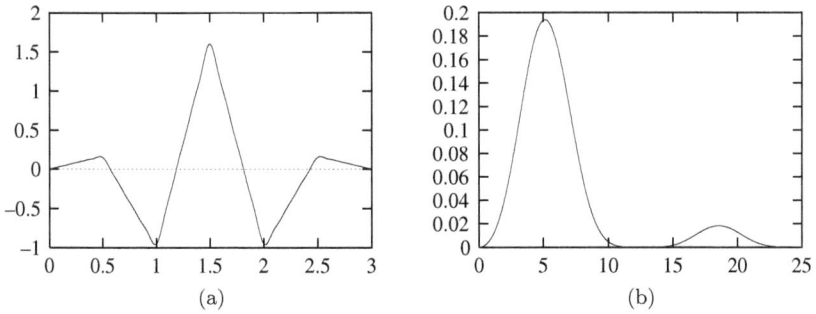

Fig. 4.34   Compactly supported spline wavelet $\psi_2$ of order 2 and its frequency spectrum: (a) Time domain; (b) frequency domain.

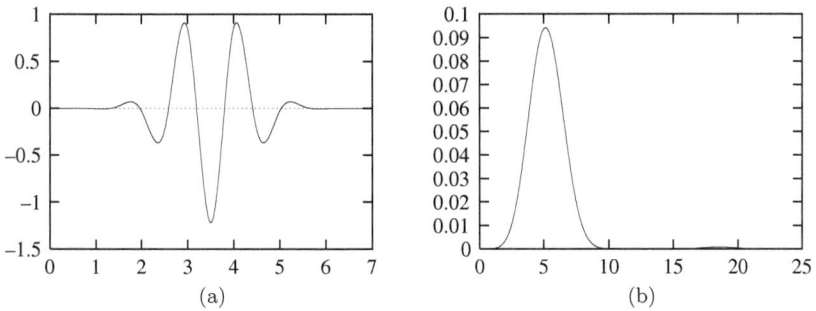

Fig. 4.35   Compactly supported spline wavelet $\psi_4$ of order 4 and its frequency spectrum: (a) Time domain; (b) frequency domain.

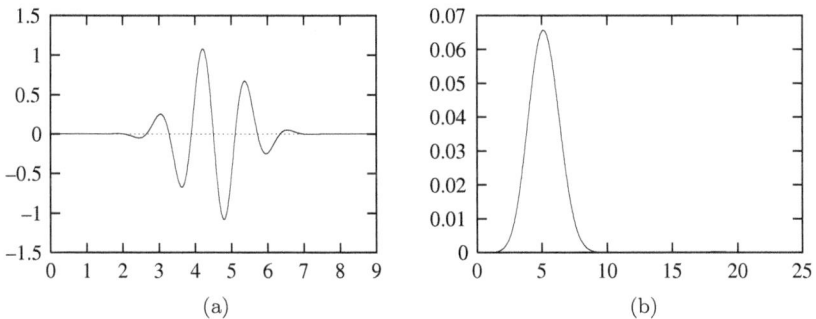

Fig. 4.36   Compactly supported spline wavelet $\psi_5$ of order 5 and its frequency spectrum: (a) Time domain; (b) frequency domain.

Table 4.12.    The centers, radii, and area of time window, frequency window, and time–frequency window of CS-spline wavelets of orders 2 to 10.

| CS-spline | Time Window Center | Radius | Freq. Window Center | Radius | Time–freq. area |
|---|---|---|---|---|---|
| 2 | 1.5 | 0.3897 | 2.6788 | 2.3717 | 3.6968 |
| 3 | 2.5 | 0.4728 | 2.5907 | 2.0004 | 3.7835 |
| 4 | 3.5 | 0.5419 | 2.5821 | 1.9410 | 4.2075 |
| 5 | 4.5 | 0.6030 | 2.5975 | 1.9162 | 4.6221 |
| 6 | 5.5 | 0.6585 | 2.5780 | 1.9004 | 5.0059 |
| 7 | 6.5 | 0.7097 | 2.5769 | 1.8891 | 5.3628 |
| 8 | 7.5 | 0.7575 | 2.5762 | 1.8805 | 5.6977 |
| 9 | 8.5 | 0.8024 | 2.5756 | 1.8738 | 6.0141 |
| 10 | 9.5 | 0.8449 | 2.5751 | 1.8684 | 6.3146 |

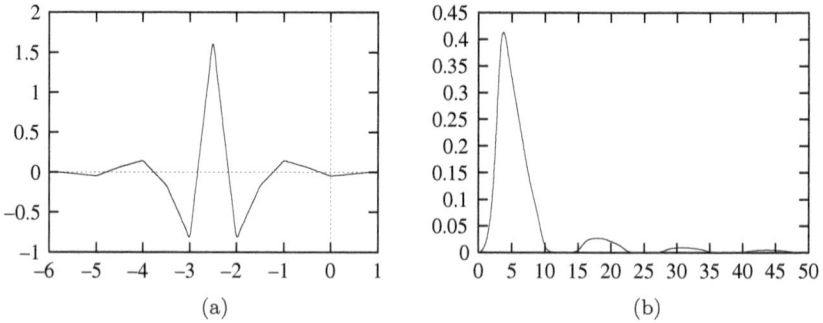

Fig. 4.37    The dual spline wavelet $\tilde{\psi}_2$ of order 2 and its frequency spectrum: (a) Time domain; (b) frequency domain.

and areas of the time windows, frequency windows, and time–frequency, windows of $\psi_m$, the compactly supported splines wavelets of orders 2–10, are given in Table 4.12.

Figures 4.37–4.39 show the dual wavelets $\tilde{\psi}_m$ of $\psi_m$ ($m = 2, 4, 5$) and their frequency property, respectively. The centers, radii and areas of the time windows, frequency windows, and time–frequency windows of the dual wavelet $\tilde{\psi}_m$ of orders 2–10 are tabulated in Table 4.13.

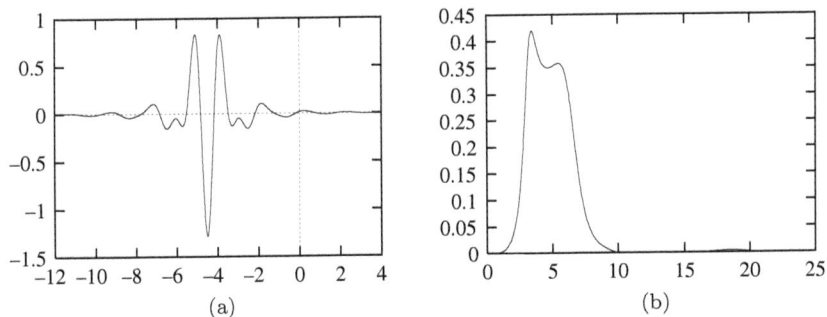

Fig. 4.38 The dual spline wavelet $\tilde{\psi}_4$ of order 4 and its frequency spectrum: (a) Time domain; (b) frequency domain.

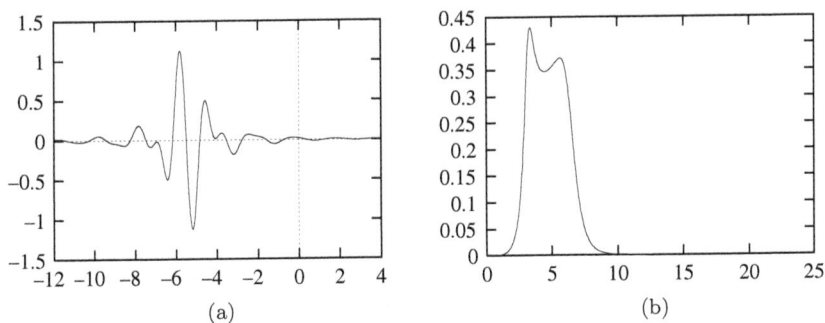

Fig. 4.39 The dual spline wavelet $\tilde{\psi}_5$ of order 5 and its frequency spectrum: (a) Time domain; (b) frequency domain.

Table 4.13. The centers, radii, and area of time window, frequency window, and time–frequency window of D-spline wavelets of orders 2–10.

| D-spline | Time Window Center | Time Window Radius | Freq. Window Center | Freq. Window Radius | Time–freq. area |
|---|---|---|---|---|---|
| 2 | −2.5 | 0.4554 | 2.3528 | 2.1734 | 3.9590 |
| 3 | −3.5 | 0.5258 | 2.3073 | 2.1734 | 3.9202 |
| 4 | −4.5 | 0.6164 | 2.3138 | 1.8301 | 4.5124 |
| 5 | −5.5 | 0.7088 | 2.3235 | 1.8239 | 5.1709 |
| 6 | −6.5 | 0.7981 | 2.3326 | 1.8829 | 6.0112 |
| 7 | −7.5 | 0.8816 | 2.3409 | 1.8230 | 6.4281 |
| 8 | −8.5 | 0.9583 | 2.3487 | 1.8233 | 6.9890 |
| 9 | −9.5 | 1.0220 | 2.3553 | 1.8224 | 7.4498 |
| 10 | −10.5 | 1.0900 | 2.3609 | 1.8207 | 7.9382 |

# Chapter 5

# Basic Principle of Deep Learning

## 5.1 Deep Learning

The term "deep learning" (DL) was first introduced to machine learning (ML) in 1986 and later used for artificial neural networks (ANNs) in 2000 [Schmidhuber, 2015]. It is a machine learning technology based on artificial neural networks. DL in "deep" refers to the use of ANN in multilayer perceptrons. The motivation for studying DL is to establish a neural network that simulates the human brain for analysis and learning, which imitates the mechanism of the human brain to interpret data, such as images, sounds, and text. The method is proposed to be closer to the initial goal of artificial intelligence. DL methods are composed of multiple layers to learn features of data with multiple levels of abstraction [LeCun et al., 2015]. DL approaches allow computers to learn complicated concepts by building them out of simpler ones [Goodfellow et al., 2016]. For ANNs, DL also known as hierarchical learning is [Deng et al., 2014] about assigning credits in many computational stages accurately, to transform the aggregate activation of the network [Schmidhuber, 2015]. To learn complicated functions, deep architectures are used with multiple levels of abstractions, i.e., nonlinear operations, e.g., ANNs with many hidden layers [Bengio et al., 2009]. To sum it up accurately, DL is a sub-field of ML, which uses many levels of nonlinear information processing and abstraction, for supervised or unsupervised feature learning and representation, classification, and pattern recognition [Deng et al., 2014].

Conventional ML techniques were limited in their ability to process natural data in their raw form. For decades, constructing a pattern-recognition or ML system required careful engineering and considerable domain expertise to design a feature extractor that transformed the raw data (such as the pixel values of an image) into a suitable internal representation or feature vector from which the learning sub-system, often a classifier, could detect or classify patterns in the input [LeCun *et al.*, 2015]. The quality of features has a crucial impact on generalization performance, and it is not easy for human experts to design good features. This can be avoided by automatically learning good features through a general-purpose learning procedure. This is the key advantage of deep learning. Representation learning is a set of methods that allows a machine to be fed with raw data and to automatically discover the representations needed for detection or classification. DL methods are representation-learning methods with multiple levels of representation, obtained by composing simple but nonlinear modules that each transform the representation at one level (starting with the raw input) into a representation at a higher, slightly more abstract level. With the composition of enough such transformations, very complex functions can be learned. For classification tasks, higher layers of representation amplify aspects of the input that are important for discrimination and suppress irrelevant variations.

DL allows computational models that are composed of multiple processing layers to learn representations of data with multiple levels of abstraction and discovers intricate structures in large datasets by using the backpropagation (BP) algorithm to indicate how a machine should change its internal parameters that are used to compute the representation in each layer from the representation in the previous layer. These methods have dramatically improved the state of the art in speech recognition, visual object recognition, object detection, and many other domains, such as drug discovery and genomics. Convolutional neural networks (CNNs) have brought about breakthroughs in processing images, video, speech, and audio, whereas recurrent neural networks (RNNs) have shone light on sequential data, such as text and speech. In addition, DL methods such as autoencoder [Deng *et al.*, 2014] , region-based CNNs [Girshick *et al.*, 2014] (R-CNN), residual network [He *et al.*, 2016] (ResNets), memory networks [Weston *et al.*, 2014], and long short-term memory networks [Hochreiter and Schmidhuber, 1997] (LSTM) have made breakthroughs in different application scenarios. Next, this chapter introduces several mainstream DL models.

## 5.2 DL Models

### 5.2.1 *Autoencoders*

An autoencoder is a type of algorithm with the primary purpose of learning an "informative" representation of the data that can be used for different applications by learning to reconstruct a set of input observations well enough [Michelucci, 2022].

In order to gain a better understanding of autoencoders, it is necessary to refer to their typical architecture. The autoencoders' main components are three: an encoder, a latent feature representation, and a decoder. The encoder and decoder are simply functions, while with the name latent feature representation, one typically intends a tensor of real numbers. Generally speaking, people want the autoencoder to reconstruct the input well enough. Still, at the same time, it should create a latent representation that is useful and meaningful. For example, latent features on hand written digits could be the number of lines required to write each number or the angle of each line and how they connect [Michelucci, 2022]. Learning how to write numbers certainly does not require to learn the gray values of each pixel in the input image. Humans do not certainly acquire the skill of writing by filling pixels with gray values. While learning, people extract the essential information that will allow them to solve a problem (writing digits, for example). This latent representation (how to write each number) can then be very useful for various tasks (for instance, feature extraction that can be then used for classification or clustering) or simply understanding the essential features of a dataset.

In most typical architectures, the encoder and the decoder are neural networks since they can be easily trained with existing software libraries such as TensorFlow or PyTorch with backpropagation. In general, the encoder can be written as a function $g$ that will depend on some parameters:

$$h_i = g(x_i), \tag{5.1}$$

where $h_i \in R^q$ (the latent feature representation) is the output of the encoder block when we evaluate it on the input $x_i$. Note that we will have $g: R^n \to R^q$. The decoder (and the output of the network that we will indicate with $\tilde{x}_i$) can be written then as a second generic function $f$ of the latent features

$$\tilde{x}_i = f(h_i) = f(g(x_i)), \tag{5.2}$$

where $\tilde{x}_i = R^n$. Training an autoencoder simply means finding the functions $g(\cdot)$ and $f(\cdot)$ that satisfy

$$\arg\min_{f,g}\langle[\Delta(\mathbf{x}_i, f(g(\mathbf{x}_i)))]\rangle, \tag{5.3}$$

where $\Delta$ indicates a measure of how the input and the output of the autoencoder differ (basically the loss function will penalize the difference between input and output) and $\langle \cdot \rangle$ indicates the average over all observations. Depending on how one designs the autoencoder, it may be possible to find $f$ and $g$ so that the autoencoder learns to reconstruct the output perfectly, thus learning the identity function.

### 5.2.2 *RNN-LSTM*

For general-purpose sequence modeling, LSTM as a special RNN structure has proven stable and powerful for modeling long-range dependencies in various previous studies [Shi *et al.*, 2015]. The major innovation of LSTM is its memory cell $c_t$ which essentially acts as an accumulator of the state information. The cell is accessed, written, and cleared by several self-parameterized controlling gates. Every time a new input comes, its information will be accumulated to the cell if the input gate $i_t$ is activated. Also, the past cell status $c_{t-1}$ could be "forgotten" in this process if the forget gate $f_t$ is on. Whether the latest cell output $c_t$ will be propagated to the final state $h_t$ is further controlled by the output gate $o_t$. One advantage of using the memory cell and gates to control information flow is that the gradient will be trapped in the cell (also known as constant error carousels) and be prevented from vanishing too quickly, which is a critical problem for the vanilla RNN model. The key equations are shown in the following where 'o' denotes the Hadamard product:

$$i_t = \sigma(W_{xi}x_t + W_{hi}h_{t-1} + W_{ci} \circ c_{t-1} + b_i), \tag{5.4}$$

$$f_t = \sigma(W_{xf}x_t + W_{hf}h_{t-1} + W_{cf} \circ c_{t-1} + b_f), \tag{5.5}$$

$$c_t = f_t \circ c_{t-1} + i_t \circ \tanh(W_{xc}x_t + W_{hc}h_{t-1} + b_c), \tag{5.6}$$

$$o_t = \sigma(W_{xo}x_t + W_{ho}h_{t-1} + W_{co} \circ c_t + b_o), \tag{5.7}$$

$$h_t = o_t \circ \tanh(c_t). \tag{5.8}$$

Multiple LSTMs can be stacked and temporally concatenated to form more complex structures. Such models have been applied to solve many real-life sequence modeling problems.

### 5.2.3 *RNN-GRU*

The gated recurrent unit (GRU) is another variant of the RNN architecture that addresses the short-term memory issue and offers a simpler structure compared to LSTM. GRU combines the input gate and forget gate of LSTM into a single update gate, resulting in a more streamlined design [Shiri *et al.*, 2023]. Unlike LSTM, GRU does not include a separate cell state.

A GRU unit consists of three main components: an update gate, a reset gate, and the current memory content. These gates enable the GRU to selectively update and utilize information from previous time steps, allowing it to capture long-term dependencies in sequences [Shiri *et al.*, 2023].

The update gate Eq. (5.9) determines how much of the past information should be retained and combined with the current input at a specific time step. It is computed based on the concatenation of the previous hidden state $h_{t-1}$ and the current input $x_t$, followed by a linear transformation and a sigmoid activation function:

$$z_t = \sigma(W_z[h_{t-1}, x_t] + b_z). \tag{5.9}$$

The reset gate Eq. (5.10) decides how much of the past information should be forgotten. It is computed in a similar manner to the update gate using the concatenation of the previous hidden state and the current input:

$$r_t = \sigma(W_r[h_{t-1}, x_t] + b_r). \tag{5.10}$$

The current memory content Eq. (5.11) is calculated based on the reset gate and the concatenation of the transformed previous hidden state and the current input. The result is passed through a hyperbolic tangent activation function to produce the candidate activation:

$$\tilde{h}_t = \tanh(W_h[r_t h_{t-1}, x_t]). \tag{5.11}$$

Finally, the final memory state $h_t$ is determined by a combination of the previous hidden state and the candidate activation Eq. (5.12). The update gate determines the balance between the previous hidden state and the candidate activation. Additionally, an output gate $o_t$ can be introduced to control the information flow from the current memory content to the output Eq. (5.13). The output gate is computed using the current memory state $h_t$ and is typically followed by an activation function, such as the sigmoid function:

$$h_t = (1 - z_t)h_{t-1} + z_t \tilde{h}_t, \tag{5.12}$$

$$o_t = \sigma_o(W_o h_t + b_o), \tag{5.13}$$

where the weight matrix of the output layer is $W_o$ and the bias vector of the output layer is $b_o$. GRU offers a simpler alternative to LSTM with fewer tensor operations, allowing for faster training.

### 5.2.4 *GAN*

Inspired by the zero-sum game in game theory, generative adversarial networks (GANs) treat the generation problem as an adversarial game between the generator and the discriminator [Zhou *et al.*, 2023].

In the standard GAN model, multilayer perceptron (MLP) structure is consisted of generator and discriminator. The purpose of generator is to learn the distribution of the real samples $x$ and generate a generated sample $G(z)$ with a high similarity with the real sample's distribution. The input of generator is a random noise $z$ obeying some simple samples distribution, such as Gaussian distribution. The output of generator is $G(z)$ with the same dimensions as real samples $x$.

Discriminator is a binary classifier with input $x$ or $G(z)$ and output discrimination results (real or fake), which is used to calculate the objective function and update the network weights by BP algorithm. The purpose of discriminator is to discriminate $x$ from $G(z)$ as accurately as possible. Discriminator expects the discrimination result to be real (Label $= 1$) when its input is $x$. And discriminator expects the discrimination result to be fake (Label $= 0$) when its input is $G(z)$, but generator expects to be able to "cheat" discriminator this time [Zhou *et al.*, 2023]. There is competition and adversarial relationship between generator and discriminator.

The training process of GAN is the binary maximin game process between generator and discriminator. The objective function is defined as

$$\min_{G} \max_{D} V(G, D) = E_{x \sim P_{\text{data}}(x)} \left[\log D(x)\right] + E_{z \sim P_z(z)} \left[\log \left(1 - D(G(z))\right)\right],$$
(5.14)

where $V(G, D)$ is a binary cross-entropy function, $E(*)$ represents the expected value of the samples distribution function, $x$ represents the real samples, $P_{\text{data}}(x)$ represents the real samples distribution, $z$ represents the random noise which input into generator, $P_z(z)$ is a random noise distribution, $G(z)$ represents the generated samples of generator, and $D(x)$ is a probability value that represents the probability that $x$ came from a real sample rather than a generated sample. $D(G(z))$ is also a probability value that represents the probability that discriminator will discriminate $G(z)$ to be a real sample.

The training process of GAN is an alternate iterative training of these two models, with one of the models being fixed during training and then updating the parameters of the other model. First, discriminator learns the distribution of $x$. When discriminator has some knowledge of the distribution of $x$, discriminator is used to discriminate the real fake of $G(z)$. Then, the ability of generator to generate samples is improved during the discrimination process of discriminator; the ability of discriminator to distinguish is increased in the processing of learning the distribution of $x$. Through continuous adversarial training, the probability value of discriminator to discriminate real-fake samples is maximized and the similarity degree between $G(z)$ and $x$ is maximized. Eventually, generator and discriminator reach the state of Nash equilibrium. At this time, it can be assumed that generator learns the distribution of the real samples [Zhou et al., 2023].

## 5.2.5 *Transformer*

The transformer was first designed in natural language processing (NLP), which mainly consists of multi head attention (MHA) and position-wise feed forward networks [Xiao et al., 2023].

In NLP, the transformer structure is composed of an encoder and a decoder, MHA is an essential part of encoder, and self-attention is the core of MHA. In addition, the MLP, normalization, skip connections, and addition operation are also important components in optimizing the transformer architecture.

The multi headed attention module is a module in which multiple attention modules learn different aspects of attention in different sub-spaces [Xin et al., 2023]. The matrix $X$ consisting of vectors was first obtained and then mapped to different sub-spaces by means of the learnable matrices $W^Q$, $W^K$, $W^V$ to obtain matrices query ($Q$), key ($K$), value ($V$), respectively. In the encoder, $Q$, $K$, and $V$ are equal, but it is worth noting that $K$ and $V$ in the second layer of the decoder come from the encoder, and $Q$ is the output of the first layer of the decoder result. $Q$, $K$, and $V$ can be written as

$$Q = XW^Q, \quad K = XW^K, \quad V = XW^V, \tag{5.15}$$

where the matrices $W^Q = R^{d_{\text{model}} \times d_k}$, $W^K = R^{d_{\text{model}} \times d_k}$, $W^V = R^{d_{\text{model}} \times d_v}$. Matrix $Q$ and the transpose of matrix $K$ are used for point multiplication, and the similarity probability is obtained through the Softmax function after scaling $\sqrt{d_k}$ times, where $d_k$ is the dimension of the matrix $K$.

Finally, multiply with $V$ to get the attention weight matrix. The attention mechanism is executed in each head. The formula is as follows:

$$Head_i = Attention(Q, K, V) = softmax\left(\frac{QK^T}{\sqrt{d_k}}\right)V, \quad (5.16)$$

$$MHA(Q, K, V) = Concat(Head_1, \ldots, Head_i)W^O. \quad (5.17)$$

Finally, project them through the learnable weights $W^O = R^{hd_v \times d_{\text{model}}}$ to obtain the final feature representation. The $i$ denotes the number of heads.

The encoder mentioned above encodes the input using a structure that combines the self-attention mechanism with the feed forward neural network, and the same structure is used in the decoder. The difference is that after the self-attention mechanism, the output of self-attention is calculated again with the output of the decoder module, and then the feed forward neural network module is entered.

## 5.3  Integration of Deep Learning with Wavelet Transform

Wavelet and convolutional network are two widely used methods in the field of image processing, which represent the processing methods of frequency domain and space domain, respectively. Wavelet transform is a multi resolution analysis method, which can decompose and reconstruct the signal on different scales, so as to extract the local and global features of the signal. Wavelet transform has many advantages, such as good time–frequency locality, adjustable resolution, and adaptive basis function, so it has a good performance in image denoising, compression, segmentation, and other tasks. Convolutional network is a deep learning model that can extract abstract features of images through multi layer convolutional layers and pooling layers, and implement tasks such as classification or regression through fully connected layers. Convolutional networks have many advantages, such as powerful expressive capabilities, efficient computing methods, and flexible network structures, so they perform well in tasks, such as image recognition, detection, and generation. The combination of wavelet and convolutional network is an idea that combines the advantages of frequency domain and space domain, aiming to improve the effect and efficiency of image processing. At present, there are two main ways of combining: one is to use wavelet transform as the activation function or filter of the convolutional network, so as to enhance the nonlinear ability and sparsity of the convolutional network. This approach can make the convolutional network more adaptable to changes in signals at different scales while reducing redundant information and noise interference. The other is to use wavelet transform as a pre-processing or post-processing step of the convolutional network, thereby reducing the amount of parameters and calculations of the convolutional network. In this way, the convolutional network can focus more on the main features of the signal while reducing memory consumption and running time.

### 5.3.1 *Fusion of Wavelet Transform and Deep Learning*

Sparse-based methods, including wavelets, have dominated the field of inverse problems for over two decades until neural networks took over [Osher *et al.*, 2005]. Among these methods, U-nets [Ronneberger *et al.*, 2015] have exhibited exceptional effectiveness. Their success can be attributed to three key factors: highly nonlinear processing, extensive learning facilitated by advanced optimization algorithms and powerful GPUs, and utilization of large-scale datasets for training. Determining the primary contributor to their performance is not straightforward. While deep learning inherently relies on multiple stages of nonlinearity, the utilization of training data for learning can also be leveraged by sparsity-based approaches. The study aims to explore the boundaries of sparsity by incorporating massive learning and large datasets, similar to U-nets, and subsequently comparing the results. A novel network architecture called "learnlets" is introduced to retain the advantageous properties of sparsity-based methods, such as exact reconstruction [Chen *et al.*, 2017] and strong generalization capabilities, while harnessing the learning power and computational efficiency of neural networks. The performance of both models is evaluated on image denoising tasks. The findings indicate that U-nets outperform learnlets in terms of image quality metrics within the given distribution, while learnlets demonstrate superior generalization properties.

#### 5.3.1.1 *Learnlets: The Model*

Let $x \in \mathcal{R}^{n \times n}$ be an image. Let $\tilde{x} = x + \varepsilon$ be the version of this image corrupted by an additive white Gaussian noise $\varepsilon \sim N(0, \sigma^2 I_{n \times n})$ whose variance $\sigma^2$ is assumed known. Let $\Sigma$ be a compact set of possible values for $\sigma$, $\sigma \sim \mathcal{U}(\Sigma)$ selected for this section. For a given number of scales $m$

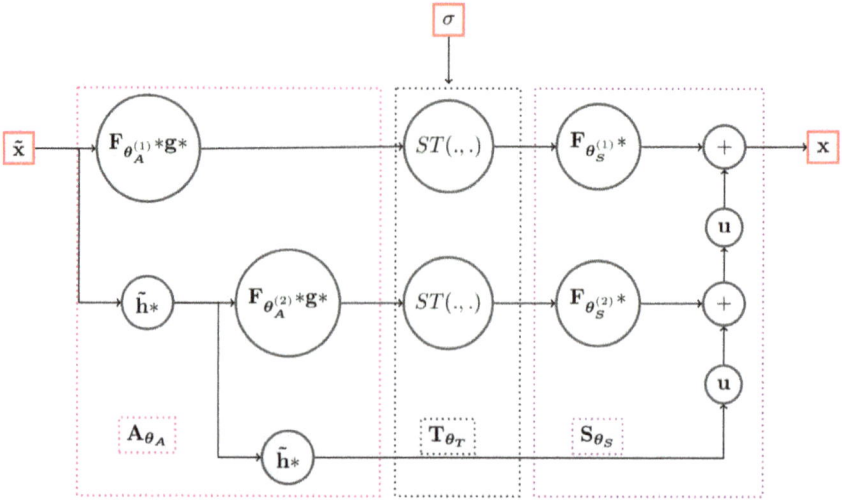

Fig. 5.1    Schematic representation of the learnlets model, with $m = 2$ scales. The red nodes are inputs/outputs. The lightly green nodes correspond to functions whose parameters can be learned. Note that the standard deviation of the noise before thresholding is not learned but rather estimated and is omitted in this diagram for clarity.

and a given set of parameters, $\theta = (\theta_S, \theta_T, \theta_A) \in \Theta_m$: This section defines the learnlets as a function $f_\theta$ from $(\mathcal{R}^{n \times n} \times \Sigma)$ to $\mathcal{R}^{n \times n}$:

$$f_\theta(\tilde{x}, \sigma) = S_{\theta_S} \left( T_{\theta_T} \left( A_{\theta_A} \left( \tilde{x} \right), \sigma \right) \right). \tag{5.18}$$

In the framework, the following components are defined: (1) $A_{\theta_A}$, the analysis function defined in analysis; (2) $T_{\theta_T}$, the thresholding function defined in Thresholding; (3) $S_{\theta_S}$, the synthesis function defined in Synthesis.

An illustration of the learnlets is given in Fig. 5.1.

## (1) **Analysis**

Intuitively, the analysis function can be perceived as the equivalent of the wavelet transform accompanied by a set of learned filters. This linear function is formally defined as follows:

$$A_{\theta_A}(\tilde{x}) = ((F_{\theta_{A^{(i)}}} * g(\tilde{h}^{i-1}(\tilde{x})))_{i=1}^{m}, \tilde{h}^m(\tilde{x})), \tag{5.19}$$

where $F_{\theta_{A^{(i)}}}$, is the filter bank at scale $i$, is the concatenation of all filters at this scale $F_{\theta_{A^{(i,j)}}}$. The convolutions are done without bias. $\theta_A{}^{(i)}$ are the $J_i$ convolution kernels all of the same square size $(k_A, k_A)$ (for now $J_i = J_m$). These filters act on the high-pass filtered versions of the image at each scale and output a corresponding feature map. Formally, $F_{\theta_{A^{(i)}}} * y = (F_{\theta_{A^{(i,j)}}} * y)_{j=1}^{J_i}$.

$\tilde{h} : y \mapsto \bar{u}(h * y)$, the low-pass filtering $(h)$ followed by a decimation $(\bar{u})$. The decimation is performed by taking one line out of 2 and one row out of 2, in line with the way it is done in wavelet transforms. $\tilde{h}^i$ is the application of $i$ times $\tilde{h}$.

$g$ the high-pass filtering is defined as $g(y) = y - u(\tilde{h}(y))$, with $u$ the upsampling operation performed with a bicubic interpolator.

For ease of manipulation, this section rewrites $A_{\theta_A}(\tilde{x}) = ((d_i)_{i=1}^m, c)$, with $d_i \in \mathcal{R}^{\frac{n}{2^{i-1}} \times \frac{n}{2^{i-1}} \times J_i}$ the detail coefficients and $c$ the coarse coefficients.

Note that low-pass and high-pass filters $(h, g)$ are fixed, and only $F_{\theta_A}^{(i)}$ filters are learned. As $g$ has a zero mean, all coefficients $d_i$ have by construction a zero mean. This wavelet property is fundamental to model the noise on the coefficients. Indeed, in the absence of signal, the coefficients follow a Gaussian distribution with a zero mean, and a standard $k\sigma$ thresholding can be applied, $\sigma$ being the noise standard deviation. With wavelets, $k$ would be chosen between 3 and 5 and would be a user parameter. In this setting, this $k$ value can be learned and can be different at each scale.

## (2) Thresholding

The nonlinearity function used for wavelet shrinkage is typically either a hard thresholding or a soft thresholding [Donoho, 1995]. The soft thresholding offers more stability. The thresholding function, in the case of a white Gaussian noise of variance $\sigma^2$, is defined as

$$T_{\theta_T}\left(((d_i)_{i=1}^m, c), \sigma\right) = \left(\left((t_{ij}(d_{ij}, \sigma))_{i=1}^{J_i}\right)_{i=1}^m, c\right), \qquad (5.20)$$

where $t_{ij}(d, \sigma) = \hat{\sigma}_{ij} ST\left(\frac{1}{\hat{\sigma}_{ij}} d_{ij}, \theta_T^{(ij)}\sigma\right)$, with

$d_{ij} \in \mathcal{R}^{\frac{n}{2^{i-1}} \times \frac{n}{2^{i-1}}}$ the output of the $j$th filter of the $i$th scale.

$\hat{\sigma}_{ij}$ is the estimated standard deviation of $d_{ij}$ when the input of the transform is set to be a white Gaussian noise of variance 1. This ensures the noise coming just before the thresholding is of variance approximately $\sigma$. The threshold is therefore truly $\theta_T^{(ij)}\sigma$.

$\theta_T^{(ij)}$ is the thresholding level applied at scale $i$ on the $j$th analysis filter. $ST(d, s)$ is the soft thresholding function applied pointwise on $d$ with threshold $s$: $ST(d, s) = \text{sign}(d) \max(|d| - s, 0)$.

It is important to note that, thanks to linearity of the analysis operator, the thresholding strategy can be very easily adapted to non-stationary Gaussian noise or to any other kind of noise, such as Poisson noise or a mixture of Gaussian and Poisson noise.

## (3) Synthesis

This section shows the synthesis function as the equivalent of the wavelet reconstruction operator, with learned filters. It is important to note that the synthesis function is linear. The synthesis function is defined recursively as

$$S_{\theta_S}((d_i)_{i=1}^m, c) = S_{\theta_S}^{(m-1)}((d_i)_{i=1}^{m-1}, u(c) + F_{\theta_{S(m)}} * d_m),\qquad(5.21)$$

where $S_{\emptyset}(\emptyset, c) = c$ and $F_{\theta_{S(i)}}$, is the filter bank at scale $i$, the concatenation of all filters at this scale $F_{\theta_{S(i,j)}}$, used for regrouping. The convolutions are done without bias and added all together. $\theta_S{}^{(i)}$ are the $J_i$ convolution kernels all of the same square size $(k_S, k_S)$. Formally, $F_{\theta_{S(m)}} * y = \sum_j F_{\theta_{S(m,j)}} * y$.

$u$, is the upsampling operation performed with a bicubic interpolator.

## (4) Constraints

Some constraints are used on the parameters of the learnlets to make them as close as possible to the wavelets and therefore make them understandable:

The analysis filters are forced to have a unit norm.
The thresholding levels are in $[0, 5]$.

## (5) Learning

The optimization problem is given as

$$\underset{\theta \in \Theta_m}{\operatorname{argmin}} E_{x,\sigma}[L_f(\theta)],\qquad(5.22)$$

where $L_f(\theta) = \|x - f_\theta(\tilde{x}, \sigma)\|_2^2$ and the expected value is computed empirically, via the empirical mean over a batch.

### 5.3.1.2 *Exact Reconstruction*

## (1) Learnlets

Exact reconstruction guarantees that if no noise is present, the signal will be perfectly reconstructed, without any error. This can be achieved using an analysis filter fixed as identity, for example, $F_{\theta_{A(i,1)}} = Id$. In particular, this section considers a single scale $i$, after the application of the $g$ filter. The operation carried out by the network, without thresholding, can be written as

$$x_{\text{out}}{}^{(i)} = \sum_{j=1}^{J_i} F_{\theta_{S(i,j)}} * F_{\theta_{A(i,j)}} * x_{\text{in}},\qquad(5.23)$$

where $N$ is the number of filters at that scale. Given the presence of $F_{\theta_{A(i,1)}} = Id$, the section proceeds to address the synthesis filter $F_{\theta_{S(i,1)}} =$

$Id - \sum_{j=2}^{N} F_{\theta_{S(i,j)}} * F_{\theta_{A(i,j)}}$ in order to rectify the situation. This trivially gives without thresholding, $x_{\text{out}} = x_{\text{in}}$. We implemented this constraint in the network, allowing us to learn a different thresholding level for this filter.

## (2) The General Case

In order to better understand the properties of exact reconstruction in the learnlets, the section studies whether it is possible to enforce it as well for black-box residual neural networks. A simple solution, given a known noise level $\sigma$, is to use the following general expression:

$$g_\theta(\tilde{\mathbf{x}}, \sigma) = \tilde{\mathbf{x}} - \sigma f_\theta(\tilde{\mathbf{x}}), \qquad (5.24)$$

where $f_\theta$ is the output of the network without exact reconstruction. It can be noted that when $\sigma$ tends to zero, then $g_\theta(\tilde{x}, \sigma) \to \tilde{x}$ and this can assure that the output will retrieve the input signal. It should be noted that this formulation might be unstable as it can amplify errors at high noise levels. This aspect is analyzed in following section.

### 5.3.1.3 *Data and Experiments*

The implementation was done in Python 3.6, using the TensorFlow 2.1 framework for model design. The training was done on the Jean Zay public supercomputer, using for each job a single GPU Nvidia Tesla V100 SXM2 with 32GB of RAM.

## (1) Data

The data used were the BSD500 dataset [Arbelaez *et al.*, 2010]. These data consist of natural images of sizes $481 \times 321$ and $321 \times 481$. The train and test subset of BSD500 were used as the training dataset. The validation subset of BSD500, containing the BSD68 [Martin *et al.*, 2001] images, was left out. Here BSD68 is used as the test dataset. This choice is motivated by the fact that many other denoising studies [Zhang *et al.*, 2017] use this dataset for comparison.

Moreover, in their experiments, Ramzi *et al.* [2023] consider the noise level to be known and do not address the task of evaluating this noise level or assessing the robustness of the different networks to errors in the noise level estimation.

## (2) Pre-processing

For training, patches of size $256 \times 256$ were randomly extracted on-the-fly. The images were then linearly mapped from $[0, 255]$ to the $[-0.5, 0.5]$

interval and converted from RGB to grayscale using the function provided by TensorFlow. In addition, data augmentation techniques such as random flipping and random $\theta$-degree ($\theta = 90°, 180°, 270°$) rotations were applied. Noise was then added by first drawing uniformly at random in the specified interval $\Sigma$ a noise level $\sigma$ and then generating a $256 \times 256$ white Gaussian noise patch $\varepsilon$ with this standard deviation. It is to be noted that during training, a single batch can feature different noise standard deviations.

At test time, the images were mirror-padded to a $352 \times 512$ size (or $512 \times 352$), in order to avoid shape mismatches when downsampling and upsampling, and the image quality metric was computed only on the original image shape. The test images were also corrupted by an additive white Gaussian noise (applied before the normalization to $[-0.5, 0.5]$) for various standard deviations $\sigma$ : $\{0.0001, 5, 15, 20, 25, 30, 50, 55, 60, 75, 85, 95, 100\}$. This allowed us to test the performance of Ramzi *et al.* [2023]'s method in different noise level settings.

## (3) Parameters

Unless specified otherwise, the learnlet parameters were chosen as follows:

$m = 5$ scales.

256 learnable analysis filters + 1 fixed analysis filter is just the identity. $F_{\theta_{A(i)}}$, of size $k_A = 11$.

$J_i = 257$ learnable synthesis filters, $F_{\theta_{S(i)}}$, of size $k_S = 13$.

The thresholding levels only depend on the scale, $\theta_T^{(ij)} = \theta_T^{(i)}$.

## (4) Evaluation metric

For the evaluation of the performance of the different models, Ramzi *et al.* [2023] utilized the peak signal-to-noise ratio (PSNR) metric. It is defined image-wise as the following (with images taken in the $[-0.5; 0.5]$ range):

$$PSNR(x, \hat{x}) = -10 \log_{10} \|x - \hat{x}\|_2^2. \tag{5.25}$$

For each test noise standard deviation $\sigma$, Ramzi *et al.* [2023] calculated the average PSNR of the denoised images across all BSD68 images.

## (5) Comparison with other methods

The work of Ramzi *et al.* [2023] compared the U-net and learnlets, both offering exact reconstruction, with algorithms that do not involve learning, specifically wavelet shrinkage. Using learning, the learnlets enhance their

Table 5.1. Runtimes of the different models for the denoising of one image.

| Model name | Wavelets | U-net 128 | Learnlets | U-net 64 |
|---|---|---|---|---|
| Denoising runtime in ms (std) | 274 (21) | 272 (18) | 106 (12) | 64 (1) |

decomposition power compared to the original wavelet model with no learning. For small noise levels, the U-net gets degraded performances compared with learnlets with exact reconstruction and wavelets. In this setting, the denoiser must act as the identity. Finally, it is observed that for unseen test noise levels, the performance of the U-net exhibits a slight drop, whereas the learnlets maintain relatively good performances.

Furthermore, it is evident from Table 5.1 that the learnlets exhibit the advantage of GPU implementation, resulting in faster execution compared to both wavelets and U-net 128.

### (6) Learnlets with exact reconstruction

As observed in Fig. 5.2, it is evident that learnlets, which exhibit exact reconstruction, present strong competition against classical methods across a broad spectrum of noise standard deviations. Figure 5.3 shows that the performance of the network with forced exact reconstruction is almost the same as the one without forced exact reconstruction (we only lose 0.1 dB at $\sigma = 30$, for example) on the majority of the test noise standard deviations. However, for low noise standard deviations, the network with forced exact reconstruction completely overpowers the other one. This is due to the fact that, at low noise standard deviations, for the $i$th scale, the term $x_{\text{out}}^{(i)}$ is practically the same as its thresholded version because the thresholds $\theta_T^{(ij)}\sigma$ are going to be low. Therefore, it is compensated in the corresponding synthesis filter used for exact reconstruction at that scale, $F_{\theta_{A^{(i,1)}}}$. This allows us to guarantee, in this case, no loss of information in the signal if it is clearly present.

### (7) Influence of the number of training samples

In a lot of computer vision (CV) problems, training data is scarce. It is reasonable to think that a small network would start performing better than a deeper one as fewer samples become available. To test this, three models were examined: two U-nets of different sizes (8 and 64) and learnlets without exact reconstruction. The first aspect that can be mentioned about Fig. 5.4

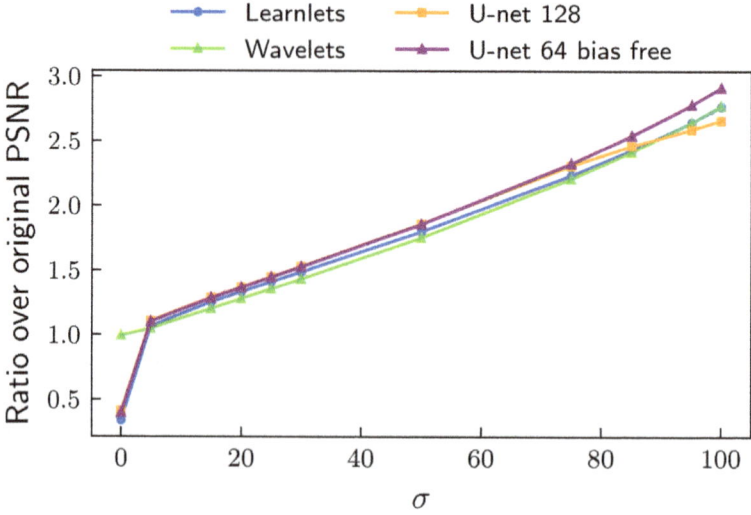

Fig. 5.2  Ratio of the denoised image PSNR compared to the original noisy image PSNR for different standard deviations of the noise added to the test images for all considered models. The train noise standard deviation range was [0; 55].

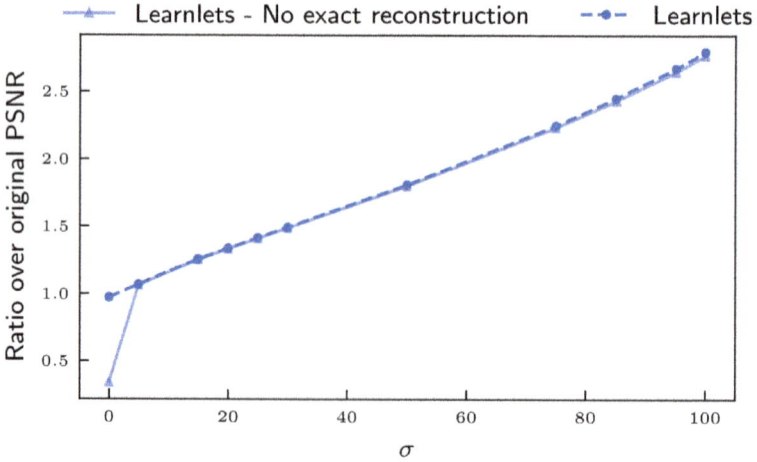

Fig. 5.3  Ratio of the denoised image PSNR compared to the original noisy image PSNR for different standard deviations of the noise added to the test images for learnlets with and without forcing exact reconstruction. The train noise standard deviation range was [0; 55]. The number of filters used was 64.

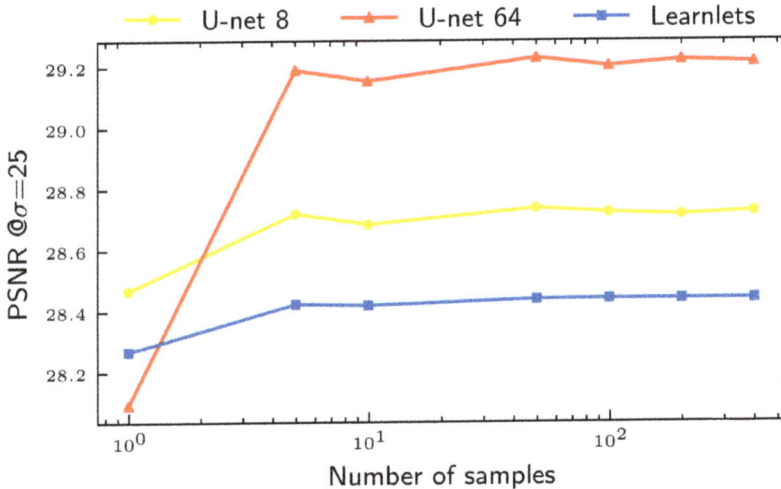

Fig. 5.4 The PSNR of the denoised image at $\sigma = 25$ added to the test images as a function of the number of training samples used during training. The train noise standard deviation range was $[0; 55]$.

is that for the three networks the PSNR does not vary significantly when reducing the original number of samples all the way down to 50. Despite learnlets overcoming U-nets with 64 base filters for the lowest number of training samples considered, they fail to outperform a U-net model with eight base filters. It can be inferred that a reduced number of parameters tends to improve the robustness of a given neural network to the number of samples. In other words, relatively few samples are required to obtain top performance.

(8) **Qualitative results**

Figure 5.5 shows that the learnlets suffer from some of the drawbacks of the wavelets like the creation of artifacts in the high-frequency parts of the image. However, the results are less blurred in comparison. Compared to the U-net, the learnlets are clearly suffering visually from a loss of contrast. This is a known effect of the soft thresholding which inherently biases the results. This could be improved by the use of reweighting [Candes *et al.*, 2008] to further approach the hard thresholding, which does not bias the results.

Original image

Noisy image

**Wavelets:** Image denoised

**Learnlets:** Image denoised

**U-net:** Image denoised

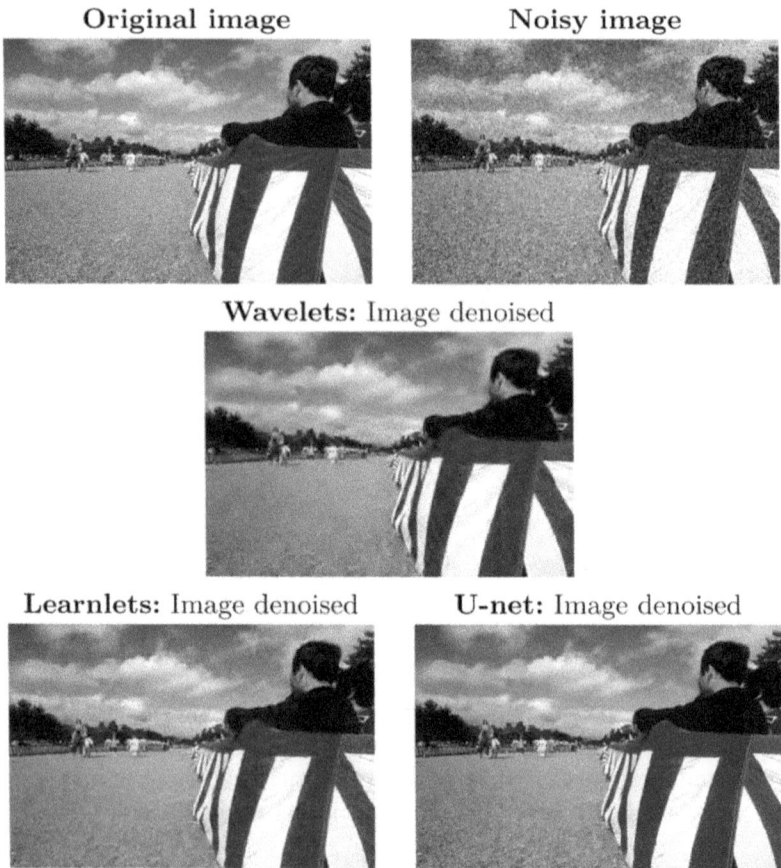

Fig. 5.5    Denoising results for a specific image in the BSD68 dataset. The noise standard deviation used was of 30.

### 5.3.2 *Deep Learning Techniques Applied in the Wavelet Domain*

This section introduces the possibility and potential advantages of learning the filters of CNNs for image analysis in the wavelet domain. Cotter and Kingsbury [2018] draw inspiration from Mallat's scattering transform [Mallat, 2012] and the concept of filtering in the Fourier domain. It is crucial to explore new learning spaces as they may offer inherent advantages not available in the pixel space. However, the scattering transform is limited in its ability to learn between scattering orders, and Fourier domain filtering is constrained by the large number of filter parameters required for localized filters. Instead, Cotter and Kingsbury [2018] propose filtering in the wavelet domain using learnable filters. The wavelet space allows for local, smooth filters with significantly fewer parameters, and the ability to learn provides flexibility.

This section introduces a novel layer that takes CNN activations into the wavelet space, learns parameters, and returns to the pixel space. This layer can easily be integrated into any neural network without affecting its structure. As part of this work, Cotter and Kingsbury [2018] demonstrate how to propagate gradients through a multirate system and present preliminary results.

### 5.3.2.1 *Method*

In a standard convolutional layer, an input with $C$ channels, $H$ rows, and $W$ columns is $X \in R^{C \times H \times W}$, which is then convolved with $F$ filters of spatial size $K - w \in R^{F \times C \times K \times K}$, giving $Y \in R^{F \times H \times W}$. In many systems, the first layer is typically a selection of bandpass filters, selecting edges with different orientations and center frequencies. In the wavelet space, this would be trivial — take a decomposition of each input channel and keep individual sub-bands (or equivalently, attenuate other bands), and then take the inverse wavelet transform. Figure 5.6 shows the frequency space for the $DTCWT$ and makes it clearer as to how this could be done practically for a two-scale transform. To attenuate all but say the 15° band at the first scale for the first input channel, the system would need to have $13C$ gains for the 13 sub-bands and $C$ input channels, $13C - 1$ of which would be zero and the remaining coefficient one. Instead of explicitly setting the gains, the gains can be randomly initialized, and backpropagation can be used to learn what they should be. This gives the system the power to learn more complex shapes rather than simple edges, as it allows the mixing of regions of the frequency space per input channel in an arbitrary way.

### (1) **Memory cost**

Instead of predefining the $w \in R^{F \times C \times K \times K}$ complex gains at the two scales and the real gain for the real lowpass filter, one considers learning them in the process as one optimizes the two-scale transform:

$$\left\{ g_1 \in C^{F \times C \times 6 \times 1 \times 1}, g_2 \in C^{F \times C \times 6 \times 1 \times 1}, g_{lp} \in R^{F \times C \times 1 \times 1} \right\}. \qquad (5.26)$$

Cotter and Kingsbury [2018] have set the spatial dimension to be $1 \times 1$ to show that this gain is identical to a $1 \times 1$ convolution over the complex

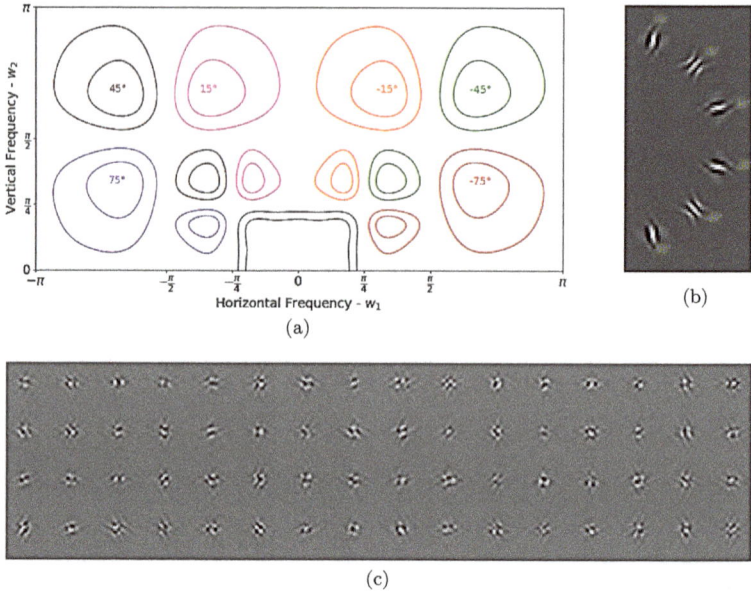

Fig. 5.6　(a) Contour plots at $-1\,\mathrm{dB}$ and $-3\,\mathrm{dB}$ showing the support in the Fourier domain of the six sub-bands of the $DTCWT$ at scales 1 and 2 and the scale 2 lowpass. (b) The pixel domain impulse responses for the second scale wavelets. (c) When $g_1$ and $g_{lp}$ are 0 and $g_2 \in C^{6 \times 1 \times 1}$, with each real and imaginary element drawn from $\mathcal{N}(0, 1)$. That is only information in the six sub-bands with $\frac{\pi}{4} < |w_1|, |w_2| < \frac{\pi}{2}$ 2 from (a) is passed through.

wavelet coefficients. If desired, larger spatial sizes can be learned to achieve more complex attenuation or magnification of the sub-bands. The number of wavelet scales can also be adjusted to be more or fewer than 2. Although the parameterization appears to have increased by a factor of 25 (13 sub-bands, with all but the lowpass being complex), each gain affects a sizable spatial region. Specifically, the effective size is approximately $5 \times 5$ pixels for the first scale and $15 \times 15$ pixels for the second scale.

## (2) Computational cost

A standard convolutional layer needs $K^2F$ multiplies per input pixel (of which there are $C \times H \times W$]). In comparison, the wavelet gain method does a set number of operations per pixel for the forward and inverse transforms and then applies gains on sub-sampled activations. For a 2-level DTCWT, the transform overhead is about 60 multiplies for both the forward and inverse transform. It is important to note that unlike the filtering operation, this does not scale with $F$. The learned gains in each sub-band do scale with the number of output channels but can have smaller spatial size (as they have larger effective sizes) as well as having fewer pixels to operate on (because of the decimation). The end result is that as $F$ and $C$ grow, the overhead of the $C$ forward and $F$ inverse transforms is outweighed by cost of $FC$ mixing processes, which should in turn be significantly less than the cost of $FC$ $K \times K$ standard convolutions for equivalent spatial sizes.

## (3) Examples

Figure 5.6(c) shows example impulse responses of the layer. These impulses were generated by randomly initializing both the real and imaginary parts of $g_2 \in C^{6 \times 1 \times 1}$ from $\mathcal{N}(0, 1)$ and $g_1$, $g_{lp}$ are set to 0. That is, each shape has 12 random variables. It is good to see that there is still a large degree of variability between shapes. Their experiments have shown that the distribution of the normalized cross-correlation between 512 of such randomly generated shapes matches the distribution for random vectors with roughly 11.5 degrees of freedom.

## (4) Forward propagation

Figure 5.7 shows the block diagram using Z-transforms for a single band of their system [Kingsbury, 2001]. To keep things simple for the rest of Section 5.2, the figure shown is for a 1D system; it is relatively straightforward to extend this to 2D [Selesnick *et al.*, 2005]. The complex analysis filter (taking us into the wavelet domain) is $P(z) = \frac{1}{2}(A(z) + jB(z))$ and the complex synthesis filter (returning us to the pixel domain) is

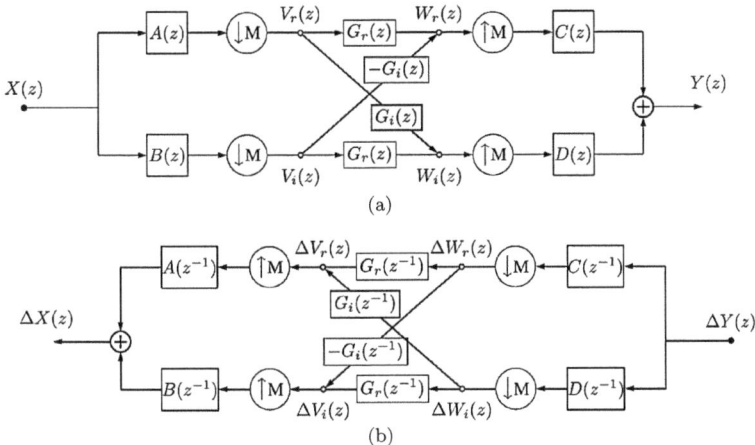

Fig. 5.7   Block diagram of a single band in the system using the Z-transform.

$Q(z) = \frac{1}{2}(C(z) - jD(z))$, where $A,B,C,D$ are real. If $G(z) = G_r(z) + jG_i(z) = 1$, then the end-to-end transfer function is (from Section 4 of the work of Kingsbury [2001])

$$\frac{Y(z)}{X(z)} = \frac{2}{M}(P(z)Q(z) + P^*(z)Q^*(z)), \tag{5.27}$$

where $P,Q$ have support only in the top half of the Fourier plane and $P^*$, $Q^*$ are $P$ and $P$ reflected in the horizontal frequency axis. Examples of $P(z)Q(z)$ for different sub-bands of a 2D *DTCWT* have spectra shown in Fig. 5.7(a); $P^*(z)Q^*(z)$ make up the missing half of the frequency space. Modifying this from the standard wavelet equations by adding the sub-band gains $G_r(z)$ and $G_i(z)$, the transfer function becomes

$$\frac{Y(z)}{X(z)} = \frac{2}{M}[G_r(z^M)(P(z)Q(z) + P^*(z)Q^*(z))$$

$$+ jG_i(z^M)(P(z)Q(z) - P^*(z)Q^*(z))]. \tag{5.28}$$

### (5) Backpropagation

They start with the commonly known property that for a convolutional block, the gradient with respect to the input is the gradient with respect to the output convolved with the time reverse of the filter. More formally, if $Y(z) = H(z)X(z)$,

$$\Delta X(z) = H(z^{-1})\Delta Y(z), \tag{5.29}$$

where $H(z^{-1})$ is the Z-transform of the time/space reverse of $H(z)$, with respect to the output, $\Delta Y(z) \triangleq \frac{\partial L}{\partial Y}(z)$ is the gradient of the loss, and $\Delta X(z) \triangleq \frac{\partial L}{\partial X}(z)$ is the gradient of the loss with respect to the input. If $H$ were complex, the first term would be $\bar{H}(1/\bar{z})$, but as each individual block in the *DTCWT* is purely real, we can use the simpler form. Assume that the work of Cotter and Kingsbury [2018] already has access to the quantity $\Delta Y(z)$ (this is the input to the backward pass). Figure 5.7(b) illustrates the backpropagation procedure. An interesting result is that the backward pass of an inverse wavelet transform is equivalent to doing a forward wavelet transform. Similarly, the backward pass of the forward transform is equivalent to doing the inverse transform. The weight update gradients are then calculated by finding $\Delta W(z) = \text{DTCWT}\{\Delta Y(z)\}$ and then convolving with the time reverse of the saved wavelet coefficients from the forward pass $-V(z)$:

$$\Delta G_r(z) = \Delta W_r(z)V_r(z^{-1}) + \Delta W_i(z)V_i(z^{-1}), \tag{5.30}$$

$$\Delta G_i(z) = -\Delta W_r(z)V_i(z^{-1}) + \Delta W_i(z)V_r(z^{-1}). \tag{5.31}$$

Table 5.2.   Comparison of LeNet with standard convolution to the methods described in this subsection which learns in the wavelet space (WaveLenet) on CIFAR-10 and CIFAR-100.

| | Train set size | 1000 | 2000 | 5000 | 10000 | 20000 | 50000 |
|---|---|---|---|---|---|---|---|
| CIFAR-10 | LeNet | 48.5 | 52.4 | 59.5 | 65.0 | 69.5 | 73.3 |
| | waveLeNet | 47.3 | 52.1 | 58.7 | 63.8 | 68.0 | 72.4 |
| CIFAR-100 | LeNet | 11.1 | 15.8 | 23.1 | 34.4 | 34.4 | 41.1 |
| | waveLeNet | 11.1 | 15.4 | 23.2 | 33.9 | 33.9 | 39.6 |

*Note*: Values reported are the average top-1 accuracy (%) rates for different train set sizes over 5 runs.

Unsurprisingly, the passthrough gradients have a similar form:

$$\Delta X(z) = \frac{2\Delta Y(z)}{M}[G_r(z^{-M})(PQ + P^*Q^*)$$
$$+ jG_i(z^{-M})(PQ - P^*Q^*)], \tag{5.32}$$

where Cotter and Kingsbury [2018] have dropped the $z$ terms on $P(z)$, $Q(z)$, $P^*(z)$, $Q^*(z)$ for brevity.

### 5.3.2.2  *Experiments and Preliminary Results*

To examine the effectiveness of our convolutional layer, Cotter and Kingsbury [2018] do is a simple experiment on CIFAR-10 and CIFAR-100. For simplicity, they compare the performance using a simple yet relatively effective convolutional architecture — LeNet [LeCun *et al.*, 1998]. LeNet has two convolutional layers of spatial size $5 \times 5$ followed by two fully connected layers and a Softmax final layer. Cotter and Kingsbury [2018] swap both these convolutional layers out for two of our proposed wavelet gain layers (keeping the ReLU between them). As CIFAR has very small spatial size, they only take a single-scale $DTCWT$. Therefore, each gain layer has six complex gains for the six sub-bands and a $3 \times 3$ real gain for the lowpass (a total of $21C$ parameters vs $25C$ for the original system). Cotter and Kingsbury [2018] train both networks for 200 epochs with Adam [Kingma and Ba, 2014] optimizer with a constant learning rate of $10^{-3}$ and a weight decay of $10^{-5}$. Table 5.2 shows the mean of the validation set accuracies for five runs. The different columns represent undersampled training set sizes (with 50,000 being the full training set). When undersampling, they keep the samples per class constant. This system performs only very slightly worse than the standard convolutional layer.

PART 3

# Chapter 6

# Step Edge Detection by Wavelet Transform

As mentioned in the previous chapters, in pattern recognition, an entire task can be divided into four phases: data acquisition, data preprocessing, feature extraction, and classification. In the first phase, the data acquisition, analog data from the physical world are acquired through a transducer and further digitized to discrete format suitable for computer processing. The physical variables, in the data acquisition phase, are converted into a set of measured data, which are then led as the input to the next phase, i.e., the data preprocessing phase. A major function of the data preprocessing phase is to modify the measured data obtained from the data acquisition phase so that those data are more suitable for further processing in the third phase (feature extraction). Many modifications are made in the data preprocessing phase. For example, some of them are listed in the following:

- gray-level histogram modification,
- smoothing and noise elimination,
- edge sharpening,
- boundary detection and contour tracing,
- thinning,
- segmentation,
- morphological processing,
- texture and object extraction from textural background,
- approximation of curves and surfaces.

In the above modifications, many items, such as noise, edges, boundaries, surfaces, textures, and curves, are of singularities in different patterns. Wavelet theory will play an important role to analyze and process such singularities. This chapter gives an example to show how the wavelet theory can be used to treat some singularities. Specifically, we study the applications of wavelet transform to edge detection.

## 6.1  Edge Detection with Local Maximal Modulus of Wavelet Transform

What is the edge of a signal? Intuitively, it is the transient component of the signal. In a 1D signal, for example, the speech signal, the edge corresponds to the abrupt change of the tune which creates a harsh noise to the ears of human. In a 2D signal, for example, an image, the edge refers to the sharp variation in color or gray. Actually, a signal is equivalent to a function, thus, the edge of the signal can be viewed as a component, whose value varies suddenly. In mathematics, it can be represented by a larger derivative on that point. Consequently, the extraction of the edge eventuates in finding the pixels with the large derivative.

In numerical applications, for instance, in pattern recognition, because the signal is in discrete form, the derivative cannot be calculated accurately. As an alternation, thus, the difference quotient can be used to approximate it. In this way, many effective and classical edge detection operators were established based on the difference quotient, for example, Robert operator, Sobel operator, Laplace operator, and so on. Unfortunately, all of these methods have a poor capability to reduce the noise due to the property of the noise itself, namely, it also is a catastrophe point and has a large value of amplitude in the signals.

As we have known, noise always distributes randomly. From the point of view of the statistics, the average value of noise is nearly constant in a certain area. In general, we can suppose the value of this constant is zero and take a weighted mean to the signal in this area. This action can be regarded as low-pass filtering. In this way, the noise will be eliminated considerably. In mathematics, this method corresponds to the smoothing of a function. The scheme of edge detection is a combination of the filtering and the derivation, in which the signal performs the filtering followed by the derivation. This idea was developed in the 1970s and 1980s. Thereafter, Marr pointed out that the best smoothing operator is the convolution of the original signal with Gaussian function. Essentially, the edge detection based on the wavelets developed in the 1990s is an extension of this method.

If we do not focus the detection on the location of the edge but the width, then only the local maximal point needs to be found out, without counting other large derivative points in this area. It corresponds to the skeleton of the edge and is effective to the isolated points (in 2D one, they are isolated in one direction). This is the basic idea of edge detection with local maximal modulus of wavelet transform.

## (1) One-dimensional signal

Let $f(x)$ be an original signal, and $\theta(x)$ be a smoothing operator. Denote

$$\theta_s(x) := \frac{1}{s}\theta\left(\frac{x}{s}\right),$$

where $s > 0$ indicates the smoothing scale. To smooth $f(x)$, we take the convolution for $f(x)$:

$$f * \theta_s(x) = \int_{\mathbb{R}} f(x - t)\theta_s(t)dt.$$

The purpose of smoothing $f(x)$ is to reduce not the noise but the edge. For this reason, $\theta(x)$ should be localized, such that $f*\theta_s(x) \sim f(x)$ when $s$ is small enough. This impliess that the smoothed signal, which was processed by a little scale, is almost the same as the original one. This conclusion can be easily proved mathematically, when $\theta(x)$ is localized and $\int_{\mathbb{R}} \theta(x)dx = 1$ (assume that the original signal is continuous).

In fact, the edge detection always conflicts with the noise reduction. The more the edge information is extracted, the more the noise will be brought out, and vice versa. In other words, the smoothing should not be too strong, although it is necessary to smooth the original signal. An example of such smoothing function is shown in Fig. 6.1, which meets the above requirement.

Suppose $f(x)$ is smoothed with $\theta(x)$ at first. Its derivative is now computed as follows:

$$\frac{d}{dx}(f * \theta_s)(x) = \int_{\mathbb{R}} f(x - t)(\theta_s)'(t)dt = \int_{\mathbb{R}} f(x - t)(\theta_s)'(t)dt. \qquad (6.1)$$

The edge pixel can be obtained by detecting the point, which possesses the local maximum of the absolute derivative.

Since

$$\int_{\mathbb{R}} \theta'(x)dx = \theta(+\infty) - \theta(-\infty) = 0.$$

Thus, $\theta'(x)$ is a wavelet function, and the right-hand side of Eq. (6.1) is $\frac{1}{s}$ times its corresponding wavelet transform.

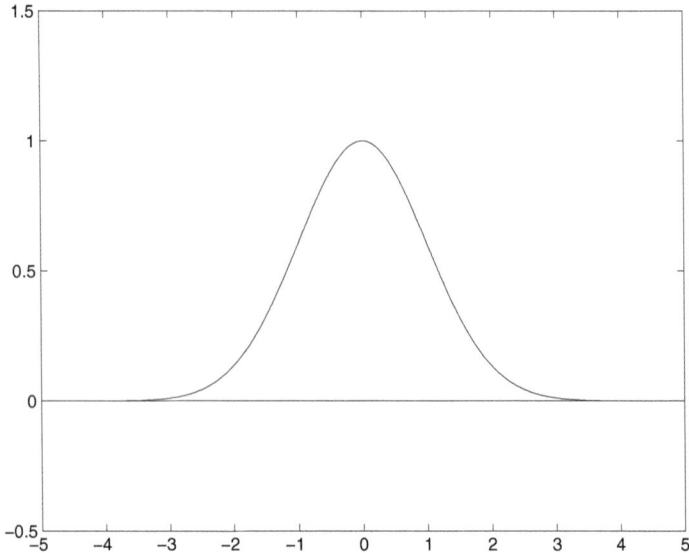

Fig. 6.1   An example of the smoothing function: Gaussian function.

Let $\psi(x) = \theta'(x)$, we have

$$W_s^\psi f(x) = \int_{\mathbb{R}} f(x-t)\psi_s(t)dt = s\frac{d}{dx}(f * \theta_s)(x).$$

This method is referred to as edge detection with local maximal modulus of wavelet transform.

It is known that the senses of audition and vision of human are limited. Hence, it is not necessary to extract all edge pixels, because some are too weak, so that they cannot reach the senses of human beings. In practice, a threshold $T > 0$ is used to decide which point will be extracted as an edge pixel that must fulfill the following two conditions:

(1)  $W_s^\psi f(x_0) \geq T$.
(2)  $|W_s^\psi f(x)|$ take the local maximum at $x_0$.

To extract the edges of a signal $f(x)$, it is required to compute its wavelet transform $\{W_s^\psi f(x)\}$. A discrete input signal $\{f(0), f(1), \ldots, f(n)\}$ can be viewed as the result of the A/D transformation from an analog signal. When applying wavelet transform to a digital signal, the discrete signal should be changed to a continuous analog signal by a suitable interpolation. Thereafter, its wavelet transform can be calculated by the following formula:

$$W_s^\psi f(x) = \int_{\mathbb{R}} f(x - t)\psi_s(t)dt.$$

After sampling the transform result, we can obtain the discrete data

$$\{W_s^\psi f(0), W_s^\psi f(1), \ldots, W_s^\psi f(n)\}.$$

Generally, different results will be achieved by different fitting methods.

It should be mentioned that the well-known zero-crossing technique for edge detection is almost the same as that with local maximal modulus of wavelet transform in principle. The former refers to the calculation of the zero-points of the second-order derivative of the smoothing function $f * \theta_s(x)$, whereas, the latter focus on the computation of the points, which possess local maxima of first-order derivative of $f * \theta_s(x)$. In mathematics, a point with local maximum of first-order derivative must be the zero-point of second-order derivative. For this reason, these two methods are the same in most cases. We should note that a zero-point of second-order derivative might not be the local maximal point of first-order derivative, and it maybe refer to the minimal point of first-order derivative of $f * \theta_s(x)$ in some cases. It implies that the pixels extracted by the zero-crossing method are probably of the false edge. From this point, the edge detection method with local maximal modulus of wavelet transform is better than that of the zero-crossing one. In addition, the first-order derivative of the smoothing function $f * \theta_s(x)$ can also be obtained by the method with local maximal modulus of wavelet transform. It has been discussed in detail by Canny [1986].

As the discrete value of wavelet transform, $\{W_s^\psi f(0), W_s^\psi f(1), \ldots, W_s^\psi f(n)\}$, has been achieved, the next step is to compute the local maximal modulus of the wavelet transform. In practice, $x = m$ is a point with local maximal modulus, if it meets the following two conditions:

(1) $|W_s^\psi f(m)| \geq |W_s^\psi f(m-1)|$, $|W_s^\psi f(m)| \geq |W_s^\psi f(m+1)|$.
(2) $|W_s^\psi f(m)| > |W_s^\psi f(m-1)|$ or $|W_s^\psi f(m)| > |W_s^\psi f(m+1)|$.

The algorithm for finding the local maximal modulus of the wavelet transform involves the following steps:

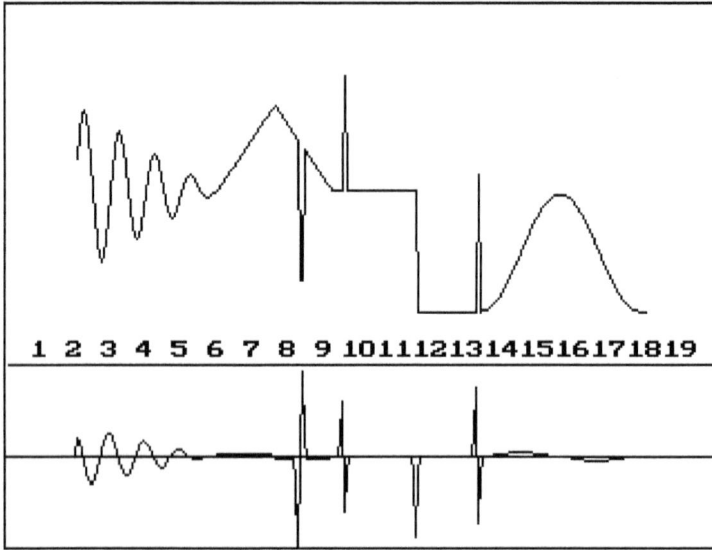

Fig. 6.2    Upper part: The original image; lower part: the result of wavelet transform.

---

**Algorithm 6.1.** Given an input discrete signal $\{f(0), f(1), \ldots, f(n)\}$, perform the following:

**Step 1:** Compute its wavelet transform $\{W_s^\psi f(0), \ W_s^\psi f(1), \ldots, W_s^\psi f(n)\}$.

**Step 2:** Take a threshold $T > 0$, for $m = 0, 1, \ldots, n$, if the following conditions satisfy:

(1) $|W_s^\psi f(m)| \geq T$,

(2) $|W_s^\psi f(m)| \geq |W_s^\psi f(m-1)|$, $|W_s^\psi f(m)| \geq |W_s^\psi f(m+1)|$,

(3) $|W_s^\psi f(m)| > |W_s^\psi f(m-1)|$, $|W_s^\psi f(m)| > |W_s^\psi f(m+1)|$, then $x = m$ is an edge pixel.

---

An example of the edge detection with the local maximal modulus of wavelet transform is graphically illustrated in Fig. 6.2.

### (2) Two-dimensional signal

We continue our discussion and extend our effort to 2D signals. For a 2D signal, for example, an image, its analysis by the wavelet theory and the algorithm can be established in a similar way to that of 1D one. In 2D signal, the input image $f(x, y)$ and smoothing function $\theta(x, y)$ are 2-variable functions. Meanwhile, $\theta(x, y)$ should have good locality and satisfy two

$$\int_{\mathbb{R}} \int_{\mathbb{R}} \theta(x, y) dx dy = 1.$$

Now, function $f(x, y)$ can be smoothed, and we shall describe this by writing $(f * \theta_s)(x, y)$:

$$(f * \theta_s)(x, y) := \int_{\mathbb{R}} \int_{\mathbb{R}} f(x - u, y - v) \theta_s(u, v) du dv,$$

where $\theta_x(u, v) := \frac{1}{s^2} \theta(\frac{u}{s}, \frac{v}{s})$ and $s > 0$ stands for a smoothing scale.

When we calculate the derivative of a 2D function, its orientation has to be considered. At each edge pixel, it reaches local maximum along the gradient direction and can be represented by

$$grad(f * \theta_s)(x, y) = \mathbf{i} \frac{\partial}{\partial x}(f * \theta_s)(x, y) + \mathbf{j} \frac{\partial}{\partial y}(f * \theta_s)(x, y),$$

where $\mathbf{i}$ and $\mathbf{j}$ correspond to $x$-axis and $y$-axis, respectively.

Now, we concentrate our discussion on the basic steps, in which, the edge detection with local maximal modulus of wavelet transform can be obtained, namely; first of all, the pixels will be extracted, such that the following modulus will reach its local maximum along the gradient direction

$$|grad(f * \theta_s)(x, y)| = \sqrt{\left|\frac{\partial}{\partial x}(f * \theta_s)(x, y)\right|^2 + \left|\frac{\partial}{\partial y}(f * \theta_s)(x, y)\right|^2},$$

and these pixels will become the components of the edge.

We can write

$$\psi^1(x, y) := \frac{\partial \theta}{\partial x}(x, y), \quad \psi^2(x, y) := \frac{\partial \theta}{\partial y}(x, y).$$

When $\theta(x, y)$ has good locality, i.e., it necessarily satisfies the conditions

$$\int_{\mathbb{R}} \int_{\mathbb{R}} \psi^1(x, y) dx dy = \int_{\mathbb{R}} [\theta(+\infty, y) - \theta(-\infty, y)] dy = 0,$$

$$\int_{\mathbb{R}} \int_{\mathbb{R}} \psi^2(x, y) dx dy = \int_{\mathbb{R}} [\theta(x, +\infty) - \theta(x, -\infty)] dx = 0,$$

then $\psi^1(x, y)$ and $\psi^2(x, y)$ become 2D wavelets.

It is easy to know that

$$|grad(f * \theta_s)(x, y)| = \frac{1}{s} \sqrt{|f * \psi_s^1)(x, y)|^2 + |f * \psi_s^2)(x, y)|^2}$$

$$= \frac{1}{s} \sqrt{|W_s^{\psi^1} f(x, y)|^2 + |W_s^{\psi^2} f(x, y)|^2},$$

where

$$W_s^{\psi^i} f(x,y) := (f * \psi^i)(x,y) = \int_{\mathbb{R}} \int_{\mathbb{R}} f(x-u, y-v) \psi_s^i(u,v) du dv \quad (i = 1, 2)$$

stand for the corresponding wavelet transforms of $\psi^1$ and $\psi^2$, respectively. The modulus of wavelet transform of $f(x,y)$ can be defined by

$$M_s f(x,y) := \sqrt{|W_s^{\psi^1} f(x,y)|^2 + |W_s^{\psi^2} f(x,y)|^2}.$$

It is clear that

$$|grad(f * \theta_s)(x,y)| = \frac{1}{s} M_s f(x,y).$$

Consequently, the calculation of the local maximal modulus of the derivative of a smoothing function along the gradient direction is equivalent to the computation of the local maximal modulus of wavelet transform.

Similar to the case of 1D, many methods can be used to compute the wavelet transform of a 2D discrete signal. The next task is to determine the gradient direction of a digital signal as well as its local maximum along this direction.

The gradient direction of a function can be accurately counted in mathematics. However, by contrast, it is difficult to express exactly if it is in discrete form. Fortunately, for the data obtained by equi-spaced sampling, only eight adjacent pixels are around a point. Hence, only the nearest eight points will be taken into account. It is said that the discrete image has eight gradient directions. Therefore, a plane can be divided into eight sectors. A gradient direction is defined by

$$\alpha_s := \arctan\left( \frac{\partial(f * \theta_s)(x,y)}{\partial y} \Big/ \frac{\partial(f * \theta_s)(x,y)}{\partial x} \right), \quad s = 1, 2, \ldots, 8.$$

When $\alpha$ falls into a sector, it will be quantified to a certain vector, which is represented by a center line of that sector. This can be found in Fig. 6.3, where the arrow shows a vector, which indicates a direction of the gradient.

Along this direction, a local maximal modulus can be achieved. The effect of any opposite gradient directions is the same. Thus, only four codes are needed to represent the gradient directions. For example, numbers 0, 1, 2, 3 can be used to code these four different directions. The tangent of each direction, $\tan \alpha_s$, falls into one of the following intervals:

$$[-1 - \sqrt{2}, 1 - \sqrt{2}), \quad [1 - \sqrt{2}, \sqrt{2} - 1),$$
$$[\sqrt{2} - 1, \sqrt{2} + 1), \quad [\sqrt{2} + 1, +\infty) \cup (-\infty, -1 - \sqrt{2}).$$

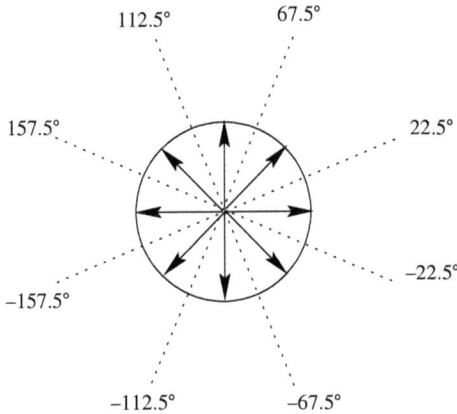

Fig. 6.3   The eight gradient directions (arrows) and the sectors divided by dashed lines.

We shall describe it by writing

$$A_s f(x,y) := \tan \alpha_s = \frac{\partial(f * \theta_s)(x,y)}{\partial y} \bigg/ \frac{\partial(f * \theta_s)(x,y)}{\partial x}.$$

When $A_s f(x,y)$ falls into the above intervals, it will be coded by 0, 1, 2, 3, respectively, They can be viewed as the functions of values 0, 1, 2, 3 and symbolized by $Code A_s f(x,y)$. Similar to the 1D signals, the algorithm of the edge detection with local maximal modulus of wavelet transform is now established as follows:

---

**Algorithm 6.2.** Given an input digital signal $\{f(k,l)|k = 0,1,\ldots,K;\ l = 0,1,\ldots,L\}$, we have the following:

**Step 1:** Calculate the modulo of its wavelet transform

$$\{M_s f(k,l)|k = 0,1,\ldots,K;\ l = 0,1,\ldots,L\}$$

as well as the codes

$$\{Code A_s f(k,l)|k = 0,1,\ldots,\ K; l = 0,1,\ldots,L\}$$

along the gradient directions.

**Step 2:** Take a threshold $T > 0$, for $k = 0,1,\ldots,K;\ l = 0,1,\ldots,L$, if

(1) $|M_s f(k,l)| \geq T$,
(2) $|M_s f(k,l)|$ reaches its local maximum along the gradient direction represented by $Code A_s f(k,l)$,

then $(k,l)$ is an edge pixel.

---

Fig. 6.4   Left: The original image; middle the edge extracted in $s = 2$; right: the edge extracted in $s = 4$.

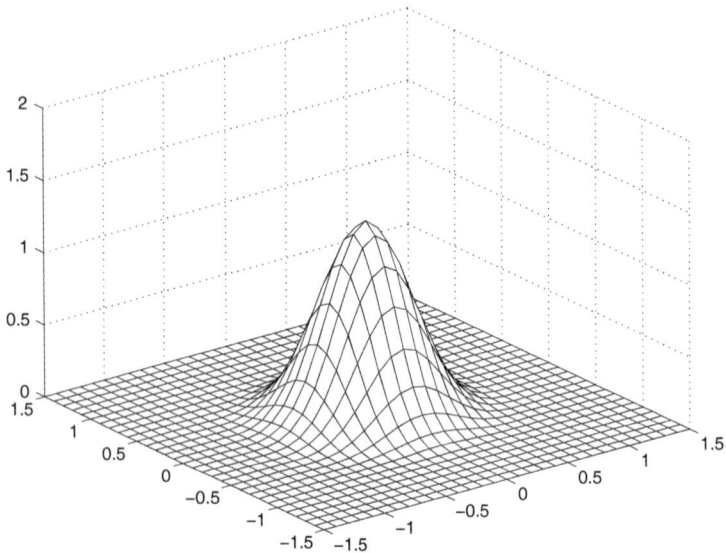

Fig. 6.5   The graphical description of smoothing function $\theta(x, y)$.

An example is given in Fig. 6.4, where a spline function is utilized as the smoothing function:

$$\theta(x, y) := \begin{cases} 8(x^2 + y^2)\left(\sqrt{x^2 + y^2} - 1\right) + \frac{4}{3} & 0 \le x^2 + y^2 \le \frac{1}{4} \\ -\frac{8}{3}\left(\sqrt{x^2 + y^2} - 1\right)^3 & \frac{1}{4} < x^2 + y^2 \le 1 \\ 0 & x^2 + y^2 > 1. \end{cases}$$

The graphical illustration of this smoothing function $\theta(x, y)$ is shown in Fig. 6.5.

## 6.2 Calculation of $W_s f(x)$ and $W_s f(x, y)$

Calculation of the wavelet transform is a key process in applications. The fast algorithm of computing $W_s f(x)$ for detecting edges and reconstructing the signal can be found in the work of Mallat and Hwang [1992] when $\psi(x)$ is a dyadic wavelet defined in that article. However, for some applications such as edge detection, the reconstruction of signals is not required. Therefore, the choice of the wavelet function will not be restricted in the conditions which were presented in the work of Mallat and Hwang [1992]. Many wavelets other than dyadic ones can be utilized. In fact, almost all the general integral wavelets satisfy this particular application. It gives much freedom to select the best wavelet $\psi$ for our task. In practice, we should digitize the wavelet transform again by the equi-spaced sampling and represent by $\{W_s^\psi f(0), W_s^\psi f(1), \ldots, W_s^\psi f(n)\}$. In general, different results are produced from the different fitting models. The simplest models are step-function fitting and line-function fitting. They are graphically displayed in Figs. 6.6–6.8. In the following sections, the wavelet transform is computed using the step-function fitting.

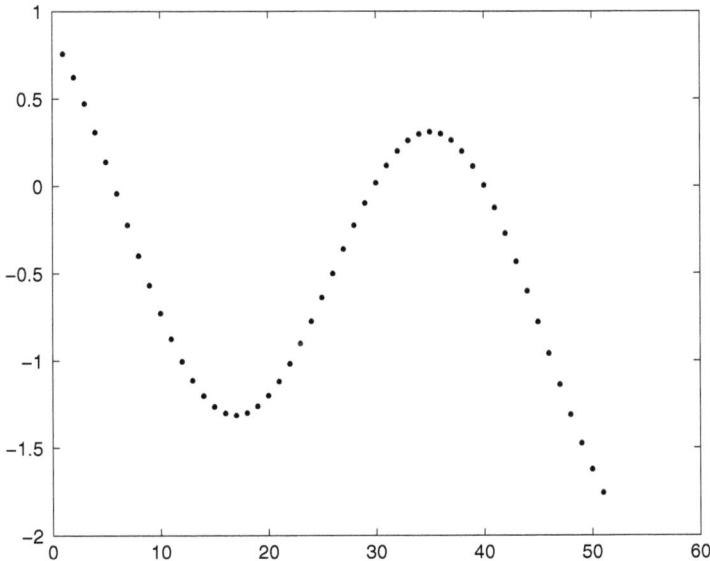

Fig. 6.6   Discrete signal.

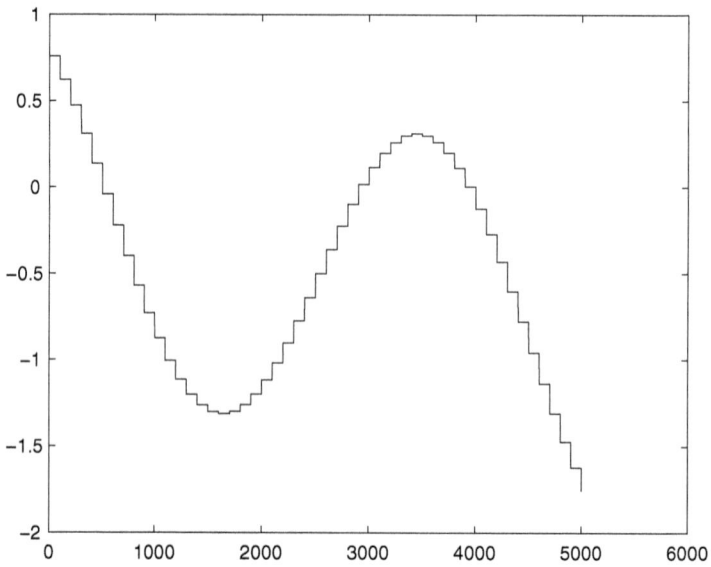

Fig. 6.7   The result fitted by step function.

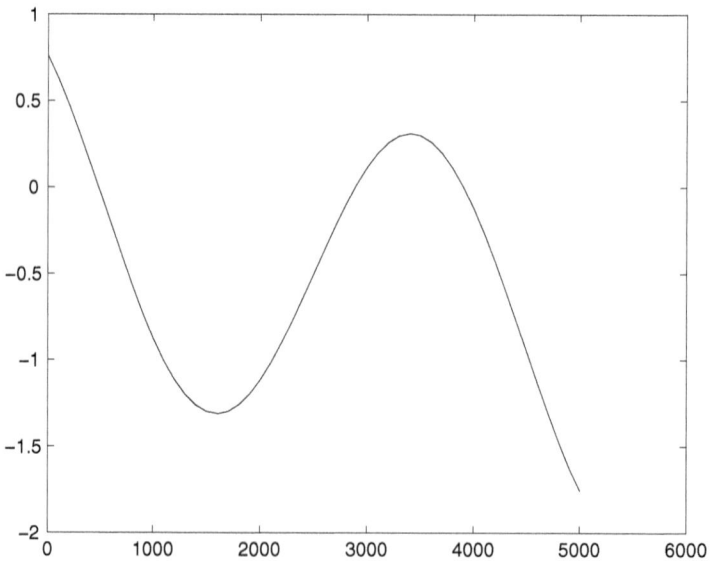

Fig. 6.8   The result from line-function fitting.

Of course, the different fitting models and basic wavelets can produce many totally different representations of a wavelet transform. The details are omitted in this book.

Instead of the algorithm stated in the work of Mallat and Hwang [1992], we calculate the corresponding wavelet transforms by the well-known trapezoidal formula of numerical integrals.

## 6.2.1 *Calculation of* $W_s f(x)$

The wavelet transform of a 1D signal $f(t)$ can be calculated as follows:

$$W_s f(k) = \int_{-\infty}^{\infty} f(t)\psi_s(k-t)dt$$

$$= \sum_{l \in \mathbb{Z}} \int_{l-1}^{l} f(l)\psi_s(k-t)dt$$

$$= \sum_{l \in \mathbb{Z}} f(l)\left[\int_{-\infty}^{l} \psi_s(k-t)dt - \int_{-\infty}^{l-1} \psi_s(k-t)dt\right]$$

$$= \sum_{l \in \mathbb{Z}} f(l)\left[\int_{(k-l)s^{-1}}^{\infty} \psi(t)dt - \int_{(k-l+1)s^{-1}}^{\infty} \psi(t)dt\right]$$

$$= \sum_{l \in \mathbb{Z}} f(l)\int_{(k-l)s^{-1}}^{\infty} \psi(t)dt - \sum_{l \in \mathbb{Z}} f(l+1)\int_{(k-l)s^{-1}}^{\infty} \psi(t)dt$$

$$= \sum_{l \in \mathbb{Z}} [f(l) - f(l+1)]\,\psi_{k-l}^{s}, \qquad (6.2)$$

where

$$\psi_k^s := \int_{k/s}^{\infty} \psi(t)dt.$$

As an example, let $\psi(x)$ be the quadric spline wavelet which is an odd function. For $x \geq 0$, it can be defined by

$$\psi(x) = \begin{cases} 8(3x^2 - 2x) & x \in [0, 1/2] \\ -8(x-1)^2 & x \in [1/2, 1] \\ 0 & x \geq 1. \end{cases}$$

Since $\psi(x)$ is an odd function, we have

$$\psi_{-k}^s = \int_{-k/s}^{k/s} \psi(t)dt + \int_{k/s}^{\infty} \psi(t)dt = \psi_k^s.$$

Hence, it is enough to calculate $\psi_k^s$ for all non-negative integer $k$:

(1) If $k \geq s$, it is obvious that $\psi_k^s = 0$.

(2) If $\frac{s}{2} \leq k < s$, $\psi_k^s$ becomes

$$\psi_k^s = \int_{k/s}^1 \psi(t)dt$$

$$= -8 \int_{k/s}^1 (t-1)^2 dt$$

$$= \frac{8}{3}\left(\frac{k}{s} - 1\right)^3.$$

(3) If $0 \leq k < \frac{s}{2}$, the following holds:

$$\psi_k^s = \int_{k/s}^1 \psi(t)dt$$

$$= 8 \int_{k/s}^{1/2} (3x^2 - 2x)dx - 8 \int_{1/2}^1 (x-1)^2 dx$$

$$= 8\left(\frac{k}{s}\right)^2\left(1 - \frac{k}{s}\right) - 1 - \frac{8}{3}\left(\frac{1}{2} - 1\right)^3$$

$$= \frac{1}{3} + 8\left(\frac{k}{s}\right)^2\left(1 - \frac{k}{s}\right).$$

The conclusions can be obtained in accordance with the above discussion, namely,

$$\psi_k^s = \begin{cases} \dfrac{8}{3}\left(\dfrac{k}{s} - 1\right)^3 & \dfrac{s}{2} \leq k < s \\[2ex] \dfrac{1}{3} + 8\left(\dfrac{k}{s}\right)^2\left(1 - \dfrac{k}{s}\right) & 0 \leq k < \dfrac{s}{2} \\[2ex] 0 & \text{otherwise.} \end{cases}$$

The filtering coefficients $\{\psi_k^s\}$ in Eq. (6.2) can be calculated when $\psi$ is given. If $\psi$ is compactly supported, only finite non-zero items $\{\psi_k^s\}$ will be obtained. Hence, the complexity of this algorithm to compute $\{W_s f(k)\}_{k=0}^N$ is $O(N)$.

Table 6.1. Filtering coefficients $\{\psi_k^{2^j}\}$ and $\{\psi_{k,l}^{2^j,l}\}$ if $\psi$ is a quadic spline wavelet.

| $k$ | $j = 1$ | $j = 2$ | $j = 3$ |
|---|---|---|---|
| 0 | −1.3333333731 | −1.3333333731 | −1.3333333731 |
| 1 | −0.3333333433 | −0.9583333135 | −1.2239583731 |
| 2 | | −0.3333333433 | −0.9583333135 |
| 3 | | −0.0416666679 | −0.6302083135 |
| 4 | | | −0.3333333433 |
| 5 | | | −0.1406250000 |
| 6 | | | −0.0416666679 |
| 7 | | | −0.0052083335 |

In the practical application, it is enough to take $s = 2^j, j = 1, 2, 3$. The algorithm presented in this chapter is $L^\infty$-stable, since

$$\|W_s f\|_{L^\infty(\mathbb{R})} \le \|f\|_{L^\infty(\mathbb{R})} \|\psi\|_{L^1(\mathbb{R})},$$

where

- $L^1(\mathbb{R})$ denotes the space of all the *Lebesgue* integrable functions on $\mathbb{R}$ and

$$\|\psi\|_{L^1(\mathbb{R})} := \int_{-\infty}^{\infty} |\psi(x)| dx,$$

- $L^\infty$ indicates the space of all the essential bounded functions on $\mathbb{R}$ and [Rudin, 1973]

$$\|f\|_{L^\infty(\mathbb{R})} := \inf_{|E|=0} \sup_{x \in \mathbb{R} \backslash E} |f(x)|.$$

The corresponding filtering coefficients $\{\psi_k^s\}$ in Eq. (6.2) for $j = 1, 2, 3$ are shown in Table 6.1.

## 6.2.2 *Calculation of $W_s f(x, y)$*

In 2D signals, the calculation of $W_{2^j}^i f(x, y)$ is more complicated than that of 2D ones. In a similar way, we have

$$W_s^1 f(n, m) = \int_{-\infty}^{\infty} \int_{-\infty}^{\infty} f(u, v) \psi_s^1(n - u, m - v) du dv$$

$$= \sum_{k,l} f(k, l) \iint_{[k,k+1] \times [l,l+1]} \psi_s^1(n - u, m - v) du dv$$

$$= \sum_{k,l} f(k,l) \iint_{[n-k-1,n-k]\times[m-l-1,m-l]} \psi_s^1(u,v)dudv$$

$$= \sum_{k,l} f(n-1-k, m-1-l)\psi_{k,l}^{s,1},$$ (6.3)

where

$$\psi_{k,l}^{s,1} = \iint_{[k,k+1]\times[l,l+1]} \psi_s^1(u,v)dv$$

$$= \int_{\frac{k}{s}}^{\frac{(k+1)}{s}} du \int_{\frac{l}{s}}^{\frac{(l+1)}{s}} \psi^1(u,v)dv.$$

The problem is, now, to lead to the calculation of $\{\psi_{k,l}^{s,1}\}$ in Eq. (6.3). Note that $\psi^1(u,v)$ is odd on $u$ and even on $v$, we have

$$\psi_{-k,l}^{s,1} = \iint_{[-k,-k+1]\times[l,l+1]} \psi_s^1(u,v)dudv$$

$$= \int_{k-1}^{k} du \int_{l}^{l+1} \psi_s^1(u,v)dudv$$

$$= -\psi_{k-1,l}^{s,1}.$$

Similarly, we can deduce that

$$\psi_{k,-l}^{s,1} = \psi_{k,l-1}^{s,1}, \quad \psi_{-k,-l}^{s,1} = -\psi_{k-1,l-1}^{s,1}.$$

Therefore, we need to calculate $\psi_{k,l}^{s,1}$ only for all $k, l \geq 0$.

Let $\psi(r)$ be an odd function, and

$$\psi^1(u,v) = \psi(r)\cos\theta,$$

where

$$r = \sqrt{u^2+v^2}, \quad \theta = arctg\frac{v}{u}.$$

Let $\phi(x)$ be

$$\phi(x) = \int_{-\infty}^{x} \psi(r)dr.$$

Then $\phi(x)$ is an even function, since $\psi(x)$ is an odd function. It is easy to find that

$$\frac{\partial}{\partial u}\phi\left(\sqrt{u^2+v^2}\right) = \phi'(r)\frac{u}{r} = \psi^1(u,v),$$

$$\frac{\partial}{\partial v}\phi\left(\sqrt{u^2+v^2}\right) = \phi'(r)\frac{v}{r} = \psi^2(u,v).$$

Hence

$$
\psi_{k,l}^{s,1} = \int_{\frac{k}{s}}^{\frac{(k+1)}{s}} du \int_{\frac{l}{s}}^{\frac{(l+1)}{s}} \frac{\partial}{\partial u} \phi\left(\sqrt{u^2 + v^2}\right) dv
$$

$$
= \int_{\frac{l}{s}}^{\frac{(l+1)}{s}} \left[ \phi\left(\sqrt{v^2 + \left(\frac{k+1}{s}\right)^2}\right) - \phi\left(\sqrt{v^2 + \left(\frac{k}{s}\right)^2}\right) \right] dv
$$

$$
= \phi_{l,k+1}^s - \phi_{l+1,k+1}^s - \phi_{l,k}^s + \phi_{l+1,k}^s,
$$

where

$$
\phi_{l,k}^s = \int_{\frac{l}{s}}^{\infty} \phi\left(\sqrt{v^2 + \left(\frac{k}{s}\right)^2}\right) dv.
$$

As for the calculation of $W_s^2 f(n,m)$, similarly, we have

$$
W_s^2 f(n,m) = \sum_{k,l} f(n-1-k, m-1-l)\psi_{k,l}^{s,2},
$$

where

$$
\psi_{k,l}^{s,2} = \int_{\frac{k}{s}}^{\frac{(k+1)}{s}} du \int_{\frac{l}{s}}^{\frac{(l+1)}{s}} \psi^2(u,v) dv
$$

$$
= \int_{\frac{k}{s}}^{\frac{(k+1)}{s}} \left[ \phi\left(\sqrt{\left(\frac{l+1}{s}\right)^2 + u^2}\right) - \phi\left(\sqrt{\left(\frac{l}{s}\right)^2 + u^2}\right) \right] du
$$

$$
= \phi_{k,l+1}^s - \phi_{k+1,l+1}^s - \phi_{k,l}^s + \phi_{k+1,l}^s
$$

$$
= \psi_{l,k}^{s,1}.
$$

Based on the above discussion, we can calculate the filtering coefficients $\{\phi_{k,l}^s\}$ only for all $k$, $l \geq 0$.

In this chapter, the quadratic spline function is served as $\psi$, then

$$
\phi(x) = \begin{cases} 8(x^3 - x^2) + \dfrac{4}{3} & 0 \leq x \leq 1/2; \\[2mm] -\dfrac{8}{3}(x-1)^3 & 1/2 < x \leq 1; \\[2mm] 0 & x > 1. \end{cases}
$$

(1) If $k^2 + l^2 \geq s^2$, we have $\phi_{k,l}^s = 0$.

(2) If $s^2 \geq k^2 + l^2 \geq \left(\frac{s}{2}\right)^2$, we have

$$\phi_{k,l}^s = -\frac{8}{3} \int_{\frac{k}{s}}^{\sqrt{1-\left(\frac{l}{s}\right)^2}} \left(\sqrt{u^2 + \left(\frac{l}{s}\right)^2} - 1\right)^3 du.$$

(3) If $k^2 + l^2 \leq \left(\frac{s}{2}\right)^2$, then

$$\phi_{k,l}^s = \int_{\frac{k}{s}}^{\sqrt{\frac{1}{4}-\left(\frac{l}{s}\right)^2}} \left[ 8\left(\left(u^2 + \left(\frac{l}{s}\right)^2\right)^{\frac{3}{2}} - \left(u^2 + \left(\frac{l}{s}\right)^2\right)\right) + \frac{4}{3}\right] du$$

$$-\frac{8}{3} \int_{\sqrt{\frac{1}{4}-\left(\frac{l}{s}\right)^2}}^{\sqrt{1-\left(\frac{l}{s}\right)^2}} \left(\sqrt{u^2 + \left(\frac{l}{s}\right)^2} - 1\right)^3 du. \tag{6.4}$$

In order to simplify the calculation, the following replacements are done:

$$\frac{k}{s} \Longrightarrow a,$$

$$\sqrt{\frac{1}{4} - \left(\frac{l}{s}\right)^2} \Longrightarrow b,$$

$$\sqrt{\frac{1}{4} - \left(\frac{l}{s}\right)^2} \Longrightarrow c,$$

$$\sqrt{1 - \left(\frac{l}{s}\right)^2} \Longrightarrow d,$$

$$\frac{l}{s} \Longrightarrow t.$$

Thus, Eq. (6.4) becomes

$$\phi_{k,l}^s = \int_a^b [8(u^2 + t^2)(\sqrt{u^2 + t^2} - 1) + \frac{4}{3}] du - \frac{8}{3} \int_c^d (\sqrt{u^2 + t^2} - 1)^3 du.$$

We denote

$$I(t, a, b) = \int_a^b \left[ 8(u^2 + t^2)(\sqrt{u^2 + t^2} - 1) + \frac{4}{3} \right] du,$$

$$J(t, c, d) = -\frac{8}{3} \int_c^d (\sqrt{u^2 + t^2} - 1)^3 du. \tag{6.5}$$

To avoid the complicated calculation, a software called *Mathematica* is utilized to solve Eq. (6.5), and the following results can be achieved:

$$I(t, a, b) = \frac{8}{3}(a^3 - b^3) - \frac{4}{3}(a - b)\left[1 - 6t^2\right]$$

$$- \sqrt{a^2 + t^2}\left[2a^3 + 5at^2\right] + \sqrt{b^2 + t^2}(2b^3 + 5bt^2)$$

$$- 3t^4 \log \frac{a + \sqrt{a^2 + t^2}}{b + \sqrt{b^2 + t^2}},$$

$$J(t, c, d) = -\frac{8}{3}(c - d)(c^2 + d^2 + cd + 3t^2 + 1)$$

$$+ \frac{1}{3}[c(2c^2 + 5t^2 + 12)\sqrt{c^2 + t^2}$$

$$- d(2d^2 + 5t^2 + 12)\sqrt{d^2 + t^2}]$$

$$+ t^2(4 + t^2) \log \frac{c + \sqrt{c^2 + t^2}}{d + \sqrt{d^2 + t^2}}.$$

Hence, for $\forall k, l \geq 0$, we have

$$\phi_{k,l}^s = \begin{cases} 0 & \text{if } k^2 + l^2 \geq s^2; \\ J\left(\frac{l}{s}, \frac{k}{s}, \sqrt{1 - \left(\frac{l}{s}\right)^2}\right) & \text{if } s^2 \geq k^2 + l^2 \geq \frac{1}{4}s^2; \\ I\left(\frac{l}{s}, \frac{k}{s}, \sqrt{\frac{1}{4} - \left(\frac{l}{s}\right)^2}\right) \\ \quad + J\left(\frac{l}{s}, \sqrt{\frac{1}{4} - \left(\frac{l}{s}\right)^2}, \sqrt{1 - \left(\frac{l}{s}\right)^2}\right) & \text{if } k^2 + l^2 \leq \frac{1}{4}s^2. \end{cases}$$

$$(6.6)$$

For $s = 2^j, j = 1, 2, 3$, the corresponding filtering coefficients $\{\psi_{k,l}^{s,1}\}$ in Eq. (6.3) are shown in Tables 6.2–6.4. For the given scale $s$, the complexity of computing $\{W_s^i f(n, m)\}_{n,m=0}^N$ is $O(N^2)$.

## 6.3 Wavelet Transform for Contour Extraction and Background Removal

In this section, the singularities are studied by Lipschitz exponent, according to the mathematical analysis of the different geometric structures of

Table 6.2. Filtering coefficients $\{\psi_{k,l}^{2^j,1}\}, (j = 1)$ if $\psi(r)$ is a quadic spline wavelet.

| $l\backslash k$ | $k = 0$ | $k = 1$ |
|---|---|---|
| $l = 0$ | $-0.3485791385$ | $-0.0351441652$ |
| $l = 1$ | $-0.1097541898$ | $-0.0085225009$ |

Table 6.3. Filtering coefficients $\{\psi_{k,l}^{2^j,1}\}, (j = 2)$ if $\psi(r)$ is a quadic spline wavelet.

| $l\backslash k$ | $k = 0$ | $k = 1$ |
|---|---|---|
| $l = 0$ | $-0.0838140026$ | $-0.0472041145$ |
| $l = 1$ | $-0.1416904181$ | $-0.0958706033$ |
| $l = 2$ | $-0.0651057437$ | $-0.0327749476$ |
| $l = 3$ | $-0.0088690016$ | $-0.0030044997$ |

| $l\backslash k$ | $k = 2$ | $k = 3$ |
|---|---|---|
| $l = 0$ | $-0.0129809398$ | $-0.0012592132$ |
| $l = 1$ | $-0.0196341425$ | $-0.0012698699$ |
| $l = 2$ | $-0.0063385521$ | $-0.0000750836$ |
| $l = 3$ | $-0.0001088651$ | |

Table 6.4. Filtering coefficients $\{\psi_{k,l}^{2^j,1}\}, (j = 3)$ if $\psi(r)$ is a quadic spline wavelet.

| $l\backslash k$ | $k = 0$ | $k = 1$ |
|---|---|---|
| $l = 0$ | $-0.0130508747$ | $-0.0107604843$ |
| $l = 1$ | $-0.0322880745$ | $-0.0277145673$ |
| $l = 2$ | $-0.0400620662$ | $-0.0347912423$ |
| $l = 3$ | $-0.0361453630$ | $-0.0306917503$ |
| $l = 4$ | $-0.0233931188$ | $-0.0196621399$ |
| $l = 5$ | $-0.0120095834$ | $-0.0100409016$ |
| $l = 6$ | $-0.0043909089$ | $-0.0034857604$ |
| $l = 7$ | $-0.0006066251$ | $-0.0003857070$ |

| $l\backslash k$ | $k = 2$ | $k = 3$ |
|---|---|---|
| $l = 0$ | $-0.0080117621$ | $-0.0051634177$ |
| $l = 1$ | $-0.0208740719$ | $-0.0181548643$ |
| $l = 2$ | $-0.0258209687$ | $-0.0154902851$ |

(*Continued*)

Table 6.4. (*Continued*)

| $l\backslash k$ | $k = 2$ | $k = 3$ |
|---|---|---|
| $l = 3$ | $-0.0216873437$ | $-0.0128720086$ |
| $l = 4$ | $-0.0139309634$ | $-0.0082158195$ |
| $l = 5$ | $-0.0069123111$ | $-0.0037158530$ |
| $l = 6$ | $-0.0020833646$ | $-0.0007867273$ |
| $l = 7$ | $-0.0001258312$ | $-0.0000085766$ |

| $l\backslash k$ | $k = 4$ | $k = 5$ |
|---|---|---|
| $l = 0$ | $-0.0025986142$ | $-0.0010915556$ |
| $l = 1$ | $-0.0065528429$ | $-0.0027379275$ |
| $l = 2$ | $-0.0097385549$ | $-0.0031414898$ |
| $l = 3$ | $-0.0063897478$ | $-0.0023643500$ |
| $l = 4$ | $-0.0037783380$ | $-0.0010769776$ |
| $l = 5$ | $-0.0013164375$ | $-0.0001667994$ |
| $l = 6$ | $-0.0001085378$ | $-0.0000003273$ |
| $l = 7$ | | |

| $l\backslash k$ | $k = 4$ | $k = 7$ |
|---|---|---|
| $l = 0$ | $-0.0003376677$ | $-0.0000403965$ |
| $l = 1$ | $-0.0008041780$ | $-0.0000769711$ |
| $l = 2$ | $-0.0008010443$ | $-0.0000415404$ |
| $l = 3$ | $-0.0004234327$ | $-0.0000038524$ |
| $l = 4$ | $-0.0000748098$ | |
| $l = 5$ | $-0.0000002738$ | |
| $l = 6$ | | |
| $l = 7$ | | |

the singularities using Lipschitz exponent. The singularities can be categorized into three basic geometric structures: step structure, roof structure, and Dirac structure. The characterization of these structures with wavelet transform is also studied in this section. A significant property will be derived, i.e., the wavelet transform of the step structure of singularity is a non-zero constant which is independent on both the gradient direction and the scale of the wavelet transform. This property leads to provide a simple and direct strategy for detecting this specific structure of singularities, including contours of the patterns. Thereafter, a novel algorithm which is referred to as scale-independent algorithm is developed, and the modular-angle-separated (MAS) wavelet transform is applied. As the applications of this new method, many examples are presented, which deal with data preprocessing in the recognition of the 2D object and document processing.

Edges are typical singularities in images, meanwhile, a contour is a specific edge, which belongs to the step structure. Extraction of singularities specifically, extraction of contours, plays an important role in the data preprocessing phase of pattern recognition, including document processing. Research on extracting the contours is one of the most significant topics in this discipline. Many methods have been developed to analyze the properties of contours and detect them from various images. For example, some works can be found in the works of Chen and Yang [1995], Deng and Lyengar [1996], Law *et al.* [1996], Matalas *et al.* [1997], Tang *et al.* [1997a; 1997c] and Thune *et al.* [1997].

The subjects of Lipschitz exponents and wavelet transform are remarkable mathematical tools to analyze the singularities including the edges and, further, to detect them effectively. A significant study related to these research topics has been done by Mallat, Hwang, and Zhong and published in *IEEE Trans. on Information Theory* [Mallat and Hwang, 1992] and *IEEE Trans. on Pattern Analysis and Machine Intelligence* [Mallat and Zhong, 1992]. Many important contributions have been made in these papers. It has been shown that the local maxima of the wavelet transform modulus can provide enough information for analyzing the singularities and can detect *all* singularities. However, it may not identify different structures of singularities. This chapter presents a new method which can handle this problem. Now, we look at an example. We consider an image which is shown in Fig. 6.9(a). Two classes of singularities are embedded in this image: (i) a contour of the aircraft which belongs to the step structure singularities and (ii) some lines and texts which belong to the Dirac structure singularities. Such an image is referred to as *multi structure singularities image*. A particular task is that we are required to extract the contour of the aircraft and to remove all lines and texts. What result will arrive when the local maxima of the wavelet transform modulus is applied?

Look at the following experiments: Figure 6.9(b) displays the modulus image, where the black pixels indicate zero values and the white ones correspond to the highest values. Figure 6.9(c) gives the angle image, in which the angle values range from 0 (black) to $2\pi$ (white). Figure 6.9(d) shows the modulus maxima. It is clear, from this result, that the method of the modulus maxima has detected all singularities without recognizing two different structures of singularities. Thus, the resulting image contains lines and texts which are required to be deleted from the image. Another example can be found in Figs. 6.10 and 6.11. A mailing address which was written on a paper with guide lines is shown in Fig. 6.10(a). In order to automatically process it by computers, a preprocessing is to remove the guide lines and extract the contours of the handwritten English letters. Figures 6.10(b)

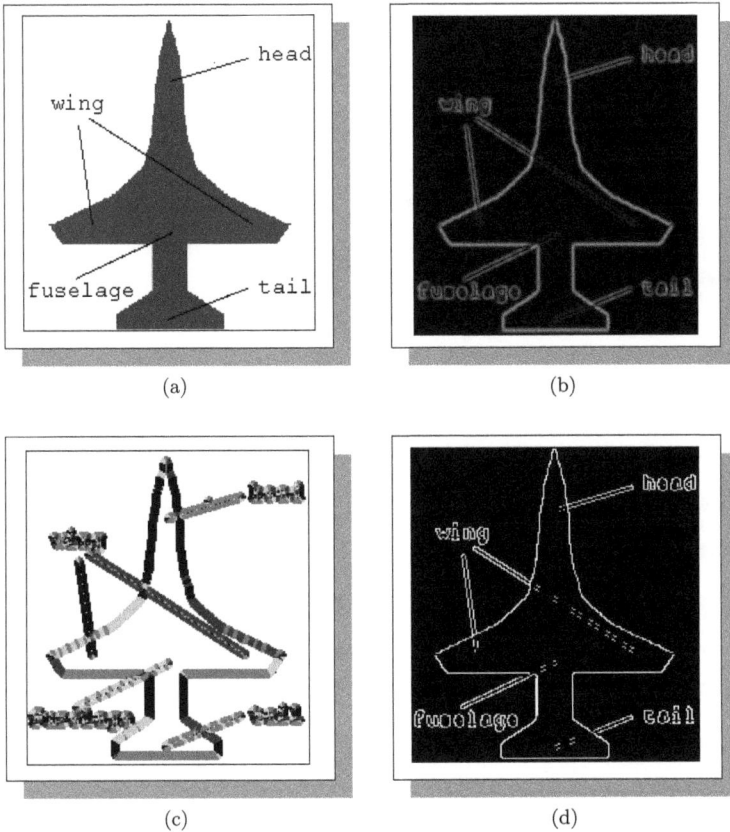

(a)

(b)

(c)

(d)

Fig. 6.9    Result of edge detection using the wavelet transform modulus maxima in data preprocessing phase of pattern recognition.

and 6.10(c) display the modulus images with scales of $s = 2, 4$ of the wavelet transform, respectively. Figures 6.10(d) and 6.10(e) state the angle images with transform scales of $s = 2, 4$, respectively. Obviously, it is still the same problem when different scales of the wavelet transforms are used, and the results are given in Figs. 6.11(a) and 6.10(b).

In order to improve the method proposed in the works of Mallat and Hwang [1992] and Mallat and Zhong [1992] so that it can be used to detect the contour of an object which is a typical step structure singularity and eliminate others, this section carefully studies the wavelet transform with respect to three basic geometric structures of singularities. A very important

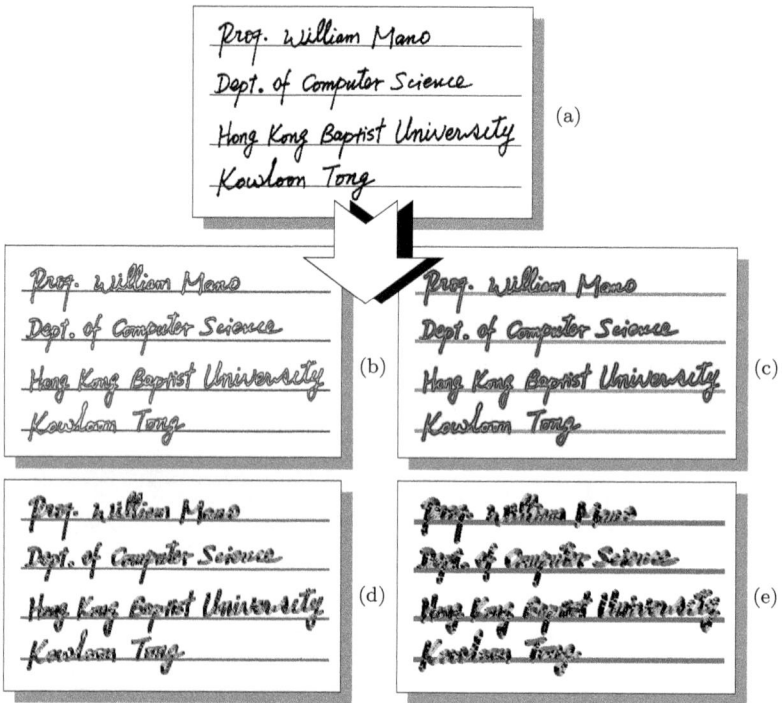

Fig. 6.10    Wavelet transform of a page of document with different scales ($s = 2, 4$).

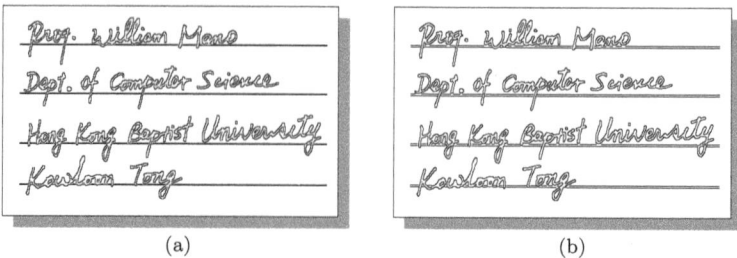

Fig. 6.11    Result of edge detection using the wavelet transform modulus maxima in document processing: (a) $s = 2$, (b) $s = 4$.

property is proven that the wavelet transform of a step singularity is a non-zero constant which is independent of the scale of the wavelet transform. Based on this result, a scale-independent wavelet algorithm is developed. This novel method can detect step structure singularities from the multi structure-singularity images.

This section is organized into the following sections:

(1) basic edge structures,
(2) analysis of the basic edge structures with wavelet transform,
(3) scale-independent algorithm,
(4) experiments.

### 6.3.1 *Basic Edge Structures*

The edge is one class of singularities in the 1D and 2D signals. Lipschitz exponent is a remarkable mathematical tool to analyze the singularities, including the edges. According to the analysis with Lipschitz exponents, the edges can be categorized into three basic geometric structures [Tang *et al.*, 1998e].

Let $0 \leq \alpha \leq 1$, function $f(x)$ is called uniformly Lipschitz $\alpha$ over $(a, b)$ if there exists a constant $K$ such that $\forall x_1, x_2 \in (a, b)$, we have

$$|f(x_1) - f(x_2)| \leq K|x_1 - x_2|^\alpha, \quad |x_1 - x_2| \to 0, \tag{6.7}$$

where $\alpha$ denotes the regularity exponent. The value of $\alpha$ plays a key role in analyzing the singularity or regularity of a signal. We are interested in the case of $0 \leq \alpha \leq 1$, which corresponds to the different geometric structures of edges. The geometric structures of the edges can be analyzed by the value of $\alpha$. The smaller the $\alpha$ the weaker the regularity (i.e., smoothness), or the smaller the $\alpha$, the stronger the singularity. Lipschitz exponent $\alpha$ can be extended to the range of $-1 \leq \alpha < 0$. $f(x)$ is called uniformly Lipschitz $\alpha$ if its primitive function is uniformly Lipschitz $\alpha + 1(-1 \leq \alpha < 0)$. This extension permits $f(x)$ to be a distribution (i.e., a generalized function). For instance, when $\alpha = -1$, the well-known *Dirac* function $\delta(x)$ is a uniform Lipschitz $-1$ in any interval containing 0. The theory of distribution can be found in the work of Rudin [1973]. In summary, three special values of $\alpha$ are considered in the concrete applications:

(1) $\alpha = 0$: This corresponds to the edges with step structure.
(2) $\alpha = 1$: It corresponds to the edges with roof structure.
(3) $\alpha = -1$: It corresponds to the edges with Dirac structure.

Any of the three structures of edges can be regarded as a superposition of an ideal edge $e(x)$ and an additional signal $f_0(x)$:

$$f(x) = e(x) + f_0(x).$$

It is proved in the following section that the latter is not important in this study. We, thus, will pay attention to the former, ideal versions, which can be formulated as follows:

(1) Ideal step edge

$$e(x) = \begin{cases} 1 & x \geq x_0; \\ 0 & x < x_0. \end{cases}$$

(2) Ideal roof edge

$$e(x) = \begin{cases} m(x - x_0) + c_0 & x_0 - h_1 < x \leq x_0; \\ n(x - x_0) + c_0 & x_0 < x < x_0 + h_2. \end{cases}$$

(3) Ideal Dirac edge

$$e(x) = \begin{cases} \infty & x = x_0; \\ 0 & x \neq x_0. \end{cases}$$

Similarly, the Lipschitz exponent of regularity for the two-dimensional images can be defined as follows:

**Definition 6.1.** Let $0 \leq \alpha \leq 1$, function $f(u, v)$ is called uniformly *Lipschitz* $\alpha$ on the interval of $(a, b) \times (c, d)$, if there exists a constant $K$, such that, $\forall (u_1, v_1) \in (a, b) \times (c, d)$, and

$$|f(u_1, v_1) - f(u_2, v_2)| \leq K(|u_1 - u_2|^2 + |v_1 - v_2|^2)^{\alpha/2},$$

$$|u_1 - u_2| \to 0, \quad |v_1 - v_2| \to 0. \tag{6.8}$$

In 2D signal, Lipschitz exponent $\alpha$ can be generalized to $-1 \leq \alpha < 0$ similar to 1D case. Tempered distribution $f(x, y)$ is an example of this generalization. Thus, $f(x, y)$ is called uniformly Lipschitz $\alpha$ over $(a, b) \times (c, d)$, if the primitive function of the distribution $f(x, y)$ is uniform Lipschitz $\alpha + 1$ ($-1 \leq \alpha < 0$) on $(a, b) \times (c, d)$.

The graphic descriptions of the basic structures of the edges are illustrated in Figs. 6.12–6.14, respectively.

## 6.3.2 *Analysis of the Basic Edge Structures with Wavelet Transform*

A remarkable property of the wavelet transform is its ability to characterize the local regularity of functions. After characterizing the singularities with Lipschitz exponents and classifying the edges into the basic structures, the

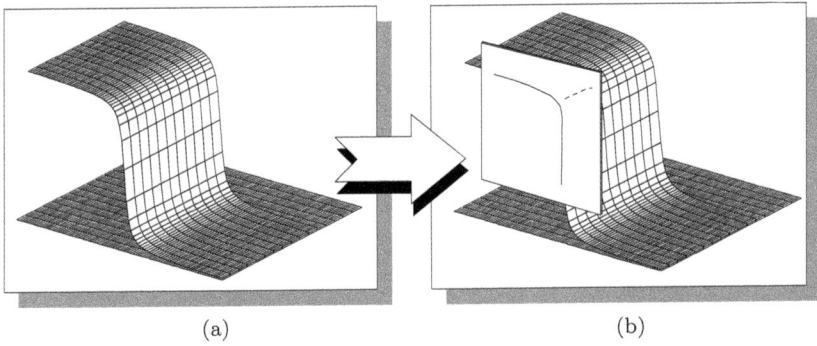

(a)　　　　　　　　　　　　(b)

Fig. 6.12　Graphic description of the step structure edge.

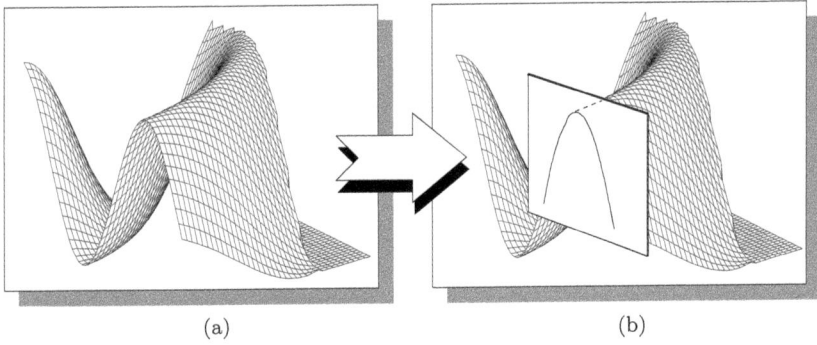

(a)　　　　　　　　　　　　(b)

Fig. 6.13　Graphic description of the roof structure edge.

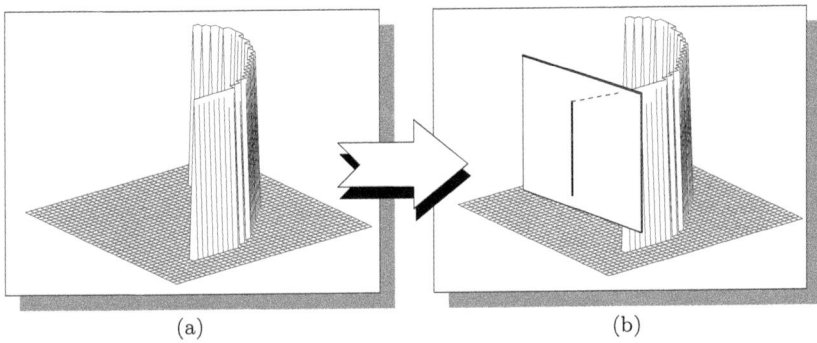

(a)　　　　　　　　　　　　(b)

Fig. 6.14　Graphic description of the Dirac structure edge.

analysis of these structures with wavelet transform is be presented in this section.

Let $L^2(\mathbb{R})$ be a *Hilbert* space of all the square-integrable functions on $\mathbb{R}$, and $\psi \in L^2(\mathbb{R})$ be a wavelet function, i.e., $\psi(x)$ decreases fast when $x$ goes to infinity and satisfies

$$\int_{\mathbb{R}} \psi(x)dx = 0.$$

In the edge detection, the speed of the decrease in $\psi(x)$ affects the result of local detection. Usually, the faster the speed the better the result. In this chapter, a compactly supported wavelet function is chosen as $\psi(x)$. $\forall f(x) \in L^2(\mathbb{R})$ and $s > 0$, the wavelet transform of $f(x)$ with the scale $s$ is defined as follows:

$$W_s f(x) := (f * \psi_s)(x) = \int_{\mathbb{R}} f(t)\frac{1}{s}\psi\left(\frac{x-t}{s}\right)dt,$$

where $*$ denotes the convolution operation and $\psi_s(x) := \frac{1}{s}\psi\left(\frac{x}{s}\right)$. Particularly, $W_{2^j} f(x)$, $(j \in \mathbb{Z})$ is called dyadic wavelet transform, where $\mathbb{Z}$ stands for the set of all integers.

The local Lipschitz singularity can be characterized with wavelet transform by the following theorem [Meyer, 1990]:

**Theorem 6.1.** *Let $\alpha$ be a real number, $\alpha \geq -1$, function $f(x)$ is said to be uniformly Lipschitz $\alpha$, if and only if there exists a constant $K$ such that*

$$|W_s f(x)| \leq Ks^\alpha, \quad \forall x \in (a, b). \tag{6.9}$$

This theorem is a theoretical basis for wavelet-based approach to edge detection. Unfortunately, it gives only an inequality, meanwhile the constant $K$ is unknown. Hence, it is difficult to apply this result to detect edges directly in practice. To contest such a deficiency, we discuss the properties of wavelet transform with respect to the three basic structures of edges, and, thereafter, find a solution to handle some specific edges. After applying wavelet transform to these edges, the following main results can be obtained:

**Theorem 6.2.** *Let $\psi(x)$ be an odd wavelet function, which is supported on $[-\delta, \delta]$, satisfying $\int_0^\infty \psi(x)dx \neq 0$, and $x_0$ be the coordinate of an ideal edge:*

(1) *The wavelet transform of a step edge is described by*

$$W_s e(x) = \int_{-\infty}^{-|x-x_0|s^{-1}} \psi(t)dt. \tag{6.10}$$

(2) *The wavelet transform of a Dirac edge can be obtained as follows:*

$$W_s e(x) = \frac{1}{s}\psi\left(\frac{x-x_0}{s}\right). \tag{6.11}$$

(3) *If $|x-x_0|+s\delta \leq h$, the wavelet transform of a roof edge can be computed in the following cases:*

(i) *If $|x - x_0| \leq s\delta$, then*

$$W_s e(x) = (x - x_0)(m - n)\int_{|x-x_0|s^{-1}}^{\infty} \psi(t)dt$$

$$- s(m+n)\int_{|x-x_0|s^{-1}}^{\infty} t\psi(t)dt$$

$$- s\int_{-|x-x_0|s^{-1}}^{|x-x_0|s^{-1}} t\psi(t)dt \cdot \begin{cases} n & \text{if } x \geq x_0; \\ m & \text{if } x < x_0. \end{cases}$$

$$\tag{6.12}$$

(ii) *If $x - x_0 > s\delta$, then*

$$W_s e(x) = -2ns\int_0^\infty t\psi(t)dt. \tag{6.13}$$

(iii) *If $x - x_0 > -s\delta$, then*

$$W_s e(x) = -2ms\int_0^\infty t\psi(t)dt. \tag{6.14}$$

Particularly, at point $x_0$, the wavelet transforms of these basic edge structures become

$$W_s e(x_0) = \begin{cases} -\int_0^\infty \psi(t)dt & \text{if } x_0 \text{ is a step edge;} \\ 0 & \text{if } x_0 \text{ is a Dirac edge;} \\ -s(m+n)\int_0^\infty t\psi(t)dt & \text{if } x_0 \text{ is a roof edge.} \end{cases} \tag{6.15}$$

**Proof.**

(1) If $x_0$ is the coordinate of a step edge, the wavelet transform becomes

$$W_s f(x) = \int f(t)\psi_s(x-t)dt$$

$$= \int_{x_0}^{\infty} \psi_s(x-t)dt$$

$$= \int_{-\infty}^{(x-x_0)s^{-1}} \psi(t)dt$$

$$= \int_{-\infty}^{-|x-x_0|s^{-1}} \psi(t)dt + \int_{-|x-x_0|s^{-1}}^{(x-x_0)s^{-1}} \psi(t)dt$$

$$= \int_{-\infty}^{-|x-x_0|s^{-1}} \psi(t)dt.$$

(2) If $x_0$ is the coordinate of a Dirac edge, the wavelet transform can be produced by

$$W_s f(x) = (f * \psi_s)(x)$$

$$= \psi_s(x-x_0)$$

$$= \frac{1}{s}\psi\left(\frac{x-x_0}{s}\right).$$

(3) If $x_0$ is the coordinate of a roof edge, it is easy to find that $x - st \in [x_0 - h, x_0 + h]$ for $\forall x : |x - x_0| \le h - s\delta$. Therefore, we have (i). If $|x - x_0| \le \delta s$, i.e., $-\delta \le (x - x_0)s^{-1} \le \delta$, the wavelet transform is computed as follows:

$$W_s e(x) = \int_{-\infty}^{\infty} e(x-t)\psi_s(t)dt$$

$$= \int_{-\delta}^{(x-x_0)s^{-1}} e(x-st)\psi(t)dt + \int_{(x-x_0)s^{-1}}^{\delta} e(x-st)\psi(t)dt$$

$$= \int_{-\delta}^{(x-x_0)s^{-1}} [n(x-x_0-st)+c_0]\psi(t)dt$$

$$+ \int_{(x-x_0)s^{-1}}^{\delta} [m(x-x_0-st)+c_0]\psi(t)dt$$

$$= n(x - x_0) \int_{-\delta}^{(x-x_0)s^{-1}} \psi(t)dt - ns \int_{-\delta}^{(x-x_0)s^{-1}} t\psi(t)dt$$

$$+ m(x - x_0) \int_{(x-x_0)s^{-1}}^{\delta} \psi(t)dt - ms \int_{(x-x_0)s^{-1}}^{\delta} t\psi(t)dt$$

$$= (m - n)(x - x_0) \int_{(x-x_0)s^{-1}}^{\delta} \psi(t)dt - ns \int_{-\delta}^{-|x-x_0|s^{-1}} t\psi(t)dt$$

$$- ns \int_{-|x-x_0|s^{-1}}^{(x-x_0)s^{-1}} t\psi(t)dt - ms \int_{(x-x_0)s^{-1}}^{|x-x_0|s^{-1}} t\psi(t)dt$$

$$- ms \int_{|x-x_0|s^{-1}}^{\delta} t\psi(t)dt$$

$$= (m - n)(x - x_0) \int_{|x-x_0|s^{-1}}^{\delta} \psi(t)dt$$

$$- (m + n)s \int_{|x-x_0|s^{-1}}^{\delta} t\psi(t)dt$$

$$- s \left( \int_{-|x-x_0|s^{-1}}^{|x-x_0|s^{-1}} t\psi(t)dt \right) \cdot \begin{cases} n & \text{if } x \geq x_0 \\ m & \text{if } x < x_0 \end{cases}$$

$$= (m - n)(x - x_0) \int_{|x-x_0|s^{-1}}^{\infty} \psi(t)dt$$

$$- (m + n)s \int_{|x-x_0|s^{-1}}^{\infty} t\psi(t)dt$$

$$- s \left( \int_{-|x-x_0|s^{-1}}^{|x-x_0|s^{-1}} t\psi(t)dt \right) \cdot \begin{cases} n & \text{if } x \geq x_0; \\ m & \text{if } x < x_0. \end{cases}$$

(ii) If $x - x_0 > s\delta$, then

$$W_s e(x) = \int_{-\delta}^{\delta} e(x - st)\psi(t)dt$$

$$= \int_{-\delta}^{\delta} [n(x - st - x_0) + c_0]\psi(t)dt$$

$$= -ns \int_{-\delta}^{\delta} t\psi(t)dt$$

$$= -2ns \int_{0}^{\infty} t\psi(t)dt.$$

(iii) If $x - x_0 < s\delta$, then

$$W_s e(x) = \int_{-\delta}^{\delta} e(x - st)\psi(t)dt$$

$$= \int_{-\delta}^{\delta} [m(x - st - x_0) + c_0]\psi(t)dt$$

$$= -2ms \int_0^{\infty} t\psi(t)dt. \qquad \square$$

It is easy to understand the following facts from Eq. (6.15) and the above analyses:

- The wavelet transform $W_s e(x_0)$ of the step structure edge is a non-zero constant which is independent of the scale of the wavelet transform. It is sign-preserving at both sides of the neighborhood of $x_0$ and the extremum is reached at $x_0$.
- For the Dirac structure edge $x_0$, its wavelet transform becomes $W_s e(x_0) = 0$ and $W_s e(x)$ has the opposite signs when $x$ belongs to the left or right side of the neighborhood of $x_0$. Moreover, the extrema are located on at both sides of the neighborhood of $x_0$. When the local extremum is used to detect the Dirac edge, ambiguity may appear, since two extrema are obtained.
- With respect to the roof structure edge $x_0$, if $m = -n$, it can be concluded that

$$W_s e(x_0) = 0,$$

$$W_s e(x) \approx 2m(x - x_0) \int_{s^{-1}|x-x_0|}^{\infty} \psi(t)dt.$$

According to the above results, one can deduce that the sign of $W_s e(x)$ at the left side of the neighborhood of $x_0$ is opposite to that at the right side. It can easily be found that if $m \neq -n$,

$$W_s e(x_0) = -s(m + n) \int_0^{\infty} t\psi(t)dt,$$

which is dependent on the scale $s$.

From the above theorem, the following conclusion are achieved:

**Corollary 6.1.** *$x_0$ is the coordinate of a step edge, if and only if the wavelet transform of $e(x)$, $W_s e(x_0)$, is a non-zero constant which is independent of the scale of the wavelet transform.*

This significant property will be utilized to extract the step structure edges from a signal with multi structure edges. The details are presented in the following section.

Similarly, for the 2D images, the characterization of the basic geometric structures of the edges with wavelet transform can also be established. However, it is more complicated to do so because the 2D image has multi directions and therefore two wavelet functions have to be applied. In this chapter, let us consider a special class of wavelet functions, $\psi^1(u, v)$ and $\psi^2(u, v)$, whose moduli and angles can be separated, such that

$$\psi^1(u, v) = \psi(r)k_1(\theta), \quad \psi^2(u, v) = \psi(r)k_2(\theta),$$

where $\psi(r)$ stands for the function with respect to the modulus $r = \sqrt{u^2 + v^2}$ and $k_1(\theta)$ and $k_2(\theta)$ denote the functions with respect to the angle $\theta = arctan(\frac{v}{u})$. They are referred to as *modulus-angle-separated wavelets* (MASWs).

In the remainder of this section, we prove that the MASWs possess two significant properties:

- The wavelet transforms of the edges are independent of the gradient directions of edges.
- The wavelet transform of the step structure edge is independent of the scales of the transform.

The first characteristic provides an ability to detect edges in different directions, while the second one can be devoted to identify the step structure edges from others. These are important in the concrete applications.

The 2D MASW transform is defined as follows:

$$W_s^i f(x, y) = \iint f(x - u, y - v)\psi_s^i(u, v)dudv, \quad (i = 1, 2),$$

where

$$\psi_s^i(u, v) = \frac{1}{s^2}\psi^i\left(\frac{u}{s}, \frac{v}{s}\right) \quad (i = 1, 2).$$

Particularly, if $s = 2^j$ $(j \in \mathbb{Z})$, they are called the dyadic wavelets.

It is easy to show that both $\psi^1$ and $\psi^2$ are 2D wavelets, if $\int_0^{2\pi} k_1(\theta)d\theta = \int_0^{2\pi} k_2(\theta)d\theta = 0$. To simplify the derivation without losing the generality, a special case is applied in this study. In this way, $k_1(\theta)$ and $k_2(\theta)$ now become

$$k_1(\theta) = \cos\theta, \quad k_2(\theta) = \sin\theta.$$

Let

$$\nabla W_s f(x, y) := \begin{pmatrix} W_s^1 f(x, y) \\ W_s^2 f(x, y) \end{pmatrix}$$

and

$$|\nabla W_s f(x, y)| := \sqrt{|W_s^1 f(x, y)|^2 + |W_s^2 f(x, y)|^2}.$$

Similarly, Theorem 6.1 can be extended to the 2D images [Mallat and Hwang, 1992] and is presented as follows:

**Theorem 6.3.** *Let $\alpha$ be a real number, $0 < \alpha < 1$, then function $f(x, y)$ is uniformly Lipschitz $\alpha$ over $(a, b) \times (c, d)$, if and only if there exists a constant $K$ such that, $\forall (x, y) \in (a, b) \times (c, d)$, we have*

$$|\nabla W_s f(x)| \leq K s^\alpha. \tag{6.16}$$

For 2D images, the geometric structures of edges are much more complicated than that of the 2D signals. Nevertheless, the step and Dirac structures of edges, in practice, are considered as contours in images. Thus, we emphasize to discuss these two structures in this section. The edge shown in Fig. 6.15(a) is said to be an ideal step structure edge, and contours of patterns are typical examples of it. The edge illustrated in Fig. 6.15(b) is called an ideal curve-Dirac edge which consists of many impulses at the gradient direction. The drawing lines in the image are examples of this edge. Another type of Dirac edge is shown in Fig. 6.15(c) which is called an ideal single-point-Dirac edge, and it possesses an impulse in any direction. It is a special case of the curve-Dirac edges, and white noise belongs to this edge. The edge can be regarded as a polygonal function, which consists of many segments. Suppose $\theta$ is an angle between any two adjacent segments, as shown in Fig. 6.15(a). If angle $\theta = \pi$, these two segments become a straight line on the edge. The main result in this section is as follows:

**Theorem 6.4.** *Let $\psi(r)$ be a compactly supported function on the interval of $[0, \infty)$. Then $\psi^1(u, v)$ and $\psi^2(u, v)$ are two 2D MASWs with the form*

$$\psi^1(u, v) = \psi(r)\cos\theta, \quad \psi^2(u, v) = \psi(r)\sin\theta, \tag{6.17}$$

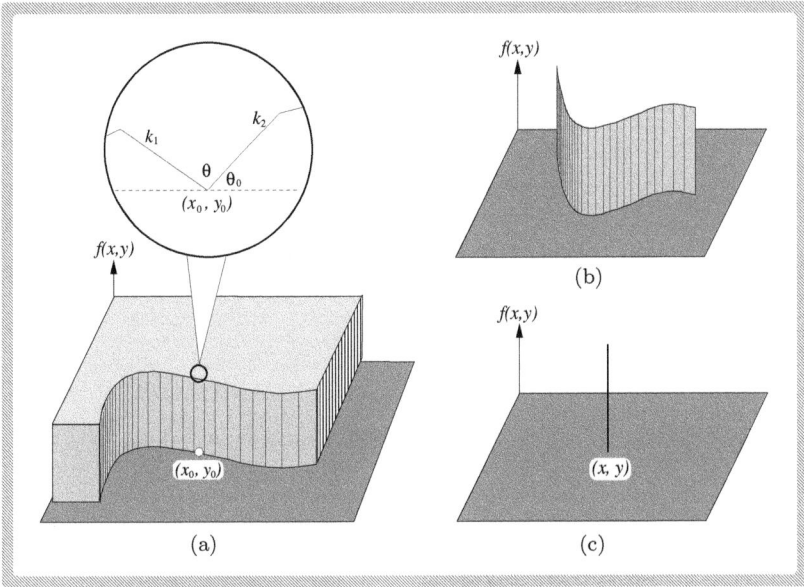

Fig. 6.15 Two-dimensional ideal edges.

*where* $r = \sqrt{u^2 + v^2}$ *and* $\theta = \arctan \frac{v}{u}$. *Furthermore, we have the following:*

(1) *If* $(x_0, y_0)$ *is the coordinate of a point on the ideal step edge with angle* $\theta$, $0 < \theta < 2\pi$, *then*

$$|\nabla W_s e(x_0, y_0)| = 2 \left| \sin \frac{\theta}{2} \right| \cdot \left| \int_0^\infty r\psi(r)dr \right|. \tag{6.18}$$

(2) *If* $(x_0, y_0)$ *is the coordinate of a point on the Dirac edge, then*

$$|\nabla W_s e(x_0, y_0)| = 0. \tag{6.19}$$

(3) *If* $(x_0, y_0)$ *is the coordinate of a point on the roof edge, then*

$$|\nabla W_s e(x_0, y_0)| \to 0, \quad (s \to 0). \tag{6.20}$$

**Proof.**

(1) First, the step structure edges are considered. Suppose that two polyg-onal lines shown in Fig. 6.15(a) are met at the coordinate $(x_0, y_0)$. Let $\theta$ be the angle between these two lines, $k_1$ and $k_2$ be the slopes of them,

and $\theta_0$ be the angle between $x$-axis and one side. It can be computed that

$$W_s^1 f(x_0, y_0) = \iint_{v \leq k_1 u \text{ and } v \leq k_2 u} \psi_s^1(u, v) du dv$$

$$= \int_0^\infty r dr \int_{\theta_0}^{\theta_0 + \theta} \frac{1}{s^2} \psi\left(\frac{r}{s}\right) \cos \alpha d\alpha$$

$$= \int_0^\infty r dr \int_{\theta_0}^{\theta_0 + \theta} \psi(r) \cos \alpha d\alpha$$

$$= \left(\int_0^\infty r\psi(r) dr\right) [\sin(\theta_0 + \theta) - \sin \theta_0]$$

$$= \left(\int_0^\infty r\psi(r) dr\right) 2\cos\left(\theta_0 + \frac{\theta}{2}\right) \sin \frac{\theta}{2}.$$

Similarly,

$$W_s^2 f(x_0, y_0) = \iint_{v \leq k_1 u, v \leq k_2 u} \psi_s^2(u, v) du dv$$

$$= \left(\int_0^\infty r\psi(r) dr\right) 2\sin\left(\theta_0 + \frac{\theta}{2}\right) \sin \frac{\theta}{2}.$$

Hence, we obtain

$$|\nabla W_s e(x_0, y_0)| = 2 \left|\sin \frac{\theta}{2}\right| \left|\int_0^\infty r\psi(r) dr\right|.$$

(2) A curve-Dirac edge can be approximated by a sequence of segments which are straight lines, if the curve is smooth. We can only consider the straight lines when the wavelet $\psi$ is compactly supported. Let the parameter equation of a straight line be the form

$$\begin{cases} u = x_0 - at; \\ v = y_0 - bt. \end{cases} \quad t \in (-\infty, \infty).$$

Then

$$W_s^1 e_l(x_0, y_0) = \int_{-\infty}^\infty \psi_s^1(at, bt) dt$$

$$= \int_{-\infty}^\infty \psi\left(\frac{\sqrt{a^2 + b^2}}{s} |t|\right) \frac{at}{\sqrt{a^2 + b^2}} dt$$

$$= 0.$$

We can also deduce that

$$W_s^2 e_l(x_0, y_0) = 0.$$

Therefore, we have

$$|\nabla W_s e(x_0, y_0)| = 0.$$

Similarly, the result for the single-point-Dirac edges can be obtained.
(3) For the roof structure edges, since $\alpha = 1$, we have

$$|\nabla W_s e(x_0, y_0)| \le Ks. \tag{6.21}$$

It is clear that when $s \to 0$, then $Ks \to 0$. Thus, $|\nabla W_s e(x_0, y_0)| \to 0$.

It is easily seen that the modulus $|\nabla W_s e(x_0, y_0)|$ of the MASW transform with respect to an ideal step structure edge is a non-zero constant which is independent of both the gradient direction of the edge and the scale of the transform, if $\int_0^\infty r\psi(r)dr \ne 0$. Based on this property, the step structure edges with various orientations can be detected effectively. The modulus $|\nabla W_s e(x_0, y_0)|$ reaches the maxima, if $\theta = \pi$. Therefore, the closer to $\pi$ the angle is, the more effective the detection will be. $\qquad \Box$

In practice, signals are usually much more complicated than the ideal ones. They can be regarded as the overlaying of the ideal edges and a background which is a smooth function, or a function with less singularity. Thus, a practical signal $f(x)$ can be written as

$$f(x) = f_0(x) + \sum_i e_i(x),$$

where $e_i(x)$ denotes an ideal edge and $f_0(x)$ is a smooth signal which has less singularity or even no singularity.

By applying wavelet transform to $f_0(x) + \sum_i e_i(x)$, we have

$$W_s f(x) = W_s f_0(x) + \sum_i W_s e_i(x).$$

Thus,

$$\left| W_s f(x) - \sum_i W_s e_i(x) \right| = |W_s f_0(x)|.$$

Theorem 6.1 shows that $|W_s f_0(x)|$ at the edge $x_0$ is usually much smaller than $|W_s e_i(x)|$, since $f_0(x)$ has smaller singularity. Hence, the following important conclusion can be obtained:

$$W_s f(x) \approx \sum_i W_s e_i(x).$$

Consequently, all results derived from the ideal case in this section can be utilized in the practical signals.

Consider a signal with a noisy function $n(x)$ which is a real, wide-sense stationary white noise of variance $\sigma^2$. Thus, its correlation function is

$$E(n(u)n(v)) = \sigma^2 \delta(u - v) = \begin{cases} \delta^2, & \text{for } u = v; \\ 0, & \text{otherwise.} \end{cases}$$

Since

$$|W_s n(x)|^2 = W_s n(x) \overline{W}_s n(x)$$
$$= \iint n(u) \overline{n}(v) \psi_s(x - u) \overline{\psi}_s(x - v) du dv,$$

we have

$$E|W_s n(x)|^2 = \iint \sigma^2 \delta(u - v) \psi_s(x - u) \overline{\psi}_s(x - v) du dv$$
$$= \sigma^2 \int |\psi_s(x - u)|^2 du$$
$$= \sigma^2 \frac{1}{s} \int |\psi_u|^2 du$$
$$= \frac{1}{s} \sigma^2 \|\psi\|_2,$$

which shows that the variance of the wavelet transform of the noise is dependent on the scale. The larger the scale the lesser the variance of the wavelet transform of the noise. Therefore, we can synthesize several scales to detect edges and to remove the noise.

### 6.3.3 *Scale-Independent Algorithm*

Based on the above analyses, the concrete algorithm can be developed. This algorithm possesses two objectives: (1) to extract the step structure edges for 1D and 2D signals; (2) to eliminate other edges, such as drawing lines, noise, and texts.

In view of Corollary 6.1, this implies that the wavelet transform of the step structure edge is a non-zero constant without considering the scales $s$. In other words, for different values of scales $s_1, s_2, \ldots, s_J$, their wavelet transforms with respect to the step structure edges, $W_{s_1} f(x)$, $W_{s_2} f(x), \ldots, W_{s_J} f(x)$, are non-zero constants and are equal each to each. At the points on the step structure edge, it is clear that

$$\frac{W_{s_j} f(x)}{W_{s_l} f(x)} = 1, \quad (j, l = 1, 2, \ldots, J). \tag{6.22}$$

In practice, the sign of equality in Eq. (6.22) may not be satisfied due to the following reasons:

- The edge is not an ideal one as mentioned in the previous subsection.
- The image to be processed is distorted by noise.
- The background of the image may bring a small error in the calculation of the wavelet transforms.
- Other factors may influence the result, for example, the accuracy of the computer system may have an effect on the computation.

Therefore, $W_{s_i} f(x)$, $i = 1, 2, \ldots, J$, may not keep the same value for different scales. Consequently, the equality sign in Eq. (6.22) should be replaced by the approximate sign, and Eq. (6.22) now becomes

$$\frac{W_{s_j} f(x)}{W_{s_l} f(x)} \approx 1, \quad (j, l = 1, 2, \ldots, J). \tag{6.23}$$

As a matter of fact, the points which satisfy Eq. (6.23) are considered to be a step structure edge.

However, the approximate sign creates uncertainty in extracting edge points. In order to avoid this uncertainty, and to facilitate the design of algorithm detecting the step structure edges, we utilize a real number $R$, instead of the integer number 1. Let $R$ be very close to 1 and greater than 1. In replacing sign "$\approx$" by sign "$\leq$", Eq. (6.23) is equivalent to

$$\frac{W_{s_j} f(x)}{W_{s_l} f(x)} \leq R, \quad \text{and} \quad \frac{W_{s_l} f(x)}{W_{s_j} f(x)} \leq R, \quad (j, l = 1, 2, \ldots, J). \tag{6.24}$$

The closer to 1 the real number $R$ the closer to 1 the $\frac{W_{s_j}f(x)}{W_{s_l}f(x)}$, i.e.,

$$R \to 1 \iff \frac{W_{s_j}f(x)}{W_{s_l}f(x)} \to 1.$$

Moreover, both $W_{s_j}f(x)$ and $W_{s_l}f(x)$ are positive values or negative ones at the same time. In other words, the following is held:

$$\frac{W_{s_j}f(x)}{W_{s_l}f(x)} > 0, \quad \text{and} \quad \frac{W_{s_l}f(x)}{W_{s_j}f(x)} > 0, \ (j, l = 1, 2, \ldots, J). \tag{6.25}$$

Combining Eqs. (6.24) and (6.25) produces

$$0 < \frac{W_{s_j}f(x)}{W_{s_l}f(x)} \leq R, \quad \text{and} \quad 0 < \frac{W_{s_l}f(x)}{W_{s_j}f(x)} \leq R, \ (j, l = 1, 2, \ldots, J), \tag{6.26}$$

which is equivalent to

$$\frac{1}{R} \leq \frac{W_{s_j}f(x)}{W_{s_l}f(x)} \leq R, \quad (j, l = 1, 2, \ldots, J), \tag{6.27}$$

in the view of mathematics. The real number $R$ in Eq. (6.27) is referred to as *Proportional Threshold.*

Because the edges are high-frequency signals, the wavelet transform of them may have large values, thus only the points, which have large values, are considered be as edges. Therefore, a threshold, $T$, is established, such that if the points satisfy

$$|W_{s_j}f(x)| \geq T, \tag{6.28}$$

they are on the edge. Number $T$ in Eq. (6.28) is referred to as *Peak Threshold.*

Finally, we can summarize that the points on the step structure edges have to satisfy

$$|W_{s_j}f(x)| \geq T, \quad \text{and} \quad \frac{1}{R} \leq \frac{W_{s_j}f(x)}{W_{s_l}f(x)} \leq R, \quad (j, l = 1, 2, \ldots, J). \tag{6.29}$$

In view of the discussion in the above, the algorithm can be designed as follows. Since the wavelet transform of the step structure edges is scale-independent, this algorithm is referred to as *Scale-Independent Algorithm.*

**Algorithm 6.3.** Detecting step structure edges:

(1) Given 1D signal $f(x)$:

**Step 1:** Take different scales $s_1, \ldots, s_J$ and calculate $W_{s_j} f(x)$, $(1 \le j \le J)$ based on Eq. (6.2) in the previous section.

**Step 2:** Select peak threshold $T$, such that

$$|W_{s_j} f(x)| \ge T.$$

**Step 3:** Select proportional threshold $R$, such that

$$\frac{1}{R} \le \frac{W_{s_j} f(x)}{W_{s_l} f(x)} \le R, \quad (1 \le j \le J).$$

Then $x$ is detected as a pixel of the step edge.

(2) For 2D signal $f(x, y)$:

**Step 1:** Take different scales $s_1, \ldots, s_J$, and calculate $W_{s_j} f(x, y)$, $(1 \le j \le J)$ based on Eq. (6.3) in the previous section.

**Step 2:** Select peak threshold $T$, such that

$$|\nabla W_{s_j} f(x, y)| \ge T.$$

**Step 3:** Select proportional threshold $R$, such that

$$\frac{1}{R} \le \frac{|\nabla W_{s_j} f(x, y)|}{|\nabla W_{s_l} f(x, y)|} \le R, \quad (1 \le j \le J).$$

Then $(x, y)$ is detected as a pixel of the step edge.

It should be pointed out that there is no deviation when the step edges are detected by the above schemes.

### 6.3.4 *Experiments*

We focus this section on the verification of the effectiveness of the proposed novel wavelet-based method by experiments. Three types of experiments have been done: (1) extraction of contours of 2D objects, (2) removal of background in images, and (3) preprocessing in document analysis and recognition. Seven examples are presented in this section.

In these experiments, Eqs. (6.3) and (6.6) have been applied to compute the filtering coefficients $\{\psi_{k,l}^{s,1}\}$ in the MASW transform. These coefficients have been used in the following examples:

First, the experiments of detecting boundaries of 2D objects are presented. Let us look back at Fig. 6.9. The particular task is that we are required to extract the contour of the aircraft and to remove all lines and texts. Unfortunately, the algorithm based on the modulus maxima of the wavelet transform [Mallat and Hwang, 1992; Mallat and Zhong, 1992] has detected all edges without identifying different structures of edges. Thus, the resulting image contains lines and texts which are required to be deleted from the image. The new method developed in this section possesses an important property, i.e., the wavelet transform of a step structure edge is scale-independent. It can improve the method of the modulus maxima, and a new result can be found in Fig. 6.16. The original image is shown in Fig. 6.16(a) containing a planner object, aircraft, with several drawing lines and texts. Figures 6.16(b) and 6.16(e) display the modulus and angle images undergone by the MASW transform with a scale of $s = 2$, respectively. Figures 6.16(c) and 6.16(f) give the modulus and angle images by the MASW transform with scale of $s = 4$, respectively. Figures 6.16(d) and 6.16(g) provide the modulus and angle ones with scale of $s = 8$, respectively. After applying Eq. (6.29) to these images, the resulting image is obtained, as shown in Fig. 6.16(h). It is clear that only the contour of the aircraft has been extracted, while all other edges including drawing lines and texts have been eliminated.

The second example of extracting the contour of the planner objects is illustrated in Fig. 6.17. The original image is given in Fig. 6.17(a). We take $s = 2, 4, 8$, thereafter, the modulus images are obtained in Figures 6.17(b)–6.17(d), the angle ones are produced in Figs. 6.17(e)–6.17(g). The result is presented in Fig. 6.17(h), where the contour of the 2-D object has been detected, while the others have been removed.

Now, we turn to other examples shown in Figs. 6.18 and 6.19. We are required to extract the boundaries of the characters, for example, English printed character "A" and Chinese handwriting "book", and to delete the drawing lines and noise. The same problem was arisen when different scales, for example, $s = 2, 4, 8$, were used in the modulus maxima of the wavelet transform. To contest this deficiency, the scale-independent wavelet method is applied to these examples. For the English printed character, the original image is shown in Fig. 6.18(a) containing a character, English letter "A", with several drawing lines and white noise. Figures 6.18(b)–6.18(d) display the modulus images of the MASW transforms with scales $s = 2, 4, 8$, respectively. Figures 6.18(e)–6.18(g) show the angle images of the MASW transforms of $s = 2, 4, 8$, respectively. After applying Eq. (6.29) to these images, the resulting image is obtained, as shown in Fig. 6.18(h). Only the

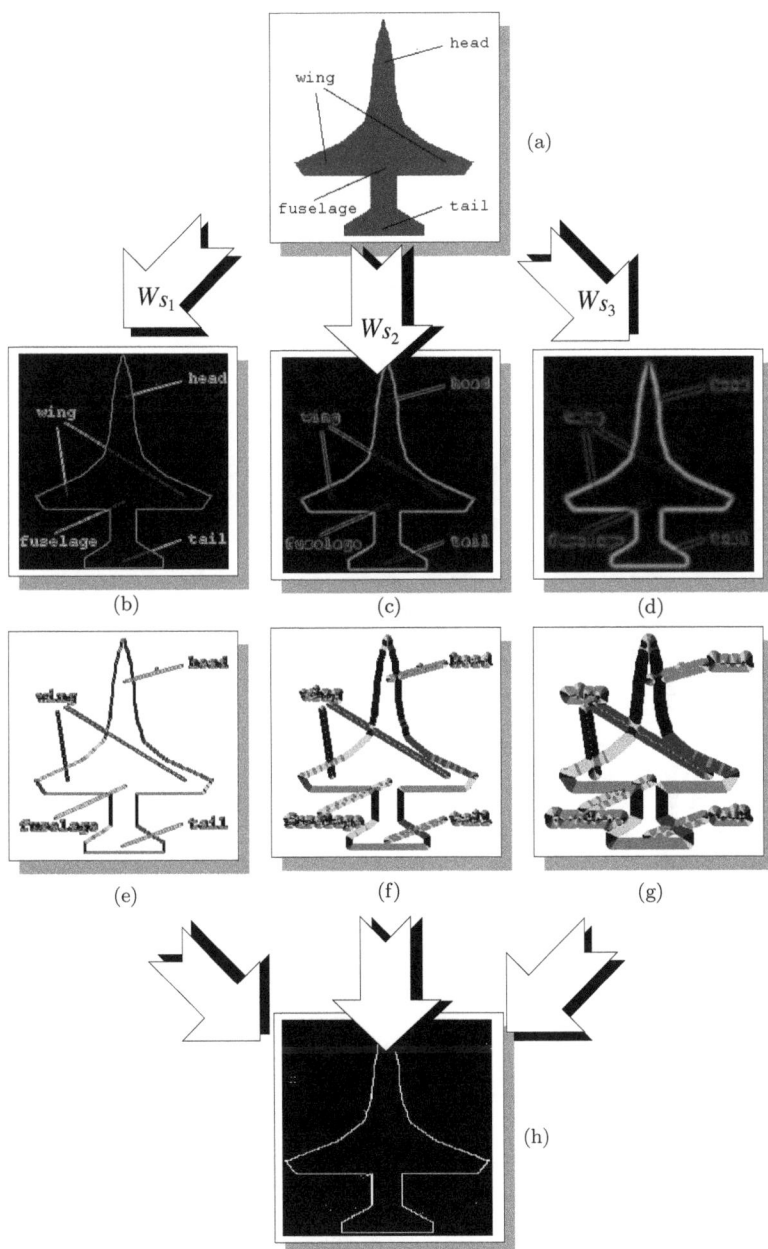

Fig. 6.16  Example 1: The contour of the aircraft is detected using the proposed algorithm.

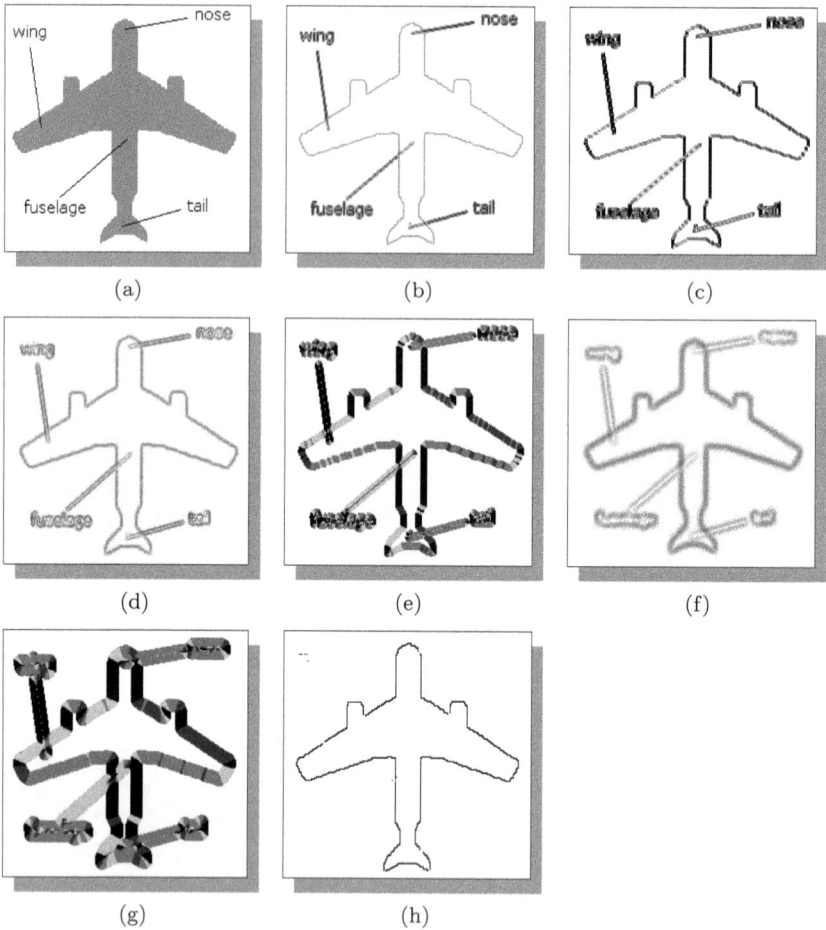

Fig. 6.17   Example 2: The contour of the aircraft is detected using the proposed algorithm.

boundary of the character has been extracted, and all other edges including drawing lines and noise have been eliminated.

Detection of the contour from a Chinese handwriting using the scale-independent algorithm can be found in Fig. 6.19. Where the original image is depicted in Fig. 6.19(a), and the resulting one is illustrated in Fig. 6.19(h).

The proposed method can be applied to document processing. An example is depicted in Fig. 6.20, which is an improvement of the results (see

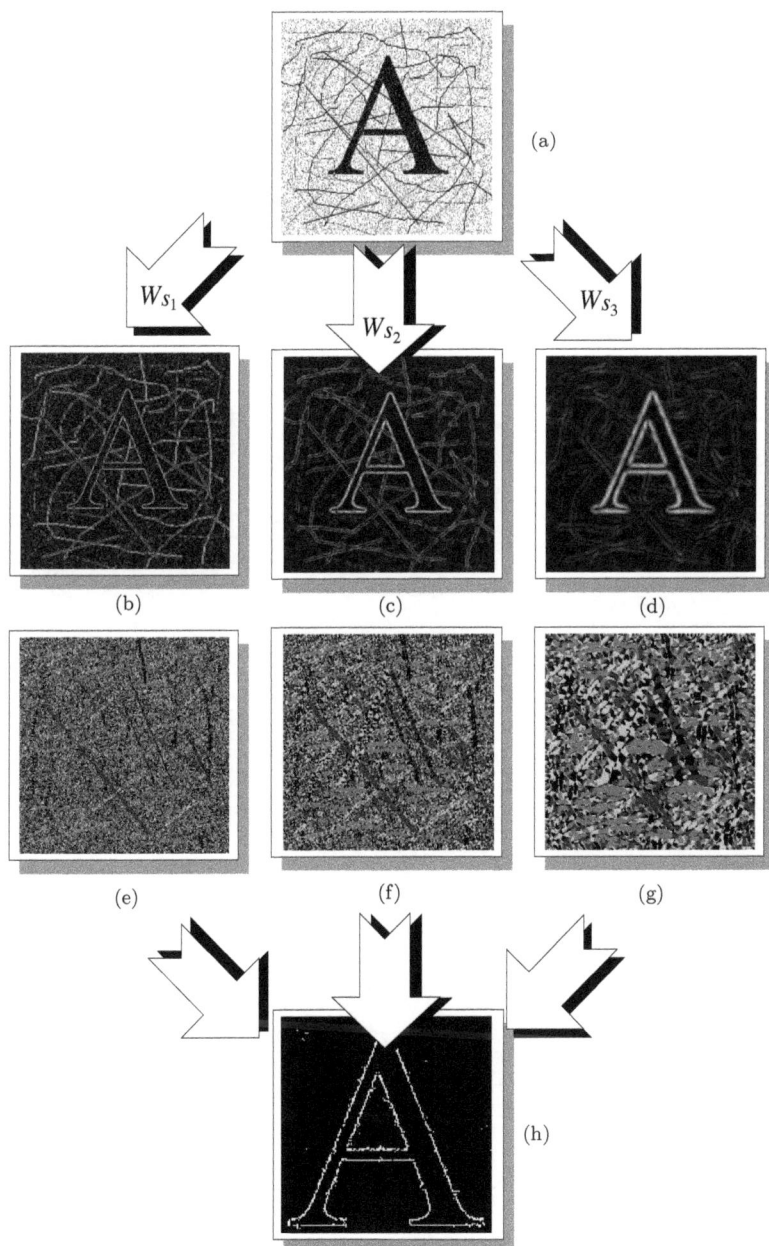

Fig. 6.18 Example 3: The boundary of English letter "A" is extracted using the proposed algorithm.

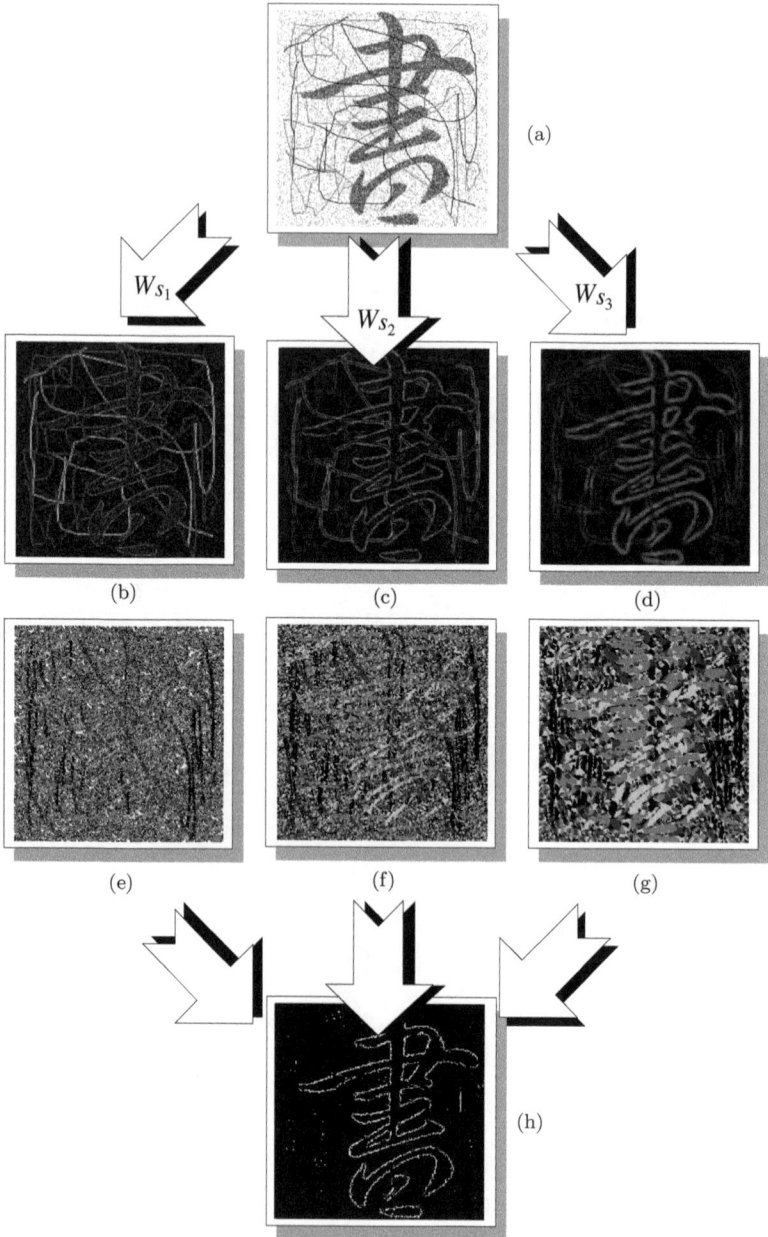

Fig. 6.19  Example 4: The contours of Chinese handwriting "book" are extracted using the proposed algorithm.

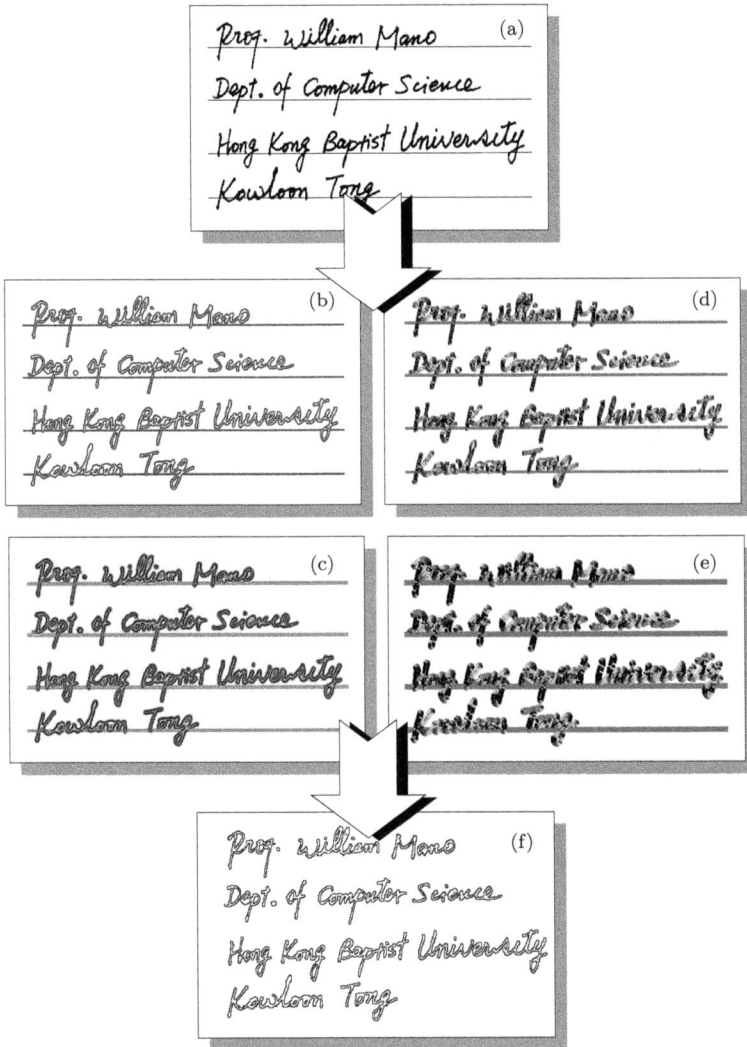

Fig. 6.20    Example 5: The boundaries of English letters are extracted using the proposed algorithm.

Figs. 6.10 and 6.11) produced by the modulus maxima. The original image is shown in Fig. 6.20(a). It is a mailing address that is written on a paper with guide lines. In order to automatically process it by computers, a pre-processing is to remove the guide lines and extract the contours of the handwriting English letters. Figures 6.20(b)–6.20(c) indicate the modulus

Fig. 6.21   Example 6: The contours of handwriting Chinese characters are extracted using the proposed algorithm.

images of MASW transform with scales $s = 2$ and $s = 4$, respectively. Figs. 6.20(d)–6.20(e), display the angle images with wavelet transform of scales $s = 2$ and $s = 4$, respectively. Using the algorithm mentioned in previous section produces an image in Fig. 6.20(f), which is what we need.

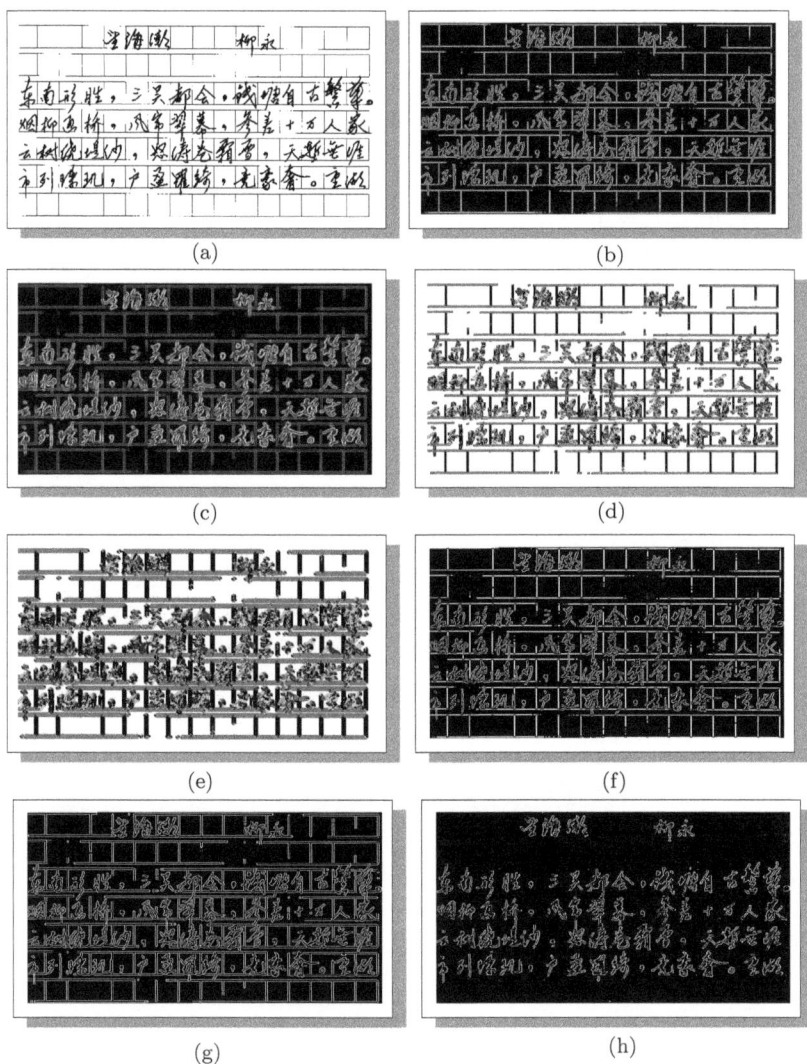

Fig. 6.22 Example 7: The boundaries of handwriting Chinese characters are extracted using the proposed algorithm.

The sixth example is shown in Fig. 6.21, where a Chinese article is written on a paper with guide mesh. Figure 6.21(a) gives the original image. Figures 6.21(b) and 6.21(c) display the modulus and angle images undergone by the MASW transform with a scale of $s = 2$, respectively. Figures 6.21(d) and 6.21(e) give the modulus and angle images by the MASW transform with a scale of $s = 4$, respectively. After applying Eq. (6.29) to these images, the resulting image is obtained, as depicted in Fig. 6.21(h).

The seventh example is depicted in Fig. 6.22. The original image is illustrated in Fig. 6.22(a), and the resulting ones in accordance with the modulus maxima are depicted in Figs. 6.22(f) and 6.22(g), which are not the desired images, since they include not only the Chinese handwriting but also the guide lines as well. Applying our method produces a good result, which is displayed in Fig. 6.22(h), in which only Chinese handwritten characters are kept and the others are removed.

# Chapter 7

# Characterization of Dirac Edges with Quadratic Spline Wavelet Transform

This chapter aims at studying the characterization of Dirac structure edges with wavelet transform and selecting the suitable wavelet functions to detect them. Three significant characteristics of the local maximum modulus of the wavelet transform with respect to the Dirac structure edges are presented: (1) Slope invariant — the local maximum modulus of the wavelet transform of a Dirac structure edge is independent of the slope of the edge. (2) Gray-level invariant — the local maximum modulus of the wavelet transform with respect to a Dirac structure edge takes place at the same points when the images with different gray levels are processed. (3) Width light-dependent — for various widths of the Dirac structure edge images, the location of maximum modulus of the wavelet transform varies lightly under certain circumscription that the scale of the wavelet transform is larger than the width of the Dirac structure edges. It is important, in practice, to select the suitable wavelet functions, according to the structures of edges. For example, Haar wavelet is better to represent brick-like images than other wavelets. A mapping technique is applied in this chapter to construct such a wavelet function. In this way, a low-pass function is mapped onto a wavelet function by a derivation operation. In this chapter, the quadratic spline wavelet is utilized to characterize the Dirac structure edges and an algorithm to extract the Dirac structure edges by wavelet transform is also developed.

A mathematical characterization of three basic geometric structures of edges (i.e., step structure, roof structure, and Dirac structure) with Lipschitz exponents is presented in Chapter 5. A significant property has been proved that the modulus of wavelet transform at each point of the step edge is a non-zero constant which is independent of both the gradient direction and the scale of the wavelet transform. This property led to provide a simple and direct strategy for detecting a specific structure of edges: the step structure edges. Thus, an algorithm called scale-independent algorithm has been developed. The method developed in Chapter 5 possesses an important property, i.e., the wavelet transform of a step structure edge is scale-independent. It can improve the method proposed in the works of Mallat and Hwang [1992] and Mallat and Zhong [1992], and the result can be found in Fig. 6.17 of Chapter 6, where the MASW has been used. The precise definition of the MASW can be found in Chapter 5 and the work of Tang *et al.* [1998c]. The original image is shown in Fig. 6.17(a) containing a planner object, aircraft, with several drawing lines and texts. Figures 6.17(b) and 6.17(c) display the modulus and angle images undergone by the MASW transform with is scale of $s = 2$, respectively. Figures 6.17(d) and 6.17(e) give the modulus and angle images by the MASW transform with a scale of $s = 4$, respectively. Figures 6.17(f) and 6.17(g) provide the modulus and angle images with a scale of $s = 8$, respectively. After applying the scale-independent algorithm to these images, the resulting image is obtained, as shown in Fig. 6.17(h). It is clear that only the contour of the aircraft has been extracted, while all other edges including drawing lines and texts have been eliminated.

On the other hand, as a complement to Chapter 5, the purpose of this chapter is to develop a method which can identify different structures of edges and, thereafter, detect the Dirac structure edges such as the drawing lines and texts in Figs. 6.17(a) and eliminate the step structure edges such as the contour of the aircraft in Figs. 6.17(a).

## 7.1  Selection of Wavelet Functions by Derivation

In practice, the selection of a suitable wavelet function in accordance with the structure of the edges is an important topic in the application of wavelet transform to image processing and pattern recognition. In this section, a method of selection of wavelet function by derivation of the low-pass function is presented.

### 7.1.1  *Scale Wavelet Transform*

Let $L^2(R^2)$ be the Hilbert space of all the square-integrable 2D functions on plane $R^2$, $\psi \in L^2(R^2)$ is called a wavelet function, if

$$\int_R \int_R \psi(x, y)dxdy = 0. \tag{7.1}$$

For $f \in L^2(R^2)$ and scale $s > 0$, the scale wavelet transform of $f(x, y)$ is defined by

$$W_s f(x, y) := (f * \psi_s)(x, y)$$

$$= \int_R \int_R f(u, v) \frac{1}{s^2} \psi \left( \frac{x - u}{s}, \frac{y - v}{s} \right) du dv, \qquad (7.2)$$

where $*$ denotes the convolution operator and

$$\psi_s(u, v) := \psi \left( \frac{u}{s}, \frac{v}{s} \right).$$

The theory dealing with the scale wavelet transform can be found in many articles, such as the works of Chui [1992a] and Daubechies [1992]. In practice, the wavelet transform can be calculated discretely using the following formula:

$$W_s f(n, m) = \iint f(u, v) \psi_s(n - u, m - v) du dv$$

$$= \sum_{k,l} f(k, l) \int_k^{k+1} \int_l^{l+1} \psi_s(n - u, m - v) du dv$$

$$= \sum_{k,l} f(k, l) \int_{n-k-1}^{n-k} \int_{m-l-1}^{m-l} \psi_s(u, v) du dv$$

$$= \sum_{k,l} f(n - k - 1, m - l - 1) \psi_{k,l}^s,$$

where

$$\psi_{k,l}^s = \int_k^{k+1} \int_l^{l+1} \psi_s(u, v) du dv = \int_{k/s}^{(k+1)/s} \int_{l/s}^{(l+1)/s} \psi(u, v) du dv.$$

Obviously, the scale wavelet transform described in Eq. (7.2) is a filter in essence. We can conclude that its Fourier transform defined by

$$\hat{\psi}(\xi, \eta) := \int_R \int_R \psi(x, y) e^{-i(\xi x + \eta y)} dx dy$$

satisfies the condition of $\hat{\psi} \in L^2(R^2)$, since $\psi \in L^2(R^2)$. Thus, both functions $\psi$ and $\hat{\psi}$ decrease at infinity. Usually, $\psi$ is chosen to meet the following two conditions at least: First, $\psi \in L^2(R^2)$ has to be held; second, both $\psi$ and $\hat{\psi}$ decrease with exponents at infinity. According to Eq. (7.1), we see that $\hat{\psi}(0, 0) = 0$, which implies that the scale wavelet transform $W_s f(x, y)$ corresponds to a band pass filter essentially. Moreover, $W_s f(x, y)$

characterizes the local properties of $f(x, y)$ on both the time and frequency domains:

- In the time domain, it is easy to see that

$$W_s f(x, y) := (f * \psi_s)(x, y) = \int_R \int_R f(x - u, y - v)\psi_s(u, v)dudv$$

  characterizes the local property of $f$ around the point $(x, y)$. The smaller the scale $s$, the more narrow the time window.
- On the frequency domain, based on the basic properties of Fourier transform, it can be concluded that

$$(W_s f)(x, y) = \left(\frac{1}{2\pi}\right)^2 \int_R \int_R (W_s f)\hat{}\,(\xi, \eta)e^{i(\xi x + \eta y)}d\xi d\eta$$

$$= \left(\frac{1}{2\pi}\right)^2 \int_R \int_R \hat{f}(\xi, \eta)(\psi_s)\hat{}\,(\xi, \eta)e^{i(\xi x + \eta y)}d\xi d\eta$$

$$= \left(\frac{1}{2\pi}\right)^2 \int_R \int_R \hat{f}(\xi, \eta)\hat{\psi}(s\xi, s\eta)e^{i(\xi x + \eta y)}d\xi d\eta.$$

Since $\hat{\psi}(s\xi, s\eta)e^{i(\xi x + \eta y)}$ decreases as $\hat{\psi}(s\xi, s\eta)$ does, thus $(W_s f)(x, y)$ characterizes the local property of $\hat{f}$. The position of the locality depends on the scale $s$ and function $\psi$. And the larger the scale $s$, the narrower the frequency window.

The reduction of $\psi$ and $\hat{\psi}$ at infinity plays an important role for characterizing the local properties of $f(x, y)$ on both the time and frequency domains. The faster the decrement of $\psi$ and $\hat{\psi}$, the better the characterization of the local properties. In other words, the shorter the support of $\psi$ and $\hat{\psi}$, the higher the quality of the localization obtained.

## 7.1.2 *Construction of Wavelet Function by Derivation of the Low-pass Function*

The Dirac structure edges such as curves are transient components with high frequencies in an image. They are highly localized in spatial positions. Such components do not resemble any of the wave basis functions, such as Fourier basis function. This makes the Fourier and other wave transforms less than optimal representations for analyzing and processing the Dirac structure edges in the images. To combat such a deficiency, wavelet analysis can be utilized. Wavelet theory is a good mathematical tool primarily used for representing such transient components more efficiently. In fact, the Dirac structure edges can be characterized by the wavelet transforms.

However, one of the key factors which obstruct the application of wavelet transform to process these Dirac structure edges is that it is still difficult, in practice, to select the suitable wavelet functions, which possess a perfect characteristic of localization.

Theoretically, the detection of the Dirac structure edges by wavelet transforms can be regarded as a particular filtering operation. The derivative function of a smooth low-pass function which decreases at infinite can become a candidate of the wavelet function. It can be considered to be a mapping, i.e., a low-pass function can be mapped onto a wavelet function by the operation of derivation, which can also be described by the following:

$$\underbrace{\text{Low} - \text{Pass Function} \overset{Mapping}{\underset{Derivation}{\Longrightarrow}} \text{Wavelet Function}}. \qquad (7.3)$$

According to Eq. (7.3), we can use such a method to produce a wavelet function in the following steps: (1) selecting a low-pass function $\theta(x, y)$ and (2) deriving this low-pass function to produce the wavelet functions $\psi^1(x, y)$ and $\psi^2(x, y)$.

Let $\theta(x, y)$ be a real function and satisfy the following:

- $\theta(x, y)$ fast decreases at infinity.
- $\theta(x, y)$ is an even function on both $x$ and $y$.
- $\hat{\theta}(0, 0) = 1$.

Consider

$$\psi^1(x, y) := \frac{\partial}{\partial x}\theta(x, y),$$

it is easy to see that $\int \int \psi^1(x, y)dxdy = 0$, which indicates that $\psi^1(x, y)$ is a 2D wavelet. Then its scale wavelet transform is

$$W_s^1 f(x, y) = (f * \psi_s^1)(x, y)$$

$$= \left(f * s\frac{\partial}{\partial x}\theta_s\right)(x, y)$$

$$= s\frac{\partial}{\partial x}(f * \theta_s)(x, y),$$

where $\theta_s(x, y) := \frac{1}{s^2}\theta(\frac{x}{s}, \frac{y}{s})$. Based on this formula, it is clear that $f * \theta_s$ is a smooth operation with scale $s$, if $\theta$ is a smooth function with fast decreasing at infinity and $\hat{\theta}(0, 0) = 1$. When a quadratic spline function is utilized as the primitive function $\theta(x, y)$, the graphical description of the

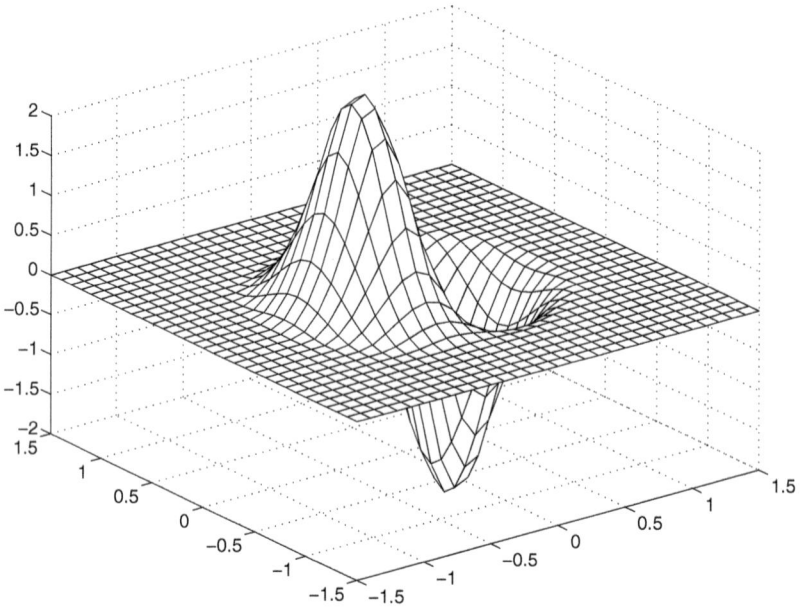

Fig. 7.1    The graphical description of derivative function $\psi^1(x, y) = \frac{\partial}{\partial x}\theta(x, y)$.

derivative function of the function $\theta(x, y)$ with respect to the horizontal axis $x$ is illustrated in Fig. 7.1.

Since $W_s^1 f(x, y)$ is the derivative function of function $\theta(x, y)$ along the horizontal axis, the characteristic of the local maxima modulus of the wavelet transform mainly influences the transient components of an image along the horizontal axis.

Similarly, let

$$\psi^2(x, y) := \frac{\partial}{\partial y}\theta(x, y).$$

It is also a 2D wavelet, and its scale wavelet transform is

$$W_s^2 f(x, y) = (f * \psi_s^2)(x, y)$$

or

$$W_s^2 f(x, y) = s\frac{\partial}{\partial y}(f * \theta_s)(x, y),$$

which is the derivative function of function $\theta(x, y)$ along the vertical axis. Therefore, the characteristic of the local maxima modulus of the wavelet transform mainly influences the transient components of an image along

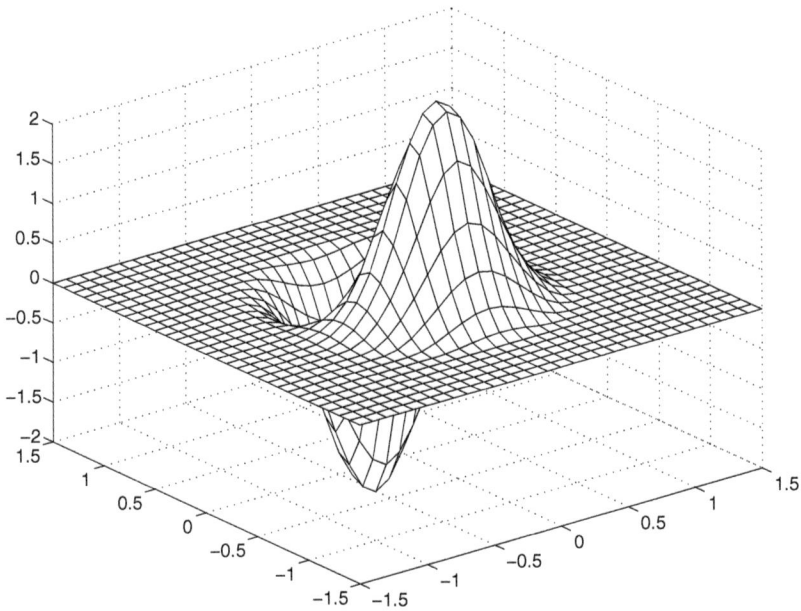

Fig. 7.2    The graphical description of derivative function $\psi^2(x, y) = \frac{\partial}{\partial y}\theta(x, y)$.

the vertical axis. The graphical description of the derivative function of the function $\theta(x, y)$ with respect to the vertical axis $y$ is illustrated in Fig. 7.2, where the primitive function is a quadratic spline function.

We denote the gradient direction and the amplitude of the wavelet transform respectively by

$$\nabla W_s f(x, y) := \begin{pmatrix} W_s^1 f(x, y) \\ W_s^2 f(x, y) \end{pmatrix} \tag{7.4}$$

and

$$|\nabla W_s f(x, y)| := \sqrt{|W_s^1 f(x, y)|^2 + |W_s^2 f(x, y)|^2}. \tag{7.5}$$

Locating the local maxima of $|\nabla W_s f(x, y)|$ along the direction of $\nabla W_s f(x, y)$ can detect the Dirac structure edges including the curves of images.

In this chapter, function $\theta(x, y)$ is defined by

$$\theta(x, y) = \phi(\sqrt{x^2 + y^2}), \tag{7.6}$$

where $\phi$ is selected to be the quadratic spline function as follows:

$$\phi(r) = \begin{cases} 8r^2(r-1) + \dfrac{4}{3} & 0 \le r < \dfrac{1}{2} \\[2mm] -\dfrac{8}{3}(r-1)^3 & \dfrac{1}{2} \le r < 1 \\[2mm] 0 & r \ge 1. \end{cases} \tag{7.7}$$

Since

$$\psi^1(x,y) = \frac{\partial}{\partial x}\theta(x,y), \quad \psi^2(x,y) = \frac{\partial}{\partial y}\theta(x,y), \tag{7.8}$$

the wavelet functions $\psi^1$, $\psi^2$, which are called the quadratic spline wavelets, can be obtained as follows:

$$\begin{cases} \psi^1(x,y) = \phi'(\sqrt{x^2+y^2})\dfrac{x}{\sqrt{x^2+y^2}} \\[3mm] \psi^2(x,y) = \phi'(\sqrt{x^2+y^2})\dfrac{y}{\sqrt{x^2+y^2}}, \end{cases} \tag{7.9}$$

which is represented graphically in Figs. 7.1 and 7.2.

## 7.2 Characterization of Dirac Structure Edges by Wavelet Transform

In this section, three significant characteristics of the local maximum modulus of the wavelet transform with respect to the Dirac structure edges in the images is presented:

- **Slope invariant:** The local maximum modulus of the wavelet transform of a Dirac structure edge is independent of the slope of the edge.
- **Gray-level invariant:** The local maximum modulus of the wavelet transforms with respect to a Dirac structure edge takes place at the same points when the images with different gray levels are to be processed.
- **Width light-dependent:** For various widths of images of the Dirac structure edges, the location of maximum modulus of the wavelet transform varies lightly under certain circumscription.

Curve is a display of the Dirac structure edge in a 2D image. To simplify the theoretical analysis, in this chapter, a segment of the curve is considered. In the remainder of this book, we do not identify the notations of the Dirac structure edge, curve, and segment of curve.

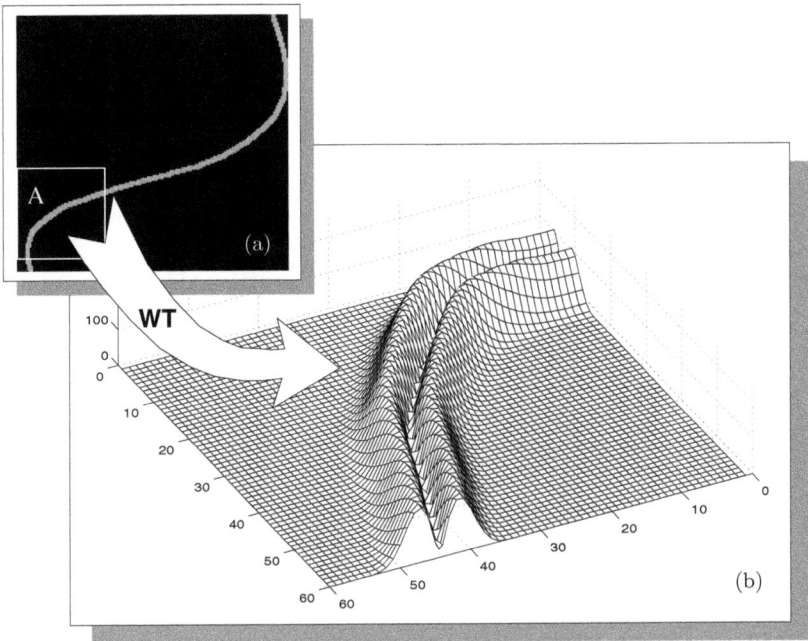

Fig. 7.3    An example of the distribution of the modulus of wavelet transform (WT) with respect to a curve.

Before mathematically analyzing the characterization of curves by wavelet transform, we look at an example as shown in Fig. 7.3. In practice, we should consider the width of the curve, especially in the application of image processing. Figure 7.3(a) gives a segment, which is marked by letter "A", of a curve with certain width. Suppose the quadratic spline function is selected to construct the wavelet functions. The distribution of the modulus of wavelet transform is illustrated in Fig. 7.3(b). Thus, the maximum modulus of wavelet transform will occur on two parallel lines around the curve image.

Now, we turn to the mathematical analysis. Suppose that the parameter equation of a curve $l_d$ can be written in form

$$l_d : \begin{cases} u = u(t) \\ v = v(t) \end{cases} \quad t \in [a, b], \tag{7.10}$$

where $[a, b]$ denotes the interval of the curve. Let $d$ be the width and $l$ be the skeleton of the curve, i.e., the central line of the curve image. A graphical description of such a curve is presented in Fig. 7.4. Note that, from the

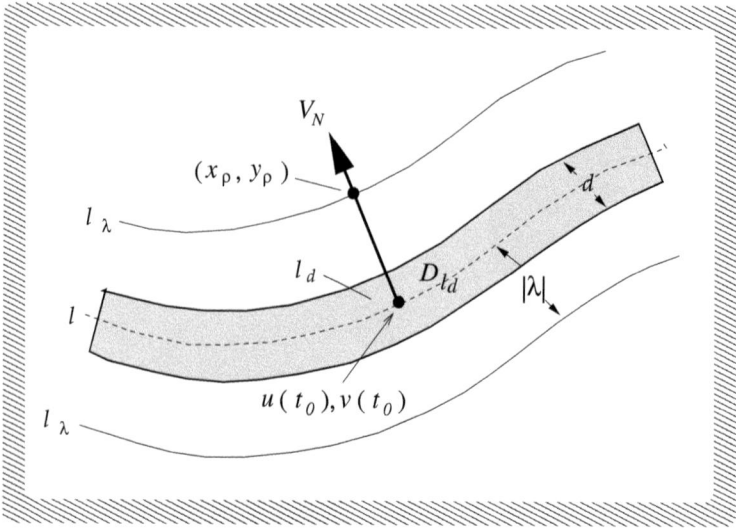

Fig. 7.4   Curve $l_d$ and its parallel line $l_\lambda$.

mathematical point of view, a curve does not have any width with it. Thus, we did not consider the width of the curve in Eq. (7.10). Therefore, we do not identify the notations of $l$ and $l_d$. In Fig. 7.4, the lines labeled by $l_\lambda$ are parallel lines, which are around the curve image, and the maximum modulus of the wavelet transform occurs on these lines. The normal unit vector of the curve $l_d$ is described by

$$V_N = \left( \frac{v'(t)}{\sqrt{u'(t)^2 + v'(t)^2}}, \; -\frac{u'(t)}{\sqrt{u'(t)^2 + v'(t)^2}} \right).$$

Therefore, the equations of the parallel line $l_\lambda$ can be represented as follows:

$$l_\lambda : \begin{cases} u_\lambda(t) = u(t) + \dfrac{\lambda v'(t)}{\sqrt{u'(t)^2 + v'(t)^2}} \\[3mm] v_\lambda(t) = v(t) - \dfrac{\lambda u'(t)}{\sqrt{u'(t)^2 + v'(t)^2}} \end{cases} \quad t \in [a, b],$$

where $|\lambda|$ is the distance between the parallel line $l_\lambda$ and the central line $l$ of the curve $l_d$, and the sign of $\lambda$ determines which side of $l_d$ the parallel line $l_\lambda$ will be located. A curve image $l_d$ can be described by

$$f_{l_d}(x, y) = c_f \chi_{D_{l_d}}(x, y),$$

where $c_f$ stands for the *gray level* of the curve and $\chi_{D_{l_d}}(x,y)$ denotes the characteristic function of the area $D_{l_d}$ which is defined by

$$D_{l_d} := \{(u_\lambda(t), v_\lambda(t)) | t \in [a,b], \ \lambda \in [-d/2, d/2]\}.$$

$\forall g(x,y)$, we have

$$\int_R \int_R f_{l_d}(u,v)g(u,v)dudv = c_f \iint_{D_{l_d}} g(u,v)dudv$$

$$= c_f \int_a^b dt \int_{-d/2}^{d/2} g(u_\lambda(t), v_\lambda(t))J(t,\lambda)d\lambda,$$

where $J(t,\lambda)$ is the Jacobi determinant of the coordinate transform

$$\begin{cases} u = u(t,\lambda) \\ v = v(t,\lambda), \end{cases}$$

i.e.,

$$J(t,\lambda) = \begin{vmatrix} \dfrac{\partial u}{\partial t} & \dfrac{\partial u}{\partial \lambda} \\[2mm] \dfrac{\partial v}{\partial t} & \dfrac{\partial v}{\partial \lambda} \end{vmatrix} = -\sqrt{u'(t)^2 + v'(t)^2} - \lambda \frac{u'(t)v''(t) - v'(t)u''(t)}{u'(t)^2 + v'(t)^2}.$$

Hence

$$\int_R \int_R f_{l_d}(u,v)g(u,v)dudv$$

$$= -c_f \int_a^b dt \int_{-d/2}^{d/2} \left[ \sqrt{u'(t)^2 + v'(t)^2} + \lambda \frac{u'(t)v''(t) - v'(t)u''(t)}{u'(t)^2 + v'(t)^2} \right]$$

$$\cdot g\left( u(t) + \frac{\lambda v'(t)}{\sqrt{u'(t)^2 + v'(t)^2}}, v(t) - \frac{\lambda u'(t)}{\sqrt{u'(t)^2 + v'(t)^2}} \right) d\lambda.$$

Let $g(u,v) = \psi_s(u,v)$, we have

$$\int_R \int_R f_{l_d}(u,v)\psi_s(u,v)dudv$$

$$= -c_f \int_a^b dt \int_{-d/2}^{d/2} \left[ \sqrt{u'(t)^2 + v'(t)^2} + \lambda \frac{u'(t)v''(t) - v'(t)u''(t)}{u'(t)^2 + v'(t)^2} \right]$$

$$\cdot \psi_s\left( u(t) + \frac{\lambda v'(t)}{\sqrt{u'(t)^2 + v'(t)^2}}, v(t) - \frac{\lambda u'(t)}{\sqrt{u'(t)^2 + v'(t)^2}} \right) d\lambda.$$

When a segment of curve $l_d$ is short enough, it can be considered to be a straight line and, thereafter, the the parameter equation of its central line can be represented as follows:

$$l : \begin{cases} u = u(t) + k_1(t - t_0) \\ v = v(t) + k_2(t - t_0) \end{cases} \quad t \in [a, b],$$

where $k_1$, $k_2$ denote the *slope* of $l$ satisfying $k_1^2 + k_2^2 = 1$. Thus, the wavelet transform of this curve can be obtained as follows:

$$W_s f_{l_d}(x, y) = -c_f \int_a^b dt \int_{-d/2}^{d/2} \psi_s(x - u(t_0) - k_1(t - t_0)$$

$$- \lambda k_2, y - v(t_0) - k_2(t - t_0) + \lambda k_1) d\lambda.$$

$\forall (u(t_0), v(t_0)) \in l$, we consider the point $(x_\rho, y_\rho)$ in the normal line:

$$\begin{cases} x_\rho = u(t_0) + k_2 \rho \\ y_\rho = v(t_0) + k_1 \rho, \end{cases}$$

where $\rho$ denotes a parameter such that its absolute value $|\rho|$ is the just distance between $(x_\rho, y_\rho)$ and $(u(t_0), v(t_0))$, i.e., $|\lambda| = |\rho|$. The local maximum modulus of wavelet transform of the point $(u(t_0), v(t_0))$ in the curve takes place at point $(x_\rho, y_\rho)$ in the normal line. As the parameter $\rho$ is determined, the point $(x_\rho, y_\rho)$ will be found. Therefore, the calculation of parameter $\rho$ plays a key role in finding the location where the maximum modulus of wavelet transform occurs. In this case, the wavelet transform becomes

$$W_s f_{l_d}(x, y) = -c_f \int_a^b dt \int_{-d/2}^{d/2} \psi_s(k_2 \rho - k_1(t - t_0) - \lambda k_2,$$

$$- k_1 \rho - k_2(t - t_0) + \lambda k_1) d\lambda$$

$$= -c_f \int_a^b dt \int_{-d/2}^{d/2} \psi_s(-k_1(t - t_0) - k_2(\lambda - \rho),$$

$$- k_2(t - t_0) + k_1(\lambda - \rho) d\lambda$$

$$= -c_f \int_{t_0-b}^{t_0-a} dt \int_{-d/2-\rho}^{d/2-\rho} \psi_s(k_1 t - k_2 \lambda, k_2 t + k_1 \lambda) d\lambda$$

$$= -c_f \int_{t_0-b}^{t_0-a} dt \int_{\rho-d/2}^{\rho+d/2} \psi_s(k_1 t + k_2 \lambda, k_2 t - k_1 \lambda) d\lambda$$

$$= -c_f \int_{(t_0-b)/s}^{(t_0-a)/s} dt \int_{(\rho-d/2)/s}^{(\rho+d/2)/s} \psi(k_1 t + k_2 \lambda, k_2 t - k_1 \lambda) d\lambda.$$

### 7.2.1  *Slope Invariant*

In this section, we prove that the local maximum modulus of the wavelet transform of the curve is independent of the slope of that curve.

For the quadratic spline wavelets $\psi^1$, $\psi^2$, since

$$\psi^1(x,y) = \frac{\partial}{\partial x}\theta(x,y), \quad \psi^2(x,y) = \frac{\partial}{\partial y}\theta(x,y),$$

where $\theta(x,y) = \phi(\sqrt{x^2 + y^2})$, and

$$\phi(r) = \begin{cases} 8r^2(r-1) + \dfrac{4}{3} & 0 \le r < \dfrac{1}{2} \\[2mm] -\dfrac{8}{3}(r-1)^3 & \dfrac{1}{2} \le r < 1 \\[2mm] 0 & r \ge 1, \end{cases}$$

therefore, $\psi^1$, $\psi^2$ can be represented as

$$\begin{cases} \psi^1(x,y) = \phi'(\sqrt{x^2+y^2})\dfrac{x}{\sqrt{x^2+y^2}} \\[3mm] \psi^2(x,y) = \phi'(\sqrt{x^2+y^2})\dfrac{y}{\sqrt{x^2+y^2}}. \end{cases}$$

The wavelet transforms using the quadratic spline wavelets $\psi^1$, $\psi^2$ are

$$W_s^1 f(x_\rho, y_\rho) = -c_f \int_{(t_0-b)/s}^{(t_0-a)/s} dt \int_{(\rho-\frac{d}{2})/s}^{(\rho+\frac{d}{2})/s} \phi'(\sqrt{t^2+\lambda^2})\frac{k_1 t + k_2\lambda}{\sqrt{t^2+\lambda^2}} d\lambda,$$

$$(7.11)$$

$$W_s^2 f(x_\rho, y_\rho) = -c_f \int_{(t_0-b)/s}^{(t_0-a)/s} dt \int_{(\rho-\frac{d}{2})/s}^{(\rho+\frac{d}{2})/s} \phi'(\sqrt{t^2+\lambda^2})\frac{k_2 t - k_1\lambda}{\sqrt{t^2+\lambda^2}} d\lambda.$$

$$(7.12)$$

It is easy to rewrite Eqs. (7.11) and (7.12) in the following forms:

$$W_s^1 f(x_\rho, y_\rho) = -c_f(k_1 w_1 + k_2 w_2)$$

$$W_s^2 f(x_\rho, y_\rho) = -c_f(k_2 w_1 - k_1 w_2),$$

by denoting

$$w_1 = \int_{(t_0-b)/s}^{(t_0-a)/s} dt \int_{(\rho-\frac{d}{2})/s}^{(\rho+\frac{d}{2})/s} \phi'(\sqrt{t^2+\lambda^2})\frac{t}{\sqrt{t^2+\lambda^2}} d\lambda$$

$$w_2 = \int_{(t_0-b)/s}^{(t_0-a)/s} dt \int_{(\rho-\frac{d}{2})/s}^{(\rho+\frac{d}{2})/s} \phi'(\sqrt{t^2+\lambda^2})\frac{\lambda}{\sqrt{t^2+\lambda^2}} d\lambda.$$

Hence, the second power of the amplitude of the wavelet transform can be obtained as follows:

$$|\nabla W_s f_{l_d}(x_\rho, y_\rho)|^2 = |W_s^1 f_{l_d}(x_\rho, y_\rho)|^2 + |W_s^2 f_{l_d}(x_\rho, y_\rho)|^2$$
$$= c_f^2[(k_1 w_1 + k_2 w_2)^2 + (k_2 w_1 - k_1 w_2)^2]$$
$$= c_f^2(w_1^2 + w_2^2).$$

Consequently, the amplitude of the wavelet transform becomes

$$|\nabla W_s f_{l_d}(x_\rho, y_\rho)| = \sqrt{c_f^2(w_1^2 + w_2^2)}. \tag{7.13}$$

In the final result of Eq. (7.13), the slope factors $k_1$ and $k_2$ disappear. That is, of course, the amplitude of the wavelet transform has the property of slope free. Consequently, the local maximum modulus of the wavelet transform also possesses this characteristic. Some graphical examples are illustrated in Fig. 7.5. Four curves, particularly, which are straight lines, with different orientations are analyzed. The original images are shown in Fig. 7.5(a). After applying the wavelet transforms to each one, the distributions of their modulus of transformations are given in Fig. 7.5(b), and the corresponding 3D graphical displays are drawn in Fig. 7.5(d). From this figure, it is clear that the highest peaks of all distributions remain in the same values. In other words, the local maximum modulus of the wavelet transform of the curve is independent of the slope of that curve. Figure 7.5(c) presents the maxima of wavelet modulus for these four curves.

### 7.2.2  *Gray-Level Invariant*

In this section, the characterization of the gray-level invariant is verified. That means, we prove that the maximum modulus of the wavelet transforms with respect to the curves takes place at the same points when the images with different gray-levels will be processed.

In fact, it is obvious to see that the wavelet transform $\nabla W_s f(x, y)$ is a linear system. Then, if the input function $f$ is scaled, the output is scaled the same. That is,

$$|\nabla W_s(cf(x, y))| := |c||\nabla W_s f(x, y)|.$$

Therefore, the factor of the gray level $c$ does not influence the location of the maximum modulus of the wavelet transform. This is the characteristic of gray-level invariant.

Some graphical examples are illustrated in Fig. 7.6. The left side of Fig. 7.6 contains two original images, which are circles. The gray levels of

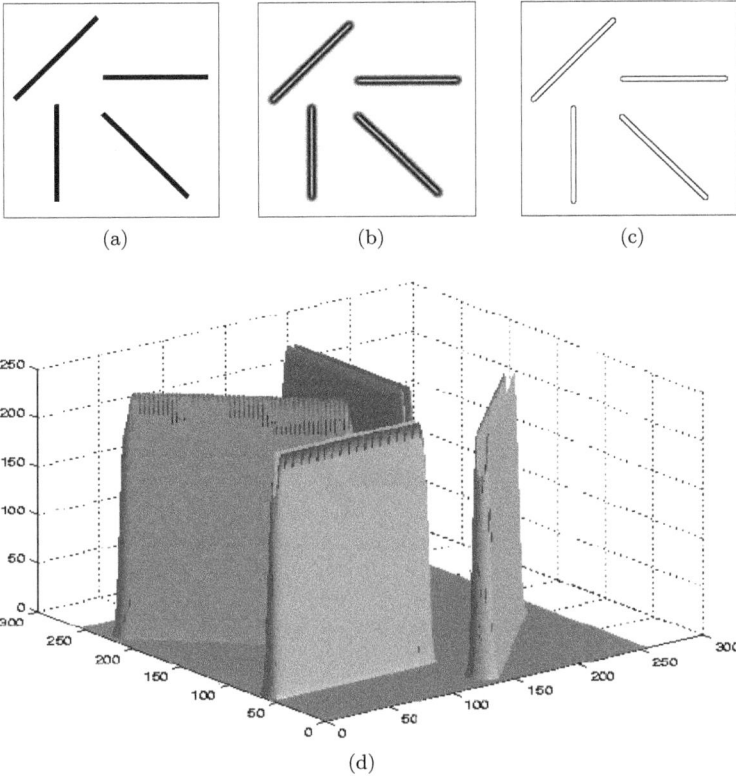

Fig. 7.5   Graphical examples of the property of the slope-invariant.

them differ from each other. The same wavelet transforms are applied to these circles. The resulting maximum moduli of the wavelet transforms with respect to them are graphically displayed in the right side of Fig. 7.6. It is obvious that they are same, i.e., the double-line circles.

## 7.2.3   *Width Light-Dependent*

The property of the width light-dependent with respect to the local maximum modulus of the wavelet transforms of curves is presented in this section. It is proved that for various widths of curves, $ds$, the location of maximum modulus of the wavelet transform varies lightly. That means parameter $\rho$ retains the same value approximately when $\rho > d$.

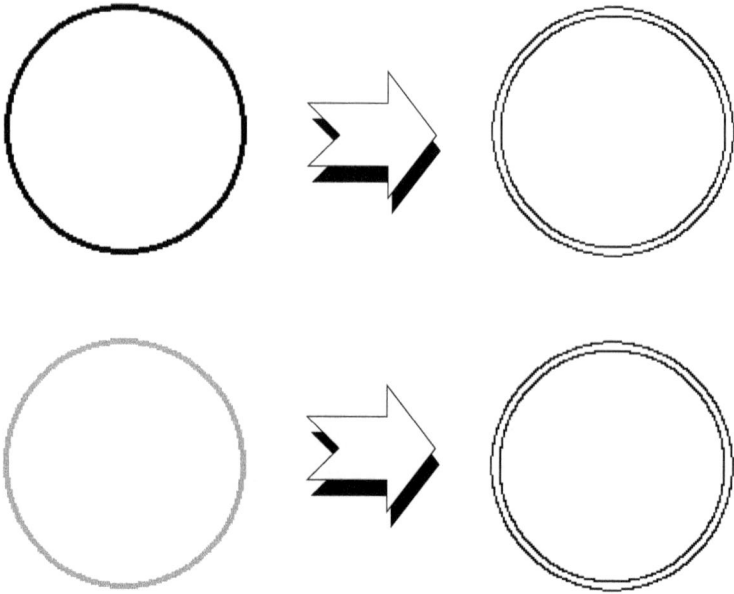

Fig. 7.6 Graphical examples of the property of the gray-level invariant.

Note that the quadratic spline function $\phi(r)$ is compactly supported, moreover the support is very short, so that for the point $t_0$, we have

$$
\begin{aligned}
w_1 &= \int_{(t_0-b)/s}^{(t_0-a)/s} dt \int_{(\rho-\frac{d}{2})/s}^{(\rho+\frac{d}{2})/s} \phi'(\sqrt{t^2+\lambda^2}) \frac{t}{\sqrt{t^2+\lambda^2}} d\lambda \\
&= \int_{-\infty}^{\infty} dt \int_{(\rho-\frac{d}{2})/s}^{(\rho+\frac{d}{2})/s} \phi'(\sqrt{t^2+\lambda^2}) \frac{t}{\sqrt{t^2+\lambda^2}} d\lambda \\
&= 0.
\end{aligned}
\tag{7.14}
$$

Substituting Eq. (7.14) into Eq. (7.13) yields

$$
\begin{aligned}
|\nabla W_s f_{l_d}(x_\rho, y_\rho)| &= \sqrt{c_f^2 w_2^2} = |c_f w_2| \\
&= |c_f| \left| \int_{-\infty}^{\infty} dt \int_{(\rho-\frac{d}{2})/s}^{(\rho+\frac{d}{2})/s} \phi'(\sqrt{t^2+\lambda^2}) \frac{\lambda}{\sqrt{t^2+\lambda^2}} d\lambda \right| \\
&= 2|c_f| \left| \int_{0}^{\infty} dt \int_{\sqrt{t^2+(\frac{\rho-d/2}{s})^2}}^{\sqrt{t^2+(\frac{\rho+d/2}{s})^2}} \phi'(\lambda) d\lambda \right|.
\end{aligned}
$$

Since $\phi'(\lambda) \leq 0$ ($\forall \lambda \geq 0$), we get

$$|\nabla W_s f_{l_d}(x_\rho, y_\rho)| = -2|c_f| \int_0^\infty dt \int_{\sqrt{t^2 + (\frac{\rho-d/2}{s})^2}}^{\sqrt{t^2 + (\frac{\rho+d/2}{s})^2}} \phi'(\lambda) d\lambda$$

$$= -2|c_f| \int_0^\infty \left[ \phi\left(\sqrt{t^2 + \left(\frac{\rho+d/2}{s}\right)^2}\right) \right.$$

$$\left. - \phi\left(\sqrt{t^2 + \left(\frac{\rho-d/2}{s}\right)^2}\right) \right] dt.$$

To find the parameter $\rho$ such that $|\nabla W_s f_{l_d}(x_\rho, y_\rho)|$ reaches the local maximum, we consider the derivative on $\rho$:

$$\frac{d}{d\rho}|\nabla W_s f_{l_d}(x_\rho, y_\rho)|$$

$$= -2|c_f| \int_0^\infty [\phi'\left(\sqrt{t^2 + (\frac{\rho+d/2}{s})^2}\right)$$

$$\frac{\left(\frac{\rho+d/2}{s}\right)\frac{1}{s}}{\sqrt{t^2 + \left(\frac{\rho+d/2}{s}\right)^2}}$$

$$- \phi'\left(\sqrt{t^2 + (\frac{\rho-d/2}{s})^2}\right) \frac{\left(\frac{\rho-d/2}{s}\right)\frac{1}{s}}{\sqrt{t^2 + \left(\frac{\rho-d/2}{s}\right)^2}}]dt.$$

Let $\frac{d}{d\rho}|\nabla W_s f_{l_d}(x_\rho, y_\rho)| = 0$, we get

$$\left(\rho + \frac{d}{2}\right) \int_0^\infty \frac{\phi'\left(\sqrt{t^2 + (\frac{\rho+d/2}{s})^2}\right)}{\sqrt{t^2 + \left(\frac{\rho+d/2}{s}\right)^2}} dt$$

$$= \left(\rho - \frac{d}{2}\right) \int_0^\infty \frac{\phi'\left(\sqrt{t^2 + (\frac{\rho-d/2}{s})^2}\right)}{\sqrt{t^2 + \left(\frac{\rho-d/2}{s}\right)^2}} dt. \tag{7.15}$$

To facilitate solving parameter $\rho$, Eq. (7.15) can be rewritten by

$$G\left(\frac{\rho + \frac{d}{2}}{s}\right) = G\left(\frac{\rho - \frac{d}{2}}{s}\right) \tag{7.16}$$

according to

$$G(\xi) := \xi \int_0^\infty \frac{\phi'(\sqrt{t^2 + \xi^2})}{\sqrt{t^2 + \xi^2}} dt. \tag{7.17}$$

Now, our question turns to find $\rho$, such that Eq. (7.16) will be held. In general, it is difficult and even impossible to solve $\rho$ from Eq. (7.16) directly, for an arbitrary wavelet $\phi'$, because of the complexity of (7.17). To simplify the estimation of $G(\xi)$ without losing its characterization for the Dirac structure edges, we utilize the quadratic spline function (7.7). First of all, we calculate $G(\xi)$, and the result is presented as follows:

(1) If $\xi \geq 1$, it is easy to see that

$$G(\xi) = 0. \tag{7.18}$$

(2) If $\frac{1}{2} \leq \xi < 1$, $G(\xi)$ is given by

$$G(\xi) = \xi \int_0^{\sqrt{1-\xi^2}} \frac{\phi'(\sqrt{t^2 + \xi^2})}{\sqrt{t^2 + \xi^2}} dt$$

$$= -8\xi \int_0^{\sqrt{1-\xi^2}} \frac{(\sqrt{t^2 + \xi^2} - 1)^2}{\sqrt{t^2 + \xi^2}} dt$$

$$= -8\xi \left[ -\frac{3}{2}\sqrt{1-\xi^2} - \frac{2+\xi^2}{2} \log\xi + \frac{2+\xi^2}{2} \log(1 + \sqrt{1-\xi^2}) \right]$$

$$= -8\xi \left[ -\frac{3}{2}\sqrt{1-\xi^2} + \frac{2+\xi^2}{2} \log\frac{1 + \sqrt{1-\xi^2}}{\xi} \right]. \tag{7.19}$$

(3) If $0 \leq \xi < \frac{1}{2}$, we obtain

$$G(\xi) = \xi \int_0^{\sqrt{\frac{1}{4}-\xi^2}} + \int_{\sqrt{\frac{1}{4}-\xi^2}}^{\sqrt{1-\xi^2}}$$

$$= 8\xi \int_0^{\sqrt{\frac{1}{4}-\xi^2}} \frac{3(t^2 + \xi^2) - 2\sqrt{t^2 + \xi^2}}{\sqrt{t^2 + \xi^2}} dt$$

$$- 8\xi \int_{\sqrt{\frac{1}{4}-\xi^2}}^{\sqrt{1-\xi^2}} \frac{(\sqrt{t^2 + \xi^2} - 1)^2}{\sqrt{t^2 + \xi^2}} dt$$

$$= 4\xi \left[ -3\sqrt{1 - 4\xi^2} + 3\sqrt{1 - \xi^2} - 3\xi^2 \log \xi \right.$$

$$+ 2 \log \frac{1 + \sqrt{1 - 4\xi^2}}{2}$$

$$+ 4\xi^2 \log \frac{1 + \sqrt{1 - 4\xi^2}}{2} - 2 \log(1 + \sqrt{1 - \xi^2})$$

$$\left. - \xi^2 \log(1 + \sqrt{1 - \xi^2}) \right]. \tag{7.20}$$

The above mathematical formulae are too complicated so that it is difficult to solve them. As an alternate, in this chapter, a graphic solution method is developed.

The graphical description of $G(\xi)$ on $[0, \infty)$ is illustrated in Fig. 7.7, based on Eqs. (7.18)–(7.20). The value of the parameter $\rho$ can be solved from this figure approximately.

From Fig. 7.7, it is easy to see that, for a constant $C$, $0 < C \leq 1$, there exists a pair of numbers $(\alpha, \beta)$ satisfying $0 \leq \alpha < \beta \leq 1$, such that

$$\begin{cases} \beta - \alpha = C \\ G(\alpha) = G(\beta). \end{cases} \tag{7.21}$$

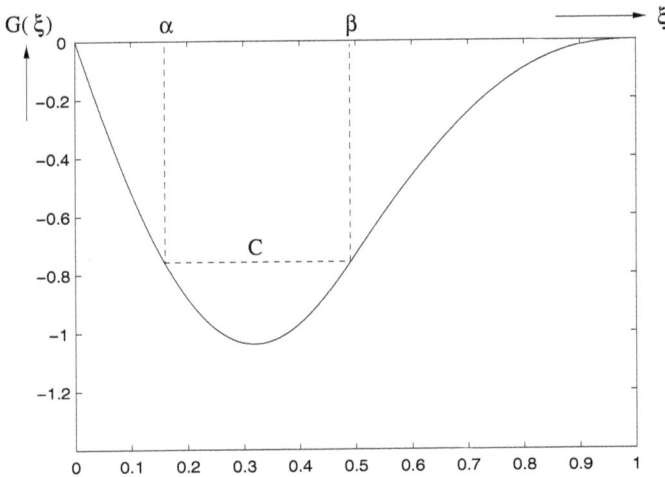

Fig. 7.7   Using the graphical description of $G(\xi)$ to find the value of the parameter $\rho$ approximately.

It is obvious that both $\alpha$ and $\beta$ depend on only the constant $C$, that means they are functions on $C$. Therefore, $\alpha$ and $\beta$ can be denoted as

$$\alpha = \alpha(C), \quad \beta = \beta(C).$$

In the following, we solve $\rho$ satisfying Eq. (7.16). Comparing Eq. (7.16) with Eq. (7.21) yields

$$\frac{\rho + d/2}{s} = \beta, \tag{7.22}$$

$$\frac{\rho - d/2}{s} = \alpha. \tag{7.23}$$

Adding Eqs. (7.22) and (7.23), we obtain

$$\rho = \frac{s}{2}(\alpha + \beta). \tag{7.24}$$

By subtracting Eq. (7.23) from Eq. (7.22), we have

$$\beta - \alpha = \frac{d}{s}. \tag{7.25}$$

Substituting Eq. (7.25) into Eq. (7.24) produces

$$\rho = \frac{s}{2}\left[\alpha\left(\frac{d}{s}\right) + \beta\left(\frac{d}{s}\right)\right] = \rho(s, d). \tag{7.26}$$

Thus, $\rho$ is dependent on $s$ and $d$. That means $\rho$ is a function on $s$ and $d$. The relationship among $\rho$, $s$, and $d$ can be displayed graphically in Fig. 7.8.

From Fig. 7.8, it is easy to understand that the dependence of $\rho$ on $d$ is very light under the condition of $0 < \frac{d}{s} \leq 1$, i.e., $s \geq d > 0$. The larger the ratio of $\frac{d}{s}$, the smaller the dependence of $\rho$ on $d$. When the scale $s$ of wavelet transform is large enough, the parameter $\rho$ is independent of the width $d$ of the curve image.

Figure 7.9 displays the relationship between $\rho$ and $d$ exactly, when some specific transform scales are given, for instance, $s = 4, 8, 12, 16$.

Another way to view Eq. (7.26) and Fig. 7.8 can also be explained in Fig. 7.10. From this figure, the following facts can be found: For fixed $ds$, the dependence of $\rho$ on $s$ is close to the linear function, when the scale $s$ of wavelet transform is large enough. That means the larger the scale $s$, the closer to the straight line the parameter $\rho$. Therefore, the relationship between $\rho$ and $d$ can be represented by a straight line, when the scale $s$ of the wavelet transform has sufficient value. The equation of this straight line

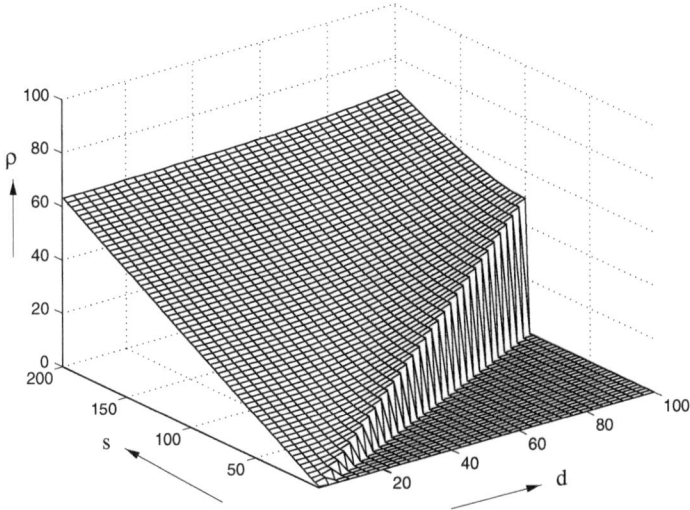

Fig. 7.8    The graphical description of $\rho = \rho(s, d)$.

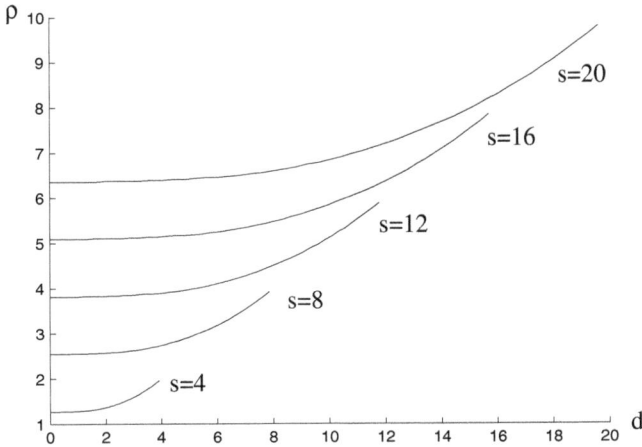

Fig. 7.9    The relationship between $\rho$ and $d$, when some specific scales are given $(s = 4, 8, 12, 16)$.

can approximately be found in Fig. 7.7, i.e., the minimum of $G(\xi)$. That is, $\rho \approx 0.318s$.

Some graphical examples are illustrated in Fig. 7.11. The left side of Fig. 7.11 contains two original images, which are circles. The width of

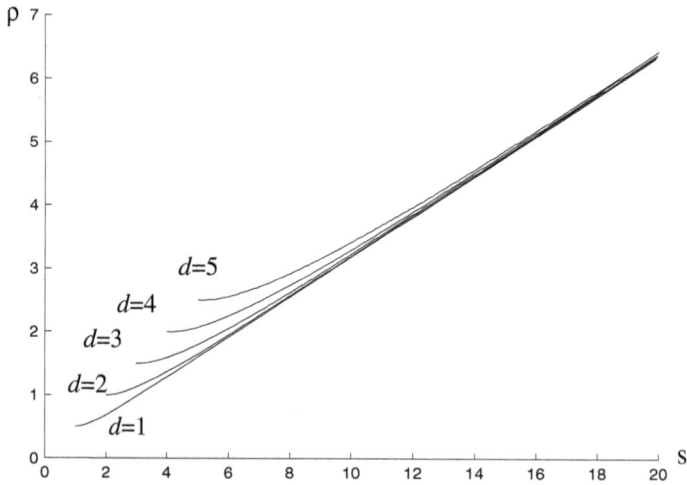

Fig. 7.10    The relationship between $\rho$ and $s$, when some specific widths are given ($d = 1 - 5$).

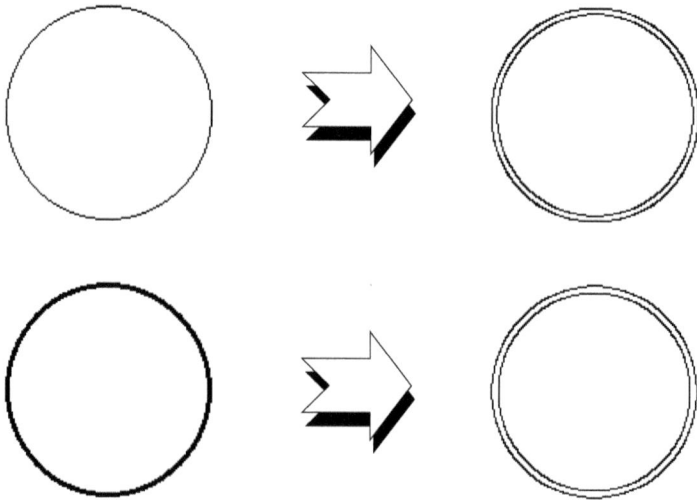

Fig. 7.11    Graphical examples of the property of the width light-invariant.

them differs from each other. The same wavelet transforms are applied to these circles. The resulting maximum moduli of the wavelet transforms with respect to them are graphically displayed on the right side of Fig. 7.11. It is obvious that the same images, i.e., the double-line circles, are obtained.

## 7.3  Experiments

Based on the theoretical analyses in the previous sections, an algorithm to detect the Dirac structure edges such as curves from a multi structure-edge image is designed as follows:

---

**Algorithm 7.1.** For a multi structure-edge image containing the Dirac structure edges, such as curves, and the wavelet transform scale $s > 0$, the edges can be detected by the following steps:

**Step 1.** Calculating all the wavelet transforms $\{W_s^1 f(x,y),\ W_s^2 f(x,y)\}$ under the quadratic spline wavelet.

**Step 2.** Calculating the local maxima $f_{\text{locmax}}$ of $|\nabla W_s f(x,y)|$ and the gradient direction $f_{\text{gradient}}$.

**Step 3.** For each point $(x,y)$ with local maximum, searching the point whose distance from $(x,y)$ to it is $0.6424s$ along the gradient direction. If it is still a point with local maximum, the center point is detected.

**Step 4.** The curves are formed by all the points detected by the above steps.

---

In Fig. 7.12, four circles with various gray levels and widths are tested using this method. The original images are illustrated on the left column of Fig. 7.12. After applying Steps 1 and 2 of the proposed wavelet transform algorithm to these circles, the local maximum modulus of the wavelet transform with respect to them can be computed and the results are given on the middle column in Fig. 7.12. Finally, the central lines of these circles are extracted using Steps 3 and 4 of the above algorithm and presented on the right column in Fig. 7.12.

Let us look at Fig. 7.13. The particular task is that we are required to identify different structures of edges and, thereafter, detect the Dirac structure edges such as drawing lines and texts in Fig. 7.13(a) and eliminate the step structure edges such as the contour of the aircraft in Fig. 7.13(a). Unfortunately, the algorithm based on the method proposed in the works of Mallat and Hwang [1992] and Mallat and Zhong [1992] has detected all edges without identifying different structures of edges. Thus, the resulting image contains the step structure edges such as the contours which are required to be deleted from the image. The method developed in this chapter possesses three significant characteristics: (1) slope invariant — the local maximum modulus of the wavelet transform of a Dirac structure edge is independent of the slope of the edge; (2) gray-level invariant — the

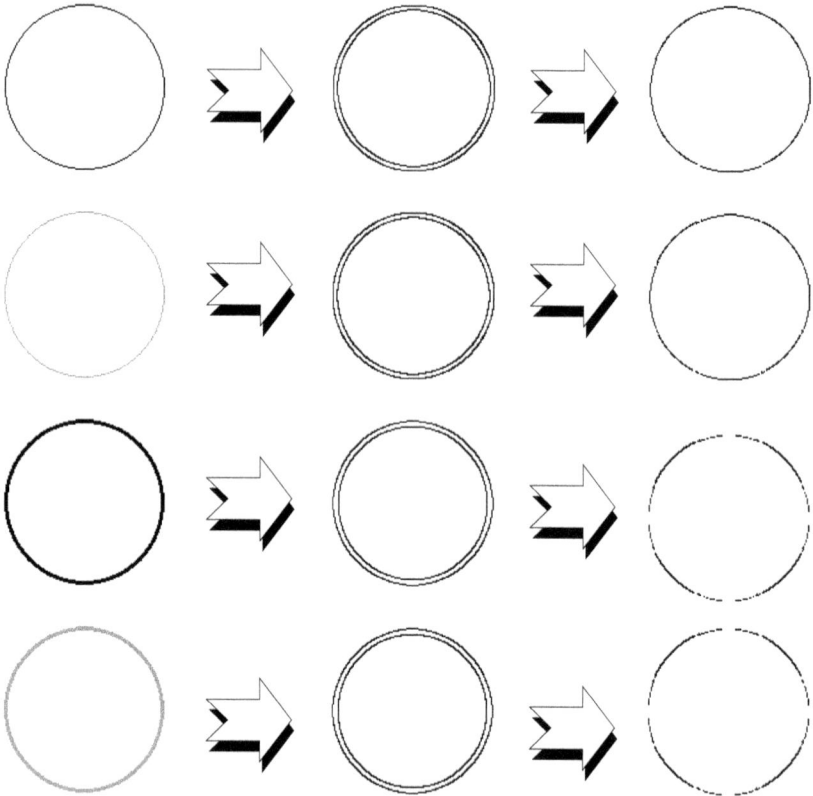

Fig. 7.12   Detection of the central lines with various gray levels and widths by proposed wavelet transform algorithm (the scale of the wavelet transform: $s = 6$). Left column: Original images, i.e., four circles with various gray levels and widths. Middle column: The local maximum modulus of the wavelet transform. Right column: The central lines of the circles are extracted.

local maximum modulus of the wavelet transform with respect to a Dirac structure edge takes place at the same points when the images with different gray levels will be processed; (3) width light-dependent — for various widths of the Dirac structure edge images, the location of maximum modulus of the wavelet transform varies lightly under the certain circumscription. According to these characteristics and our early work [Tang *et al.*, 1998c], we can recognize three basic structures of edges and further, extract the Dirac structure ones. After applying the above algorithm, new result can be found in Fig. 7.13(b). It extracts all lines and texts and removes the contour of the aircraft.

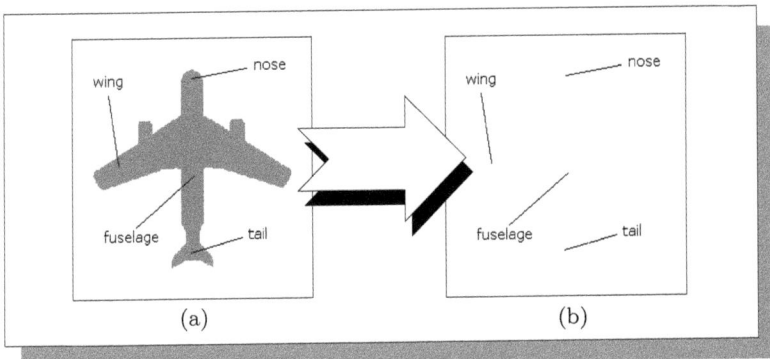

Fig. 7.13 Result using the method proposed in this chapter. (a) Original image: Two classes of edges are embedded in this image, i.e., a contour of the aircraft which belongs to the step structure edge, and some lines and texts which belong to the Dirac structure edges. (b) The result which is obtained from the new algorithm, all lines and texts are extracted, and the contour of the aircraft is removed.

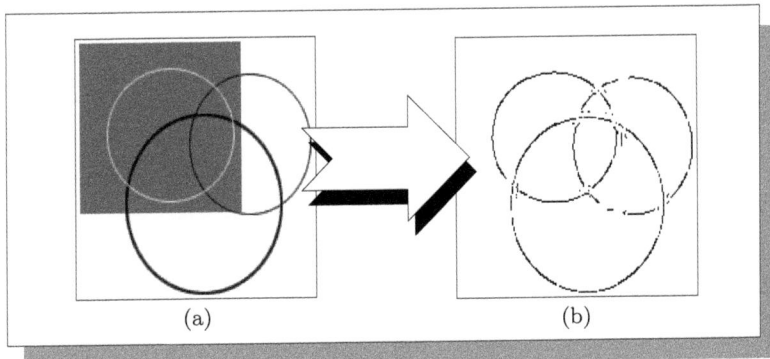

Fig. 7.14 Another result using the method proposed in this chapter: (a) Original image which has two classes of structures of edges, i.e., Dirac structure edges (three circles with different gray levels and widths), and step structure edge (a planner object, square which has a contour). (b) Resulting image which has only three circles, and the contour of the square is removed.

Now, turn to another example shown in Fig. 7.14. The original image is illustrated in Fig. 7.14(a) which has two classes of structures of edges: (1) Dirac structure edges, i.e., three circles with different gray levels and widths; (2) step structure edge, that is, a planner object, square, which has a contour. Upon utilizing the method proposed in this chapter, we can produce a new image, as shown in Fig. 7.14(b). It contains only three

Fig. 7.15 Lines extracted from image of Lena by proposed wavelet transform algorithm (the scale of the wavelet transform: $s = 6$). (a) Original image which is the image Lena with several characters. (b) Local maximum modulus of the wavelet transform of image of Lena. (c) The characters which are embedded in Lena image are extracted.

circles which belong to the Dirac structure edges, and the step structure edge, contour of the square, is removed.

The final experiment is presented in Fig. 7.15. Original image is the image Lena plus several characters as illustrated in Fig. 7.15(a). It is a multi structure-edge one, where several characters are embedded in it. Upon performing Steps 1 and 2 of the proposed wavelet transform algorithm, its local maximum modulus of the wavelet transform can be computed and the result is given in Fig. 7.15(b). Next, the characters which are embedded in the image are extracted by Steps 3 and 4 of the above algorithm and presented in Fig. 7.15(c). Finally, the noise is removed and a fine image which contains only the characters "lena" is obtained and displayed in Fig. 7.15(d).

# Chapter 8

# Construction of New Wavelet Function and Application to Curve Analysis

In Chapter 6, a very important characterization of the Dirac-structure edges by wavelet transform was provided. Three significant characteristics of the local maximum modulus of the wavelet transform with respect to the Dirac-structure edges were presented: (1) Slope invariant — the local maximum modulus of the wavelet transform of a Dirac-structure edge is independent of the slope of the edge. (2) Gray-level invariant — the local maximum modulus of the wavelet transform with respect to a Dirac-structure edge takes place at the same points when the images with different gray levels are to be processed. (3) Width light-dependent — for various widths of the Dirac-structure edge images, the location of the maximum modulus of the wavelet transform varies slightly when the scale $s$ of the wavelet transform is larger than the width $d$ of the Dirac structure edge images.

Based on the characteristics, an algorithm to detect the Dirac-structure edges from an image has been developed. An example of applying this algorithm to detect the Dirac-structure edge can be found in Fig. 8.1. The original image shown in Fig. 8.1(a) is a multi structure-edge one, which contains an image of the Chinese traditional word "book" where several drawing curves are embedded on it. Upon performing the algorithm proposed in Chapter 6, its local maximum modulus of the wavelet transform has been computed and is shown in Fig. 8.1(b), and the result is presented in Fig. 8.1(c). Finally, the noise is removed and a fine image which contains only the drawing curves is obtained and displayed in Fig. 8.1(d).

However, note the third property in Chapter 6, it says "width light-dependent" and does not say "width-invariant". This means that for various widths of the Dirac-structure edge images, the location of maximum

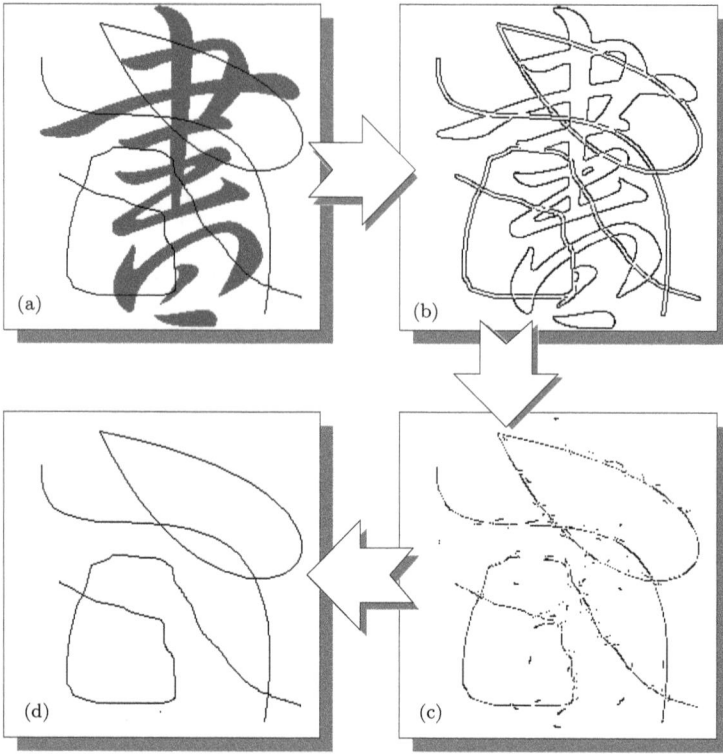

Fig. 8.1 Lines extracted from an image by the proposed algorithm in Chapter 6: (a) Original image; (b) local maximum modulus of the wavelet transform of the original image; (c) the drawing Lines extracted by the algorithm; (d) the resulting image.

modulus of the wavelet transform may change (these changes are small). What we want is that the location of maximum modulus does not change, i.e., the location of maximum modulus has the property of width-invariant. Let us look at Fig. 8.2. The first row of Fig. 8.2 has three circles. The left image is the original one which contains a circle with various widths. The middle one is the location of the maximum modulus of the wavelet transform with scale $s = 6$, which depends on the width of the circle in some way. Finally, by utilizing the algorithm proposed in Chapter 6, the skeleton of the circle is extracted and displayed on the right of Fig. 8.2. We can find that the skeleton of the circle is broken. The second row of Fig. 8.2 has trees, where the sizes of the branches vary, some are thick and some are thin. The left image is the original one. The middle one illustrates the location of the maximum modulus of the wavelet transform with scale $s = 6$. It is clear

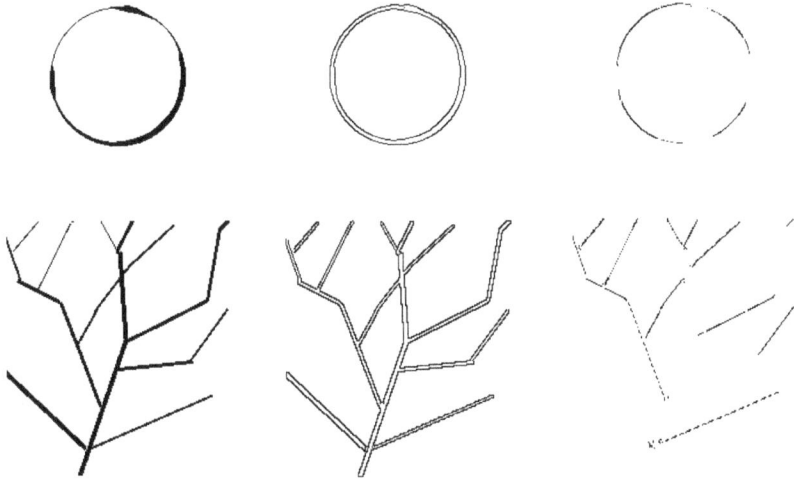

Fig. 8.2 (a) The original images; (b) The location of maximum modulus of the wavelet transform with $s = 6$; (c) The skeleton images extracted by the algorithm of Chapter 6.

that the location of the maximum modulus of the different branches has slight changes. The right of Fig. 8.2 is the skeleton extracted utilizing the algorithm proposed in Chapter 6. It is easy to see that some branches of the tree are lost.

To overcome such a defect, a natural question is whether there exists a more suitable wavelet function such that the location of the maximum modulus of the wavelet transform of a curve is width-invariant as well as slope-invariant and gray-invariant.

In this chapter, we present a development [Yang *et al.*, 2003a], where a novel wavelet is constructed, so that the above "width light-dependent" properties can be improved to "width invariant" without losing the "slope invariant" and "gray-level invariant". Due to this improvement, the detection of Dirac-structure edge is more accurate.

## 8.1 Construction of New Wavelet Function: Tang–Yang Wavelet

The scale wavelet transform described in Eq. (7.2) is a filter, and since $\psi \in L^2(R^2)$, its Fourier transform can be defined by

$$\hat{\psi}(\xi, \eta) := \int_R \int_R \psi(x, y) e^{-i(\xi x + \eta y)} \, dx \, dy,$$

which satisfies the condition of $\hat{\psi} \in L^2(R^2)$. Thus, both functions $\psi$ and $\hat{\psi}$ decrease at infinity. Usually, we choose $\psi \in L^2(R^2)$ such that both $\psi$ and $\hat{\psi}$ decrease at infinity with exponents at least. We know that $\hat{\psi}(0,0) = 0$ from Eq. (7.1), which implies that the scaled wavelet transform $W_s f(x, y)$ corresponds to a band pass filter essentially. Moreover, since $\psi$ and $\hat{\psi}$ decrease at infinity, $W_s f(x, y)$ characterizes the local properties of $f$ both on the time domain and frequency domain. In fact, two points are considered:

- On one hand, due to the fast decrease of $\psi$ at infinity, formula (7.2) characterizes the local property of $f$ around point $(x, y)$. To understand this, one can consider that $\psi$ vanishes outside $[-1, 1]$. In this case, we have

$$\psi\left(\frac{x - u}{s}, \frac{y - v}{s}\right) = 0,$$

for every $(u, v)$ satisfying

$$-1 \le \frac{x - u}{s} \le 1 \quad \text{or} \quad -1 \le \frac{y - v}{s} \le 1,$$

i.e.,

$$x - s \le u \le x + s \quad \text{or} \quad y - s \le v \le y + s.$$

This concludes that

$$W_s f(x, y) = \int_R \int_R f(u, v) \frac{1}{s^2} \psi\left(\frac{x - u}{s}, \frac{y - v}{s}\right) dudv$$

$$= \int_{x-s}^{x+s} \int_{y-s}^{y+s} f(u, v) \frac{1}{s^2} \psi\left(\frac{x - u}{s}, \frac{y - v}{s}\right) dudv,$$

which tells us that $W_s f(x, y)$ characterizes the local properties of $f$ on the interval $[x - s, x + s] \times [y - s, y + s]$ centering at $(x, y)$. The smaller the $s$ the narrower the interval, i.e., the time window.

- On the other hand, based on the basic properties of Fourier transform, we can obtain

$$(W_s f)(x, y) = \left(\frac{1}{2\pi}\right)^2 \int_R \int_R (W_s f)\widehat{}\,(\xi, \eta) e^{i(\xi x + \eta y)} d\xi d\eta$$

$$= \left(\frac{1}{2\pi}\right)^2 \int_R \int_R \hat{f}(\xi, \eta)(\psi_s)\widehat{}\,(\xi, \eta) e^{i(\xi x + \eta y)} d\xi d\eta$$

$$= \left(\frac{1}{2\pi}\right)^2 \int_R \int_R \hat{f}(\xi, \eta)\hat{\psi}(s\xi, s\eta) e^{i(\xi x + \eta y)} d\xi d\eta.$$

Since $\hat{\psi}(s\xi, s\eta)e^{i(\xi x + \eta y)}$ decreases as $\hat{\psi}(s\xi, s\eta)$ does at infinity, thus $(W_s f)(x, y)$ characterizes the local property of $\hat{f}$. The position of the locality depends on scale $s$ and function $\hat{\psi}$. The larger the $s$ the narrower the frequency window. A detailed analysis can be done similarly to the above.

Theoretically, Eq. (7.1), i.e., $\hat{\psi}(0, 0) = 0$, implies that $\psi(x, y)$ is a band-pass filter but a high-pass one because of the decrease in its Fourier transform at infinity. It is easy to see that the partial derivatives of a low-pass function can become the candidates of the wavelet functions. Here, we consider such kinds of wavelets, i.e.,

$$\psi^1(x, y) := \frac{\partial}{\partial x}\theta(x, y), \quad \psi^2(x, y) := \frac{\partial}{\partial y}\theta(x, y),$$

where $\theta(x, y)$ denotes a real function satisfying the following:

- $\theta(x, y)$ fast decreases at infinity.
- $\theta(x, y)$ is an even function on both $x$ and $y$.
- $\hat{\theta}(0, 0) = 1$.

For wavelet $\psi^1(x, y)$ defined above, its scale wavelet transform is

$$W_s^1 f(x, y) = (f * \psi_s^1)(x, y)$$

$$= \left( f * s\frac{\partial}{\partial x}\theta_s \right)(x, y)$$

$$= s\frac{\partial}{\partial x}(f * \theta_s)(x, y),$$

where $\theta_s(x, y) := \frac{1}{s^2}\theta(\frac{x}{s}, \frac{y}{s})$.

This formula tells us that $W_s^1 f(x, y)$ is essentially the derivative of the smoothness function along the horizontal axis and then the local maxima of the derivative function correspond to the points of the smoothness image with sharp variation along the horizontal axis. It is equivalent to the classical multi scale edge detection [Canny, 1986; Marr and Hildreth, 1980], if $\theta(x, y)$ is set to be a Gaussian, which is defined by

$$G_\sigma(x, y) = \frac{1}{2\pi\sigma^2} \exp\left( -\frac{x^2 + y^2}{2\sigma^2} \right). \tag{8.1}$$

That is why $\theta(x, y)$ is assumed to satisfy $\hat{\theta}(0, 0) = 1$.

A similar explanation for wavelet $\psi^2(x, y)$ defined above can be made. However, the partial derivative is along the vertical direction instead of the horizontal one.

Gaussian function has been employed in image processing. It possesses some excellent properties, such as the locality in both the time domain and frequency domain, the same widths in both the time window and frequency window, and so on. All these properties make it applied extensively and deeply in the area of filtering, and it already almost becomes the best candidate for low-pass filters in practice. Unfortunately, the Gaussian function is not always the best one for all applications. In fact, we have shown that it is not the best candidate for characterizing a Dirac-structure edge [Tang *et al.*, 2000]. Even the quadratic spline wavelet is better than it, although, the quadratic spline wavelet is still not a perfect one for such applications. Tang *et al.* [2000], proved that the location of the maximum modulus of the wavelet transform with respect to a Dirac-structure edge is not width invariant. It still depends on the width of the edge even though it depends lightly. To avoid such dissatisfaction, a new wavelet called the Tang–Yang wavelet is constructed, and its definition is described in the following [Yang *et al.*, 2003a]:

Let

$$
\begin{cases}
\psi_1(x) = -\dfrac{2}{\pi}\left(-8x\ln\dfrac{1+\sqrt{1-16x^2}}{4x} + \dfrac{1}{2x}\sqrt{1-16x^2}\right) \\[3mm]
\psi_2(x) = -\dfrac{2}{\pi}\left(8x\ln\dfrac{3+\sqrt{9-16x^2}}{4x} - \dfrac{3}{2x}\sqrt{9-16x^2}\right) \\[3mm]
\psi_3(x) = -\dfrac{2}{\pi}\left(-4x\ln\dfrac{1+\sqrt{1-x^2}}{x} + \dfrac{4}{x}\sqrt{1-x^2}\right).
\end{cases}
$$

Then, the 1D wavelet $\psi(x)$ is an odd function defined on $(0, \infty)$ by

$$
\psi(x) := \begin{cases}
\psi_1(x) + \psi_2(x) + \psi_3(x) & x \in (0, \tfrac{1}{4}) \\
\psi_2(x) + \psi_3(x) & x \in [\tfrac{1}{4}, \tfrac{3}{4}) \\
\psi_3(x) & x \in [\tfrac{3}{4}, 1) \\
0 & x \in [1, \infty).
\end{cases}
\tag{8.2}
$$

Let $\phi(x) := \int_0^x \psi(x)dx$. Then $\phi(x)$ is an even function, compactly supported on $[-1, 1]$, and $\phi'(x) = \psi(x)$. Figure 8.3 displays functions $\psi(x)$ and $\phi(x)$ graphically.

The smoothness function $\theta(x, y)$ is then defined by

$$
\theta(x, y) := \phi(\sqrt{x^2 + y^2}),
$$

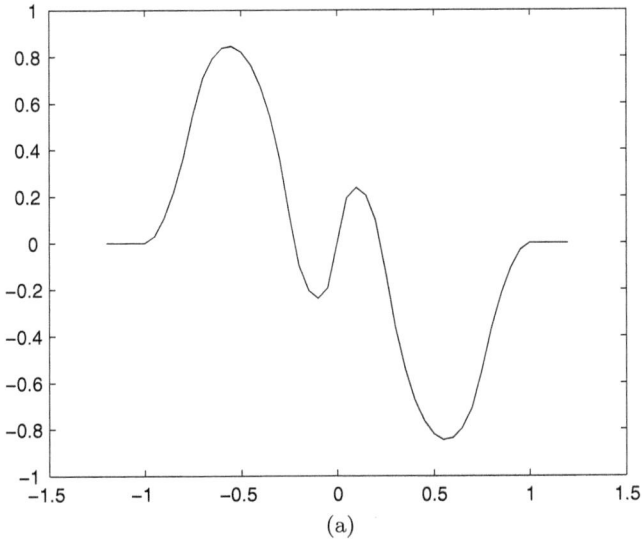

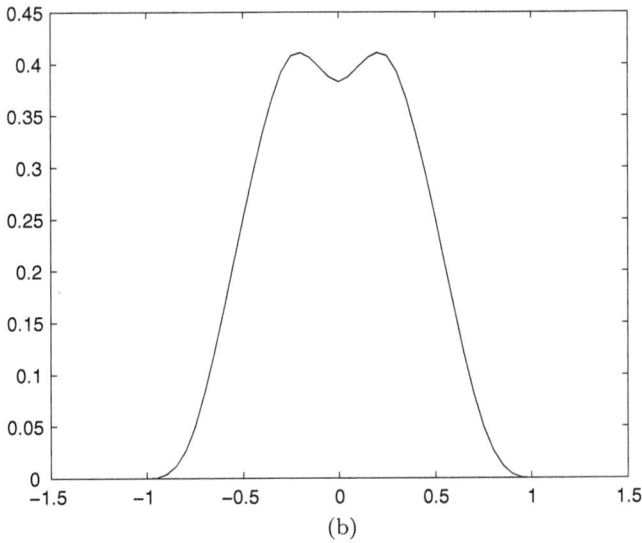

Fig. 8.3   The graphical descriptions of $\psi(x)$ and $\phi(x)$: (a) The graphical descriptions of function $\psi(x)$; (b) its primitive function $\phi(x)$.

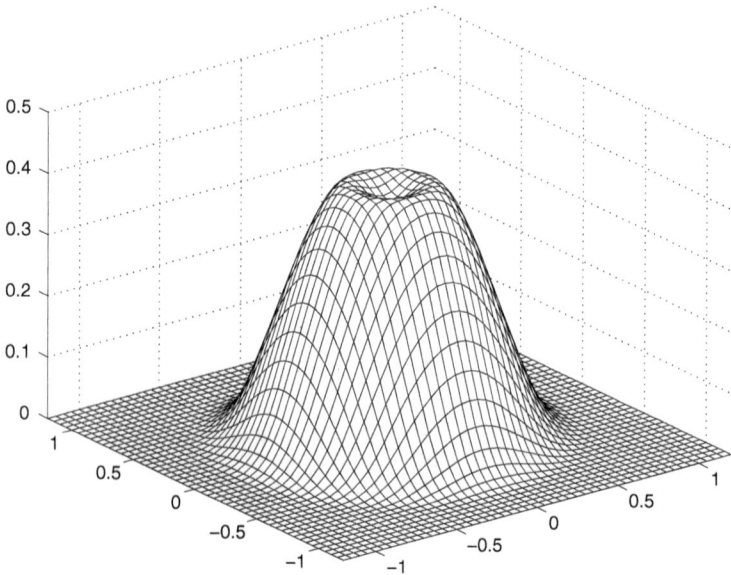

Fig. 8.4   The graphical descriptions of function $\theta(u, v)$.

which is graphically shown in Fig. 8.4, and the 2D wavelets are defined by

$$
\begin{cases}
\psi^1(x, y) := \dfrac{\partial}{\partial x}\theta(x, y) = \phi'(\sqrt{x^2 + y^2})\dfrac{x}{\sqrt{x^2 + y^2}}, \\[4mm]
\psi^2(x, y) := \dfrac{\partial}{\partial y}\theta(x, y) = \phi'(\sqrt{x^2 + y^2})\dfrac{y}{\sqrt{x^2 + y^2}}
\end{cases}
\tag{8.3}
$$

and are illustrated in Fig. 8.5.

The gradient direction and the amplitude of the wavelet transform are denoted, respectively, by

$$
\nabla W_s f(x, y) := \begin{pmatrix} W_s^1 f(x, y) \\ W_s^2 f(x, y) \end{pmatrix}
\tag{8.4}
$$

and

$$
|\nabla W_s f(x, y)| := \sqrt{|W_s^1 f(x, y)|^2 + |W_s^2 f(x, y)|^2}.
\tag{8.5}
$$

By locating the local maxima of $|\nabla W_s f(x, y)|$, we can detect the edges of the images.

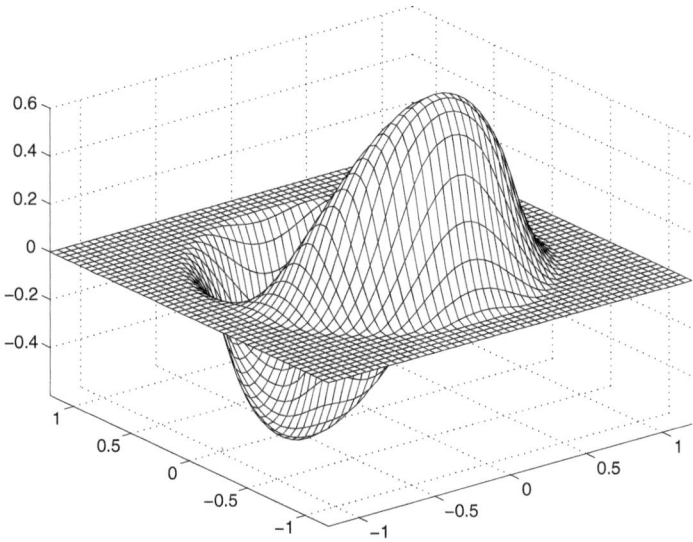

(a)

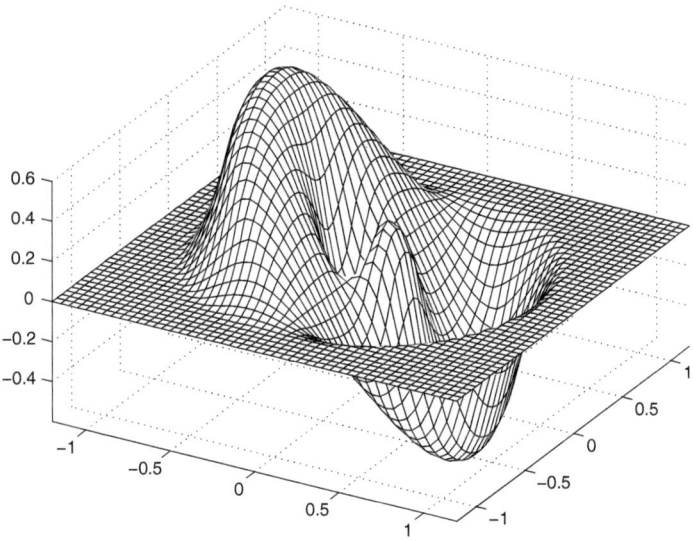

(b)

Fig. 8.5    The graphical descriptions of 2D wavelet functions: (a) Function $\psi^1(x, y)$; (b) function $\psi^2(x, y)$.

## 8.2 Characterization of Curves through New Wavelet Transform

Three significant characteristics of the local maximum modulus of the wavelet transform with respect to the Dirac-structure edges in images are held:

- **Gray-level invariant:** The local maximum modulus of the wavelet transform with respect to a Dirac-structure edge takes place at the same points when the images with different gray levels are to be processed.
- **Slope invariant:** The local maximum modulus of the wavelet transform of a Dirac-structure edge is independent of the slope of the edge.
- **Width invariant:** For various widths of the Dirac-structure edges in an image, the location of the maximum modulus of the wavelet transform does not vary under certain circumstances.

The discussion of the Gray-level invariant and slope invariant is the same as that in Chapter 6. In the following, we present the characteristics of the width invariant.

In this section, is proved that for various widths of curves, $ds$, the location of the maximum modulus of the wavelet transform does not vary. It means that parameter $\rho$ retains the same value when $\rho \geq d$ and the distance from the central line is just $\rho = \frac{s}{2}$.

Now that $\phi(r)$ is compactly supported and the support is very short, for the point $t_0$ far away from $a$ and $b$, we have

$$w_1 = \int_{(t_0-b)/s}^{(t_0-a)/s} dt \int_{(\rho-\frac{d}{2})/s}^{(\rho+\frac{d}{2})/s} \phi'(\sqrt{t^2+\lambda^2}) \frac{t}{\sqrt{t^2+\lambda^2}} d\lambda$$

$$= \int_{-\infty}^{\infty} dt \int_{(\rho-\frac{d}{2})/s}^{(\rho+\frac{d}{2})/s} \phi'(\sqrt{t^2+\lambda^2}) \frac{t}{\sqrt{t^2+\lambda^2}} d\lambda = 0,$$

which concludes that

$$|\nabla W_s f_{l_d}(x_\rho, y_\rho)|$$

$$= \sqrt{c_f^2 w_2^2} = |c_f w_2|$$

$$= |c_f| \left| \int_{-\infty}^{\infty} dt \int_{(\rho-\frac{d}{2})/s}^{(\rho+\frac{d}{2})/s} \phi'(\sqrt{t^2+\lambda^2}) \frac{\lambda}{\sqrt{t^2+\lambda^2}} d\lambda \right|$$

$$= 2|c_f| \left| \int_0^\infty dt \int_{(\sqrt{t^2+(\frac{\rho-d/2}{s})^2}}^{(\sqrt{t^2+(\frac{\rho+d/2}{s})^2}} \phi'(\lambda) d\lambda \right|$$

$$= 2 \left| c_f \int_0^\infty \left[ \phi \left( \sqrt{t^2 + \left(\frac{\rho+d/2}{s}\right)^2} \right) - \phi \left( \sqrt{t^2 + \left(\frac{\rho-d/2}{s}\right)^2} \right) \right] dt \right|.$$

To find $\rho$ such that $|\nabla W_s f_{l_d}(x_\rho, y_\rho)|$ reaches the local maximum, we consider its derivative on $\rho$. We denote

$$G(x) := x \int_0^\infty \frac{\phi'(\sqrt{t^2+x^2})}{\sqrt{t^2+x^2}} dt = x \int_0^\infty \frac{\psi(\sqrt{t^2+x^2})}{\sqrt{t^2+x^2}} dt, \qquad (8.6)$$

then have

$$\frac{\partial}{\partial x} \int_0^\infty \phi(\sqrt{t^2+x^2}) dt = G(x).$$

If $G(x) \leq 0$, we have

$$|\nabla W_s f_{l_d}(x_\rho, y_\rho)| = -2|c_f| \int_0^\infty \left[ \phi \left( \sqrt{t^2 + \left(\frac{\rho+d/2}{s}\right)^2} \right) \right.$$

$$\left. - \phi \left( \sqrt{t^2 + \left(\frac{\rho-d/2}{s}\right)^2} \right) \right] dt. \qquad (8.7)$$

Therefore,

$$\frac{d}{d\rho} |\nabla W_s f_{l_d}(x_\rho, y_\rho)|$$

$$= -2|c_f| \int_0^\infty \left[ \phi' \left( \sqrt{t^2 + \left(\frac{\rho+d/2}{s}\right)^2} \right) \frac{\left(\frac{\rho+d/2}{s}\right) \frac{1}{s}}{\sqrt{t^2 + \left(\frac{\rho+d/2}{s}\right)^2}} \right.$$

$$\left. - \phi' \left( \sqrt{t^2 + \left(\frac{\rho-d/2}{s}\right)^2} \right) \frac{\left(\frac{\rho-d/2}{s}\right) \frac{1}{s}}{\sqrt{t^2 + \left(\frac{\rho-d/2}{s}\right)^2}} \right] dt$$

$$= -2|c_f| \frac{1}{s} \left[ G \left( \frac{\rho}{s} + \frac{d}{2s} \right) - G \left( \frac{\rho}{s} - \frac{d}{2s} \right) \right].$$

To guarantee that $|\nabla W_s f_{l_d}(x_\rho, y_\rho)|$ reaches the local maximum, we let $\frac{d}{d\rho}|\nabla W_s f_{l_d}(x_\rho, y_\rho)| = 0$, and we can get

$$G\left(\frac{\rho}{s} + \frac{d}{2s}\right) = G\left(\frac{\rho}{s} - \frac{d}{2s}\right). \tag{8.8}$$

By Eq. (8.7), it is easily seen that $\rho$ must satisfy the following condition:

$$\left|\frac{\rho}{s} - \frac{d}{2s}\right| \le 1, \quad \left|\frac{\rho}{s} + \frac{d}{2s}\right| \le 1 \tag{8.9}$$

so that $|\nabla W_s f_{l_d}(x_\rho, y_\rho)|$ reaches the local maximum.

Hence, our question turns to solve $\rho$ satisfying Eqs. (8.8) and (8.9). To do this, we can evaluate $G(x)$ first. For $\psi$ defined by Eq. (8.2), the following is clear:

(1) For $x \in (0, \frac{1}{4})$,

$$
\begin{aligned}
G(x) = x &\int_x^{\frac{1}{4}} \frac{\psi_1(t) + \psi_2(t) + \psi_3(t)}{\sqrt{t^2 - x^2}} dt \\
&+ x \int_{\frac{1}{4}}^{\frac{3}{4}} \frac{\psi_2(t) + \psi_3(t)}{\sqrt{t^2 - x^2}} dt + x \int_{\frac{3}{4}}^{1} \frac{\psi_3(t)}{\sqrt{t^2 - x^2}} dt \\
= x &\int_x^{\frac{1}{4}} \frac{\psi_1(t)}{\sqrt{t^2 - x^2}} dt + x \int_x^{\frac{3}{4}} \frac{\psi_2(t)}{\sqrt{t^2 - x^2}} dt + x \int_x^{1} \frac{\psi_3(t)}{\sqrt{t^2 - x^2}} dt.
\end{aligned}
$$

(2) For $x \in [\frac{1}{4}, \frac{3}{4})$,

$$
\begin{aligned}
G(x) = x &\int_x^{\frac{3}{4}} \frac{\psi_2(t) + \psi_3(t)}{\sqrt{t^2 - x^2}} dt + x \int_{\frac{3}{4}}^{1} \frac{\psi_3(t)}{\sqrt{t^2 - x^2}} dt \\
= x &\int_x^{\frac{3}{4}} \frac{\psi_2(t)}{\sqrt{t^2 - x^2}} dt + x \int_x^{1} \frac{\psi_3(t)}{\sqrt{t^2 - x^2}} dt.
\end{aligned}
$$

(3) For $x \in [\frac{3}{4}, 1)$,

$$G(x) = x \int_x^{1} \frac{\psi_3(t)}{\sqrt{t^2 - x^2}} dt.$$

Thus, our problem now turns to evaluate the following:

$$
\begin{cases}
J_1(x) := x \int_x^{\frac{1}{4}} \dfrac{\psi_1(t)}{\sqrt{t^2 - x^2}} dt, \\[2mm]
J_2(x) := x \int_x^{\frac{3}{4}} \dfrac{\psi_2(t)}{\sqrt{t^2 - x^2}} dt, \\[2mm]
J_3(x) := x \int_x^{1} \dfrac{\psi_3(t)}{\sqrt{t^2 - x^2}} dt.
\end{cases}
$$

According to the definition of $\psi$ described by Eq. (8.2), we have

$$
-\frac{\pi}{2} J_1(x) = x \int_x^{\frac{1}{4}} \frac{1}{\sqrt{t^2 - x^2}} \left( -8t \ln \frac{1 + \sqrt{1 - 16t^2}}{4t} + \frac{\sqrt{1 - 16t^2}}{2t} \right) dt
$$

$$
= -4x \int_x^{\frac{1}{4}} \frac{1}{\sqrt{t^2 - x^2}} \ln \frac{1 + \sqrt{1 - 16t^2}}{4t} dt^2
$$

$$
+ x \int_x^{\frac{1}{4}} \frac{\sqrt{1 - 16t^2}}{4t^2 \sqrt{t^2 - x^2}} dt^2
$$

$$
= -8x \int_x^{\frac{1}{4}} \ln \frac{1 + \sqrt{1 - 16t^2}}{4t} d\sqrt{t^2 - x^2} + \frac{x}{4} \int_{x^2}^{\frac{1}{16}} \frac{\sqrt{1 - 16t}}{t\sqrt{t - x^2}} dt
$$

$$
= -8x \left[ \sqrt{t^2 - x^2} \ln \frac{1 + \sqrt{1 - 16t^2}}{4t} \Big|_{t=x}^{\frac{1}{4}} + \int_x^{\frac{1}{4}} \frac{\sqrt{t^2 - x^2}}{t\sqrt{1 - 16t^2}} dt \right]
$$

$$
+ \frac{x}{4} \int_{x^2}^{\frac{1}{16}} \frac{\sqrt{1 - 16t}}{t\sqrt{t - x^2}} dt
$$

$$
= -4x \int_x^{\frac{1}{4}} \frac{\sqrt{t^2 - x^2}}{t^2\sqrt{1 - 16t^2}} dt^2 + \frac{x}{4} \int_{x^2}^{\frac{1}{16}} \frac{\sqrt{1 - 16t}}{t\sqrt{t - x^2}} dt
$$

$$
= -4x \int_{x^2}^{\frac{1}{16}} \frac{\sqrt{t - x^2}}{t\sqrt{1 - 16t}} dt + \frac{x}{4} \int_{x^2}^{\frac{1}{16}} \frac{\sqrt{1 - 16t}}{t\sqrt{t - x^2}} dt
$$

$$
= -\frac{x}{4} \int_{x^2}^{\frac{1}{16}} \frac{1}{t} \left( \frac{16\sqrt{t - x^2}}{\sqrt{1 - 16t}} - \frac{\sqrt{1 - 16t}}{\sqrt{t - x^2}} \right) dt.
$$

Hence,

$$
J_1(x) = \frac{x}{2\pi} \int_{x^2}^{\frac{1}{16}} \frac{1}{t} \left( \frac{16\sqrt{t - x^2}}{\sqrt{1 - 16t}} - \frac{\sqrt{1 - 16t}}{\sqrt{t - x^2}} \right) dt.
$$

$J_2(x)$ and $J_3(x)$ can be calculated similarly. We omit the details here and give the results as follows:

$$J_2(x) = -\frac{3x}{2\pi} \int_{x^2}^{\frac{9}{16}} \frac{1}{t} \left( \frac{16\sqrt{t - x^2}}{\sqrt{9 - 16t}} - \frac{\sqrt{9 - 16t}}{\sqrt{t - x^2}} \right) dt,$$

$$J_3(x) = \frac{x}{\pi} \int_{x^2}^{1} \frac{1}{t} \left( \frac{16\sqrt{t - x^2}}{\sqrt{16 - 16t}} - \frac{\sqrt{16 - 16t}}{\sqrt{t - x^2}} \right) dt.$$

For $k = 1, 3, 4$, we have

$$\int_{x^2}^{\frac{k^2}{16}} \frac{1}{t} \left( \frac{16\sqrt{t - x^2}}{\sqrt{k^2 - 16t}} - \frac{\sqrt{k^2 - 16t}}{\sqrt{t - x^2}} \right) dt$$

$$= \int_{x^2}^{\frac{k^2}{16}} \frac{32t - 16x^2 - k^2}{t\sqrt{-16t^2 + (16x^2 + k^2)t - k^2x^2}} dt$$

$$= 32 \int_{x^2}^{\frac{k^2}{16}} \frac{dt}{\sqrt{-16t^2 + (16x^2 + k^2)t - k^2x^2}}$$

$$-(16x^2 + k^2) \int_{x^2}^{\frac{k^2}{16}} \frac{dt}{t\sqrt{-16t^2 + (16x^2 + k^2)t - k^2x^2}}$$

$$= 8 \arcsin \frac{32t - (16x^2 + k^2)}{k^2 - 16x^2} \Bigg|_{t=x^2}^{\frac{k^2}{16}}$$

$$- \frac{16x^2 + k^2}{kx} \arcsin \frac{(16x^2 + k^2)t - 2k^2x^2}{t(k^2 - 16x^2)} \Bigg|_{t=x^2}^{\frac{k^2}{16}}$$

$$= \pi \left( 8 - \frac{16x^2 + k^2}{kx} \right).$$

Hence,

$$J_1(x) = \frac{x}{2\pi} \pi \left( 8 - \frac{16x^2 + 1}{x} \right) = -8x^2 + 4x - \frac{1}{2},$$

$$J_2(x) = -\frac{3x}{2\pi} \pi \left( 8 - \frac{16x^2 + 9}{3x} \right) = 8x^2 - 12x + \frac{9}{2},$$

$$J_3(x) = \frac{x}{\pi} \pi \left( 8 - \frac{16x^2 + 16}{4x} \right) = -4(1 - x)^2.$$

Consequently,

(1) For $x \in (0, \frac{1}{4})$,

$$G(x) = J_1(x) + J_2(x) + J_3(x) = -4x^2,$$

(2) For $x \in [\frac{1}{4}, \frac{3}{4})$,

$$G(x) = J_2(x) + J_3(x) = 4x^2 - 4x + \frac{1}{2},$$

(3) for $x \in [\frac{3}{4}, 1)$,

$$G(x) = J_3(x) = -4(x - 1)^2,$$

i.e.,

$$G(x) = \begin{cases} -4x^2, & x \in (0, \frac{1}{4}), \\ 4x^2 - 4x + \frac{1}{2}, & x \in [\frac{1}{4}, \frac{3}{4}), \\ -4(x-1)^2, & x \in [\frac{3}{4}, 1), \\ 0, & x = 0 \text{ or } x \geq 1, \\ -G(-x), & x < 0. \end{cases}$$

It is clear that $G(x) \leq 0$ and is symmetric about $x = \frac{1}{2}$, therefore, Eqs. (8.8) and (8.9) are equivalent to the followings:

$$\begin{cases} 0 \leq \dfrac{\rho}{s} - \dfrac{d}{2s} \leq \dfrac{\rho}{s} + \dfrac{d}{2s} \leq 1, \\ \dfrac{1}{2} - \left( \dfrac{\rho}{s} - \dfrac{d}{2s} \right) = \left( \dfrac{\rho}{s} + \dfrac{d}{2s} \right) - \dfrac{1}{2}, \end{cases}$$

or

$$\begin{cases} -1 \leq \dfrac{\rho}{s} - \dfrac{d}{2s} \leq \dfrac{\rho}{s} + \dfrac{d}{2s} \leq 0, \\ -\dfrac{1}{2} - \left( \dfrac{\rho}{s} - \dfrac{d}{2s} \right) = \left( \dfrac{\rho}{s} + \dfrac{d}{2s} \right) + \dfrac{1}{2}, \end{cases}$$

i.e.,

$$\begin{cases} 0 \leq \rho - \dfrac{d}{2} \leq \rho + \dfrac{d}{2} \leq s, \\ \rho = \dfrac{s}{2}, \end{cases} \quad \text{or} \quad \begin{cases} -s \leq \rho - \dfrac{d}{2} \leq \rho + \dfrac{d}{2} \leq 0, \\ \rho = -\dfrac{s}{2}, \end{cases}$$

i.e.,

$$\begin{cases} s \geq d, \\ \rho = \dfrac{s}{2} \end{cases} \quad \text{or} \quad \begin{cases} s \geq d, \\ \rho = -\dfrac{s}{2}. \end{cases}$$

Hence, Eqs. (8.8) and (8.9) are solvable if and only if $s \geq d$. There are two solutions:

$$\rho_1 = -\frac{s}{2}, \quad \rho_2 = \frac{s}{2}.$$

These two solutions realize that the local maxima of $|\nabla W_s f_{l_d}(x_\rho, y_\rho)|$ arrive at both sides of the central line $l$ of $l_d$ and the distance from $l$ is $\frac{s}{2}$, which is independent of the width $d$.

In summary, The above three invariance properties can be rewritten as the following theorem:

**Theorem 8.1.** *Let $l_d$ be a Dirac-structure edge with width $d$ and $l$ be its central line. The local maxima modulus of the wavelet transform corresponding to the wavelets of Eq. (8.3) forms two new lines which are located symmetrically on both sides of the central line and have the same direction with it. If scale $s \geq d$, then the distance between the two new ones equals $s$.*

This theorem describes the width invariant property, which is important. It improves our former results in Chapter 6 and the work of Tang *et al.* [2000]. A couple of graphical examples are shown in Fig. 8.6.

Such an invariance property can be illustrated in Figs. 8.7 and 8.8 clearly. The graph in Fig. 8.7(a) displays function $G(\alpha)$ defined by Eq. (8.6). The graph in Fig. 8.7(b) indicates the positions of local maximum modulus by the wavelets defined by Eq. (8.2), where the three horizontal straight segments correspond to the widths of three curves. Point 0 at the horizontal axis corresponds to the center of the curves. The vertical axis corresponds to the *scale s* of the wavelet transform. The slant straight lines correspond to the positions of the local maximum moduli of the wavelet transforms. For each scale $s$, the local maximum moduli of the wavelet transforms with respect to the curves of different widths are located at the same positions.

To be more clear, we look at Fig. 8.8. There are three curves with different widths of $d = 1, d = 2$, and $d = 3$, as shown in Figs. 8.8(a)–8.8(c). The three horizontal straight segments in Fig. 8.8(d) correspond to these three curves. This width invariance can guarantee that for each scale $s$,

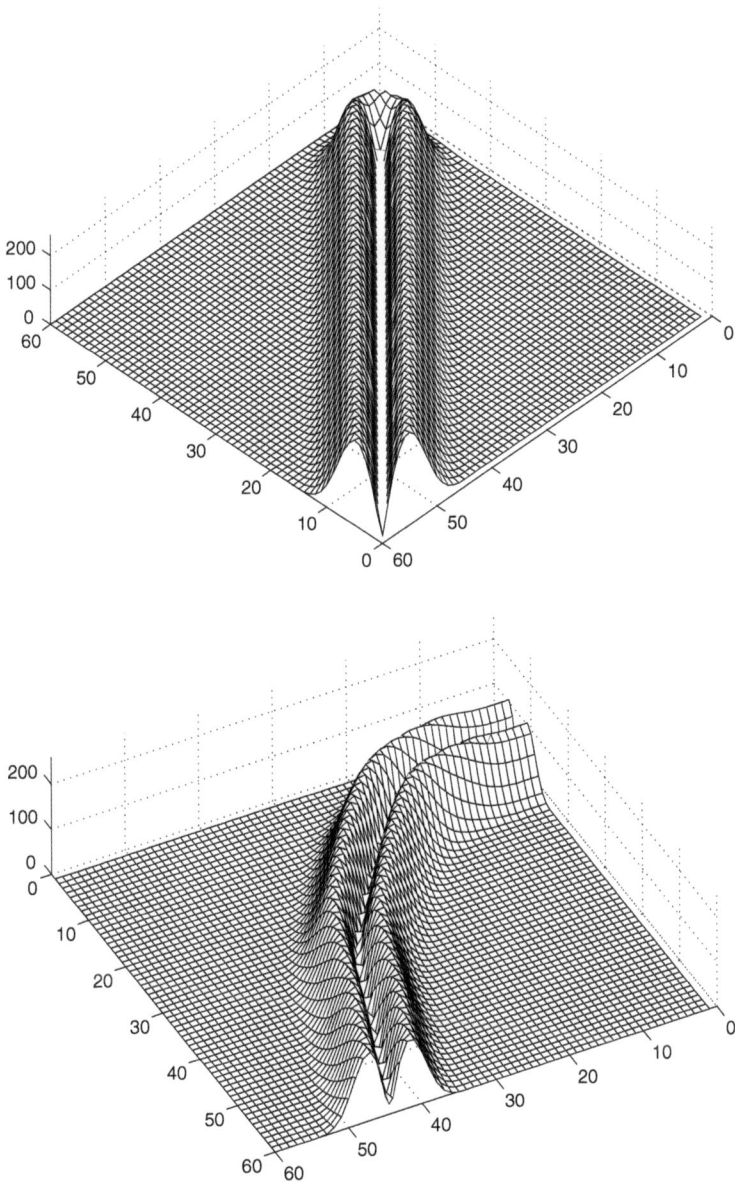

Fig. 8.6   Modulus of wavelet transforms corresponding to a segment of straight line and a curve.

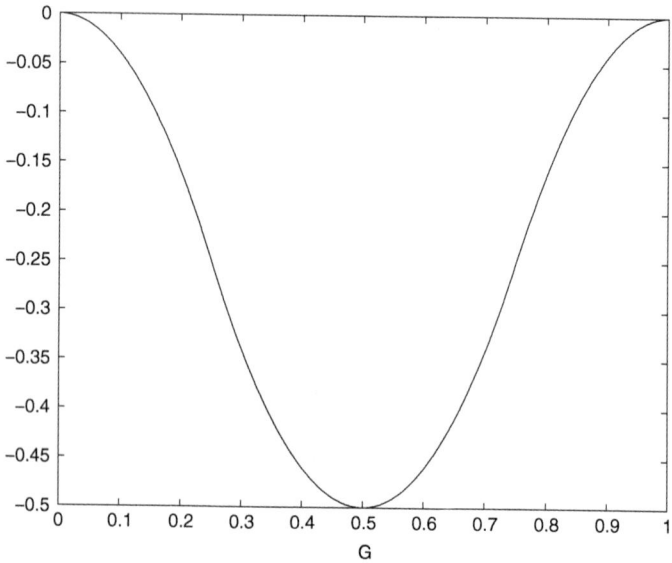

(a)

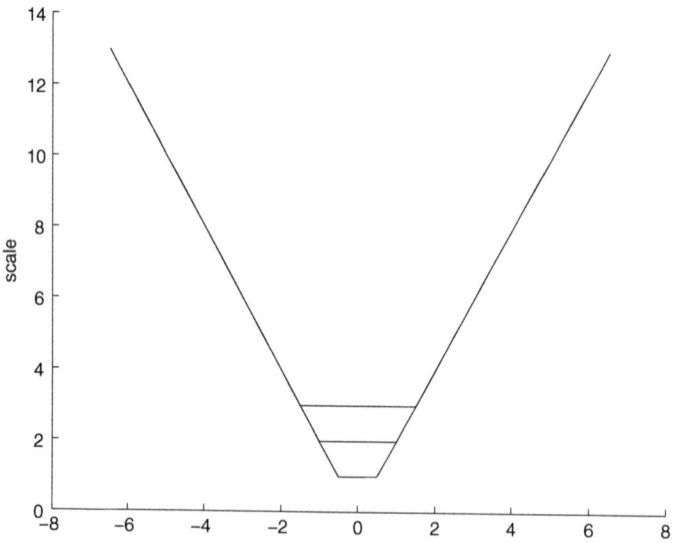

(b)

Fig. 8.7   (a) Function $G(\alpha)$ defined by Eq. (8.6); (b) positions of local maximum modulus by the wavelets defined by Eq. (8.2).

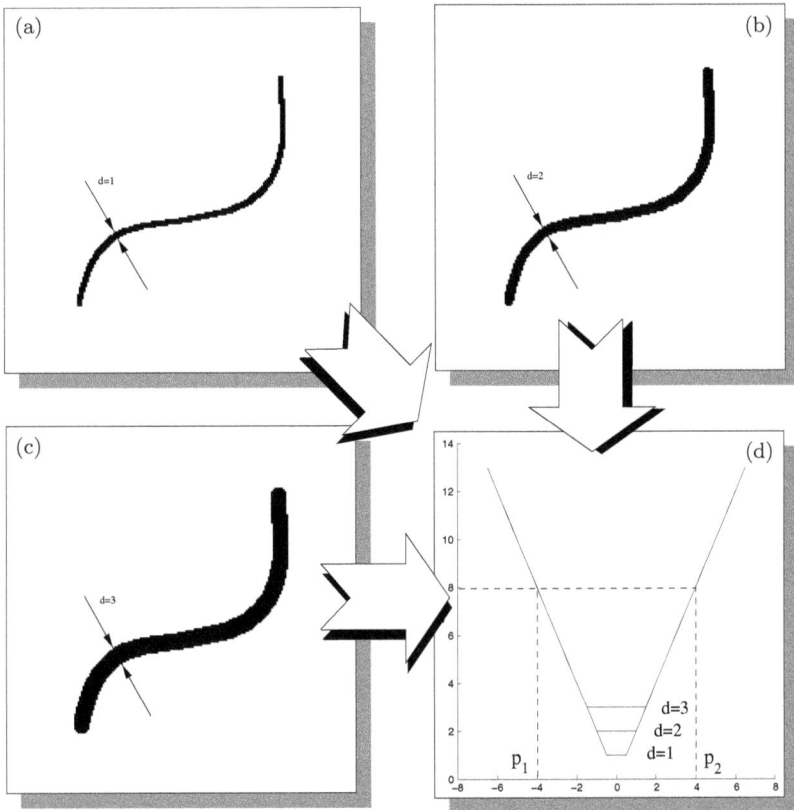

Fig. 8.8   Three curves with different widths of (a) $d = 1$, (b) $d = 2$, (c) $d = 3$, and (d) the positions of local maximum modulus by the wavelets defined by Eq. (8.2).

the local maximum moduli of the wavelet transforms with respect to the curves of different widths are located at the same positions. For example, when we apply the wavelet transform with scale $s = 8$ to these three curves, the locations of local maximum moduli of the wavelet transform for three curves, which have different widths of $d = 1, 2$, and 3, are the same. They are located on points $-4$ and 4 and denoted by $p_1$ and $p_2$, respectively, in Fig. 8.8(d).

## 8.3   Comparison with Other Wavelets

In this section, we compare our new wavelet with two other kinds of wavelets:

- **Gaussian wavelets:** they are derived from the smoothness functions $\theta(x)$, which are Gaussian functions as Eq. (8.1),
- The quadratic spline wavelets [Tang *et al.*, 2000].

### 8.3.1 *Comparison with Gaussian wavelets*

For the first case, $\theta(x)$ is a Gaussian function, we let $\sigma := 1/4$ for simplicity. Obviously, its derivative is a wavelet which is called a Gaussian wavelet, as shown in Fig. 8.9(a). A comparison between the new wavelet defined by Eq. (8.2) and the Gaussian wavelet is described in the following:

Let $\phi(x)$ in Eq. (8.6) be the Gaussian function of $\sigma := 1/4$ and $\psi(x)$ be its derivative, i.e.,

$$\phi(x) = \frac{4}{\sqrt{2\pi}} e^{-8x^2}, \quad \psi(x) = -\frac{64}{\sqrt{2\pi}} x e^{-8x^2}.$$

Then the function $G(x)$ defined by (8.6) can be calculated as follows:

$$G(x) := x \int_0^\infty \frac{\psi(\sqrt{t^2 + x^2})}{\sqrt{t^2 + x^2}} dt$$

$$= -\frac{64}{\sqrt{2\pi}} x \int_0^\infty e^{-8(x^2 + t^2)} dt$$

$$= \left( \int_0^\infty e^{-8t^2} dt \right) \psi(x).$$

Hence, Eq. (8.8) is equivalent to

$$\psi\left( \frac{\rho}{s} + \frac{d}{2s} \right) = \psi\left( \frac{\rho}{s} - \frac{d}{2s} \right), \tag{8.10}$$

i.e.,

$$\left( \frac{\rho}{s} + \frac{d}{2s} \right) \exp\left[ -8 \left( \frac{\rho}{s} + \frac{d}{2s} \right)^2 \right] = \left( \frac{\rho}{s} - \frac{d}{2s} \right) \exp\left[ -8 \left( \frac{\rho}{s} - \frac{d}{2s} \right)^2 \right].$$

Although this equation cannot be solved analytically, a numerical solution can be obtained. According to Fig. 8.9 which describes function $\psi(x)$, for a constant $C: 0 < C \leq 1$, there exists a pair of numbers $(\alpha, \beta)$ satisfying

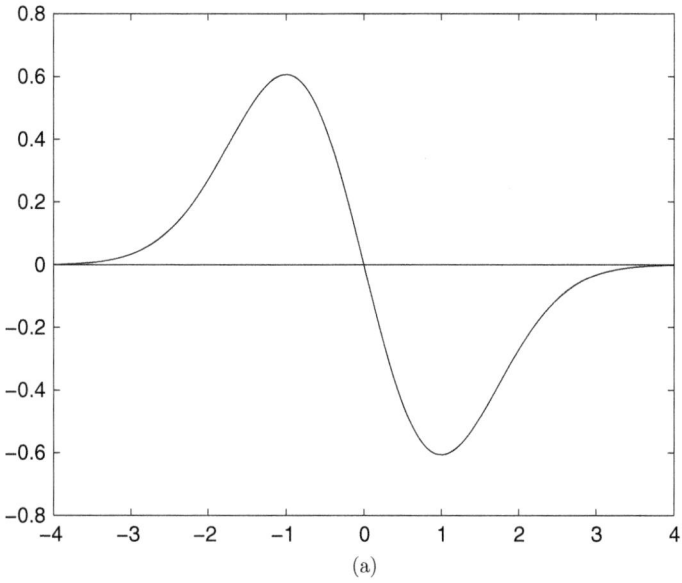

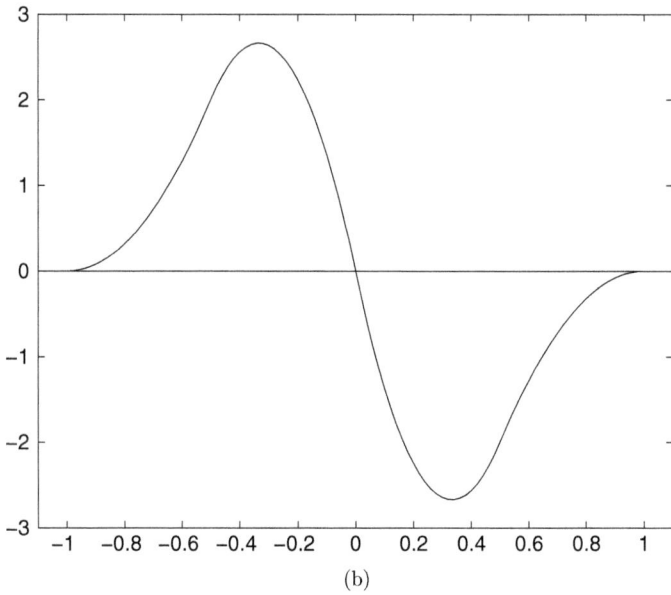

Fig. 8.9 (a) Gaussian wavelet; (b) quadratic spline wavelet.

$0 \leq \alpha < \beta \leq 1$, such that

$$\begin{cases} \beta - \alpha = C \\ \psi(\alpha) = \psi(\beta). \end{cases} \tag{8.11}$$

It is obvious that both $\alpha$ and $\beta$ depend on only the constant $C$, which means that they are functions on $C$. Therefore, $\alpha$ and $\beta$ can be denoted by

$$\alpha = \alpha(C), \quad \beta = \beta(C). \tag{8.12}$$

In the following, we solve $\rho$ to satisfy Eq. (8.10). Comparing Eq. (8.10) and Eq. (8.11) yields

$$\frac{\rho}{s} + \frac{d}{2s} = \beta, \quad \frac{\rho}{s} - \frac{d}{2s} = \alpha. \tag{8.13}$$

Therefore, we obtain

$$\rho = \frac{s}{2}(\alpha + \beta), \quad C = \beta - \alpha = \frac{d}{s}.$$

By Eq. (8.12), we can deduce that

$$\rho = \frac{s}{2}(\alpha + \beta) = \frac{s}{2}\left(\alpha\left(\frac{d}{s}\right) + \beta\left(\frac{d}{s}\right)\right).$$

Then, for a fixed $d$, the parameter $\rho$ is a function of $s$. Figure 8.10(a) shows function $G(\alpha)$ with respect to the Gaussian wavelet, and Fig. 8.10(b) illustrates the positions of local maximum modulus by Gaussian wavelet for $d = 1, 2, 3$, respectively. Figure 8.10(e) shows function $G(\alpha)$ with respect to the new wavelet, and Fig. 8.10(f) illustrates the positions of local maximum modulus by the new wavelet for $d = 1, 2, 3$, respectively. The readers are suggested to compare Figs. 8.10(b) and 8.10(f). It is obvious that the locations of the local maximum of the wavelet transform with respect to the Gaussian wavelet depend nonlinearly on the curve width $d$ heavily, while the locations of the local maximum of the wavelet transform using the new wavelet are fixed.

### 8.3.2  *Comparison with quadratic spline wavelets*

$\theta(x)$ is set to be the quadratic spline as discussed in Chapter 6 and the work of Tang *et al.* [2000], and a graphical display is present in Fig. 8.9(b). Figure 8.10(c) presents function $G(\alpha)$ with respect to the quadratic spline wavelet. Figure 8.10(d) illustrates the corresponding locations of the maximum moduli for $d = 2, 4, 8$, respectively. Similarly, it can be shown that $\rho$ depends nonlinearly on $d$ lightly.

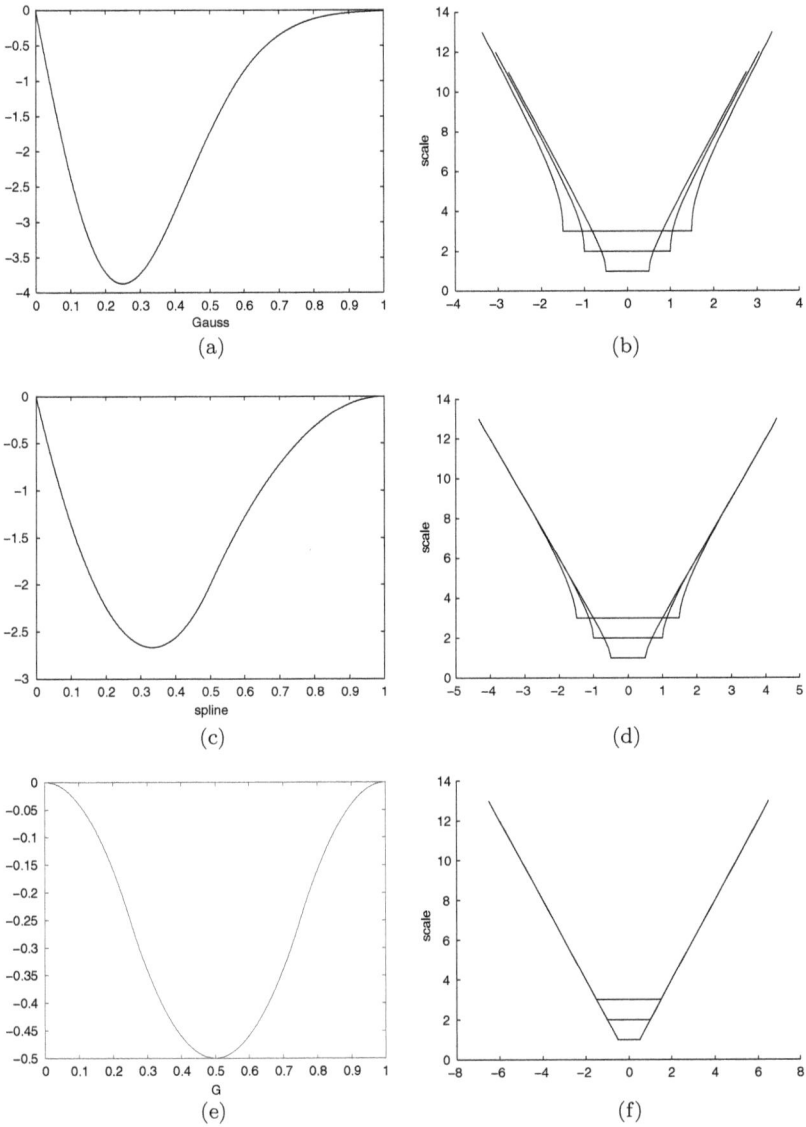

Fig. 8.10   (a), (c), and (e) The functions $G(\alpha)$ with respect to the Gaussian wavelet, quadratic spline wavelet, and the new wavelet defined by Eq. (8.2), respectively. (b), (d), and (f) The positions of local maximum moduli by Gaussian wavelet, quadratic spline wavelet, and the new wavelet defined by Eq. (8.2).

## 8.4 Algorithm and Experiments

In this section, the algorithm for extracting the Dirac-structure edges is present, including the calculation of the non-zero coefficients $\{\phi_{k,l}^s\}$ for different scales. Several experiments are also conducted.

### 8.4.1 *Algorithm*

In practice, the wavelet transform should be calculated discretely. We have the following formula:

$$
\begin{aligned}
W_s^i f(n, m) &= \iint f(u, v)\psi_s^i(n - u, m - v)\,du\,dv \\
&= \sum_{k,l} f(k, l) \int_k^{k+1} \int_l^{l+1} \psi_s^i(n - u, m - v)\,du\,dv \\
&= \sum_{k,l} f(k, l) \int_{n-k-1}^{n-k} \int_{m-l-1}^{m-l} \psi_s^i(u, v)\,du\,dv \\
&= \sum_{k,l} f(n - k - 1, m - l - 1)\psi_{k,l}^{s,i}, \quad (i = 1, 2),
\end{aligned}
$$

where

$$
\begin{aligned}
\psi_{k,l}^{s,i} &= \int_k^{k+1} \int_l^{l+1} \psi_s^i(u, v)\,du\,dv \\
&= \int_{k/s}^{(k+1)/s} \int_{l/s}^{(l+1)/s} \psi^i(u, v)\,du\,dv, \quad (i = 1, 2).
\end{aligned}
$$

In fact, the computation of the above formula is a discrete convolution. For a small scale $s$, it can be performed directly. But for a large scale $s$, considering the amount of computation, FNT (fast numerous transform) or other fast algorithms can be utilized instead. The details can be found in the works of Mallat and Hwang [1992] and Tang *et al.* [1998b]. In the following, we discuss the calculation of the coefficients $\{\psi_{k,l}^{s,i}\}$.

For the wavelets defined by Eq.(8.3), since

$$
\psi^1(u, v) = \phi'(\sqrt{u^2 + v^2})\frac{u}{\sqrt{u^2 + v^2}} = \psi^2(v, u),
$$

we deduce that

$$
\begin{aligned}
\psi_{k,l}^{s,1} &= \int_{k/s}^{(k+1)/s} du \int_{l/s}^{(l+1)/s} \psi^1(u,v)dv \\
&= \int_{k/s}^{(k+1)/s} du \int_{l/s}^{(l+1)/s} \psi^2(v,u)dv \\
&= \int_{k/s}^{(k+1)/s} dv \int_{l/s}^{(l+1)/s} \psi^2(u,v)du \\
&= \psi_{k,l}^{s,2}.
\end{aligned}
$$

Therefore, we need to calculate only $\psi_{k,l}^{s,1}$ for all $k,l \in Z$. Note that $\psi^1(u,v)$ is odd on $u$ and even on $v$, thus we have

$$
\begin{aligned}
\psi_{-k,l}^{s,1} &= \int_{-k/s}^{(-k+1)/s} \int_{l/s}^{(l+1)/s} \psi^1(u,v)dudv \\
&= -\int_{(k-1)/s}^{k/s} \int_{l/s}^{(l+1)/s} \psi^1(u,v)dudv \\
&= -\psi_{k-1,l}^{s,1}.
\end{aligned}
$$

Similarly,

$$
\psi_{k,-l}^{s,1} = \psi_{k,l-1}^{s,1}, \quad \psi_{-k,-l}^{s,1} = -\psi_{k-1,l-1}^{s,1}.
$$

Consequently, we further need to calculate only $\psi_{k,l}^{s,1}$ for all positive integers $k,l$. By $\psi^1(u,v) = \frac{\partial}{\partial u}[\phi(\sqrt{u^2+v^2})]$, we have

$$
\begin{aligned}
\psi_{k,l}^{s,1} &= \int_{k/s}^{(k+1)/s} \int_{l/s}^{(l+1)/s} \frac{\partial}{\partial u}[\phi(\sqrt{u^2+v^2})]dudv \\
&= \int_{l/s}^{(l+1)/s} \left[ \phi\left(\sqrt{v^2 + \left(\frac{k+1}{s}\right)^2}\right) - \phi\left(\sqrt{v^2 + \left(\frac{k}{s}\right)^2}\right) \right] dv \\
&= \phi_{l,k+1}^s + \phi_{l+1,k}^s - \phi_{l+1,k+1}^s - \phi_{l,k}^s,
\end{aligned}
$$

where

$$
\phi_{k,l}^s = \int_{k/s}^{\infty} \phi\left(\sqrt{v^2 + \left(\frac{l}{s}\right)^2}\right) dv.
$$

Hence, our question turns to calculate $\phi_{k,l}^s$ for all non-negative integers $k$ and $l$. Since

$$\psi^1(u,v) = \phi'(\sqrt{u^2 + v^2})\frac{u}{\sqrt{u^2 + v^2}} = \psi(\sqrt{u^2 + v^2})\frac{u}{\sqrt{u^2 + v^2}},$$

where $\psi$ is defined by Eq. (8.2), we have for any non-negative integers $k, l$ satisfying $k^2 + l^2 < s^2$,

$$
\begin{aligned}
\phi_{k,l}^s &= \int_{k/s}^{\infty} \phi(\sqrt{v^2 + (l/s)^2}\,)dv \\
&= \int_{\frac{\sqrt{k^2+l^2}}{s}}^{\infty} \phi(v)d\sqrt{v^2 - (l/s)^2} \\
&= \sqrt{v^2 - (l/s)^2}\phi(v)\Big|_{v=\frac{\sqrt{k^2+l^2}}{s}}^{\infty} - \int_{\frac{\sqrt{k^2+l^2}}{s}}^{\infty} \phi'(v)\sqrt{v^2 - (l/s)^2}dv \\
&= -\frac{l}{s}\phi\left(\frac{\sqrt{k^2+l^2}}{s}\right) - \int_{\frac{\sqrt{k^2+l^2}}{s}}^{1} \psi(v)\sqrt{v^2 - (l/s)^2}dv \\
&= -\frac{l}{s}\int_{-\infty}^{\frac{\sqrt{k^2+l^2}}{s}} \psi(v)dv - \int_{\frac{\sqrt{k^2+l^2}}{s}}^{1} \psi(v)\sqrt{v^2 - (l/s)^2}dv \\
&= -\frac{l}{s}\int_{0}^{\frac{1}{s}\sqrt{k^2+l^2}} \psi(v)dv - \frac{l}{s}\int_{-1}^{0} \psi(v)dv \\
&\quad - \int_{\frac{1}{s}\sqrt{k^2+l^2}}^{1} \psi(v)\sqrt{v^2 - (l/s)^2}dv \\
&= -\frac{l}{s}\int_{0}^{\frac{1}{s}\sqrt{k^2+l^2}} \psi(v)dv + \frac{l}{s}\int_{0}^{1} \psi(v)dv \\
&\quad - \int_{\frac{1}{s}\sqrt{k^2+l^2}}^{1} \psi(v)\sqrt{v^2 - (l/s)^2}dv \\
&= \frac{l}{s}\int_{\frac{1}{s}\sqrt{k^2+l^2}}^{1} \psi(v)dv - \int_{\frac{1}{s}\sqrt{k^2+l^2}}^{1} \psi(v)\sqrt{v^2 - (l/s)^2}dv \\
&= \int_{\frac{1}{s}\sqrt{k^2+l^2}}^{1} \left[\frac{l}{s} - \sqrt{v^2 - (l/s)^2}\right] \psi(v)dv.
\end{aligned}
$$

On the other hand, it is easy to see that $\phi_{k,l}^s = 0$ for all integers $k, l$ satisfying $k^2 + l^2 \geq s^2$ due to the compact support $[-1, 1]$ of $\phi(x)$. According to the composite trapezoidal formula of numerical quadrature [Chaohao, 1992]

$$\int_a^b f(x)dx \approx \frac{b-a}{2n}\left[f(a)+f(b)+2\sum_{i=1}^{n-1}f\left(a+i\frac{b-a}{n}\right)\right],$$

we can calculate all the coefficients $\phi_{k,l}^s$ numerically for non-negative integers $k, l$. The positive integer $n$ in the formula can be set to be so large that the error is smaller than any prior number. The possible non-zero items of $\phi_{k,l}^s$ for $s = 2, 4, 8$ are listed in Tables 8.1–8.4.

Based on the characterization of a straight line in an image developed in Section 8.2, an algorithm to detect straight lines in an image can be designed. The result is also valid for general curves since a short segment

Table 8.1.   The non-zero coefficients $\{\phi_{k,l}^s\}$ for $s = 2$.

| $l\backslash k$ | $k = 0$ | $k = 1$ |
|---|---|---|
| $l = 0$ | 0.2500 | 0.1617 |
| $l = 1$ | 0.1241 | 0.0656 |

Table 8.2.   The non-zero coefficients $\{\phi_{k,l}^s\}$ for $s = 4$.

| $l\backslash k$ | $k = 0$ | $k = 1$ | $k = 2$ | $k = 3$ |
|---|---|---|---|---|
| $l = 0$ | 0.2500 | 0.2138 | 0.1170 | 0.0435 |
| $l = 1$ | 0.1520 | 0.1171 | 0.0612 | 0.0236 |
| $l = 2$ | 0.0602 | 0.0477 | 0.0260 | 0.0084 |
| $l = 3$ | 0.0207 | 0.0169 | 0.0078 | 0 |

Table 8.3.   The non-zero coefficients $\{\phi_{k,l}^s\}$ for $s = 8$, $k = 0 - 3$.

| $l\backslash k$ | $k = 0$ | $k = 1$ | $k = 2$ | $k = 3$ |
|---|---|---|---|---|
| $l = 0$ | 0.2500 | 0.2254 | 0.1871 | 0.1343 |
| $l = 1$ | 0.1844 | 0.1698 | 0.1374 | 0.0963 |
| $l = 2$ | 0.1202 | 0.1110 | 0.0882 | 0.0603 |
| $l = 3$ | 0.0683 | 0.0630 | 0.0495 | 0.0330 |
| $l = 4$ | 0.0335 | 0.0308 | 0.0239 | 0.0158 |
| $l = 5$ | 0.0144 | 0.0133 | 0.0104 | 0.0071 |
| $l = 6$ | 0.0060 | 0.0056 | 0.0046 | 0.0033 |
| $l = 7$ | 0.0025 | 0.0023 | 0.0018 | 0.0009 |

Table 8.4. The nonzero coefficients $\{\phi^s_{k,l}\}$ for $s = 8$, $k = 4 - 7$.

| $l\backslash k$ | $k = 4$ | $k = 5$ | $k = 6$ | $k = 7$ |
|---|---|---|---|---|
| $l = 0$ | 0.0821 | 0.0423 | 0.0190 | 0.0079 |
| $l = 1$ | 0.0577 | 0.0292 | 0.0133 | 0.0056 |
| $l = 2$ | 0.0350 | 0.0173 | 0.0081 | 0.0032 |
| $l = 3$ | 0.0187 | 0.0094 | 0.0046 | 0.0013 |
| $l = 4$ | 0.0091 | 0.0049 | 0.0023 | 0 |
| $l = 5$ | 0.0045 | 0.0024 | 0.0005 | 0 |
| $l = 6$ | 0.0020 | 0.0005 | 0 | 0 |
| $l = 7$ | 0 | 0 | 0 | 0 |

of a curve can be regarded as a straight line approximately. In fact, wavelet transforms are essentially local analysis. Therefore, the result of Theorem 8.1 can be applied to the general curves in an image. Our algorithm to detect curves in an image is designed as follows:

---

**Algorithm 8.1.** Let $f(x, y)$ be an image containing curve. For a scale $s > 0$. We have the following:

**Step 1:** Calculate all the wavelet transforms $\{W^1_s f(x, y),\ W^2_s f(x, y)\}$ with respect to the wavelets defined by Eq. (8.3).

**Step 2:** Calculate the local maxima $f_{\text{locmax}}$ of $|\nabla W_s f(x, y)|$ and the gradient direction $f_{\text{gradient}}$.

**Step 3:** For each point $(x, y)$ with local maximum, search the point whose distance along the gradient direction from $(x, y)$ is $s$. If it is a point of local maxima, the center point is detected.

**Step 4:** The curves formed by all the points detected in Step 3 are what we need.

---

## 8.4.2 *Experiments*

Four circles with various gray levels and widths as shown in Fig. 8.11 are tested using the new method. The original images are illustrated on the left column of Fig. 8.11. After applying Steps 1 and 2 of the proposed wavelet transform algorithm to these circles, the local maximum modulus of the wavelet transform with respect to them can be computed and the results are given on the middle column in Fig. 8.11. Finally, the central lines of

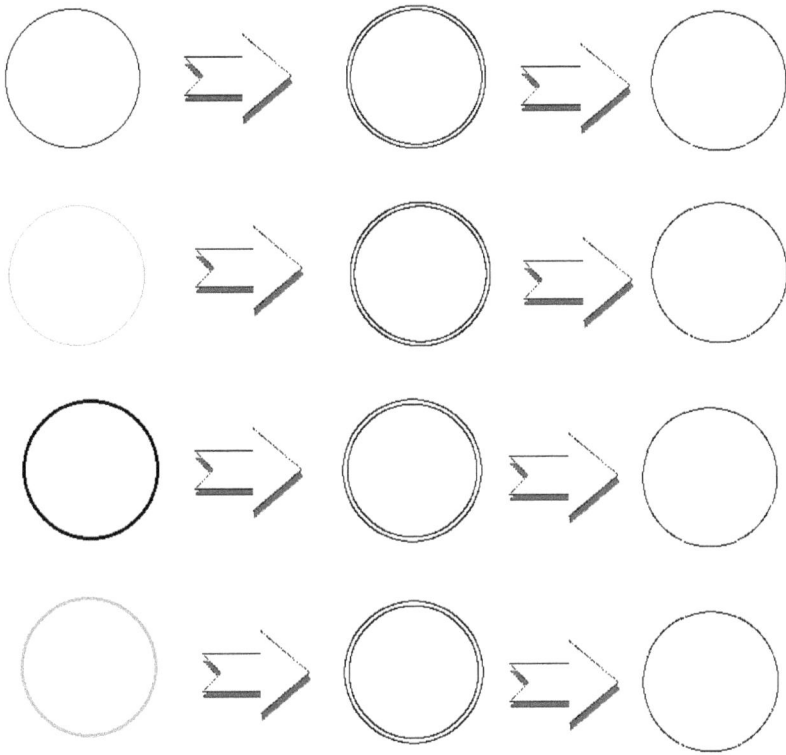

Fig. 8.11   Detection of lines by new wavelet transform, $s = 6$.

these circles are extracted using Steps 3 and 4 of the above algorithm and presented on the right column in Fig. 8.11.

Next, let us turn back to the beginning of this chapter and look at Fig. 8.12. The particular task is that we are required to extract the skeleton of the circle with various widths. Unfortunately, as we have shown in Fig. 8.12, the algorithm based on the spline wavelet in Chapter 6 cannot work well due to the width dependence of the detection. Fortunately, as described in detail in Section 8.2, the method developed in this chapter possesses the width invariant, gray-level invariant, as well as slope invariant. According to these properties, the skeleton of the circle and tree in Fig. 8.12 can be extracted. After applying Steps 1 and 2 of the above algorithm to the original image as displayed on the left column of Fig. 8.12, the local maximum modulus of the wavelet transform with respect to them can

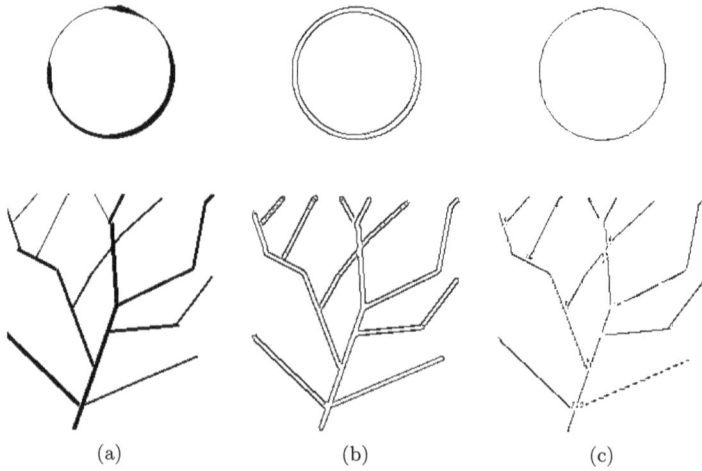

Fig. 8.12 (a) The original image; (b) The location of maximum modulus of the wavelet transform corresponding to $s = 6$; (c) The skeleton extracted by the algorithm in this chapter.

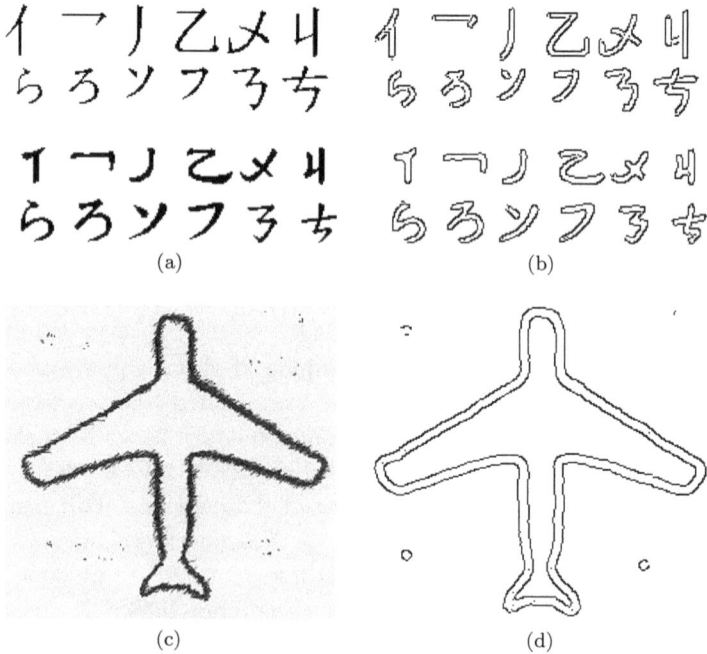

Fig. 8.13 (a) An original image containing some basic strokes of Chinese and Japanese characters in different fonts; (b) the location of the MMWTs of (a) with the new wavelet corresponding to $s = 5$; (c) an original image of a noisy contour of an airplane; (d) the location of the MMWTs of (c) with the new wavelet corresponding to $s = 5$.

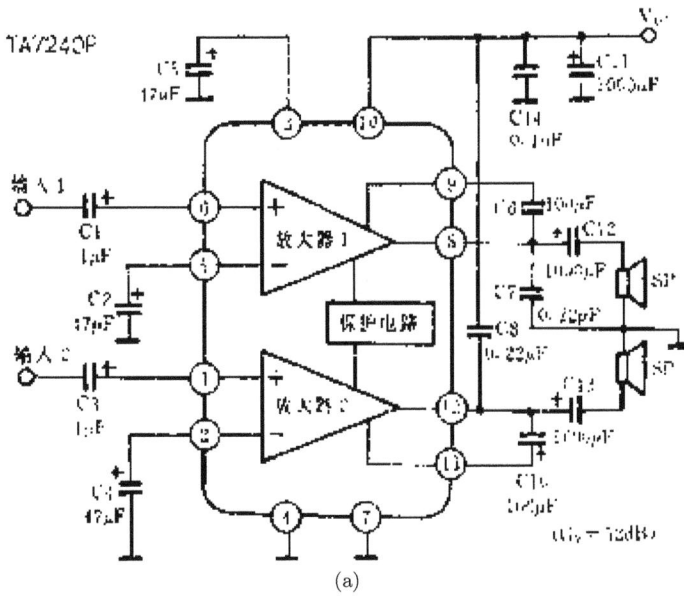

(a)

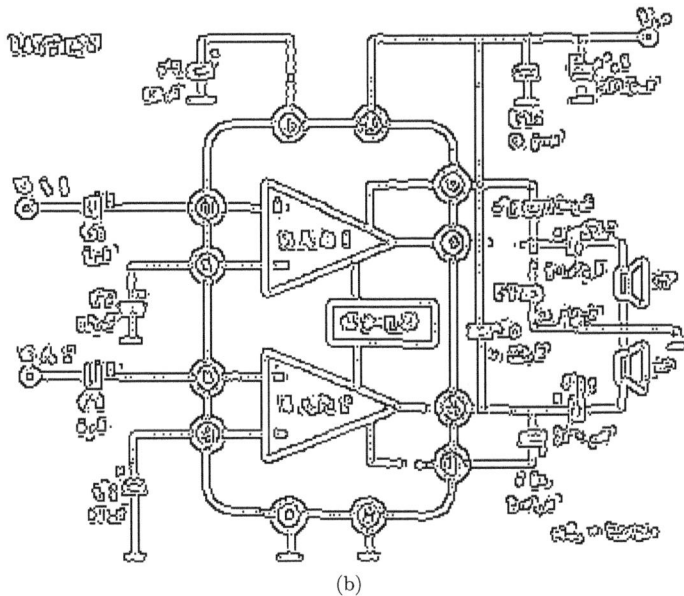

(b)

Fig. 8.14 (a) A circuit diagram; (b) the locations of the MMWTs of (a) with the new wavelet corresponding to $s = 4$.

(a)

(b)

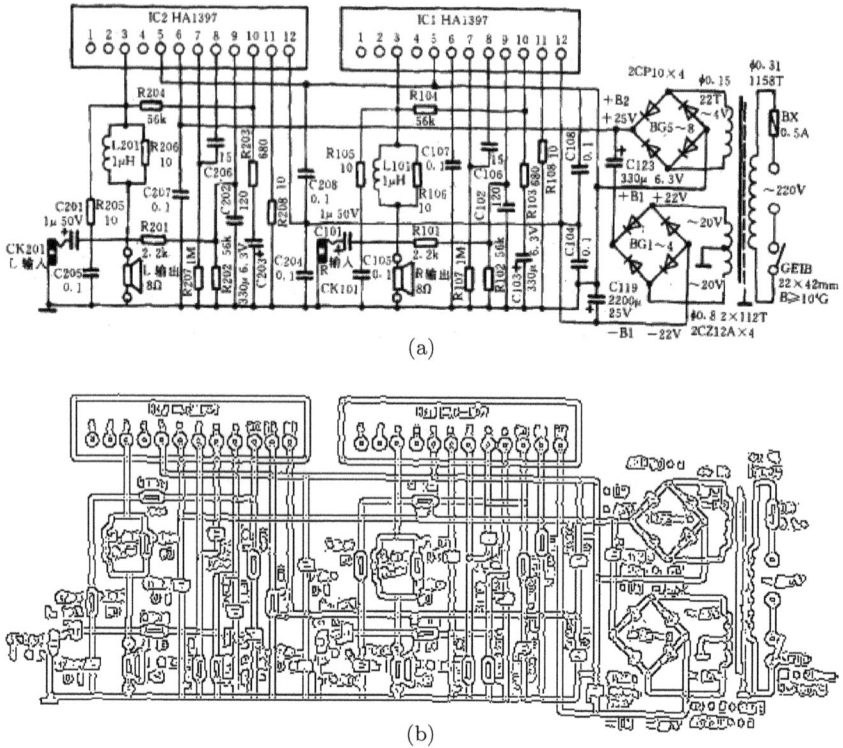

Fig. 8.15   (a) A circuit diagram; (b) the locations of the MMWTs of (a) with the new wavelet corresponding to $s = 4$.

be computed and presented on the middle column in Fig. 8.12. At last, the central lines are extracted using Steps 3 and 4 of the above algorithm and presented on the right column in Fig. 8.12.

Figures 8.13–8.15 show experimental results for four other images. Figure 8.13(a) is the original image which consists of some basic strokes of Chinese and Japanese characters in different fonts. Its (maximum moduli of the wavelet transforms MMWTs) are displayed in Fig. 8.13(b). It is easy to see that the locations of its MMWTs are independent of the width of the strokes. Figure 8.13(c) is the original image of a noisy contour of an airplane. There are many anomalistic blurs along the contour and the widths along the contour are erratically different. It is interesting to see that its MMWTs, which are shown in Fig. 8.13(d), are smooth with the same widths along the contour. Figures 8.14(a) and 8.15(a) are two circuit diagrams. Some lines of the schematic diagrams are eroded because of aging. Some pixels on the lines of the schematic diagrams disappear and some other noises appear instead. The widths of each line are thus not the same. With the novel wavelet presented in this chapter, their MMWTs, which are shown in Figs. 8.14(b) and 8.15(b), respectively, remain the same.

Chapter 9

# Skeletonization of Ribbon-Like Shapes with New Wavelet Function

One of the most significant topics in pattern recognition is the analysis of Ribbon-like shapes. It can be applied to character recognition, signature verification, understanding of paper-based graphics including maps and engineering drawings, computer-assisted cartoons, fingerprint analysis, etc. [Janssen, 1997; Tombre, 1998]. Representation of a shape using a suitable form is essential in these recognition systems. An appropriate representation of a Ribbon-like shape is its skeleton [Zou and Yan, 2001]. The objective of this study is to extract skeletons from planar Ribbon-like shapes, which have the following properties: (1) conforming to human perceptions of the original shapes, (2) being centered inside the original shapes, (3) being efficiently computable, and (4) being robust against noise and geometric transformation.

Skeletons have been defined in various ways in different literatures. Generally, the skeleton of a shape is referred to as the locus of the symmetric points or symmetric axes of the local symmetries of the shape, in other words, different local symmetry analyses may result in different symmetric points, and hence different skeletons are produced. There are three methods of local symmetry analysis, which are well known to the shape analysis community:

1. Blum's symmetric axs tansform (SAT) [Blum, 1967],
2. Brady's smoothed local symmetry (SLS) [Brady, 1983],
3. Leyton's process-inferring symmetry analysis (PISA) [Leyton, 1988].

Each method contributes to the skeletonization of the shapes greatly. The difference in the selection of the location of a symmetric point makes these methods distinguishable from each other. Nevertheless, the major problem of SAT and PISA is that the symmetric points of a local symmetry and hence a skeleton segment may lie in a perceptually distinct part of the underlying shape. Although the skeleton obtained from the SLS has a pleasing visual appearance, the major shortcoming of the SLS is that some perceptually irrelevant symmetric axes may be created. Some grouping rules can be used [Connell and Brady, 1987]. However, these rules may not be appropriate for a wide range of shapes. An alternative way is to divide the shape into several parts, and thereafter, the SLS axes are computed within each part [Rom and Medioni, 1984]. Dividing a shape into suitable parts is crucial in this method, unfortunately, it is a difficult task. Another powerful measure of the symmetry proposed by Kovesi [1995, 1997] is based on the analysis of phase information. The mathematics of phase congruency using wavelets has been studied by Kovesi [1995], where the phase information can be used to construct a contrast-invariant measure of the symmetry that does not require any prior recognition or segmentation. Although it is possible to calculate the phase congruency directly [Venkatesh and Owens, 1990], it is a rather awkward quantity to calculate. Moreover, it provides limited information about the overall shape of an object, and it is not shown further whether it is capable of skeleton extraction.

The computation of the skeleton of a shape is another challenge for skeletonization. In other words, the determination of the symmetric points of a shape by using the above symmetry analysis is a key issue. Computing the skeleton of a shape in the continuous domain is not an easy task. Suppose the boundaries of a shape are represented by two curves, generally, it is difficult to determine the symmetric points from those curves. Especially in the discrete domain, it is even more difficult to compute the skeleton of a shape using the definitions of the skeleton given above, since the shape is represented by a set of discrete pixels, not in a continuous form. The constrained Delaunay triangulation technique [Zou and Yan, 2001] is a sound solution for this problem. However, it has to suffer from complicated computation and it costs too much computation time.

Therefore, many approximation methods have been proposed to compute skeletons in the discrete world, which can be divided into two groups: pixel-based method and non-pixel-based method.

In the pixel-based method, each foreground pixel is utilized for computation in the skeletonization process. Techniques used in the pixel-based method include thinning [Lam *et al.*, 1992; Smith, 1987] and distance transform [Smith, 1987]. The former is applied to most of the existing skeletonization methods. Their basic principle is to repetitively remove the pixels of the outside layer of an image until only the central pixels remain, which establishes the approximated skeleton of the shape. It can be regarded as the grassfire process, that produces a homotopic skeleton, does not alter the topology of the shape, and is easy to implement. The distance transform possesses the following advantages: (1) A skeleton is produced in a fixed number of passes through the image regardless of the size of the object. (2) It may operate faster than the thinning method does, since it requires $O(n^2)$ to $O(n^3)$ time to compute the skeleton of an $n \times n$ image [Smith, 1987] rather than the latter, which usually takes $O(n^3)$ to do so [Smith, 1987].

The pixel-based method often suffers from one or more of the following drawbacks:

First, the generated skeletons are generally in discrete forms, where the skeleton points are discrete pixels. However, such a skeleton is not helpful for recognizing the underlying shape unless skeleton pixels are linked by lines or curves implicitly or explicitly to form a graph.

Second, even if the skeleton pixels are linked, the resulting skeleton may not be centered inside the underlying shape due to the use of discrete data.

Third, the computational complexity is quite high, since all foreground pixels are needed for the computation in the skeletonization process.

In the non-pixel-based method, the skeleton of a shape is analytically derived from the border of the image. There are two types of non-pixel-based methods, which are based on either cross-section [Pavlidis, 1986] or Voronoi diagrams [Ogniewicz and Kubler, 1995]. These methods attempt to determine the symmetric points of a shape without the intermediate step of the grassfire propagation. The fundamental concept of these methods is that the local symmetric axes of a shape are derived from pairs of contour pixels or a contour segment representing a sequence of the contour pixels. The mid-points or center lines of the pairs of the contour elements are connected to generate the skeleton. An obvious advantage of these methods over the pixel-based methods is that fewer data (only the contour pixels) are used for the skeletonization. Hence, they are expected to be faster. Especially, compared with the thinning technique, another advantage of the non-pixel-based method is that the real coordinates are available for the skeleton points, which are not restricted by the discrete grids. Thus, the generated skeleton can be centered inside the underlying shapes.

On the other hand, the accurate identification of the local symmetries of the underlying shape is the major problem in the non-pixel-based methods. In fact, given a contour pixel of a digital shape, it may be impossible to find another pixel on the opposite contour, such that they are exactly mathematically symmetrical, due to the digitization. Thus, a fundamental problem that needs to be solved is the definition of symmetries in the discrete domain, so that the skeleton of a digital shape can be computed correctly.

All in all, although more than 300 skeletonization algorithms have been proposed, improvement is still required, since the existing approximation algorithms of skeletonization often suffer from one or more of the following drawbacks [Chang and Yan, 1999; Ge and Fitzpatrick, 1996; Lam *et al.*, 1992; Smith, 1987; You and Tang, 2007]:

(1) It may take a long time to skeletonize a high-resolution image.
(2) Skeletons may not be centered inside the underlying shapes.
(3) Skeletons are sensitive to noise and shape variations, such as rotation and scaling.
(4) A shape and its skeleton may have different numbers of connected components.
(5) Skeleton may contain artifacts such as noisy spurs and spurious short branches between split junction points.
(6) Skeleton branches may seriously erode.
(7) In addition, most methods are suitable for only the binary images rather than the gray images.

To overcome the above problems, a novel wavelet-based method is presented in Tang and You [2003] and You and Tang [2007]. In this way, a new wavelet function called Tang–Yang wavelet, which has been constructed by our research group [Yang *et al.*, 2001, 2003a], is applied; a new symmetry analysis, which is called symmetry analysis of maximum moduli of the wavelet transform, is also proposed. This approach benefits from the following desired characteristics of the new wavelet function investigated:

• The position of the local maximum moduli of the wavelet transform with respect to the Ribbon-like shape is independent of the gray levels of the image.
• The local maximum moduli of the wavelet transform of the Ribbon-like shape form two new contours, when the appropriate scale of the wavelet transform is selected according to the width of the shape. Meantime,

the new contours are located symmetrically on both sides of the original image and keep the same topological and geometric properties.

- The distance from one contour to the opposite one, which is completely independent of the width of the shape, equals the scale of the wavelet transform.

Based on the above symmetry analyses, a new skeleton called wavelet skeleton can be defined, which not only maintains the good properties provided by the existing methods but also improves the technique in the following respects: (1) Generally, it is easy to determine the symmetric points from the symmetric curves proposed and a skeleton may be centered exactly inside the underlying shape. (2) The computation of the skeleton of a shape is readily simple, moreover, it takes less time to perform the skeleton. (3) A skeleton representation is robust against noise and insensitive to linear geometric transformations, such as translation, rotation, and scaling. (4) The image to be processed may be extended to any gray levels.

This chapter is organized as follows. In Section 9.1, the Tang–Yang wavelet function is introduced. In Section 9.2, a corresponding symmetry analysis will be developed based on three important properties of the maximum moduli of the wavelet transform. The characterization of the skeleton produced by the proposed method will also be compared with the existing ones. In Section 9.3, the wavelet skeleton and its implementation will further be discussed. A set of techniques for modifying the artifacts of the primary wavelet skeleton will also be given. An algorithm to perform the proposed new scheme to extract the skeleton of the Ribbon-like shapes as well as experiments will be illustrated in Section 9.4.

## 9.1 Tang–Yang Wavelet Function

It's well known that settling the contour or boundary of the shape is key to extracting the skeleton. With the growth of wavelet theory, the wavelet transform has been found to be a remarkable mathematical tool to analyze the singularities, including the edges, in particular, the sharp boundary of the shape, and further to detect them effectively [Canny, 1986; Mallat, 1998; Mallat and Hwang, 1992; Mallat and Zhong, 1992; Tang *et al.*, 2000]. Although lots of wavelet functions have been found so far, the construction of an appropriate one according to the application in practice is still a great challenge for researchers worldwide. In order to extract the skeleton of a shape, detecting the boundary or contour of the shape is required.

In this chapter, a novel wavelet function called Tang–Yang wavelet, which is constructed in our work [Yang *et al.*, 2001; 2003a], is utilized.

As we have known, the edge points of a digital image are the pixels, which are as close as possible to the center of the true edge but absolute. Hence, when we extract the central point of the shape via its edges, the location of edge pixels may be shifted around the center of the true edge without losing the human vision. In our motivation, when we detect the edges from the local maxima of the wavelet transform modulus, the desired operator not only detects the singular points of the signal but also adjusts properly the location of the edge points around the center of the true edge, which depends strongly on the scale of the wavelet transform. Therefore, a novel wavelet function $\psi(x, y)$ is a considerable candidate, which satisfies the mentioned extra requirement besides those conditions that Gaussian function and quadratic spline satisfy, such as when it is used as the wavelet function, the location of the maximum moduli of the wavelet transform is relatively dependent on the scale of the wavelet transform. Obviously, it is preferable for the boundary detection of the Ribbon-like shapes. An odd function $\psi(x)$ (for the sake of simplicity, we discuss the one-dimension case first) is considered as the new wavelet function. Consequently, we consider an odd function $\psi(x)$, which is defined on $(0, \infty)$ by

$$\psi(x) := \begin{cases} \psi_1(x) + \psi_2(x) + \psi_3(x) & x \in (0, \frac{1}{4}), \\ \psi_2(x) + \psi_3(x) & x \in [\frac{1}{4}, \frac{3}{4}), \\ \psi_3(x) & x \in [\frac{3}{4}, 1), \\ 0 & x \in [1, \infty), \end{cases} \tag{9.1}$$

where

$$\begin{cases} \psi_1(x) = -\frac{2}{\pi}(-8x \ln \frac{1+\sqrt{1-16x^2}}{4x} + \frac{1}{2x}\sqrt{1-16x^2}), \\ \psi_2(x) = -\frac{2}{\pi}(8x \ln \frac{3+\sqrt{9-16x^2}}{4x} - \frac{3}{2x}\sqrt{9-16x^2}), \\ \psi_3(x) = -\frac{2}{\pi}(-4x \ln \frac{1+\sqrt{1-x^2}}{x} + \frac{4}{x}\sqrt{1-x^2}), \end{cases}$$

as the candidate of 1D wavelet function. Apparently, function $\phi(x) := \int_0^x \psi(x)dx$ is an even function compactly supported on $[-1, 1]$, and $\phi'(x) = \psi(x)$ holds as well. The graphical description of functions $\psi(x)$ and $\phi(x)$ can be found in Fig. 8.3 in Chapter 8. The 2D wavelet functions are given by

$$\begin{cases} \psi^1(x,y) := \frac{\partial}{\partial x}\theta(x,y) = \phi'(\sqrt{x^2+y^2})\frac{x}{\sqrt{x^2+y^2}}, \\ \psi^2(x,y) := \frac{\partial}{\partial y}\theta(x,y) = \phi'(\sqrt{x^2+y^2})\frac{y}{\sqrt{x^2+y^2}}, \end{cases} \tag{9.2}$$

and are graphically displayed in Fig. 8.5 in Chapter 8.

The gradient direction and the amplitude of the wavelet transform are denoted respectively by

$$\nabla W_s f(x,y) := \begin{pmatrix} W_s^1 f(x,y) \\ W_s^2 f(x,y) \end{pmatrix} \tag{9.3}$$

and

$$|\nabla W_s f(x,y)| := \sqrt{|W_s^1 f(x,y)|^2 + |W_s^2 f(x,y)|^2}. \tag{9.4}$$

Here, $|\nabla W_s f(x,y)|$ is called the modulus of the wavelet transform at point $(x,y)$. The boundary of the shape can be detected by locating the local maxima of the wavelet transform modulus. The details of the selection of the new function and analysis of its property as well as the comparisons with Gaussian function and quadratic spline are presented in Yang *et al.* [2003a].

## 9.2 Characterization of the Boundary of a Shape by Wavelet Transform

In this section, some significant characteristics of the Ribbon-like shape with respect to the local maximum moduli of the wavelet transform based on the above wavelet function are presented as follows:

- **Gray-level-invariant:** The local maximum moduli of the wavelet transform of the Ribbon-like shape take place at the same points regardless of the different gray levels of the image.
- **Slope invariant:** The local maximum moduli of the wavelet transform of the Ribbon-like shape are independent of the slope of the object.
- **Width-invariant:** The maximum moduli of the wavelet transform of the Ribbon-like shape produce two new lines around the original boundary. Moreover, the location of maximum moduli is independent of the width of the shape, namely, the distance between the two new lines depends on only the scale of the wavelet transform rather than width of the shape under certain circumstances.

- **Symmetry:** The location of the maximum moduli of the wavelet transform cover exactly the points of the original boundary if the transform scale $s$ equals the width of the shape. Moreover, when the scale is bigger than or equal the width of a shape, the two new lines formed by the maximum moduli of the wavelet transform are exactly symmetrical with respect to the central line of the shape.

The above properties can be summarized in the following theorem, and the proofs can be found in Yang *et al.* [2003a].

**Theorem 9.1.** *Let $l_d$ be a straight segment of the Ribbon-like shape with width $d$ and central line $l$. If the scale of the wavelet transform $s \geq d$, then the local maximum moduli of the wavelet transform using wavelet function of Eq. (9.2) generate two new periphery lines around the original segment, which have the following properties: The two new lines are exactly symmetric with respect to the central line of the segment; the distance between the new lines equals scale $s$, in other words, the location of the maximum moduli of the wavelet transform depends completely on scale $s$ and the location of the central line of the segment. Moreover, the two new lines possess the same gradient direction as the central line does. Furthermore, if and only if scale $s$ equals the width of the shape, the locations of points of the maximum moduli lie exactly on the boundaries of the shape.*

Therefore, the following definitions are natural outcomes:

**Definition 9.1.** In wavelet transform, if scale $s$ is bigger than or equal to the width of the Ribbon-like shape, the points of the maximum moduli will generate two new lines located on the periphery of the shape. Moreover, they are locally symmetrical with respect to the central line of the shape. This symmetry is called *maximum moduli symmetry (MMS)*.

**Definition 9.2.** The *wavelet skeleton* of the Ribbon-like shape is defined as the curve of all connective midpoints of the segment lines, which are connected by all pairs of the symmetrical maximum moduli of the wavelet transform of the shape.

To illustrate the above definitions more clearly, now a brief comparison of our new symmetry analysis with the traditional ones, i.e., Blum's SAT [Blum, 1967], Brady's SLS [Brady, 1983], and Leyton's PISA [Leyton, 1988], will be presented.

The graphical descriptions of four symmetry analyses are shown in Figs. 9.1 and 9.2. Here, $l_1$ and $l_2$ are two opposite boundaries of the shape.

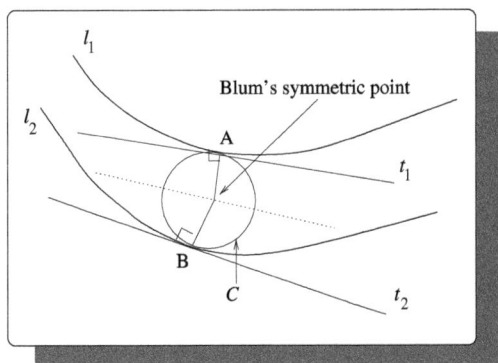

(a)

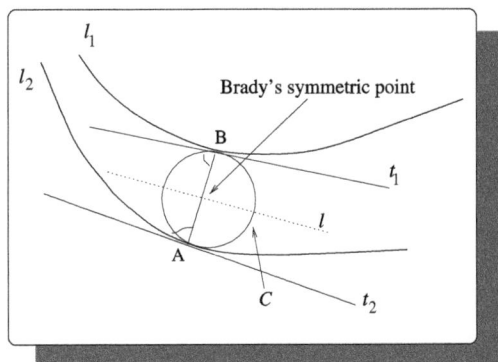

(b)

Fig. 9.1   The illustration of symmetry analyses: (a) Blum's symmetric axis transform and (b) Brady's smoothed local symmetry.

A circle labeled by $C$ is placed between $l_1$ and $l_2$ such that it is simultaneously tangential to both boundaries at points $A$ and $B$ in Figs. 9.1(a) and (b) and 9.2(a). Blum's SAT is shown in Fig. 9.1(a), and its symmetric point of the local symmetry formed by $A$ and $B$ is defined as the center of circle $C$. Hence, its corresponding skeleton is defined as the locus of the central points of the maximal inscribed symmetric circles of the shape. It is obvious that the interior of circle $C$ must lie entirely inside either the foreground or the background of the image. Most of the existing skeletonization algorithms are based on the concept of this symmetry analysis, such as the grassfire technique.

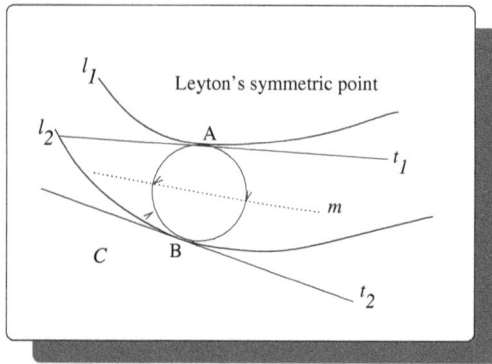

(a)

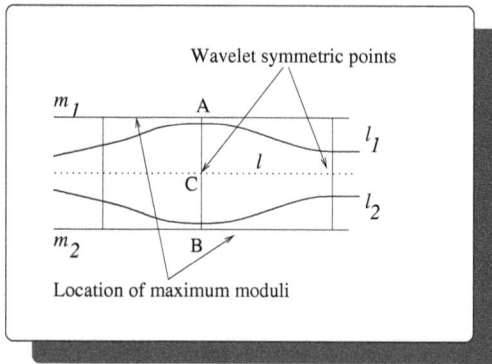

(b)

Fig. 9.2   The illustration of symmetry analyses: (a) Leyton's process–inferring symmetry and (b) maximum moduli symmetry of wavelet transform.

In SLC symmetry analysis, Brady defines the symmetric point of the local symmetry by the intersecting point of two lines, the mirror $l$ and segment $\overline{AB}$, i.e., the midpoint of $\overline{AB}$. Its corresponding skeleton is the locus of the mid-points of the symmetric lines of the symmetric circles, as shown in Fig. 9.1(b). In Brady's original proposal, a basic criterion, called the equal-angle criterion, asks that angle $(\theta)$ between symmetric line $\overline{AB}$ and a tangent line $t_1$ equals the angle between $\overline{AB}$ and another tangent line $t_2$. More precisely, two points, $A$ and $B$, on a planar curve form a local symmetry if the angle between vector $\overrightarrow{AB}$ and the normal of the curve at $B$ equals the angle between vector $\overrightarrow{BA}$ and the normal of the curve at $A$. However, the interior of circle $C$ is not required to stay inside the shape completely. This characteristic can be either an advantage or a disadvantage compared with SAT and PISA methods. The advantage is that a wider variety of situations are considered. For instance, SAT fails to produce the minor axis of an ellipse, which can be generated by SLS [Brady and Asada, 1993]. The disadvantage is that a symmetric axis from SLS may not lie inside the shape entirely [Saint-Marc *et al.*, 1993]. The additional measures have to be used while the perceptually salient axes are selected to characterize the shape. In addition, the question of whether the locus of symmetries lies within the shape or not depends not only on the choice of locality of the symmetry point (e.g., center of circle, mid-chord) but also on whether the bitangent circle is forced to be inscribed (entirely contained within the shape). Typically SLS does not enforce this constraint while the SAT of Blum does.

Leyton's symmetric point is defined as the intersecting point of "mirror" $m$ and symmetric circle $C$, i.e., the midpoint of the arc $\widehat{AB}$, and the corresponding skeleton is defined by the locus of the midpoints of the symmetric arcs of the circles, which can be shown in Fig. 9.2(a). Since there are two symmetric arcs, and two symmetric points associated with the symmetric pair $A$ and $B$, PISA becomes ambiguous when it comes to decide which symmetric point is appropriate to represent the symmetry property of the object.

In our proposed symmetry analysis, which is shown in Fig. 9.2(b), the symmetric point is defined as the midpoint of segment $\overline{AB}$. Meanwhile, the corresponding skeleton of a shape is defined as the locus of the midpoints of the segments connecting any pair of points, which lie on two opposite symmetric maximum moduli $m_1$ and $m_2$. The first advantage is that a symmetric axis must be inside the shape entirely. Moreover, the skeleton of the shape can be mathematically centered inside the underlying shape. Another obvious advantage is that its implementation is relatively easy, in other words, the computation of the skeleton of the shape is simple and direct because its two symmetric maximum moduli lines are easily obtained by applying the wavelet transform to the shape. Moreover, the distance between the two symmetric maximum moduli lines completely depends on the scale of the wavelet transform, thus, it is relatively easy to locate the midpoint of the segment in practice. As far as the symmetry analyses in either SAT or SLS or PISA are concerned, the implementation is not straightforward and relatively difficult. Since it is hard to fix circle $C$ between $l_1$ and $l_2$ such that it is simultaneously tangential to both contours at $A$ and $B$, which can be found in Figs. 9.1(a) and (b) and 9.2(a), especially in the discrete domain. Moreover, it also implies that the above indirect approaches (SAT, SLS, and PISA) suffer from the higher computational cost. Our approach is a direct method, in which the points of the modulus maxima of the wavelet transform are just contour points of the shape, therefore, the computational cost is mainly determined by this phase itself. The extra computational cost of searching the tangent circles of the contours in SAT, SLS, and PISA can be saved.

## 9.3  Wavelet Skeletons and Its Implementation

In this section, a set of schemes for extracting wavelet skeletons of the Ribbon-like shapes will be presented, including two parts: (1) implementation of the wavelet transform in the discrete domain to generate a primary wavelet skeleton and (2) modification of the primary wavelet skeleton.

### 9.3.1  *Wavelet Transform in the Discrete Domain*

In practice, the images to be processed are in a discrete world. In this section, we will give the wavelet transform formula in the discrete domain and calculate the corresponding wavelet coefficients. In fact, the wavelet transform formula can be rewritten as follows:

$$W_s^i f(n,m) = \iint f(u,v)\psi_s^i(n-u, m-v)dudv$$

$$= \sum_{k,l} f(k,l) \int_k^{k+1} \int_l^{l+1} \psi_s^i(n-u, m-v)dudv$$

$$= \sum_{k,l} f(n-k-1, m-l-1)\psi_{k,l}^{s,i}, \quad (i = 1, 2),$$

where

$$\psi_{k,l}^{s,i} = \int_k^{k+1} \int_l^{l+1} \psi_s^i(u,v)dudv$$

$$= \int_{k/s}^{(k+1)/s} \int_{l/s}^{(l+1)/s} \psi^i(u,v)dudv, \quad (i = 1, 2).$$

It is clear that the computation of the above formula is a discrete convolution. For a small scale $s$, it can be performed directly. But for a large scale $s$, considering the amount of computation, fast numerous transform (FNT) or other fast algorithms can be utilized instead. The details can be found in Tang *et al.* [2000]. In the following, we will discuss the calculation of the coefficients $\{\psi_{k,l}^{s,i}\}$.

For the wavelet functions defined in Eq. (9.2), since

$$\psi^1(u,v) = \phi'(\sqrt{u^2 + v^2})\frac{u}{\sqrt{u^2 + v^2}} = \psi^2(v,u),$$

we can deduce that

$$\psi_{k,l}^{s,1} = \psi_{l,k}^{s,2}, \quad \psi_{-k,l}^{s,1} = -\psi_{k-1,l}^{s,1}, \quad \psi_{k,-l}^{s,1} = \psi_{k,l-1}^{s,1}, \quad \psi_{-k,-l}^{s,1} = -\psi_{k-1,l-1}^{s,1}.$$

Consequently, we further need to calculate only $\psi_{k,l}^{s,1}$ for all positive integers $k$ and $l$. By $\psi^1(u,v) = \frac{\partial}{\partial u}[\phi(\sqrt{u^2 + v^2})]$, we have

$$\psi_{k,l}^{s,1} = \phi_{l,k+1}^s + \phi_{l+1,k}^s - \phi_{l+1,k+1}^s - \phi_{l,k}^s,$$

where

$$\phi_{k,l}^s = \int_{k/s}^\infty \phi\left(\sqrt{v^2 + \left(\frac{l}{s}\right)^2}\right) dv.$$

Hence, our question turns to calculate $\phi_{k,l}^s$ for all non-negative integers $k$ and $l$. Since

$$\psi^1(u,v) = \phi'(\sqrt{u^2 + v^2})\frac{u}{\sqrt{u^2 + v^2}} = \psi(\sqrt{u^2 + v^2})\frac{u}{\sqrt{u^2 + v^2}},$$

where $\psi$ is defined by Eq. (9.1), for any non-negative integers $k$ and $l$ satisfying $k^2 + l^2 < s^2$, we have

$$\phi_{k,l}^s = \int_{k/s}^\infty \phi(\sqrt{v^2 + (l/s)^2})dv = \int_{\frac{1}{s}\sqrt{k^2 + l^2}}^1 \left[\frac{k}{s} - \sqrt{v^2 - (l/s)^2}\right]\psi(v)dv.$$

On the other hand, it is easy to see that $\phi_{k,l}^s = 0$ for all integers $k$ and $l$ satisfying $k^2 + l^2 \geq s^2$ due to the compact support $[-1,1]$ of $\phi(x)$.

We can calculate all coefficients $\phi_{k,l}^s$ numerically for non-negative integers $k$ and $l$. The positive integer $n$ in the formula can be set to be so large that the error is smaller than any prior number. The possible non-zero items for $s = 2, 4, 6, 8$ are listed in Tables 9.1–9.5.

## 9.3.2 Generation of Wavelet Skeleton in the Discrete Domain

Based on the discussion in Section 9.2, a set of schemes to detect the segment of shape can be designed. The result is also valid for general Ribbon-like shapes since a short segment of a shape can be regarded as a straight

Table 9.1. The non-zero coefficients $\{\phi_{k,l}^s\}$ for $s = 2$.

| $l\backslash k$ | $l = 0$ | $l = 1$ |
|---|---|---|
| $k = 0$ | 0.2500 | 0.1250 |
| $k = 1$ | 0.0497 | 0.0111 |

Table 9.2. The non-zero coefficients $\{\phi_{k,l}^s\}$ for $s = 4$.

| $k\backslash l$ | $l = 0$ | $l = 1$ | $l = 2$ | $l = 3$ |
|---|---|---|---|---|
| $k = 0$ | 0.2500 | 0.2292 | 0.1250 | 0.0208 |
| $k = 1$ | 0.1468 | 0.1206 | 0.0552 | 0.0060 |
| $k = 2$ | 0.0497 | 0.0366 | 0.0111 | 0.0003 |
| $k = 3$ | 0.0047 | 0.0026 | 0.0002 | 0 |

Table 9.3. The non-zero coefficients $\{\phi_{k,l}^s\}$ for $s = 6$.

| $k\backslash l$ | $l = 0$ | $l = 1$ | $l = 2$ | $l = 3$ | $l = 4$ | $l = 5$ |
|---|---|---|---|---|---|---|
| $k = 0$ | 0.2500 | 0.2438 | 0.2022 | 0.1250 | 0.0478 | 0.0062 |
| $k = 1$ | 0.1831 | 0.1718 | 0.1333 | 0.0767 | 0.0257 | 0.0025 |
| $k = 2$ | 0.1106 | 0.1003 | 0.0723 | 0.0367 | 0.0094 | 0.0005 |
| $k = 3$ | 0.0497 | 0.0436 | 0.0281 | 0.0111 | 0.0017 | 0.0000 |
| $k = 4$ | 0.0133 | 0.0109 | 0.0056 | 0.0014 | 0.0000 | 0 |
| $k = 5$ | 0.0011 | 0.0008 | 0.0002 | 0.0000 | 0 | 0 |

Table 9.4. The non-zero coefficients $\{\phi_{k,l}^s\}$ for $s = 8, l = 0 - 3$.

| $k\backslash l$ | $l = 0$ | $l = 1$ | $l = 2$ | $l = 3$ |
|---|---|---|---|---|
| $k = 0$ | 0.2500 | 0.2474 | 0.2292 | 0.1849 |
| $k = 1$ | 0.2006 | 0.1950 | 0.1741 | 0.1358 |
| $k = 2$ | 0.1468 | 0.1403 | 0.1206 | 0.0902 |
| $k = 3$ | 0.0935 | 0.0882 | 0.0733 | 0.0517 |
| $k = 4$ | 0.0497 | 0.0462 | 0.0366 | 0.0236 |
| $k = 5$ | 0.0199 | 0.0180 | 0.0132 | 0.0072 |
| $k = 6$ | 0.0047 | 0.0041 | 0.0026 | 0.0011 |
| $k = 7$ | 0.0004 | 0.0003 | 0.0001 | 0 |

line. In fact, wavelet transform is essentially a local analysis. Therefore, the result of Theorem 9.1 can be applied to the general Ribbon-like object.

Table 9.5.   The non-zero coefficients $\{\phi_{k,l}^s\}$
for $s = 8$, $l = 4 - 7$.

| $k\backslash l$ | $l = 4$ | $l = 5$ | $l = 6$ | $l = 7$ |
|---|---|---|---|---|
| $k = 0$ | 0.1250 | 0.0651 | 0.0208 | 0.0026 |
| $k = 1$ | 0.0884 | 0.0433 | 0.0126 | 0.0013 |
| $k = 2$ | 0.0552 | 0.0244 | 0.0060 | 0.0004 |
| $k = 3$ | 0.0287 | 0.0107 | 0.0020 | 0 |
| $k = 4$ | 0.0111 | 0.0032 | 0.0003 | 0 |
| $k = 5$ | 0.0026 | 0.0004 | 0 | 0 |
| $k = 6$ | 0.0002 | 0 | 0 | 0 |
| $k = 7$ | 0 | 0 | 0 | 0 |

Maximum moduli symmetry is defined in the continuous domain as mentioned in the previous discussion. In fact, the exact symmetries between contour pixels may not exist in the discrete domain. When a continuous shape is digitized, its boundary is sampled by discrete points. For some contour pixels of a digital shape, it may not be possible to find their symmetrical counterparts on the opposite contour to form local symmetries. Even if a pair of symmetric contour pixels can be found precisely, their corresponding symmetrical central point may not exist.

In addition, in the continuous domain, the gradient direction of a function can be accurately counted in mathematics. However, in contrast, it is difficult to express it exactly in the discrete form. Consequently, we will devote our efforts to finding the best way to locate the symmetric maximum moduli points and the corresponding central points.

First, we determine the gradient direction of a digital signal and its corresponding local maximum. Fortunately, in the equispaced sampling, only eight adjacent pixels are around a point. Hence, only the nearest eight points will be taken into account. It is said that the discrete image has eight gradient directions. Therefore, a plane can be divided into eight sectors. A gradient direction is defined by

$$\alpha_s := \arctan \left( \frac{\partial (f * \theta_s)(x, y)}{\partial y} \bigg/ \frac{\partial (f * \theta_s)(x, y)}{\partial x} \right),$$

where $s$ denotes the wavelet transform scale.

When $\alpha_s$ falls into a sector, it will be quantified to a certain vector, which is represented by a central line of that sector. It indicates the direction of the gradient, and along this direction, the local maximal moduli can be achieved. The effect of any opposite gradient direction is the same. Thus, only four codes are needed to be used to code these different directions. The tangent of each direction, which is described by $tg\alpha_s = \frac{\partial(f*\theta_s)(x,y)}{\partial y} / \frac{\partial(f*\theta_s)(x,y)}{\partial x}$, falls into one of the following intervals:

$$[-1 - \sqrt{2}, 1 - \sqrt{2}), \quad [1 - \sqrt{2}), \sqrt{2} - 1), \quad [\sqrt{2} - 1, \sqrt{2} + 1),$$
$$[\sqrt{2} + 1, +\infty) \cup (-\infty, -1 - \sqrt{2}).$$

The above gradient code is called *gradient code of wavelet transform* (GCWT). It will play an important role not only in locating the central points (primary skeleton points) but also in modifying artifacts of the primary wavelet skeleton whereafter.

On the other hand, for every maximum moduli point, even though the corresponding symmetric counterpart cannot be achieved or does not exist along its gradient direction, one or more maximum moduli points may appear in a sector, which contains certain GCWT or gradient direction. Hence, from these points, we can select one, where the distance from this point to the original maximum moduli point approximates to scale $s$ of the wavelet transform as the symmetric counterpart of the original maximum moduli point. Meantime, the midpoint or approximate midpoint of the segment, which is connected through the pair of maximum moduli symmetric points, is regarded as the symmetric central point, namely, the skeleton point. In practice, this approach can be implemented accurately and easily in mathematics.

An important issue in skeletonization is how to reduce the noise. As we have know, the noise always distributes randomly. From the point of view of the statistics, the average value of the noise is nearly a constant in a certain area. In general, we can suppose the value of this constant is zero and take a weighted mean to the signal in this area. This action can be regarded as a low-pass filtering. In this way, the noise will be eliminated considerably. Virtually, $\theta(x, y)$ is a smoothing operator, which convolutes with $f(x, y)$, thus, the noise can be reduced. In fact, we often have some prior information to identify the difference between the singularities of signal and noise. The technique of reducing noise based on the detection of the wavelet transform maxima has been developed [Canny, 1986; Mallat, 1998; Mallat and Hwang, 1992; Mallat and Zhong, 1992; Tang *et al.*, 2000; You and Tang, 2007]. In our practice, threshold processing is applied to reduce the noise and retain the edge information. In this process, after calculating the modulus of the wavelet transform, a threshold is used to decide which point is a noise one to be eliminated. If the modulus of the wavelet transform is less than the threshold, then their modulus will be reset as 0. Even though the threshold cannot be automatically computed, it can be selected by the experiments. In our practice, we have $T = C_T \times M$, where $C_T$ is a constant to be adjusted manually according to the amount of noise or the distracting background (in our experiments, $C_T \approx 0.35$) and $M$ is the maximum modulus of the wavelet transform. Based on the threshold processing, the prospective edge or contour of the shape can be detected effectively and exactly.

### 9.3.3  *Modification of Primary Wavelet Skeleton*

Obviously, a few points on the central line produced by the above technique may be lost. Hence, the primary skeleton obtained from the above approach does not resemble somewhat human perceptions. The lost points may be resumed through a slight modification. Namely, for each lost point, we examine the neighbor points along the normal of the gradient; if they have the same GCWT, then the lost points will be considered to be the skeleton points. Meantime, the typical junction points and intersections of the shape, such as T-pattern, K-pattern, and cross-pattern intersections as shown in Fig. 9.3, are applied to the modification of the primary wavelet skeleton.

Let us look at an example, as shown in Fig. 9.5(c), by using the proposed foregoing central line searching technique to the letter "H", some skeleton lines in the junctions are lost. In fact, this is a key problem, which is needed to be solved in any skeletonization method. Fortunately, the GCWT plays an extremely important role in our new technique. Six typical junctions are listed in Fig. 9.3, which may cause the pixel-losing in the primary skeleton. Some modifying techniques are proposed in this chapter to reduce the pixel-losing. It will be noted that they still depend completely on the relevant results of the above wavelet transform.

For the cross-patterns as shown in Fig. 9.3, the points between $A$ and $B$ of the wavelet skeleton are truncated. Fortunately, it is easy to find these points through the above wavelet transform since they have exactly the same gradient direction or the same GCWT. Based on the foregoing definition of the wavelet symmetry, one can easily accept the fact that the distance between points $A$ and $B$ equals (approximate) scale $s$ of the wavelet transform. Consequently, in practice, from a terminal point $A$, we can find another point $B$ along its planar normal direction or the same GCWT. A similar process can be performed between $C$ and $D$. It is obvious that this modification approach is suitable for the shape with X-pattern intersection as well.

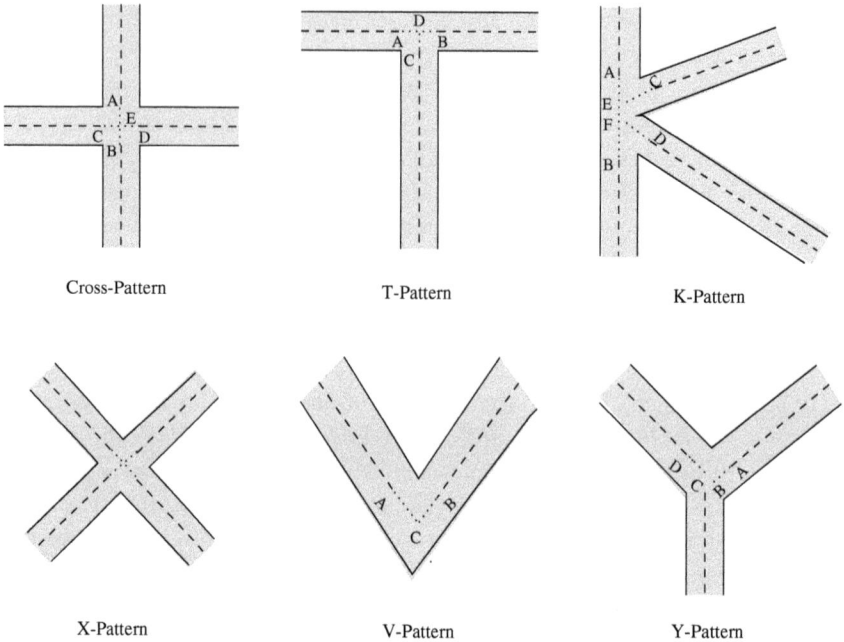

| | | |
|---|---|---|
| Cross-Pattern | T-Pattern | K-Pattern |
| X-Pattern | V-Pattern | Y-Pattern |

Fig. 9.3    All sorts of junction points.

The primary wavelet skeleton of the shape with a T-pattern junction is also shown in Fig. 9.3. Here, the points between $A$ and $B$ as well as $C$ and $D$ are lost. Likewise, the lost points between $A$ and $B$ in the primary skeleton possess the same gradient direction as that of points $A$ and $B$. Moreover, along either its gradient direction or opposite direction, for every such point, one and only one corresponding point has the maximum moduli, and the distance between them is half of scale $s$. Consequently, we only need to search for the lost points, which satisfy the above two conditions, and then retrieve these points from the primary skeleton line to find the modifying skeleton locus. It is noted that if the distance from some lost point to the location of the maximum moduli along the gradient direction or normal direction at the lost point is less than half of the scale $s$ of the wavelet transform, then the retrieve process needs to be stopped. For example, since the distance from the next neighbor of point $D$ along its gradient direction is less than half of scale $s$ for the T-pattern of Fig. 9.3, the retrieve process from point $C$ to $D$ has to end at point $D$.

Virtually, the analogical process can be done for K-pattern, V-pattern, and Y-pattern, which are also shown in Fig. 9.3. Apparently, the modified rules depend strongly on the gradient direction and location of the local maximum modulus points. In fact, all retrieve and extend processes are strictly limited within the area enclosed by the local maximum modulus pixels. Therefore, it does not result in the incorrect connection of unrelated contours or create a new artificial branch in even closely neighboring shapes.

Generally, if the terminal or end point lying on the locus of the primary skeleton satisfies one of the following conditions, it will be called unstable terminal point:

- **Condition 1:** For the terminal or end point lying on the locus of the primary skeleton, along the vertical direction of its gradient, there exists a neighbor point with the same GCWT such that the distance from itself to the nearest location of the maximum moduli is more than scale $s$.
- **Condition 2:** Along the vertical direction of the gradient, there exists another terminal or end point with the same GCWT lying on another segment of the locus of the primary skeleton, every point between these two end points lies completely inside the shape and the distance between the two terminal points equals scale $s$ of the wavelet transform.
- **Condition 3:** For the next neighbor point of a terminal point along the vertical direction of its gradient, there exists only one location of maximum moduli point along its gradient direction or opposite direction such that the distance from this point to the location of maximum moduli is half of scale $s$; further, along the planar normal of the gradient direction at the point, no maximum moduli point exists or the distance from this point to the location of maximum moduli is more than half of scale $s$ of the wavelet transform.

As a result, the above modifying process for the primary wavelet skeleton can be rewritten as the following algorithm:

---

**Input:** An image contains the primary wavelet skeleton with the corresponding maximum moduli resource.
**Output:** An image contains the modified wavelet skeleton.


**REPEAT**
**FOR** *Every unstable terminal point in the image*
**IF** *It satisfies Condition 1*

---

**THEN** *Retrieve its next neighbor point located at the vertical direction of its gradient as a modified skeleton;*
**ELSE IF** *It satisfies Condition 2*
**THEN** *Retrieve s points between two terminal points as a part of the modified skeleton;*
**ELSE IF** *It satisfies Condition 3*
**THEN** *Retrieve its next neighbor point located at the vertical direction of its gradient as one modified skeleton.*
**END IF;**
**END FOR;**
**UNTIL** *No unstable terminal point is detected.*
**END.**

In addition, it has been noted that a primary wavelet skeleton representation can also be used to recover the contours of the underlying shape. This is because the contour information of the shape is recorded in its primary representation. In other words, the wavelet skeleton preserves its width and direction information of the shape. This property is often desirable since a shape is completely specified by its contours and further by its wavelet skeletons.

## 9.4 Algorithm and Experiment

To implement the proposed method to extract the wavelet skeletons of the Ribbon-like shapes, in this section, an algorithm will be provided, followed by several experiments.

### 9.4.1 *Algorithm*

The algorithm based on the maximum moduli analysis of the wavelet transform (MMAWT) method to extract the wavelet skeleton of the Ribbon-like shape in an image is designed as follows:

**Algorithm 9.1.** Let $f(x, y)$ be an image containing Ribbon-like shapes and scale $s > 0$,

**Step 1** Select the suitable scale of the wavelet transform according to the width of the Ribbon-like shape.
**Step 2** Calculate wavelet transforms $\{W_s^1 f(x, y), W_s^2 f(x, y)\}$ using the wavelets defined by Eq. (9.2).

---

**Step 3** Calculate modulus $|\nabla W_s f(x,y)|$ of the wavelet transform and the gradient direction $f_{\text{gradient}}$.

**Step 4** Take threshold $T$ according to the amount of noise and background in the original image and proceed with threshold on the modulus image (if necessary).

**Step 5** For each point $(x,y)$, the modulus of the wavelet transform is compared with one of its neighboring points along its gradient direction; if its modulus arrives at the maximum, it will be recorded as the local modulus maximum $f_{\text{locmax}}$.

**Step 6** For each point $(x,y)$ with local maximum, search the point whose distance to $(x,y)$ along the gradient direction is $s$. If it is a point with the local maximum, the central point is detected.

**Step 7** The primary skeleton is formed by all the points detected in Step 4.

**Step 8** Modify the above primary wavelet skeleton according to the foregoing modification program to obtain the final wavelet skeleton.

---

Obviously, the implementation of the above algorithm is easy, simple, and fast due to the following reasons: It is well known that performing wavelet transform for an image is easy as long as the discrete transform formula is given. Secondly, only the points of the maximum moduli of the wavelet transform, i.e., contour pixels are considered in Steps 3–5. Hence, it is expected to be fast. Finally, the skeletons produced from the proposed algorithm are not sensitive to noise and shape variations such as rotation and scaling.

However, the selection of the proper scale according to the width of the shape is tough and depends on the experience in practice. It is emphasized that the properties of the modulus maximum of the wavelet transform still hold as long as the scale $s \geq d$. Therefore, in practice, the scale can be selected to be bigger than the true value, which is accepted within a certain degree.

## 9.4.2 *Experiments*

In this section, we focus on the verification of the effectiveness of the skeletonization based on the MMAWT method by experiments. Three types of experiments have been carried out: (1) production of maximum moduli image of a shape, (2) extraction of primary wavelet skeleton of the Ribbon-like shape, and (3) modification of the primary wavelet skeleton. The images

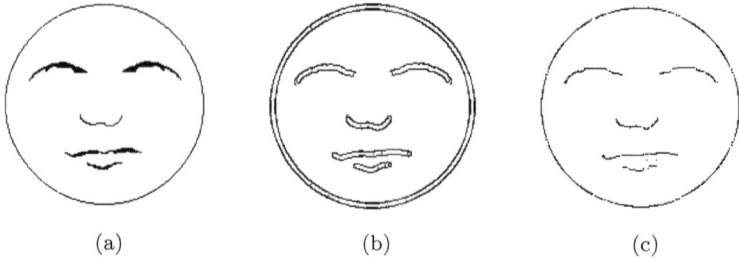

Fig. 9.4   (a) The original image; (b) the location of maximum moduli of the wavelet transform corresponding to $s = 6$; (c) the skeletons extracted by the proposed algorithm.

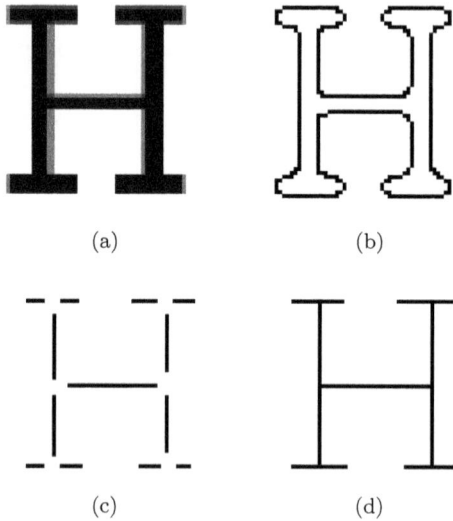

Fig. 9.5   (a) The original image; (b) the location of maximum moduli of the wavelet transform corresponding to $s = 4$; (c) the primary skeletons extracted by the proposed algorithm; (d) the final skeletons obtained by applying the modification algorithm.

used in the experiments vary in type, noise, gray levels, etc. Some examples will be presented in the following.

An interesting example is shown in Fig. 9.4. The original image consists of a face drawing with various widths. By carrying out the algorithm of this chapter, the skeleton is extracted, which is shown graphically in Fig. 9.4(c). The second example is illustrated in Fig. 9.5. The original image, the English letter "H" with varied gray levels, is given in Fig. 9.5(a). We take scale $s = 4$, the maximum moduli image is presented in Fig. 9.5(b).

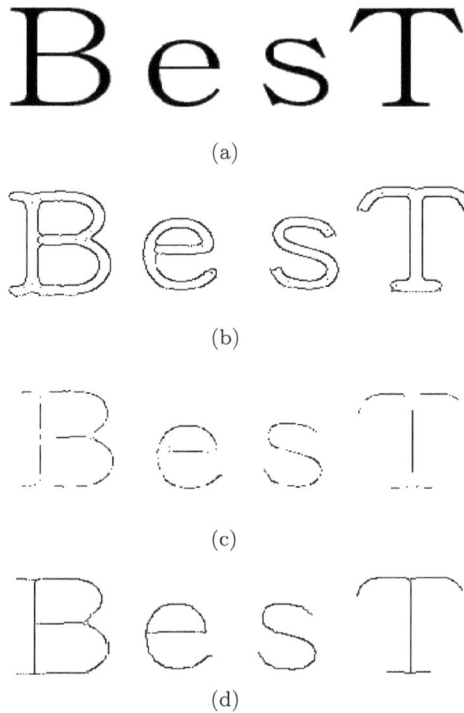

Fig. 9.6   (a) The original image of English letters; (b) the location of maximum moduli of the wavelet transform corresponding to $s = 8$; (c) the primary skeletons extracted by the proposed algorithm; (d) the final skeletons modified by applying the modification program.

The primary wavelet skeleton is obtained from MMAWT and shown in Fig. 9.5(c). Obviously, the skeleton loci which are located in some intersections of the shape are lost. The modified skeleton of the English letter "H" is shown in Fig. 9.5(d), where the disappeared skeleton loci which are located in these intersections of the shape are retrieved successfully.

Some other examples are shown in Figs. 9.6–9.8, where English printed characters "BesT", and handwritings "$C$" and "$D$" are presented respectively. We take the scales of the wavelet transform, $s = 8$ and $s = 4$, respectively. The original images are shown in Figs. 9.6(a), 9.7(a), and 9.8(a), where each letter contains some strokes with a variety of widths. Fortunately, their locations of the maximum moduli have exactly the same width as shown in Figs. 9.6(b), 9.7(b), and 9.8(b). Some truncations in the primary skeleton loci are apparent in the intersections of the shapes; in

Fig. 9.7 (a) The original image of English handwriting; (b) the location of maximum moduli of the wavelet transform corresponding to $s = 4$; (c) the primary skeletons extracted by the proposed algorithm; (d) the final skeletons modified by applying the modification program.

Fig. 9.8 (a) The original image; (b) the location of maximum moduli of the wavelet transform corresponding to $s = 4$; (c) the primary skeletons extracted by the proposed algorithm; (d) the final skeletons modified by applying the modification program.

addition, some truncations are also discovered in the non-intersection parts, which can be found in Figs. 9.6(c), 9.7(c), and 9.8(c). After applying the modification processes to these images, those truncated points are retrieved successfully in final wavelet skeletons as illustrated in Figs. 9.6(d), 9.7(d), and 9.8(d), which closely resemble human perceptions of the underlying shapes.

Chinese printed characters "Tian'an gate" and "material", "glory", and "silk" are displayed, respectively, in Figs. 9.9 and 9.10. Obviously, each Chinese character contains some strokes with a variety of widths as shown in Figs. 9.9(a) and 9.10(a), respectively. Virtually, the width in the same stroke maybe different. The locations of the maximum moduli of the wavelet transform with respect to these Chinese characters can be computed, and the results are given in Figs. 9.9(b) and 9.10(b). Likewise, their locations of the maximum moduli of the wavelet transform have the same width. The primary skeletons of these Chinese characters are extracted utilizing Steps 3 and 4 of the above algorithm and presented in Figs. 9.9(c) and 9.10(c), respectively. Final skeletons are obtained by MMAWT and shown in Figs. 9.9(d) and 9.10(d). Apparently, all final skeletons closely resemble human perceptions of the underlying shapes as well as preserve their original topological and geometric properties.

Fig. 9.9  (a) The original image of Chinese characters; (b) the location of maximum moduli of the wavelet transform corresponding to $s = 4$; (c) the primary skeletons extracted by the proposed algorithm; (d) the final skeletons modified by applying the modification program.

Fig. 9.10  (a) The original image of Chinese characters; (b) the location of maximum moduli of the wavelet transform corresponding to $s = 4$; (c) the primary skeletons extracted by the proposed algorithm; (d) the final skeletons modified by applying the modification program.

The image of a raw chest X-ray with varying gray level distribution and noise is illustrated in Fig. 9.11(a). The raw modulus image after performing the wavelet transform is presented in Fig. 9.11(b). The corresponding output of edge detection obtained from non-threshold and threshold processing are shown, respectively, in Figs. 9.11(c) and 9.11(d). Applying MMAWT to Figs. 9.11(c) and 9.11(d) produces the primary skeleton as shown in Figs. 9.11(e) and 9.11(f). By comparing the result of the non-threshold

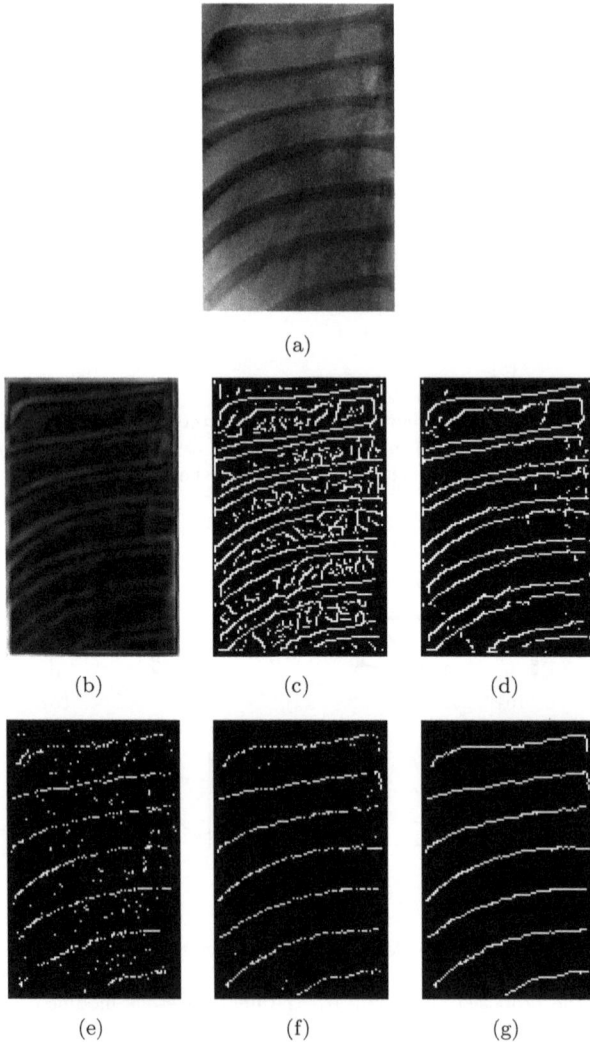

Fig. 9.11     (a) The chest X-ray image; (b) raw of modulus of wavelet transform; (c) non-threshold processing on modulus image, raw output of edge obtained by MMAWT; (d) the output of edge obtained after threshold processing on modulus image; (e) raw skeleton obtained from (b); (f) the primary skeleton obtained from (c); (g) the final modified skeleton from (e).

processing shown in Fig. 9.11(c) with that of the threshold one shown in Fig. 9.11(d), we can conclude that the threshold processing of the wavelet transform modulus is robust against the noise. Finally, it is clear that the

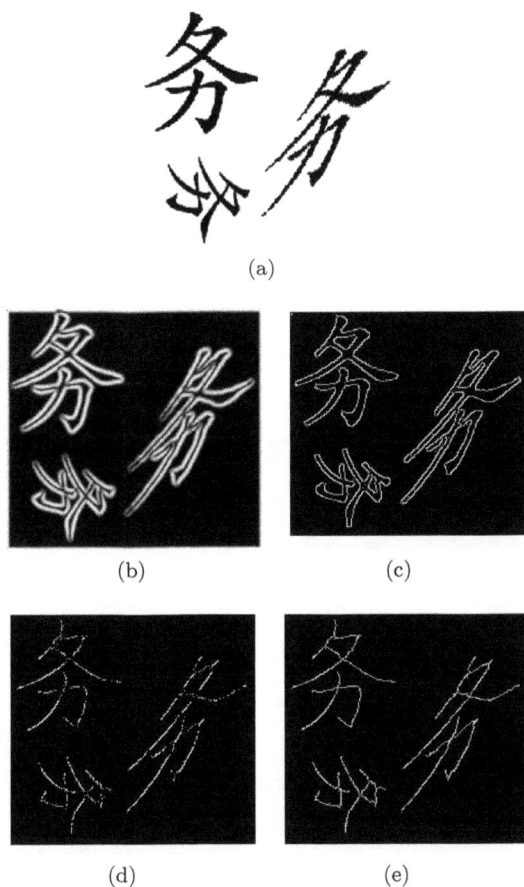

Fig. 9.12 (a) The original image of Chinese characters with affine transforms; (b) raw of modulus of wavelet transform; (c) raw output of edge detection through the modulus maximum of wavelet transform; (d) the primary skeleton extracted by the proposed algorithm; (e) the final modified skeletons.

modified result shown in Fig. 9.11(g) is much better than the primary skeletons shown in Fig. 9.11(f), which can verify the desired effect.

The experiment results in Figs. 9.12 and 9.13 show that the proposed approach is robust against both the noise and affine transformation. The image in Fig. 9.12(a) contains three affine transformed patterns of the same Chinese character. In Fig. 9.13(a), the image is harmed by both the affine transformation and white "salt and pepper" noise. The wavelet transform with scale $s = 4$ is applied to the images in Figs. 9.12(a) and

(a)

(b)

(c)

(d)

(e)

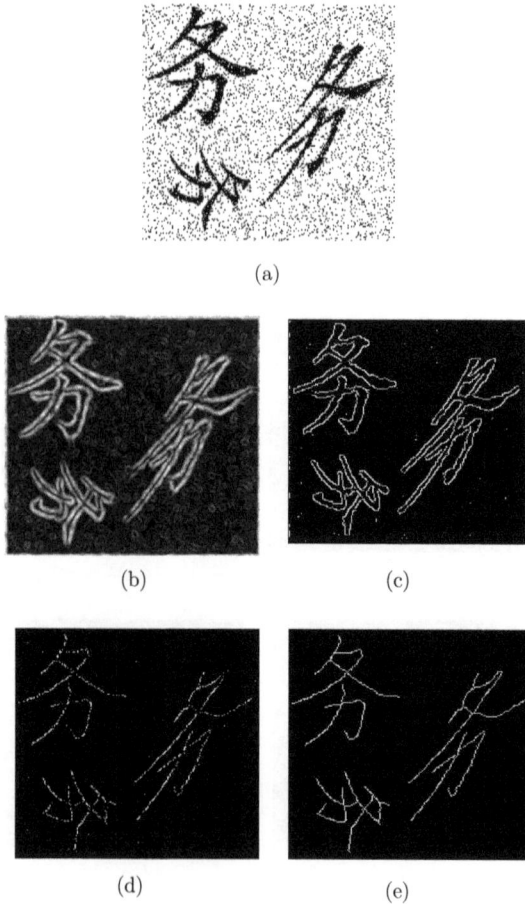

Fig. 9.13 (a) The original image which is harmed by both the affine transform and noise; (b) raw of modulus of wavelet transform; (c) raw output of edge detection obtained by the modulus maximum of wavelet transform; (d) the output of primary skeleton extracted by the proposed algorithm; (e) the modified skeletons.

9.13(a) and their corresponding raw outputs of modulus images are placed in Figs. 9.12(b) and 9.13(b), respectively. To reduce the noise affection to the edge detection, we set threshold $T = 0.39 \times M$ for eliminating the points, which have relatively weak modulus values and were caused by the noise or the distracting background, namely, for each point, if the modulus value of the wavelet transform is less than 0.39 times the maxima modulus, its modulus value will be reset to 0. After the threshold processing, the raw edge for further skeleton extraction can be detected and shown

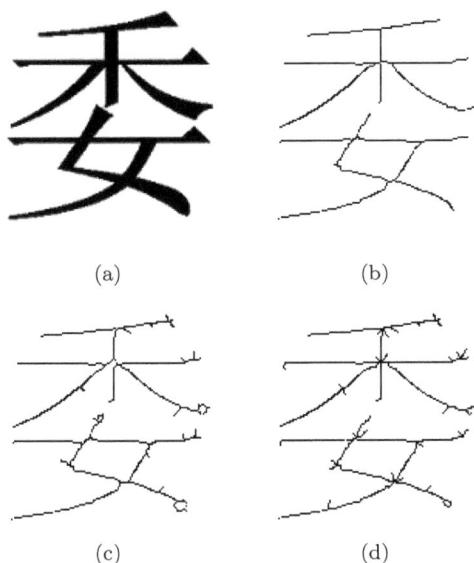

Fig. 9.14   (a) The original image of a Chinese character; (b) the skeleton extracted by the proposed algorithm; (c) the skeleton obtained from ZSM; (d) the skeleton obtained from CYM.

in Figs. 9.12(c) and 9.13(c). The primary and modified skeletons of three Chinese characters are extracted and shown, respectively, in Figs. 9.12(d) and (e) and 9.13(d) and (e).

Finally, to evaluate the performance of our method, we compare it with two typical skeletonization methods: the ZSM method [Lam *et al.*, 1992] and the CYM method [Chang and Yan, 1999]. Our method is suitable for removing both periphery and intersection artifacts, whereas the ZSM and CYM methods can remove only one type of artifact. The results of the comparison are illustrated in Fig. 9.14. The original image, which is a Chinese character, is shown in Fig. 9.14(a). The skeletons obtained from the proposed method, ZSM and CYM are shown in Figs. 9.14(b)–(d), respectively. Apparently, the result obtained from the proposed method is much better than those from ZSM and CYM, and it resembles closer human perception than those obtained from ZSM and CYM.

# Chapter 10

# Feature Extraction by Wavelet Sub-Patterns and Divider Dimensions

Feature extraction is the heart of a pattern recognition system. In pattern recognition, features are utilized to identify one class of pattern from another. The pattern space is usually of high dimensionality. The objective of the feature extraction is to characterize the object and, further, to reduce the dimensionality of the measurement space to a space suitable for the application of pattern classification techniques. In the feature extraction phase, only the salient features necessary for the recognition process are retained such that the classification can be implemented on a vastly reduced feature space.

Feature extraction can be viewed as a mapping, which maps a pattern space into a feature space. Pattern space $\mathbf{X}$ may be described by a vector of $m$ pattern vectors such that

$$\mathbf{X} = \begin{vmatrix} X_1^T \\ X_2^T \\ \vdots \\ X_m^T \end{vmatrix} = \begin{vmatrix} x_{11} & x_{12} & \cdots & x_{1n} \\ x_{21} & x_{22} & \cdots & x_{2n} \\ \vdots & & & \vdots \\ x_{m1} & x_{m2} & \cdots & x_{mn} \end{vmatrix}, \tag{10.1}$$

where the superscript $T$ for each vector stands for its transpose, the $X_i^T = (x_{i1}, x_{i2}, \ldots, x_{in})$, $i = 1, \ldots, m$, represent pattern vectors.

The objective of the feature extraction functions is dimensionality reduction. It maps the pattern space (i.e., original data) into the feature space (i.e., feature vectors). The dimensionality of the feature space has to be smaller than that of the pattern space. Obviously, the feature vectors can be represented by

$$X_i^T = (x_{i1}, x_{i2}, \ldots, x_{ir}), \quad i = 1, \ldots, m, r < n. \tag{10.2}$$

Note that since the feature space is in a smaller dimension, the maximum value of $r$ has to be smaller than that of $n$, i.e., $r < n$.

There are many methods to feature selection, however, they can be categorized into the following [Tou and Gonzalez, 1974]:

- **Entropy minimization:** Entropy is a statistical measure of uncertainty. It can be used as a suitable criterion in the design of optimum feature selection. The entropy minimization approach is based on the assumption that the pattern classes under consideration are normally distribution.
- **Orthogonal expansion:** When the assumption of the normal distribution is not valid, the method of orthogonal expansion offers an alternative approach to feature extraction. The principal advantage of this method is that it does not require knowledge of the various probability densities.
- **Functional approximation:** If the features of a class of objects can be characterized by a function that is determined on the basis of observed data, the feature extraction can be considered to be a problem of functional approximation.
- **Divergence:** It can be used to determine feature ranking and to evaluate the effectiveness of class discrimination. The divergence can also be used as a criterion function for generating an optimum set of features.

In this chapter, we present a novel approach to extract features in pattern recognition that utilizes ring-projection-wavelet-fractal signatures (RPWFSs). This approach can be categorized into the second method listed in the preceding paragraph. In particular, this approach reduces the dimensionality of a 2D pattern by way of a ring-projection method and, thereafter, performs Daubechies' wavelet transform on the derived 1D pattern to generate a set of wavelet transformed sub-patterns, namely, curves that are non-self-intersecting. Further from the resulting non-self-intersecting curves, the divider dimensions are readily computed. These divider dimensions constitute a new feature vector for the original 2D pattern, defined over the curves' fractal dimensions.

Wavelet analysis and its applications have become the fastest growing research areas in recent years. Advanced research and development in wavelet analysis have found numerous applications in such areas as signal processing, image processing, and pattern recognition with many encouraging results [Daubechies, 1990; IEEE, 1992, 1993; Mallat and Hwang, 1992; Tang *et al.*, 1997a, 1998a]. During this fast growth in theories and applications, the theoretical development of high-dimensional wavelet analysis is somewhat lagging behind compared to that of the 1D wavelet. As has been shown in several real-life applications, the 2D wavelet analysis has not been as effectively applied as the 1D analysis.

The goal of this chapter is to, through mathematically sound derivations, reduce the problem of 2D patterns into that of 1D ones and, thereafter, utilize the well-established 1D wavelet transform coupled with fractal theory to extract the 1D patterns' feature vectors for the purpose of pattern recognition.

It is a well-known fact that in many real-life pattern recognition situations, such as optical character recognition (OCR), patterns are often found to be rotated due to experimentation constraints or errors. This implies that a new pattern recognition method must be developed that is invariant to rotations. In 1991, Tang *et al.* [1991] first proposed a method of transforming 2D patterns into 1D patterns through so-called *ring projections*. As the projections are done in the form of rings, the 1D pattern obtained from ring projection is invariant to rotations.

Drawing on the aforementioned work by Tang *et al.* [1991], this work further explores the use of ring projection in reducing the dimensionality of 2D patterns such as alpha-numeric symbols into 1D ones. Consequently, we can perform Daubechies' wavelet transform on the derived 1D patterns to generate a set of *wavelet-transformed sub-patterns* which are curves with non-self-intersecting. We then compute the *divider dimensions* for the individual curves. This allows us to map the set of non-self-intersecting curves into a feature vector defined over the *curves' fractal dimensions*, which is also the feature vector corresponding to the original 2D patterns. An overall description of this approach can be illustrated by a diagram shown in Fig. 10.1.

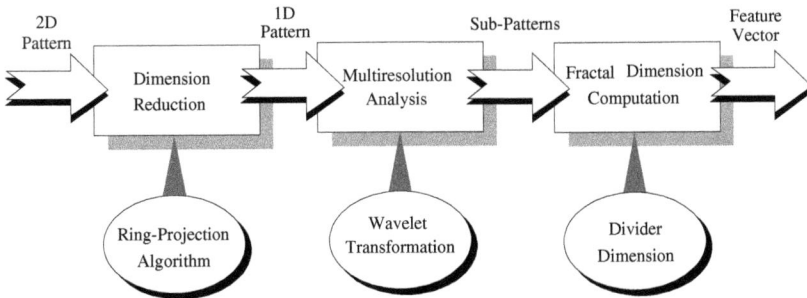

Fig. 10.1   Diagram of ring-projection-wavelet-fractal method.

Following the descriptions of the theoretical constructs of the RPWFSs method, we shall also present the results of experiments in which 2D patterns were tested. The experiment data consist of a subset of Chinese characters and a set of printed alphanumeric symbols, including 26 alphabets in both upper and lower cases, 10 numeric digits, and an additional 10 ASCII symbols. Our method yielded a 100% correct classification rate, even when the orientations or fonts of these characters were altered.

The overall sequence of presentation of this chapter is illustrated as follows:

- dimensionality reduction of 2D patterns with a ring-projection,
- wavelet orthonormal decomposition to produce sub-patterns,
- wavelet-fractal scheme:

  — basic concepts of fractal dimension,
  — the divider dimension of one-dimensional patterns.

- experiments:

  — experimental procedure,
  — experimental results.

## 10.1 Dimensionality Reduction of 2D Patterns with Ring-Projection

This section provides an overview of the ring-projection method for reducing the dimensionality of 2D patterns. First, suppose that a 2D pattern such as an alphanumeric symbol has been represented into a binary image. Taking the letter $A$ as an example, its gray scale image, $p(x, y)$, can be discretized into binary values as follows:

$$p(x, y) = \begin{cases} 1 & \text{if } (x, y) \in \mathcal{D} \\ 0 & \text{otherwise,} \end{cases} \tag{10.3}$$

where domain $\mathcal{D}$ corresponds to the white region of letter $A$, as shown in Fig. 10.2. The above multivariate function $p(x, y)$ can also be viewed as a 2D density function of mass distribution on the plane. From Eq. (10.3), it is readily noted that the corresponding density function is a uniform distribution. That is also to say, the mass is homogeneously distributed over the region $\mathcal{D}$. From this uniform mass distribution, we can derive the centroid of the mass for the region $\mathcal{D}$, as denoted by $m(x_0, y_0)$ and,

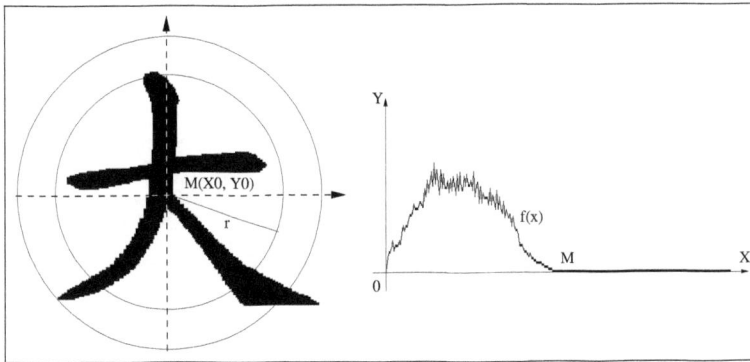

Fig. 10.2 A binary image of letter "A" whose centroid is used to place the origin of a new reference frame.

subsequently, translate the origin of our reference frame to this centroid, as has been illustrated in Fig. 10.2.

Next, we let

$$M = \max_{N \in \mathcal{D}} \mid N(x, y) - m(x_0, y_0) \mid,$$

where $\mid N(x, y) - m(x_0, y_0) \mid$ represents the Euclidean distance between two points, $N$ and $m$, on the plane. Further, we transform the original reference Cartesian frame into a polar frame based on the following relations:

$$\begin{cases} x = \gamma \cos \theta \\ y = \gamma \sin \theta. \end{cases} \tag{10.4}$$

Hence,

$$p(x, y) = p(\gamma \cos \theta, \gamma \sin \theta).$$

where $\gamma \in [0, \infty)$, $\theta \in (0, 2\pi]$.

For any fixed $\gamma \in [0, M]$, we then compute the following integral:

$$f(\gamma) = \int_0^{2\pi} p(\gamma \cos \theta, \gamma \sin \theta) d\theta. \tag{10.5}$$

The resulting $f(\gamma)$ is in fact equal to the total mass as distributed along circular rings, as shown in Fig. 10.3. Hence, the derivation of $f(\gamma)$ is also termed as a ring projection of the planar mass distribution. The single-variate function $f(\gamma)$, $\gamma \in [0, M]$, sometimes also denoted as $f(x)$, $x \in [0, M]$, can be viewed as a 1D pattern that is directly transformed from the original 2D pattern through a ring projection. Owing to the facts

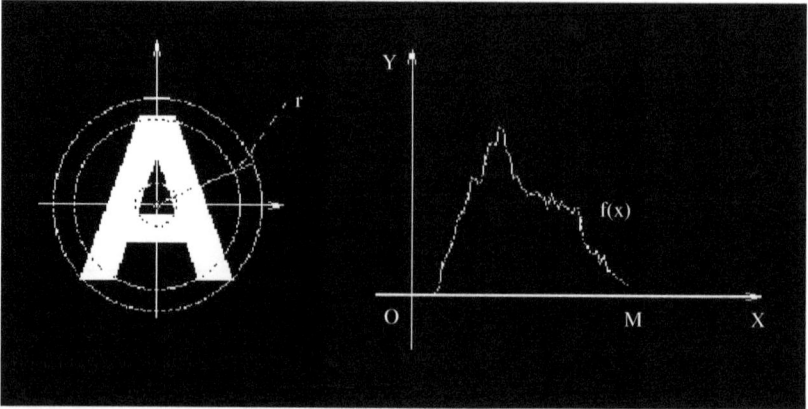

Fig. 10.3 An illustration of the ring projection for letter *A*.

that the centroid of the mass distribution is invariant to rotation and that the projection is done along circular rings, the derived 1D pattern will be invariant to the rotations of its original 2D pattern. In other words, the ring projection is rotation-invariant.

From a practical point of view, the images to be analyzed by a recognition system are most often stored in discrete formats. Catering to such discretized 2D patterns, we shall modify Eq. (10.5) into the following expression (see Fig. 10.4):

$$f(\gamma) = \sum_{k=0}^{M} p(\gamma \cos \theta_k, \gamma \sin \theta_k). \tag{10.6}$$

Two examples of dimension reduction for 2D patterns can be found in Figs. 10.5 and 10.6. In the first example, the image of the Chinese character "big" is a 2D pattern as shown on the left side of Fig. 10.5. After applying the operation of the ring projection, a 1D signal is obtained and illustrated on the right side of Fig. 10.5. In the second example, the original pattern is a 2D object, an aircraft, which is displayed on the left side of Fig. 10.6. As the ring-projection operation is applied to it, a 1D signal is obtained and presented on the right side of Fig. 10.6.

Interested readers are referred to the work of Tang *et al.* [1991] for a thorough discussion on ring projection. In the succeeding sections, we are concerned mainly with how to extract as much information as possible from the obtained ring projection, i.e., a 1D pattern, by way of the

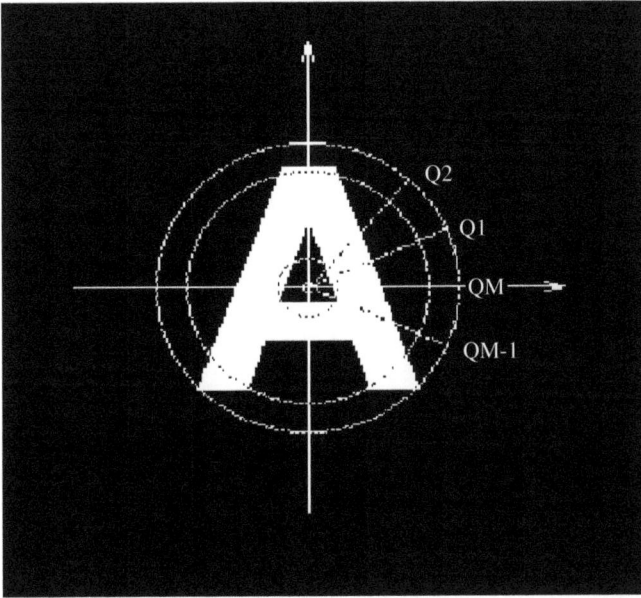

Fig. 10.4   Projection of a 2D pattern along rings in a discrete manner.

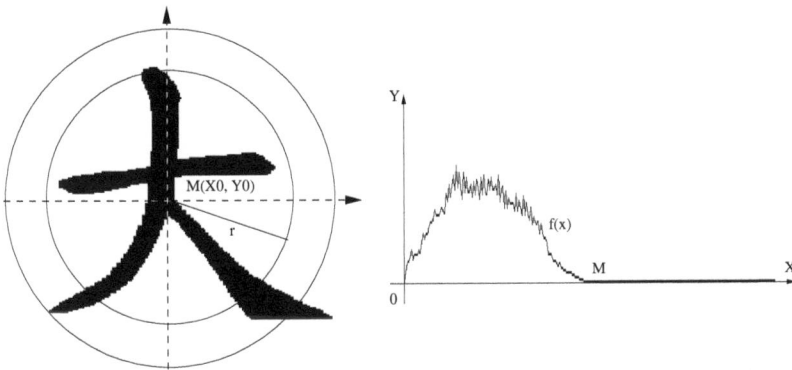

Fig. 10.5   An example of dimension reduction for Chinese character "big".

wavelet transform. This, as is described later, enables us to obtain a set of wavelet-transformed sub-patterns — curves that are non-self-intersecting, from which feature vectors defined over the curves' fractal dimensions can easily be computed.

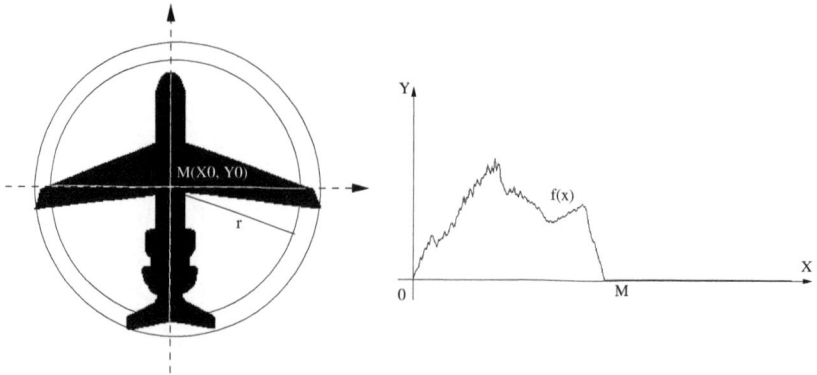

Fig. 10.6   An example of dimension reduction for a 2D pattern.

## 10.2  Wavelet Orthonormal Decomposition to Produce Sub-Patterns

Let $\{V_j\}$ be an orthonormal MRA, then $V_{j-1}$ can be decomposed orthogonally as follows:

$$V_{j-1} = W_j \oplus V_j. \tag{10.7}$$

Note that the order of $V_j$ and $V_{j-1}$ here, in this chapter, differs from that in the previous chapters:

- we use the subspaces in $\{V_j\}$ by the order of

$$\cdots \subset V_{j+1} \subset V_j \subset V_{j-1}$$

  in the present chapter,
- we use the order of

$$\cdots \subset V_{j-1} \subset V_j \subset V_{j+1}$$

  in the previous chapters.

Actually, these two representations are no different.

As can be realized, any real-world instruments for measuring physical data are capable of acquiring information with limited precision. In other words, the signals obtained using such instruments will have limited resolution. The same is also true for the 1D pattern that we have derived in the preceding sections, $f(x)$; it must belong to a closed space $V_j$. Mathematically, we can express this as follows:

$$P_j : L^2(\mathbb{R}) \Longrightarrow V_j \text{ projective operator from } L^2(\mathbb{R}) \text{ to } V_j,$$

$$Q_j : L^2(\mathbb{R}) \Longrightarrow W_j \text{ projective operator from } L^2(\mathbb{R}) \text{ to } W_j.$$

Since $f(x) \in V_j \subset L^2(\mathbb{R})$, we arrive at

$$f(x) = P_j f(x) = \sum_{k \in \mathbb{Z}} c_{j,k} \varphi_{j,k}(x)$$

$$= P_{j+1} f(x) + Q_{j+1} f(x)$$

$$= \sum_{m \in \mathbb{Z}} c_{j+1,m} \varphi_{j+1,m}(x) + \sum_{m \in \mathbb{Z}} d_{j+1,m} \psi_{j+1,m}(x), \qquad (10.8)$$

where

$$c_{j+1,m} = \langle P_{j+1} f, \varphi_{j+1,m} \rangle = \sum_{k \in \mathbb{Z}} c_{j,k} \hat{h}_{k-2m}, \qquad (10.9)$$

$$d_{j+1,m} = \langle Q_{j+1} f, \psi_{j+1,m} \rangle = \sum_{k \in \mathbb{Z}} c_{j,k} \hat{g}_{k-2m}, \qquad (10.10)$$

and $\hat{h}_{k-2m}$ and $\hat{g}_{k-2m}$ denote the conjugate vectors of $h_{k-2m}$ and $g_{k-2m}$, respectively. When both $h_{k-2m}$ and $g_{k-2m}$ are real, we have $\hat{h}_{k-2m} = h_{k-2m}$ and $\hat{g}_{k-2m} = g_{k-2m}$.

It should be mentioned that both Eqs. (10.9) and (10.10) compute the sum of an infinite number of terms. However, from a computational point of view, we wish to see only a finite number of non-zero terms, $h_k$, hence reducing the problem of infinite summation to that of finite summation. In doing so, we carry out the following procedure, in order to find the expression

$$m_0(\omega) = \sum_{k \in \mathbb{Z}} \frac{1}{\sqrt{2}} h_k e^{-ik\omega},$$

such that the number of non-zero $h_k$ is finite:

First, let us consider

$$1 = \left( \cos^2 \frac{\omega}{2} + \sin^2 \frac{\omega}{2} \right)^3$$

$$= \cos^6 \frac{\omega}{2} + 3 \cos^4 \frac{\omega}{2} \sin^2 \frac{\omega}{2} + 3 \cos^2 \frac{\omega}{2} \sin^4 \frac{\omega}{2} + \sin^6 \frac{\omega}{2}. \qquad (10.11)$$

Suppose that

$$| m_0(\omega) |^2 = \cos^6 \frac{\omega}{2} + 3 \cos^4 \frac{\omega}{2} \sin^2 \frac{\omega}{2}, \qquad (10.12)$$

hence,

$$| m_0(\omega + \pi) |^2 = \sin^6 \frac{\omega}{2} + 3 \sin^4 \frac{\omega}{2} \cos^2 \frac{\omega}{2}. \qquad (10.13)$$

Since

$$\frac{1 + e^{-i\omega/2}}{2} = \frac{e^{-i\omega/2}(e^{i\omega/2} + e^{-i\omega/2})}{2} = e^{-i\omega/2} \cos \frac{\omega}{2}, \qquad (10.14)$$

$$\frac{1 - e^{-i\omega/2}}{2} = \frac{e^{-i\omega/2}(e^{i\omega/2} - e^{-i\omega/2})}{2} = ie^{-i\omega/2} \sin \frac{\omega}{2}, \qquad (10.15)$$

we have

$$
\begin{aligned}
m_0(\omega) &= (e^{-i\omega/2} \cos \frac{\omega}{2})^2 e^{-i\omega/2} \left( \cos \frac{\omega}{2} + i\sqrt{3} \sin \frac{\omega}{2} \right) \\
&= \left( \frac{1 + e^{-i\omega/2}}{2} \right)^2 \left( \frac{1 + e^{-i\omega/2}}{2} + \sqrt{3} \frac{1 - e^{-i\omega/2}}{2} \right) \\
&= \frac{1}{8} \left[ (1 + \sqrt{3}) + (3 + \sqrt{3})e^{-i\omega} \right] \\
&\quad + \frac{1}{8} \left[ (3 - \sqrt{3})e^{-i\omega 2} + (1 - \sqrt{3})e^{-i\omega 3} \right],
\end{aligned}
\qquad (10.16)
$$

at the same time, we know

$$m_0(\omega) = \sum_{k=0}^{3} \frac{1}{\sqrt{2}} h_k e^{-ik\omega}. \qquad (10.17)$$

By comparing Eqs. (10.16) and (10.17), we can immediately obtain the following:

$$h_0 = \frac{1 + \sqrt{3}}{4\sqrt{2}} = 0.4829629131445341,$$

$$h_1 = \frac{3 + \sqrt{3}}{4\sqrt{2}} = 0.8365163037378077,$$

$$h_2 = \frac{3 - \sqrt{3}}{4\sqrt{2}} = 0.2241438680420134,$$

$$h_3 = \frac{1 - \sqrt{3}}{4\sqrt{2}} = -0.1294095225512603.$$

In other words, if the above terms are chosen as $h_k$, $k = 0, 1, 2, 3$,

$$m_0(\omega) = \sum_{k=0}^{3} \frac{1}{\sqrt{2}} h_k e^{-ik\omega}$$

will satisfy the following conditions:

$$|m_0(\omega)|^2 + |m_0(\omega + \pi)|^2 = 1,$$

$$m_0(0) = 1.$$

Now, if we define that

$$\hat{\varphi}(\omega) = \prod_{j=1}^{\infty} m_0(\omega/2^j),$$

then the resulting scaling function, $\varphi(x)$, can be said to have a compact support [Daubechies, 1988]. Thus, we can simplify the wavelet transform expressions of Eqs. (10.9) and (10.10) into the following:

$$c_{j+1,m} = \sum_{k=0}^{3} h_k c_{j,k+2m}, \tag{10.18}$$

$$d_{j+1,m} = \sum_{k=0}^{3} g_k c_{j,k+2m}. \tag{10.19}$$

Next, $m_0(\omega)$ is also determined following the above-mentioned steps, except that we examine the expression of

$$1 = \left( \cos^2 \frac{\omega}{2} + \sin^2 \frac{\omega}{2} \right)^5$$

to obtain

$$m_0(\omega) = \sum_{k=0}^{5} \frac{1}{\sqrt{2}} h_k e^{-ik\omega}.$$

In this case, we can simplify the wavelet transform expressions of Eqs. (10.9) and (10.10) into the following:

$$c_{j+1,m} = \sum_{k=0}^{5} h_k c_{j,k+2m}, \tag{10.20}$$

$$d_{j+1,m} = \sum_{k=0}^{5} g_k c_{j,k+2m}. \tag{10.21}$$

Daubechies [1992] used the following notations:

$$nM_0(\xi) = \frac{1}{\sqrt{2}} \sum_{n=0}^{2N-1} Nh_n e^{-in\xi}, \tag{10.22}$$

where $nM_0(\xi)$ is equivalent to $m_0(\omega)$ in our case and $Nh_n$ is equivalent to our $h_k$. Daubechies also provided the exact values of $h_k$ (or $Nh_n$) for $N = 2$ to 10 in tabular forms. The tables might be of interest to those readers who want to use Eqs. (10.9) and (10.10) but not know how $h_k$s are computed. When $N$ is large enough, the computation of $h_k$ could become problem-dependent. Generally speaking, the larger the $N$ value, the higher the resolution of the wavelet orthonormal decomposition will be, and at the same time, the more costly the computation will become.

According to the wavelet orthonormal decomposition as shown in Eq. (10.7), first $V_j$ is decomposed orthogonally into a high-frequency sub-space $W_{j+1}$ and a low-frequency one $V_{j+1}$ using the wavelet transform Eqs. (10.20) and (10.21). The low-frequency sub-space $V_{j+1}$ is further decomposed into $W_{j+2}$ and $V_{j+2}$. Afterward, $V_{j+2}$ is broken into $W_{j+3}$ and $V_{j+3}$, and these processes can be continued. The above wavelet orthonormal decomposition can be represented by

$$
\begin{aligned}
V_j = \quad & W_{j+1} \quad \oplus \quad V_{j+1} \\
& \qquad\qquad\qquad \Downarrow \\
& \qquad (W_{j+2} \oplus V_{j+2}) \\
& \qquad\qquad\qquad\quad \Downarrow \\
& \qquad\qquad (W_{j+3} \oplus V_{j+3}). \\
& \qquad\qquad\qquad\qquad\quad \Downarrow
\end{aligned}
$$

$$\cdots\cdots\cdots$$

In the view of pattern recognition, $V_j$ can be considered to be an original pattern, while $V_{j+1}, V_{j+2}, \ldots, W_{j+1}, W_{j+2}, \ldots$ can be regarded as the sub-patterns which are referred to as *wavelet transform sub-patterns*.

In Fig. 10.7, we have shown the 1D pattern resulting from the ring projection [Tang *et al.*, 1991] of a 2D pattern, for example, a Chinese character, and its corresponding wavelet transform sub-patterns. In this figure, $V_0$ denotes the resulting 1D pattern from the ring projection, which is a dimension reduction operation. $V_j$ denotes the wavelet transform sub-pattern resulting from the $j$th wavelet transform based on Eq. (10.20). $W_j$ denotes the wavelet transform sub-pattern from the $j$th wavelet transform

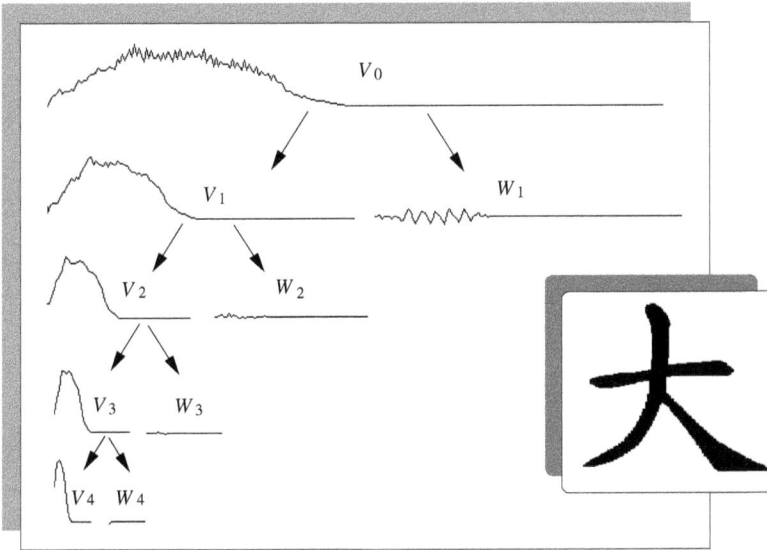

Fig. 10.7 Examples of the wavelet transform sub-patterns from a Chinese character.

based on Eq. (10.21). Since

$$V_j = W_{j+1} \oplus V_{j+1}$$
$$= W_{j+1} \oplus W_{j+2} \oplus V_{j+2}$$
$$= W_{j+1} \oplus W_{j+2} \oplus W_{j+3} \oplus V_{j+3}$$
$$= \ldots \ldots \ldots,$$

it can be seen that in Fig. 10.7,

$$V_0 = W_1 \oplus V_1$$
$$= W_1 \oplus W_2 \oplus V_2$$
$$= W_1 \oplus W_2 \oplus W_3 \oplus V_3$$
$$= W_1 \oplus W_2 \oplus W_3 \oplus W_4 \oplus V_4.$$

Another example of the wavelet transform sub-patterns from a 2D object using the wavelet orthonormal decomposition is illustrated in Fig. 10.8. The original pattern is an aircraft, which is a 2D object. The dimension reduction operation is applied to it, and a 1D pattern is produced which is a curve labeled by $V_0$ in Fig. 10.8. After utilizing the wavelet orthonormal

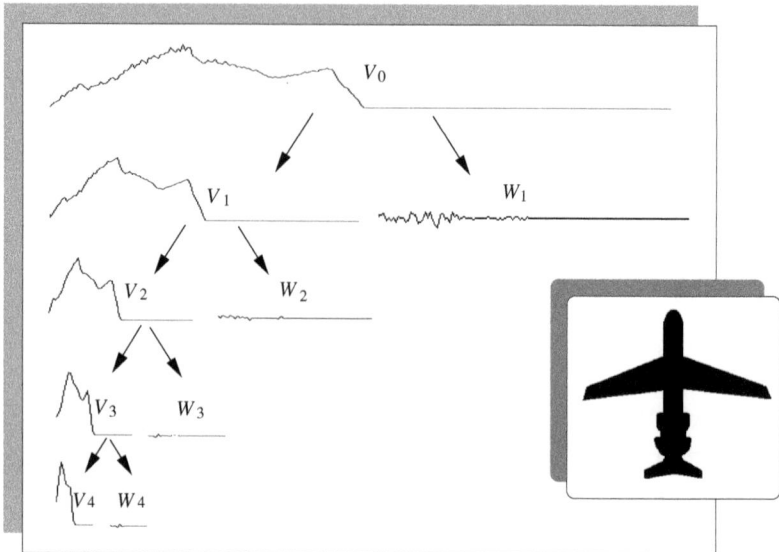

Fig. 10.8   Examples of the wavelet transform sub-patterns from a pattern.

decomposition, eight wavelet transform sub-patterns are obtained, i.e., $V_1$–$V_4$ and $W_1$–$W_4$ in Fig. 10.8.

In Fig. 10.9, we have shown the 1D pattern resulting from the ring projection of an alphabet, $A$, and its corresponding wavelet transformation sub-patterns. In the figure, $S_{20}A$ denotes the resulting 1D pattern. $S_{2j}A$ denotes the wavelet sub-pattern resulting from the $j$th wavelet transformation of $S_{20}A$ based on Eq. (10.20). $W_{2j}A$ denotes the wavelet sub-pattern resulted from the $j$th wavelet transformation of $S_{20}A$ based on Eq. (10.21).

The above wavelet orthonormal decomposition can be represented by

$$S_{20}A = W_{21}A \oplus S_{21}A$$
$$\Downarrow$$
$$(W_{22}A \oplus V_{22}A)$$
$$\Downarrow$$
$$(W_{23}A \oplus V_{23}A)$$
$$\Downarrow$$
$$(W_{24}A \oplus V_{24}A).$$

In the view of pattern recognition, $S_{20}A$ can be considered to be an original pattern, while $V_{2j}A$ and $W_{2j}A$ can be regarded as the sub-patterns.

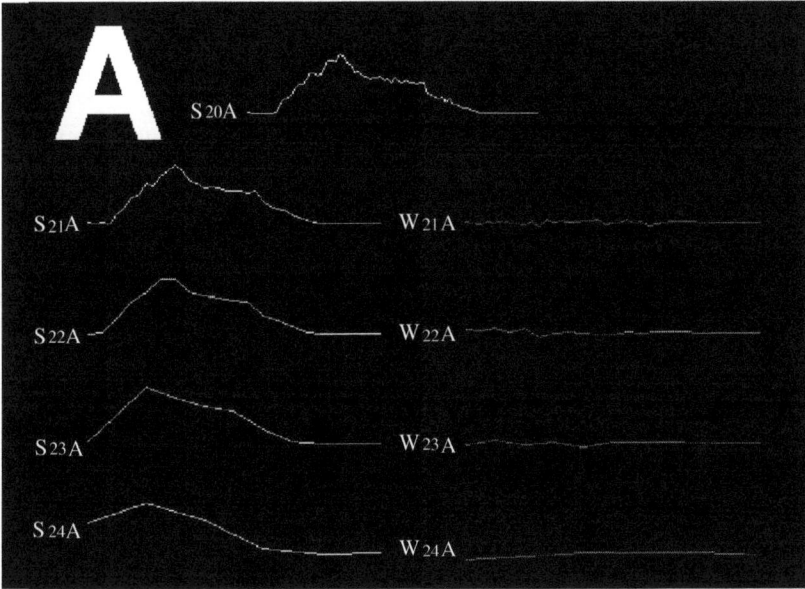

Fig. 10.9   The 1D pattern obtained after ring projection and its corresponding wavelet transformation sub-patterns.

## 10.3   Wavelet-Fractal Scheme

In 1975, B. Mandelbrot first developed the notion of *fractal* [Mandelbrot, 1982]. For the past decades, the theories of fractal geometry have been widely applied in a variety of domains, such as computer science, physics, chemistry, biology, material science, geography, geology, and even in social science and humanities. Of equal significance is its applications in pattern recognition. Among others, in 1994, Tang *et al.* examined the essence of Minkowski fractional dimension and applied the Minkowski blanket technique in deriving fractal signatures for document processing, as part of a new method for page layout analysis [Ma *et al.*, 1995; Tang *et al.*, 1995b, 1997b].

### 10.3.1   *Basic Concepts of Fractal Dimension*

In this section, the basic concepts of *fractals* are introduced.

**Definition 10.1.** A collection $\Re$ of subsets of the space $X$ is said to have order $m + 1$ if some point of $X$ lies in $m + 1$ elements of $\Re$ and no point of

$X$ lies in more than $m + 1$ elements of $\Re$. Given a collection $\Re$ of subsets of $X$, a collection $\Im$ is said to *refine* $\Re$, or to be a *refinement* of $\Re$, if each element $B$ of $\Im$ is contained in at least one element of $\Re$.

Now we define what we mean by the *topological dimension* of a space $X$.

**Definition 10.2.** A space $X$ is said to be *finite-dimensional* if there is some integer $m$ such that for every open covering $\Re$ of $X$, there is an open covering $\Im$ of $X$ that refines $\Re$ and has order at most $m + 1$. The *topological dimension* of $X$ is defined to be the smallest value of $m$ for which this statement holds. We use the notation $dim_T X$ to represent the topological dimension of $X$.

The topological dimension is a complicated and advanced mathematical topic; more details about it can be found in the work of Munkres [1975].

What are fractals? There are many definitions, because it is very difficult to define fractal strictly. B. Mandelbrot gave two definitions in 1982 and 1986, respectively:

(1) The first definition from his original essay (1982) says the following:

**Definition 10.3.** A set $F$ is called a fractal set if its *Hausdorff dimension* $(\dim_H F)$ is greater than the *Topological dimension* $(\dim_T F)$:

$$\dim_H F > \dim_T F.$$

(2) In 1986, B. Mandelbrot defined the fractal as follows:

**Definition 10.4.** Fractal is a compound object, which contains several sub-objects. The global characteristic of this object is similar to the local characteristics of each sub-object.

(3) A more precise definition of the fractal set $F$ can be provided in the following:

**Definition 10.5.** A set $F$ is called a fractal set if the following conditions are satisfied:

- The global characteristic of the set $F$ is self-similar to the local characteristics of each sub set:

$$\Im(F) \sim \Im(f_i), \quad f_i \supset F,$$

where $\Im(.)$ stands for the characteristic of $(.)$.

- The set $F$ is infinitely separable, i.e.,

$$F = \{f_1^1, f_2^1, \ldots, f_i^1, \ldots, f_n^1\},$$
$$f_i^1 = \{f_1^2, f_2^2, \ldots, f_k^2, \ldots, f_n^2\},$$

$$\cdots\cdots\cdots$$

$$f_k^m = \{f_1^{m+1}, f_2^{m+1}, \ldots, f_k^{m+1}, \ldots, f_n^{m+1}\}, \quad m+1 \to \infty.$$

- Usually, the fractal dimension of the set $F$ is a fraction and greater than the *Topological dimension* $\dim_T F$:

$$\dim_H F > \dim_T F.$$

- In many cases, the definition of $F$ is recursive.

Let $U$ be a non-empty subset of $n$-dimensional Euclid space $\mathbb{R}^n$, and the diameter of $U$ is defined as

$$|U| = \sup\{|x - y| : x, y \in U\},$$

where $sup\{.\}$ stands for the supremum of $\{.\}$. Thus, the diameter of $U$ is the greatest distance apart of any pair of points in $U$. If $\{U_i\}$ is a countable collection of sets of diameter at most $\delta$ that cover $F$, such that

$$\Re(\delta) = \{U_i\} = \{U_i : i = 1, 2, \ldots\}$$

and

$$F \subset \bigcup_{i=1}^{\infty} U_i, \quad 0 < |U_i| \leq \delta,$$

we say that $\{U_i\}$ is a $\delta$-cover of $F$.

Suppose that $F \subset \mathbb{R}^n$ and $s$ is a real number and $s \geq 0$. For any $\delta > 0$, we define

$$H_\delta^s(F) = \inf_{\Re(\delta)} \left\{ \sum_{i=1}^{\infty} |U_i|^s : \{U_i\} \text{ is a } \delta\text{-cover of } F \right\}, \tag{10.23}$$

where the symbol $\inf\{.\}$ indicates the infimum of $\{.\}$. As $\delta$ decreases, the class of permissible covers of $F$ in Eq. (10.23) is reduced. Consequently, the infimum $H_\delta^s(F)$ increases and so approaches a limit as $\delta \to 0$. We have the following definition:

**Definition 10.6.** When $\delta \to 0$, the limit of $H_\delta^s(F)$ exists for any subset $F$ of $I\!R^n$, and the limiting value can be (and usually) 0 or $\infty$. The *s-dimensional Hausdorff measure* of $F$ can be defined by

$$H^s(F) = \lim_{\delta \to 0} H_\delta^s(F)$$

$$= \lim_{\delta \to 0} \left[ \inf_{\Re(\delta)} \left\{ \sum_{i=1}^{\infty} |U_i|^s : \{U_i\} \text{ is a } \delta\text{-cover of } F \right\} \right]. \quad (10.24)$$

We can clearly prove that $H^s$ is a measure. Specifically, $H^s(\phi) = 0$, and if $E \subset F$, then $H^s(E) \leq H^s(F)$. If $\{F_i\}$ is any countable collection of *Borel set*, such that

$$\bigcap_{i=1}^{\infty} F_i = \phi,$$

Then we have

$$H^s\left(\bigcup_{i=1}^{\infty} F_i\right) = \sum_{i=1}^{\infty} H^s(F_i).$$

Furthermore, if $F$ is a Borel subset of $I\!R^n$, then the $n$-dimensional *Hausdorff measure* of $F$ can be deduced as

$$H^n(F) = \frac{\pi^{\frac{n}{2}}}{2^n \Gamma(\frac{n+2}{2})} vol^n(F),$$

where

- $H^n(F)$ stands for the $n$-dimensional *Hausdorff measure* of $F$,
- $vol^n(F)$ represents the $n$-dimensional volume of $F$ which is called the *Lebesgue measure* of $F$:
  - $vol^1$ is length,
  - $vol^2$ is area,
  - $vol^3$ is the usual 3D volume.

Consequently, Hausdorff's measures generalize the familiar ideas of length, area, and volume.

Let us review Eq. (10.23). For any set $F$ and $\delta < 1$, $H_\delta^s(F)$ is a non-increasing function of $s$. According to Eq. (10.24), it can be shown that $H^s(F)$ is also a non-increasing function of $s$. In fact, the stronger conclusion is that if $t > 0$ and $\{U_i\}$ is a $\delta$-cover of $F$, we have

$$H^t_\delta(F) \leq \sum_i |U_i|^t \leq \delta^{t-s} \sum_i |U_i|^s. \tag{10.25}$$

We take the infimum, that is,

$$H^t_\delta(F) \leq \delta^{t-s} H^s_\delta(F).$$

**Definition 10.7.** Let $\delta \to 0$; if $H^s(F) < \infty$, then $H^t(F) = 0$ for $s < t$. Therefore, there exists a critical value of $s$, such that $H^s(F)$ jumps from $\infty$ to 0 at this point. This critical value is called the *Hausdorff Dimension* of $F$, and it is symbolized by $\dim_H F$.

Formally, we have

$$\dim_H F = \inf \{s : H^s(F) = 0\} = \sup \{s : H^s(F) = \infty\} \tag{10.26}$$

and

$$H^s(F) = \begin{cases} \infty & \text{if } s < \dim_H F \\ 0 & \text{if } s > \dim_H F. \end{cases}$$

If $s = \dim_H F$, probably $H^s(F)$ is 0 or $\infty$, or may satisfy

$$0 < H^s(F) < \infty.$$

A Borel set is called an *s-set* if the latter condition as shown in the above is satisfied. More details about the Hausdorff dimension can be found in the works of Falconer [1990, 1985]. Hausdorff dimension is the oldest and probably the most important one of the fractal dimensions. It has the following advantages: (1) Hausdorff dimension can be defined for any set. (2) It is mathematically convenient. (3) It is based on measures, which are relatively easy to manipulate. However, the major disadvantage of the Hausdorff dimension is that it is difficult to compute or estimate in many cases. In practice, *box-computing dimension* is convenient to apply. Therefore, our study focuses on the box-computing dimension.

Fundamental to most definitions of dimension is the idea of measurement at scale $\delta$. For each $\delta$, a set can be measured in a way that ignores irregularities of size less than $\delta$, and we see how these measurements behave as $\delta \to 0$ [Falconer, 1990].

Suppose $F$ is a plane curve, the measurement $M_\delta(F)$ denotes the number of sets (with length $\delta$) which divide the set $F$. A dimension of $F$ is determined by the power law obeyed by $M_\delta(F)$ as $\delta \to 0$. If

$$M_\delta(F) \sim \mathcal{K}\delta^{-s}, \tag{10.27}$$

where $\mathcal{K}$ and $s$ are constants, we might say that $F$ has dimension $s$, and $\mathcal{K}$ can be considered as the "$s$-dimensional length" of $F$.

Taking the logarithm of both sides in Eq. (10.27) yields the following formula:

$$\log_2 M_\delta(F) \simeq \log_2 \mathcal{K} - s \log_2 \delta,$$

in the sense that the difference between the two sides tends to be 0 with $\delta$, we have

$$s = \lim_{\delta \to 0} \frac{\log_2 M_\delta(F)}{-\log_2 \delta}. \tag{10.28}$$

From the above equation, $s$ can be regarded as a slope on a log–log scale [Falconer, 1990].

Box-computing dimension or box dimension is one of the most widely used fractal dimensions. The popularity of the box-computing dimension is largely due to its relative ease of mathematical calculation and empirical estimation.

Let $F$ be a non-empty and bounded subset of $\mathbb{R}^n$ and $\xi = \{\omega_i : i = 1, 2, 3, \ldots\}$ be covers of the set $F$, as shown in Fig. 10.10. $N_\delta(F)$ denotes the number of covers, such that

$$N_\delta(F) = |\xi : d_i \le \delta|,$$

where $d_i$ stands for the diameter of the $i$th cover. This equation means that $N_\delta(F)$ is the smallest number of subsets which cover the set $F$, and their diameters $d_i$s are not greater than $\delta$ (Fig. 10.10).

The upper and lower bounds of the box-computing dimension of $F$ can be defined by the following formulas:

$$\underline{\dim}_B F = \liminf_{\delta \to 0} \frac{\log_2 N_\delta(F)}{-\log_2 \delta}, \tag{10.29}$$

$$\overline{\dim}_B F = \overline{\lim_{\delta \to 0}} \frac{\log_2 N_\delta(F)}{-\log_2 \delta}, \tag{10.30}$$

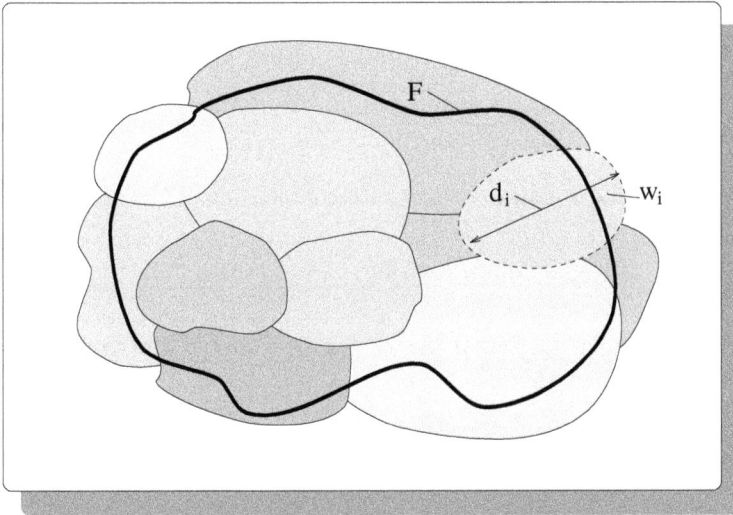

Fig. 10.10    Opening covers with diameters $d_i$s covering $F$.

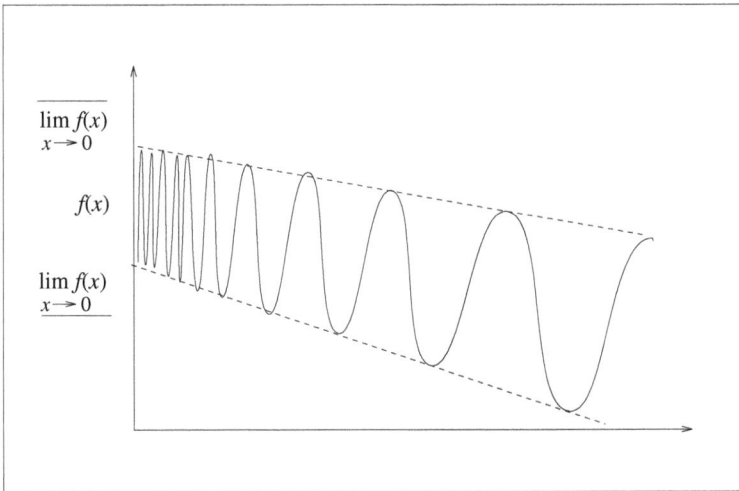

Fig. 10.11    Upper and lower bounds of a function.

where the over line of $\dim_B F$ stands for the upper bound of dimension while the under line of $\dim_B F$ is for the lower bound. An example of the upper bound and the lower bound is shown in Fig. 10.11

**Definition 10.8.** If both the upper bound $\overline{\dim_B}F$ and the lower bound $\underline{\dim_B}F$ are equal, i.e.,

$$\liminf_{\delta \to 0} \frac{\log_2 N_\delta(F)}{-\log_2 \delta} = \overline{\lim_{\delta \to 0}} \frac{\log_2 N_\delta(F)}{-\log_2 \delta},$$

the common value is called *box-computing dimension* or *box dimension* of $F$:

$$\dim_B F = \lim_{\delta \to 0} \frac{\log_2 N_\delta(F)}{-\log_2 \delta}. \tag{10.31}$$

Further discussions on fractal theory can be found in the works of Edgar [1990] and Falconer [1990].

## 10.3.2 *The Divider Dimension of 1D Patterns*

In the preceding sections, we have shown how to carry out ring projections to reduce an original 2D pattern, such as alphanumeric symbols, into a 1D pattern and furthermore described how to apply wavelet transform to the resulting 1D patterns in order to obtain a set of wavelet transformed sub-patterns. Such sub-patterns are in fact non-self-intersecting curves. In this section, we address the problem of computing the divider dimension of those curves and, thereafter, use the computed divider dimension to construct a feature vector for the original 2D pattern in question for pattern recognition. In what follows, we shall first formally define the notion of the divider dimension of a non-self-intersecting curve.

**Definition 10.9.** Suppose that $C$ is a non-self-intersecting curve, and $\delta > 0$. Let $M_\delta(C)$ be the maximum number of ordered sequence of points $x_0, x_1, \ldots, x_M$ on curve $C$, such that $\mid x_k - x_{k-1} \mid = \delta$ for $k = 1, 2, \ldots, M$. The divider dimension, $\dim_D C$, of curve $C$ is defined as follows:

$$\dim_D C = \lim_{\delta \to 0} \frac{\log M_\delta(C)}{-\log \delta}, \tag{10.32}$$

where $\mid x_k - x_{k-1} \mid$ represents the magnitude of the difference between two vectors $x_k$ and $x_{k-1}$, as illustrated in Fig. 10.12.

It should also be mentioned that $x_M$ is not necessarily the end point of curve $C$, $x_T$, but $\mid x_T - x_M \mid < \delta$. Furthermore, $(M_\delta(C) - 1)\delta$ may be viewed as the "length" of curve $C$ as measured using a pair of dividers that are set $\delta$ distance apart.

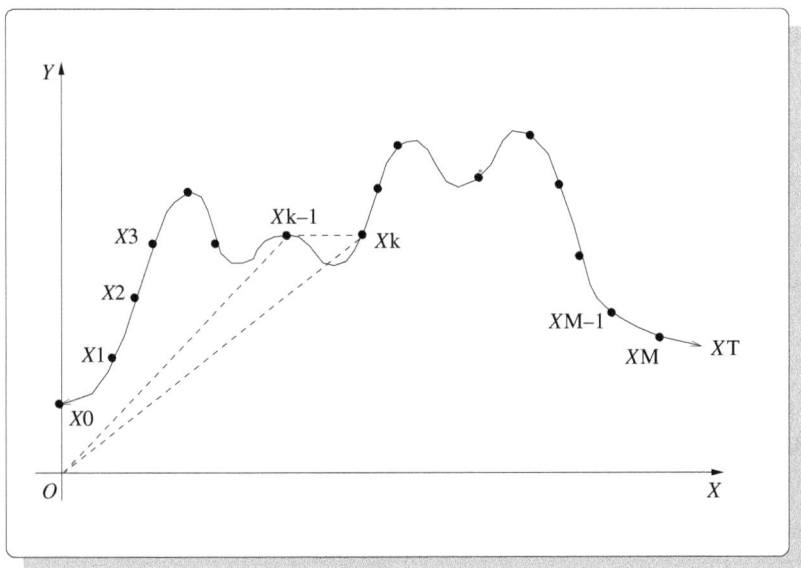

Fig. 10.12 The difference between two vectors on curve $C$.

If the readers are interested in the details of the divider dimension, they can refer to the work of Falconer [1990].

## 10.4 Experiments

This section presents the procedure as well as the results of our experiments that aim at recognizing a set of 2D patterns.

### 10.4.1 *Experimental Procedure*

The key steps of the experimental procedure consist of the following:

**Step 1. Ring projection of 2D patterns:**
We denote each of the 2D patterns in question by $p(x, y)$. Thus, the ring projection of $p(x, y)$ can be expressed as follows:

$$f(x_k) = \sum_{i=0}^{M} p(x_k \cos \theta_i, x_k \sin \theta_i). \tag{10.33}$$

**Step 2. Wavelet transform of the 1D patterns:**

Let $f(x_k) = c_{j,k}$, where $k = 0, 1, \ldots, 2N - 1$, and

$$S_{20}A = c_{j,k} = \{c_{j,0}, c_{j,1}, \ldots, c_{j,2N-1}\}.$$

Thus, the expressions for the wavelet transform of $S_{20}A$ can be written as follows:

$$c_{j+1,m} = \sum_{k=0}^{5} h_k c_{j,k+2m},$$

$$d_{j+1,m} = \sum_{k=0}^{5} g_k c_{j,k+2m}, \qquad (10.34)$$

where $m = 0, 1, \ldots, N - 1$.

Hence, the wavelet transformed sub-patterns of $S_{20}A$ obtained using Eq. (10.34) will become

$$S_{21}A = \{c_{j+1,0}, c_{j+1,1}, \ldots, c_{j+1,N-1}\},$$

$$W_{21}A = \{d_{j+1,0}, d_{j+1,1}, \ldots, d_{j+1,N-1}\}.$$

**Step 3. Computation of divider dimensions for wavelet transformed sub-patterns:**

Since the divider dimensions of non-self-intersecting curves are asymptotic values, we can derive their approximations based on the following expression:

$$\frac{\log M_\delta(C)}{-\log \delta},$$

when $\delta$ is set small enough.

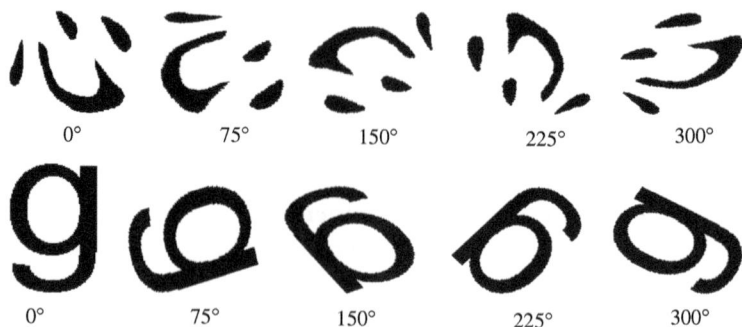

Fig. 10.13 Examples of the Chinese character "Xin" (Heart) and English letter "g" rotated at 0°, 75°, 150°, 225°, and 300°.

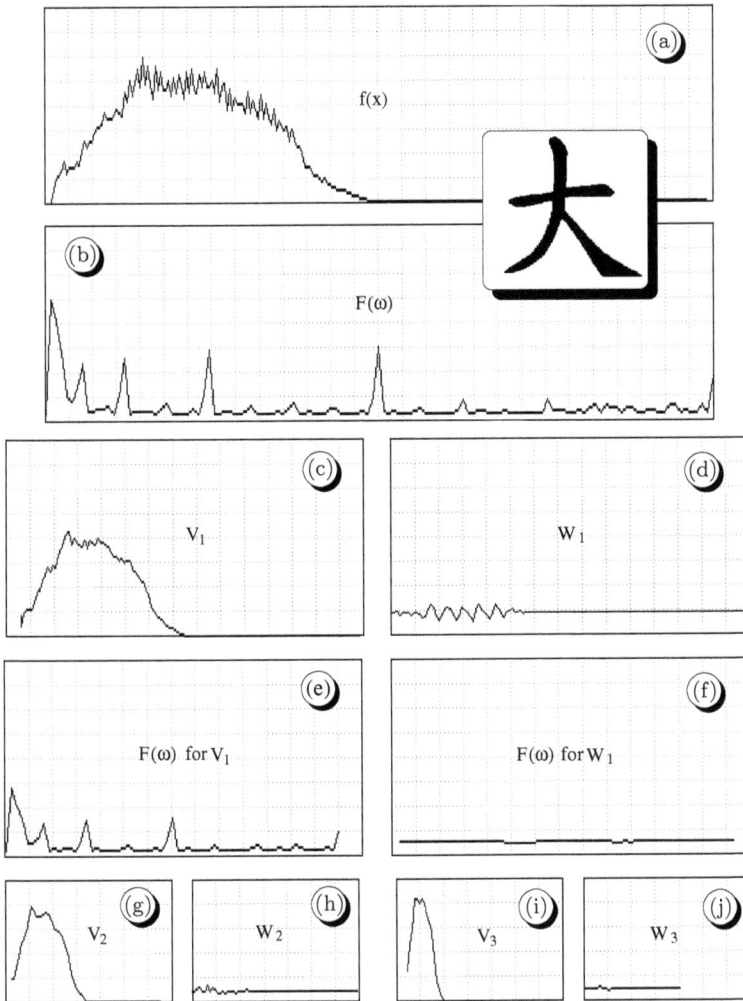

Fig. 10.14 Wavelet transform sub-patterns of Chinese character "Da" (Big) and their frequency analyses.

In our experiments, we performed three consecutive wavelet transformations for each 1D pattern. Hence, the 1D pattern, such as the ring projection of letter $A$, will yield seven non-self-intersecting curves:

$$S_{20}A, \ S_{21}A, \ S_{22}A, \ S_{23}A, \ W_{21}A, \ W_{22}A, \ W_{23}A.$$

For each of the seven curves, we further compute its divider dimension and, therefore, relate each symbol with a feature vector.

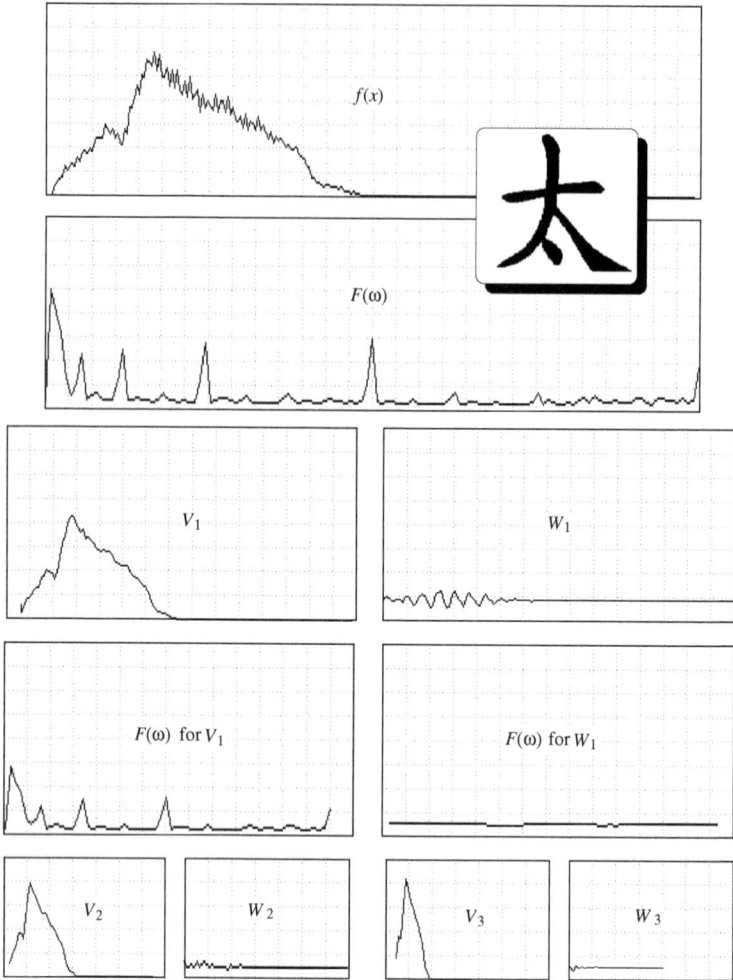

Fig. 10.15　Wavelet transform sub-patterns of Chinese character "Tai" (Too much) and their frequency analyses.

## 10.4.2 *Experimental Results*

This section presents the results of our experiments that aim at recognizing a set of 2D patterns, including a subset of Chinese characters, 52 upper and lower case English letters, and 10 numeric digits plus additional 10 ASCII. All samples were rotated at different angles. Examples of the Chinese character "Xin" (in English it means heart) and English letter "g"

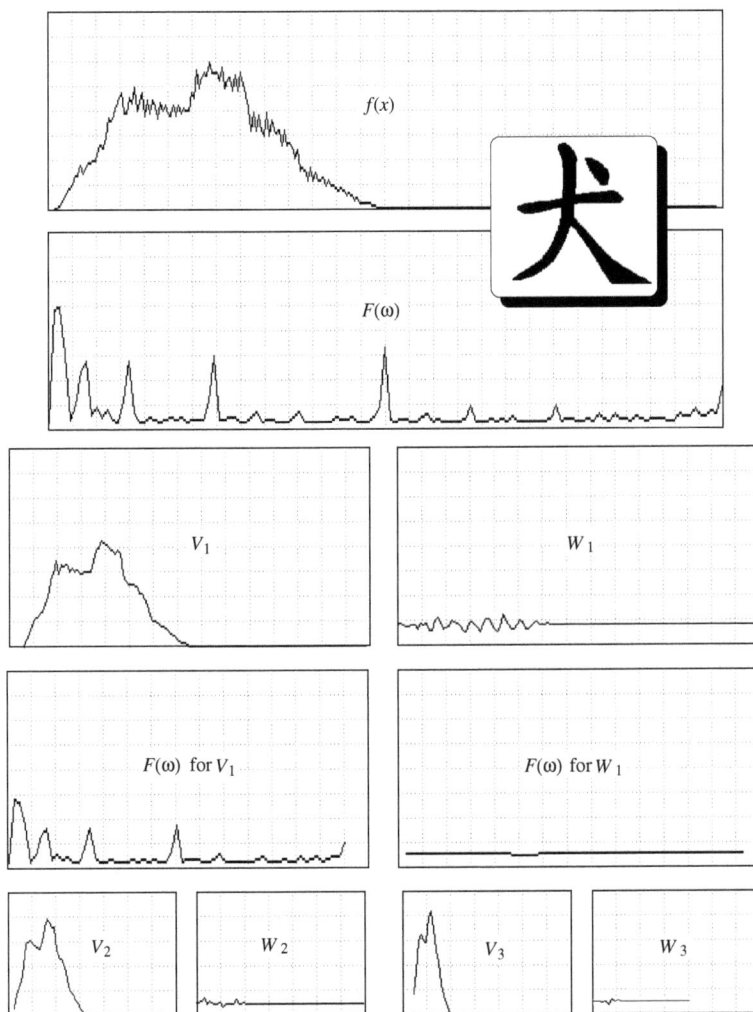

Fig. 10.16    Wavelet transform sub-patterns of the Chinese character "Quan" (Dog) and their frequency analyses.

rotated at 0°, 75°, 150°, 225°, and 300° are illustrated in Fig. 10.13. The 52 upper and lower case alphabets with 5 different fonts were considered in the feature-vector computation.

In our experiments, we performed three consecutive wavelet transforms for each 1D pattern. Hence, the one-dimensional pattern, such as the ring projection of the Chinese character "Da" (in English it means big), will

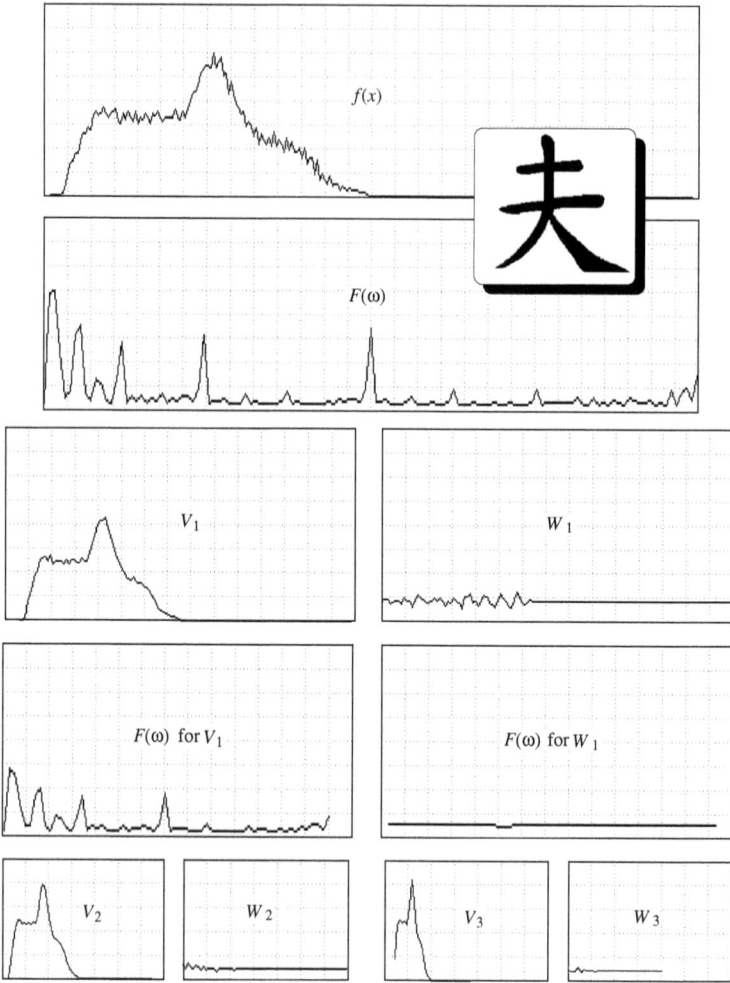

Fig. 10.17   Wavelet transform sub-patterns of the Chinese character "Fu" (Husband) and their frequency analyses.

yield seven non-self-intersecting curves:

$$V_0, \ V_1, \ V_2, \ V_3, \ W_1, \ W_2, \ W_3,$$

since

$$V_0 = W_1 \oplus V_1$$
$$= W_1 \oplus W_2 \oplus V_2$$
$$= W_1 \oplus W_2 \oplus W_3 \oplus V_3.$$

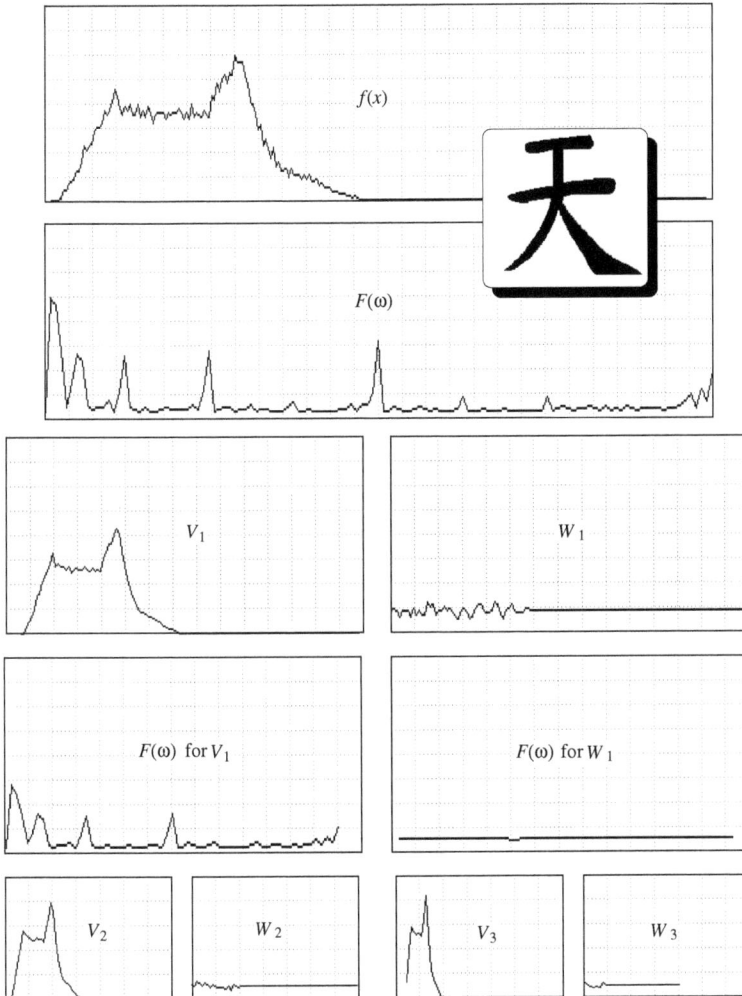

Fig. 10.18 Wavelet transform sub-patterns of the Chinese character "Tian" (Sky) and their frequency analyses.

The graphical illustration is shown in Fig. 10.14, where the curves (a), (c), (g), (i) and (d), (h), (j) represent $V_0$–$V_3$ through $W_1$–$W_3$, respectively. Note that $f(x)$ is the 1D pattern of the Chinese character "Da" (Big), and $V_0$ is $f(x)$ itself in this case. Figure 10.14(b) is the frequency spectrum of $V_0$, while Fig. 10.14(e) is the frequency spectrum of $V_1$. By comparing these two

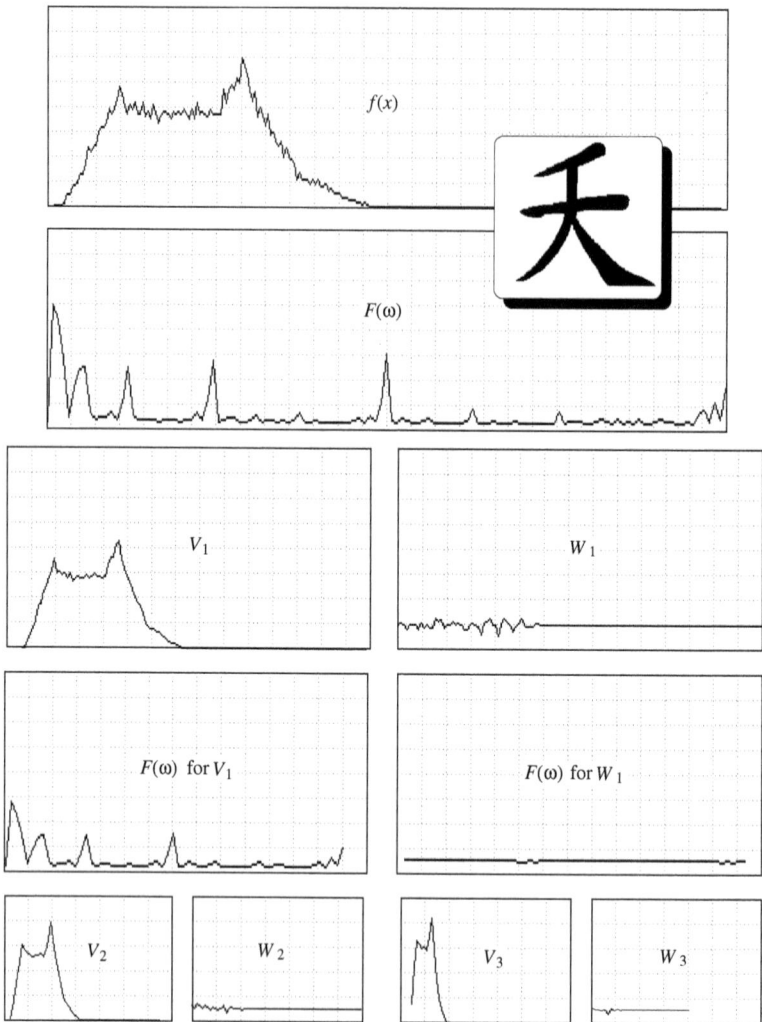

Fig. 10.19    Wavelet transform sub-patterns of the Chinese character "Yao" (Die prematurely) and their frequency analyses.

frequency spectra, we can find that $V_0$ contains all frequency components of $f(x)$, while $V_1$ contains only one-half of the frequency components of $f(x)$. The reason is that $V_0$ is divided into two parts:

$$V_0 = V_1 \oplus W_1.$$

Table 10.1.  Divider dimensions computed for 72 printed symbols.

| Char | $S_{20}(*)$ | $S_{21}(*)$ | $W_{21}(*)$ |
|------|-------------|-------------|-------------|
| A | 1.1139433384 | 1.1047174931 | 1.0648593903 |
| B | 1.1238559484 | 1.1136001348 | 1.0730702877 |
| C | 1.1167527437 | 1.1132092476 | 1.0624803305 |
| D | 1.1197948456 | 1.1136578321 | 1.0700432062 |
| E | 1.1139442921 | 1.1020903587 | 1.0629045963 |
| F | 1.1100301743 | 1.0874330997 | 1.0602608919 |
| G | 1.1257215738 | 1.1170243025 | 1.0674185753 |
| H | 1.1142077446 | 1.1060614586 | 1.0664664507 |
| I | 1.0840704441 | 1.0549546480 | 1.0441081524 |
| J | 1.1017495394 | 1.0808098316 | 1.0476298332 |
| K | 1.1139196157 | 1.0974267721 | 1.0708941221 |
| L | 1.1026504040 | 1.0669544935 | 1.0534558296 |
| M | 1.1212527752 | 1.1126092672 | 1.0775213242 |
| N | 1.1176242828 | 1.1083389521 | 1.0729970932 |
| O | 1.1214458942 | 1.1196649075 | 1.0770056248 |
| P | 1.1121571064 | 1.1039747000 | 1.0661952496 |
| Q | 1.1247720718 | 1.1199330091 | 1.0716108084 |
| R | 1.1206896305 | 1.1102559566 | 1.0806628466 |
| S | 1.1185953617 | 1.1158298254 | 1.0743304491 |
| T | 1.1036618948 | 1.0789839029 | 1.0521930456 |
| U | 1.1125218868 | 1.1064934731 | 1.0647976398 |
| V | 1.1079084873 | 1.0890588760 | 1.0515108109 |
| W | 1.1243741512 | 1.1173660755 | 1.0764572620 |
| X | 1.1142315865 | 1.0871822834 | 1.0648663044 |
| Y | 1.1017326117 | 1.0773953199 | 1.0506635904 |
| Z | 1.1157565117 | 1.1043380499 | 1.0640560389 |
| 0 | 1.1168053150 | 1.1078009605 | 1.0619374514 |
| 1 | 1.1028971672 | 1.0694137812 | 1.0519375801 |
| 2 | 1.1089236736 | 1.0893353224 | 1.0689998865 |
| 3 | 1.1099339724 | 1.0979336500 | 1.0579220057 |
| 4 | 1.1061652899 | 1.0899257660 | 1.0559248924 |
| 5 | 1.1143358946 | 1.0999945402 | 1.0730379820 |
| 6 | 1.1117491722 | 1.0993975401 | 1.0590016842 |
| 7 | 1.1040900946 | 1.0803054571 | 1.0495930910 |
| 8 | 1.1123523712 | 1.1005206108 | 1.0584080219 |
| 9 | 1.1148401499 | 1.1010701656 | 1.0597313643 |

As for the first part, $V_1$, only the low-frequency components of $V_0$ are kept, and the high-frequency components are lost. In addition, only the high-frequency components of $V_0$ are kept, and the low-frequency components are lost in $W_1$. Similarly, $V_1$ is further divided into two parts:

$$V_1 = V_2 \oplus W_2,$$

Table 10.2.  Divider dimensions computed for 72 printed symbols.

| Char | $S_{22}(*)$ | $W_{22}(*)$ | $S_{23}(*)$ | $W_{23}(*)$ |
|------|-------------|-------------|-------------|-------------|
| A | 1.0620796680 | 1.0203253031 | 1.0321298838 | 1.0096461773 |
| B | 1.0757583380 | 1.0170320272 | 1.0424953699 | 1.0116459131 |
| C | 1.0778726339 | 1.0138599873 | 1.0455238819 | 1.0122945309 |
| D | 1.0791871548 | 1.0267117023 | 1.0460337400 | 1.0130724907 |
| E | 1.0550003052 | 1.0216575861 | 1.0250078440 | 1.0069218874 |
| F | 1.0459311008 | 1.0240161419 | 1.0195240974 | 1.0061128139 |
| G | 1.0895853043 | 1.0255262852 | 1.0553485155 | 1.0214713812 |
| H | 1.0645886660 | 1.0131975412 | 1.0322223902 | 1.0091987848 |
| I | 1.0211712122 | 1.0151667595 | 1.0079869032 | 1.0055073500 |
| J | 1.0463556051 | 1.0200892687 | 1.0200742483 | 1.0079109669 |
| K | 1.0503554344 | 1.0194057226 | 1.0224356651 | 1.0070123672 |
| L | 1.0315659046 | 1.0158586502 | 1.0114238262 | 1.0061960220 |
| M | 1.0746159554 | 1.0180855989 | 1.0394244194 | 1.0094848871 |
| N | 1.0676525831 | 1.0181691647 | 1.0326002836 | 1.0105568171 |
| O | 1.0961915255 | 1.0249220133 | 1.0584391356 | 1.0235477686 |
| P | 1.0630793571 | 1.0239624977 | 1.0319234133 | 1.0084927082 |
| Q | 1.0963153839 | 1.0238838196 | 1.0569688082 | 1.0244978666 |
| R | 1.0684889555 | 1.0262739658 | 1.0370421410 | 1.0095887184 |
| S | 1.0754898787 | 1.0259306431 | 1.0420234203 | 1.0142017603 |
| T | 1.0403193235 | 1.0152760744 | 1.0144587755 | 1.0082896948 |
| U | 1.0712003708 | 1.0221848488 | 1.0397174358 | 1.0074933767 |
| V | 1.0530000925 | 1.0157818794 | 1.0261570215 | 1.0078532696 |
| W | 1.0759001970 | 1.0372149944 | 1.0386097431 | 1.0088177919 |
| X | 1.0430511236 | 1.0131111145 | 1.0191410780 | 1.0054205656 |
| Y | 1.0399550200 | 1.0197407007 | 1.0164992809 | 1.0062215328 |
| Z | 1.0620642900 | 1.0264353752 | 1.0310798883 | 1.0084336996 |
| 0 | 1.0728080273 | 1.0246663094 | 1.0417411327 | 1.0079085827 |
| 1 | 1.0326262712 | 1.0114061832 | 1.0116872787 | 1.0071448088 |
| 2 | 1.0527602434 | 1.0210211277 | 1.0254948139 | 1.0066089630 |
| 3 | 1.0563023090 | 1.0109890699 | 1.0268583298 | 1.0093988180 |
| 4 | 1.0551825762 | 1.0126544237 | 1.0278050900 | 1.0051388741 |
| 5 | 1.0546742678 | 1.0228652954 | 1.0254955292 | 1.0088582039 |
| 6 | 1.0559306145 | 1.0179107189 | 1.0281671286 | 1.0088255405 |
| 7 | 1.0430243015 | 1.0153788328 | 1.0195813179 | 1.0059862137 |
| 8 | 1.0582846403 | 1.0142199993 | 1.0291059017 | 1.0085783005 |
| 9 | 1.0582153797 | 1.0221956968 | 1.0298372507 | 1.0067170858 |

where the low-frequency components of $V_1$ remain in $V_2$, while the high-frequency components of $V_1$ are kept in $W_2$. Hence, $V_2$ contains only one-half of the frequency components of $V_1$, that is, only one-quarter of the frequency components of $f(x)$. Again for the same reason, $V_3$ contains only one-eighth of the frequency components of $f(x)$, etc.

Table 10.3. Divider dimensions computed for 72 printed symbols.

| Char | $S_{20}(*)$ | $S_{21}(*)$ | $W_{21}(*)$ |
|---|---|---|---|
| a | 1.1124289036 | 1.1017379761 | 1.0647013187 |
| b | 1.1129053831 | 1.1025065184 | 1.0577483177 |
| c | 1.1096138954 | 1.1018588543 | 1.0524585247 |
| d | 1.1138308048 | 1.1023379564 | 1.0613058805 |
| e | 1.1173223257 | 1.1131840944 | 1.0712224245 |
| f | 1.0966352224 | 1.0739761591 | 1.0401916504 |
| g | 1.1138812304 | 1.1035263538 | 1.0677573681 |
| h | 1.1104953289 | 1.0909165144 | 1.0611625910 |
| i | 1.0898797512 | 1.0654842854 | 1.0428867340 |
| j | 1.0985819101 | 1.0733737946 | 1.0505572557 |
| k | 1.1036610603 | 1.0771027803 | 1.0567890406 |
| l | 1.0866196156 | 1.0642234087 | 1.0423823595 |
| m | 1.1117404699 | 1.1028811932 | 1.0665664673 |
| n | 1.1099185944 | 1.1014550924 | 1.0498541594 |
| o | 1.1197787523 | 1.1124036312 | 1.0888708830 |
| p | 1.1132911444 | 1.1032136679 | 1.0608453751 |
| q | 1.1141122580 | 1.1080610752 | 1.0727815628 |
| r | 1.0962741375 | 1.0674076080 | 1.0423336029 |
| s | 1.1148400307 | 1.1055955887 | 1.0563476086 |
| t | 1.0966463089 | 1.0725744963 | 1.0399602652 |
| u | 1.1121745110 | 1.1033401489 | 1.0550986528 |
| v | 1.1025810242 | 1.0756613016 | 1.0482001305 |
| w | 1.1197830439 | 1.1085958481 | 1.0760765076 |
| x | 1.1038902998 | 1.0788321495 | 1.0512022972 |
| y | 1.1015509367 | 1.0733027458 | 1.0484067202 |
| z | 1.1092675924 | 1.0997101068 | 1.0636947155 |
| + | 1.0951038599 | 1.0682361126 | 1.0504192114 |
| - | 1.0777190924 | 1.0501787663 | 1.0364004374 |
| * | 1.0885618925 | 1.0625029802 | 1.0379626751 |
| / | 1.0820037127 | 1.0522413254 | 1.0374629498 |
| = | 1.1034380198 | 1.0821037292 | 1.0473917723 |
| > | 1.0986634493 | 1.0686403513 | 1.0433781147 |
| < | 1.1005434990 | 1.0753053427 | 1.0443359613 |
| ( | 1.0821937323 | 1.0506196022 | 1.0377899408 |
| ) | 1.0795989037 | 1.0480804443 | 1.0329642296 |
| ? | 1.1018882990 | 1.0856221914 | 1.0461981297 |

A series of experiments have been conducted to verify the proposed method. The results are shown in Figs. 10.14–10.19.

After wavelet decomposition, a pattern has produced seven sub-patterns. For each of the seven curves, we further computed its divider dimension and, therefore, related each pattern with a feature vector. The experiment data consist of a subset of Chinese characters.

Table 10.4. Divider dimensions computed for 72 printed symbols (*Continued*).

| Char | $S_{22}(*)$ | $W_{22}(*)$ | $S_{23}(*)$ | $W_{23}(*)$ |
|---|---|---|---|---|
| a | 1.0639376640 | 1.0190188885 | 1.0330584049 | 1.0155755281 |
| b | 1.0619244576 | 1.0232038498 | 1.0326099396 | 1.0063673258 |
| c | 1.0650064945 | 1.0149598122 | 1.0355489254 | 1.0124291182 |
| d | 1.0654503107 | 1.0228716135 | 1.0351454020 | 1.0064482689 |
| e | 1.0756344795 | 1.0332411528 | 1.0424871445 | 1.0176811218 |
| f | 1.0408747196 | 1.0150277615 | 1.0157804489 | 1.0071839094 |
| g | 1.0618500710 | 1.0178625584 | 1.0322690010 | 1.0100055933 |
| h | 1.0562689304 | 1.0192553997 | 1.0275653601 | 1.0072581768 |
| i | 1.0308871269 | 1.0149402618 | 1.0106828213 | 1.0073467493 |
| j | 1.0384023190 | 1.0228306055 | 1.0125319958 | 1.0095375776 |
| k | 1.0393686295 | 1.0122373104 | 1.0165767670 | 1.0052092075 |
| l | 1.0321862698 | 1.0211771727 | 1.0108287334 | 1.0083882809 |
| m | 1.0571626425 | 1.0170327425 | 1.0283771753 | 1.0088099241 |
| n | 1.0639086962 | 1.0118888617 | 1.0350456238 | 1.0062968731 |
| o | 1.0819115639 | 1.0584974289 | 1.0469623804 | 1.0224372149 |
| p | 1.0617285967 | 1.0226594210 | 1.0317831039 | 1.0102237463 |
| q | 1.0720965862 | 1.0300858021 | 1.0401749611 | 1.0091934204 |
| r | 1.0329773426 | 1.0121468306 | 1.0128583908 | 1.0058519840 |
| s | 1.0681239367 | 1.0374056101 | 1.0327115059 | 1.0203156471 |
| t | 1.0337879658 | 1.0192551613 | 1.0106018782 | 1.0087591410 |
| u | 1.0648975372 | 1.0165529251 | 1.0357654095 | 1.0063078403 |
| v | 1.0425969362 | 1.0102100372 | 1.0186051130 | 1.0057344437 |
| w | 1.0638923645 | 1.0303624868 | 1.0287992954 | 1.0120517015 |
| x | 1.0377725363 | 1.0120788813 | 1.0158277750 | 1.0059341192 |
| y | 1.0364419222 | 1.0107214451 | 1.0150331259 | 1.0061452389 |
| z | 1.0546982288 | 1.0181393623 | 1.0260168314 | 1.0094790459 |
| + | 1.0358840227 | 1.0237274170 | 1.0138664246 | 1.0080053806 |
| - | 1.0249313116 | 1.0114963055 | 1.0089241266 | 1.0068209171 |
| * | 1.0311700106 | 1.0164752007 | 1.0129623413 | 1.0083116293 |
| / | 1.0214143991 | 1.0115058422 | 1.0073471069 | 1.0066345930 |
| = | 1.0502055883 | 1.0121105909 | 1.0243492126 | 1.0072437525 |
| ¿ | 1.0380367041 | 1.0120087862 | 1.0165723562 | 1.0057575703 |
| ¡ | 1.0435506105 | 1.0124135017 | 1.0198223591 | 1.0060656071 |
| ( | 1.0224382877 | 1.0079113245 | 1.0076332092 | 1.0052670240 |
| ) | 1.0189806223 | 1.0097336769 | 1.0072559118 | 1.0047732592 |
| ? | 1.0515812635 | 1.0236015320 | 1.0239137411 | 1.0104774237 |

and a set of printed alphanumeric symbols, including 26 alphabets in both upper and lower cases, 10 numeric digits, and an additional 10 ASCII symbols. For the printed alphanumeric symbols, their feature vectors are presented in Tables 10.1–10.4, where the values in each row indicate the divider dimensions of seven sub-patterns for each symbol.

# Chapter 11

# Document Analysis by Reference Line Detection with 2D Wavelet Transform

Document processing has become a very active topic in areas of pattern recognition, office automation, artificial intelligence, and knowledge engineering for a decade Tang *et al.* [1994]. The acquisition of knowledge from documents by an information system can involve an extensive amount of hand-crafting which can be time-consuming. Actually, it is a bottleneck of information systems. Thus, automatic knowledge acquisition from documents is an important subject, and many researchers are trying to find new techniques for processing documents. In reality, it is very difficult to develop a general system that can process all kinds of documents, such as technical reports, government files, newspapers, books, journals, magazines, letters, and bank cheques. As the first step, many researchers concentrated their study on developing specific ones to treat some specific types of documents. After carefully studying the major characteristics of different types of documents, the specific properties of form documents have been analyzed in our earlier work [Tang *et al.*, 1995c]. According to these properties, we have taken the approach of building a simpler system for processing form documents instead of a complex one for all sorts of documents.

First, let us consider the major characteristics of forms:

- In general, a form consists of straight lines, which are oriented mostly in horizontal and vertical directions. These lines are referred to as *reference lines*, which can be found in an example, a Canadian cheque, shown in Fig. 11.1(a).
- The reference lines are pre-printed to guide the users to complete the form. A typical example is a bank cheque, where reference lines are printed to guide the users to write in the name of the payee, amount, and date in the appropriate places.

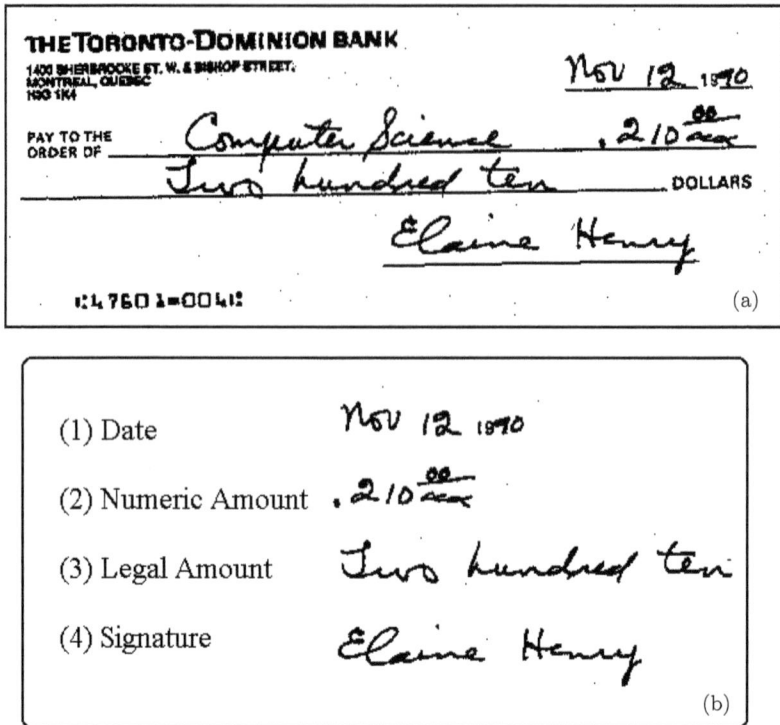

Fig. 11.1    (a) A Canadian cheque image and (b) the item images extracted from (a).

- Not all the information in a form is useful. The information that should be entered into the computer and processed is usually the filled data. For example, in Fig. 11.1(a), the date, numeric amount, legal amount, and signature are the filled data. These data shown in Fig. 11.1(b) need to be extracted from the cheque image in Fig. 11.1(a).
- In order to indicate the filling position, the reference lines can be used and the filled information usually appears either above, beneath, or beside these reference lines. Thus, in form processing, the reference lines have to be detected first, then we can find the useful information from a form based on these, and thereafter, enter into the computers.

Consequently, extraction of such reference lines plays a very important role in form processing. However, the extraction of the reference lines from a complex-background document is a difficult task. A traditional method to detect lines from images is the Hough transform. In 1994, the journal

*IEEE PAMI* published a new method called Subspace-Based Line Detection (SLIDE) proposed by Aghajan and Kailath [1994]. The SLIDE yields closed-form and high-resolution estimates for line parameters, and its computational complexity and storage requirement are far less than those of the Hough transform.

In this chapter, a novel wavelet-based method will be presented. In this method, 2D multiresolution analysis (MRA), wavelet decomposition algorithm, and compactly supported orthonormal wavelets are used to transform a document image into several sub-images. Based on these sub-images, the reference lines of a complex-background document can be extracted, and knowledge about the geometric structure of the document can be acquired. Particularly, this approach appears to be more efficient in processing form documents with multi-gray-level backgrounds.

## 11.1 Two-Dimensional MRA and Mallat Algorithm

The principle of the 1D MRA presented in Chapter 3 can be extended directly to 2D multiresolution by replacing $L^2(\mathbb{R})$ with $L^2(\mathbb{R}^2)$.

First of all, we declare that, as in the last chapter, we use the subspaces in $\{V_j\}$ by the order of

$$\cdots \subset V_{j+1} \subset V_j \subset V_{j-1},$$

when we deal with MRAs in this chapter.

**Definition 11.1.** Let $\{V_j\}_j \in \mathbb{Z}$ be a sequence of closed subspaces in $L^2(\mathbb{R})$. The 2D MRA can be constructed by *tensor product space* $\{V_j^2\}_{j\in\mathbb{Z}}$ if and only if $\{V_j\}_{j\in\mathbb{Z}}$ is a 1D MRA in $L^2(\mathbb{R})$ and

$$V_j^2 = V_j \otimes V_j. \tag{11.1}$$

The scaling function $\Phi(x,y)$ for 2D MRA has the form of

$$\Phi(x,y) = \varphi(x)\varphi(y),$$

where $\varphi$ is the real scaling function of the 1D multiresolution analysis $\{V_j\}_{j\in\mathbb{Z}}$. For each $j \in \mathbb{Z}$, the orthonormal bases of $V_j^2$ can be produced by the function system of

$$\{\Phi_{j,k_1,k_2} = \varphi_{j,k_1}(x)\varphi_{j,k_2}(y)|(k_1,k_2) \in \mathbb{Z}^2\}.$$

Such MRA $\{V_j^2\}_{j\in\mathbb{Z}}$ in space $L^2(\mathbb{R}^2)$ is called *divisible MRA*.

We define a wavelet space $W_j^2 = (V_j^2)^{\perp}$, i.e., $V_j^2 \oplus W_j^2 = V_{j-1}^2$. Thus, the wavelet function consists of three basic wavelet functions: $\psi^1$, $\psi^2$, and $\psi^3$.

The orthonormal bases of the wavelet space $W_j^2$ can be obtained from the "single wavelets" $\psi^1$, $\psi^2$, and $\psi^3$ by a *binary dilation*, i.e., dilation by $2^j$ and *dyadic translation* (of $k/2^j$). More precisely, we have the following theorem.

**Theorem 11.1.** *Let* $\{V_j^2\}_{j\in\mathbb{Z}}$ *be a divisible MRA in* $L^2(\mathbb{R}^2)$: $V_j^2 = V_j \otimes V_j$, *where* $\{V_j\}_{j\in\mathbb{Z}}$ *is a 1D MRA in the space* $L^2(\mathbb{R})$ *with scaling function* $\varphi$ *and wavelet function* $\psi$. *We define the following three functions:*

$$\begin{cases} \psi^1(x,y) = \varphi(x)\psi(y), \\ \psi^2(x,y) = \psi(x)\varphi(y), \\ \psi^3(x,y) = \psi(x)\psi(y). \end{cases}$$

For any $j \in \mathbb{Z}$, the orthonormal bases of the space $W_j^2$ can be obtained from the following function system:

$$\begin{cases} \Psi_{j,k,m}^1 = \varphi_{j,k}(x)\psi_{j,m}(y), \\ \Psi_{j,k,m}^2 = \psi_{j,k}(x)\varphi_{j,m}(y), \\ \Psi_{j,k,m}^3 = \psi_{j,k}(x)\psi_{j,m}(y). \end{cases}$$

Therefore, the function system

$$\{\Psi_{j,k,m}^e | e = 1, 2, 3; j, k, m \in \mathbb{Z}\} \tag{11.2}$$

becomes a set of orthonormal bases of $L^2(\mathbb{R}^2)$.

**Proof.** From Eq. (11.1), a very significant formula can be produced:

$$\begin{aligned} V_{j-1}^2 &= V_{j-1} \otimes V_{j-1} \\ &= (V_j \oplus W_j) \otimes (V_j \oplus W_j) \\ &= (V_j \otimes V_j) \oplus (V_j \otimes W_j) \oplus (W_j \otimes V_j) \oplus (W_j \otimes W_j). \end{aligned} \tag{11.3}$$

It can be rewritten as

$$\begin{aligned} V_{j-1}^2 &= V_{j-1} \otimes V_{j-1} \\ &= (V_j \otimes V_j) \oplus (V_j \otimes W_j) \oplus (W_j \otimes V_j) \oplus (W_j \otimes W_j) \\ &= V_j^2 \oplus \underbrace{[(V_j \otimes W_j) \oplus (W_j \otimes V_j) \oplus (W_j \otimes W_j)]}_{W_j^2}. \end{aligned} \tag{11.4}$$

Since $\{\varphi_{j,k}|k \in \mathbb{Z}\}$ is an orthonormal base for $V_j$, the set $\{\psi_{j,k}|k \in \mathbb{Z}\}$ becomes the orthonormal bases of the space $W_j$. Therefore,

- $\{\Psi^1_{j,k,m}|m \in \mathbb{Z}\}$ is an orthonormal base of $V_j \otimes W_j$;
- $\{\Psi^2_{j,k,m}|m \in \mathbb{Z}\}$ is an orthonormal base of $W_j \otimes V_j$;
- $\{\Psi^3_{j,k,m}|m \in \mathbb{Z}\}$ is an orthonormal base of $W_j \otimes W_j$.

Consequently, Eq. (11.4) indicates that the function represented by Eq. (11.2) has constituted the orthonormal bases of $W^2_j$.   □

Let $P_j$, $D^1_j$, $Q^2_j$ and $Q^3_j$ be projection operators from $L^2(\mathbb{R}^2)$ to its subspaces $(V_j \otimes V_j)$, $(V_j \otimes W_j)$, $(W_j \otimes V_j)$ and $(W_j \otimes W_j)$, respectively. In practice, the document image $f(x,y) \in V^2_{j_1}$ has a limited resolution, namely, $j_1$ is a certain integer. We have

$$f(x,y) = P_{j_1} f(x,y)$$

$$= \sum_{k_1 \in \mathbb{Z}} \sum_{k_2 \in \mathbb{Z}} c_{j_1,k_1,k_2} \Phi_{j_1,k_1,k_2}$$

$$= \sum_{k_1 \in \mathbb{Z}} \sum_{k_2 \in \mathbb{Z}} c_{j_1,k_1,k_2} \varphi_{j_1,k_1}(x)\varphi_{j_1,k_2}(y)$$

$$= P_{j_1+1}f + Q^1_{j_1+1}f + Q^2_{j_1+1}f + Q^3_{j_1+1}f, \qquad (11.5)$$

where

$$c_{j_1,k_1,k_2} = \langle P_{j_1+1}f(x,y), \varphi_{j_1,k_1}(x)\varphi_{j_1,k_2}(y)\rangle,$$

and

- $P_{j_1} f \in V_{j_1} \otimes V_{j_1}$;
- $P_{j_1+1} f \in V_{j_1+1} \otimes V_{j_1+1}$;
- $Q^1_{j_1+1} f \in V_{j_1+1} \otimes W_{j_1+1}$;
- $Q^2_{j_1+1} f \in W_{j_1+1} \otimes V_{j_1+1}$;
- $Q^3_{j_1+1} f \in W_{j_1+1} \otimes W_{j_1+1}$.

In Eq. (11.5), $P_{j_1+1}f$ and $Q^\beta_{j_1+1}f$ ($\beta = 1,2,3$) can be computed by

$$P_{j_1+1}f = \sum_{m_1 \in \mathbb{Z}} \sum_{m_2 \in \mathbb{Z}} c_{j_1+1,m_1,m_2} \Phi_{j_1+1,m_1,m_2}$$

$$= \sum_{m_1 \in \mathbb{Z}} \sum_{m_2 \in \mathbb{Z}} c_{j_1+1,m_1,m_2} \varphi_{j_1+1,m_1}(x)\varphi_{j_1+1,m_2}(y)$$

and

$$Q^{\beta}_{j_1+1}f = \sum_{m_1\in\mathbb{Z}}\sum_{m_2\in\mathbb{Z}} d^{\beta}_{j_1+1,m_1,m_2}\,\Psi^{\beta}_{j_1+1,m_1,m_2}$$

$$= \sum_{m_1\in\mathbb{Z}}\sum_{m_2\in\mathbb{Z}} d^{\beta}_{j_1+1,m_1,m_2}\,\psi^{\beta}_{j_1+1,m_1}(x)\psi^{\beta}_{j_1+1,m_2}(y) \quad (\beta=1,2,3).$$

Equation (11.5) can be written as follows:

$$f(x,y) = P_{j_1+1}f + Q^1_{j_1+1}f + Q^2_{j_1+1}f + Q^3_{j_1+1}f$$

$$= \sum_{m_1\in\mathbb{Z}}\sum_{m_2\in\mathbb{Z}} c_{j_1+1,m_1,m_2}\varphi_{j_1+1,m_1}(x)\varphi_{j_1+1,m_2}(y)$$

$$= \sum_{m_1\in\mathbb{Z}}\sum_{m_2\in\mathbb{Z}} d^1_{j_1+1,m_1,m_2}\varphi_{j_1+1,m_1}(x)\psi_{j_1+1,m_2}(y)$$

$$= \sum_{m_1\in\mathbb{Z}}\sum_{m_2\in\mathbb{Z}} d^2_{j_1+1,m_1,m_2}\psi_{j_1+1,m_1}(x)\varphi_{j_1+1,m_2}(y)$$

$$= \sum_{m_1\in\mathbb{Z}}\sum_{m_2\in\mathbb{Z}} d^3_{j_1+1,m_1,m_2}\psi_{j_1+1,m_1}(x)\psi_{j_1+1,m_2}(y).$$

An iterative algorithm called the *Mallat algorithm* is presented as follows:

$$\begin{cases} c_{j_1+1,m_1,m_2} = \displaystyle\sum_{k_1\in\mathbb{Z}}\sum_{k_2\in\mathbb{Z}} h_{k_1-2m_1}h_{k_2-2m_2}c_{j_1,k_1,k_2}, \\[2mm] d^1_{j_1+1,m_1,m_2} = \displaystyle\sum_{k_1\in\mathbb{Z}}\sum_{k_2\in\mathbb{Z}} h_{k_1-2m_1}g_{k_2-2m_2}c_{j_1,k_1,k_2}, \\[2mm] d^2_{j_1+1,m_1,m_2} = \displaystyle\sum_{k_1\in\mathbb{Z}}\sum_{k_2\in\mathbb{Z}} g_{k_1-2m_1}h_{k_2-2m_2}c_{j_1,k_1,k_2}, \\[2mm] d^3_{j_1+1,m_1,m_2} = \displaystyle\sum_{k_1\in\mathbb{Z}}\sum_{k_2\in\mathbb{Z}} g_{k_1-2m_1}g_{k_2-2m_2}c_{j_1,k_1,k_2}. \end{cases} \quad (11.6)$$

Let $H_r = (H_{k_1,m_1})$, $H_c = (H_{k_2,m_2})$, $G_r = (G_{k_1,m_1})$, and $H_c = (G_{k_2,m_2})$ be matrices. The subscript "r" indicates an operation on the rows of the matrix, while the subscript "c" is for an operation on the column.

Thus, Eq. (11.6) can be represented by a simple form:

$$\begin{cases} C_{j_1+1} = H_r H_c C_{j_1}, \\ Q^1_{j_1+1} = H_r G_c C_{j_1}, \\ Q^2_{j_1+1} = G_r H_c C_{j_1}, \\ Q^3_{j_1+1} = G_r G_c C_{j_1}. \end{cases} \tag{11.7}$$

$P_{j_1+1}f$ will be decomposed in the same way, and $P_{j_1+2}f$ and $Q^\beta_{j_1+2}f$ will be produced. This iteration procedure will be continued. After $j_2 - j_1$ steps, we arrive at

$$f(x, y) = P_{j_2} f(x, y) + \sum_{j=j_1+1}^{j_2} \sum_{\beta=1}^{3} Q^\beta_j f(x, y). \tag{11.8}$$

## 11.2 Detection of Reference Line from Sub-Images by the MRA

In this section, the basic idea of form analysis by the MRA will be presented. First, we will introduce the properties of the wavelet-transformed sub-images. A document image can be transformed into four sub-images by applying the Mallat algorithm: (1) LL sub-image, (2) LH sub-image, (3) HL sub-image, and (4) HH sub-image. According to Eq. (11.3), these sub-images possess the following properties:

- **LL sub-image:** Both horizontal and vertical directions have low-frequencies. It corresponds to $(V_j \otimes V_j)$, and its orthonormal basis is $\{\Phi_{j,k,m} | k, m \in \mathbb{Z}\}$.
- **LH sub-image:** The horizontal direction has low frequencies, and the vertical one has high frequencies. It corresponds to $(V_j \otimes W_j)$, and its orthonormal basis is $\{\Psi^1_{j,k,m} | k, m \in \mathbb{Z}\}$.
- **HL sub-image:** The horizontal direction has high frequencies, and the vertical one has low frequencies. It corresponds to $(W_j \otimes V_j)$, and its orthonormal basis is $\{\Psi^2_{j,k,m} | k, m \in \mathbb{Z}\}$.
- **HH sub-image:** Both horizontal and vertical directions have high frequencies. It corresponds to $(W_j \otimes W_j)$, and its orthonormal basis is $\{\Psi^3_{j,k,m} | k, m \in \mathbb{Z}\}$.

An example is illustrated in Fig. 11.2. The original image is a square with gray level shown in Fig. 11.2(a). It has been transformed into four sub-images by the MRA, as illustrated in Fig. 11.2(b). The LL sub-image is the result from a filter, which allows lower frequencies to pass through along the horizontal direction as well as the vertical direction. That is a "smoothing" effect in both directions. The HH sub-image comes from a filter where the higher frequency components can cross it along both directions. That is an "enhancing" effect on the horizontal and vertical directions.

We are interested in the LH and HL sub-images. The LH sub-image is achieved from a filter which allows lower frequency components to reach across along the horizontal direction as well as the higher frequencies along the vertical direction. That is an "enhancing" effect on the vertical and a "smoothing" effect on the horizontal. The result of the HL sub-image is opposite to that of the LH one. In this way, the horizontal direction of the filter opens for the higher frequencies and the vertical direction for lower frequency components. That is an "enhancing" effect on the horizontal and a "smoothing" effect on the vertical.

Another example can be shown in Figs. 11.3 and 11.4. The input image is illustrated in Fig. 11.3, which contains several lines with complex-background including texts. Its LH and HL sub-images are depicted in Figs. 11.4(a) and 11.4(b), respectively.

From Fig. 11.4(a), it is clear that only horizontal lines remain in the LH sub-image, while only vertical ones remain in the HL sub-image, as shown in Fig. 11.4(b). Both the LH and HL sub-images keep only straight lines, the gray-level background is removed. We are interested in Figs. 11.4(b) and 11.4(c) because these important properties can be used to extract the horizontal and vertical lines in a form document, which has complex background. In form documents, the information that should be entered into the computer and processed is usually the filled data. In order to indicate the filling position, some pre-printed reference lines should be extracted. The useful information, in general, is either above, beneath, or beside these specific lines. In our work, the LH and HL sub-images have been used to extract such lines.

To perform the above, proposed method, an MRA algorithm and compactly supported orthonormal wavelets are used.

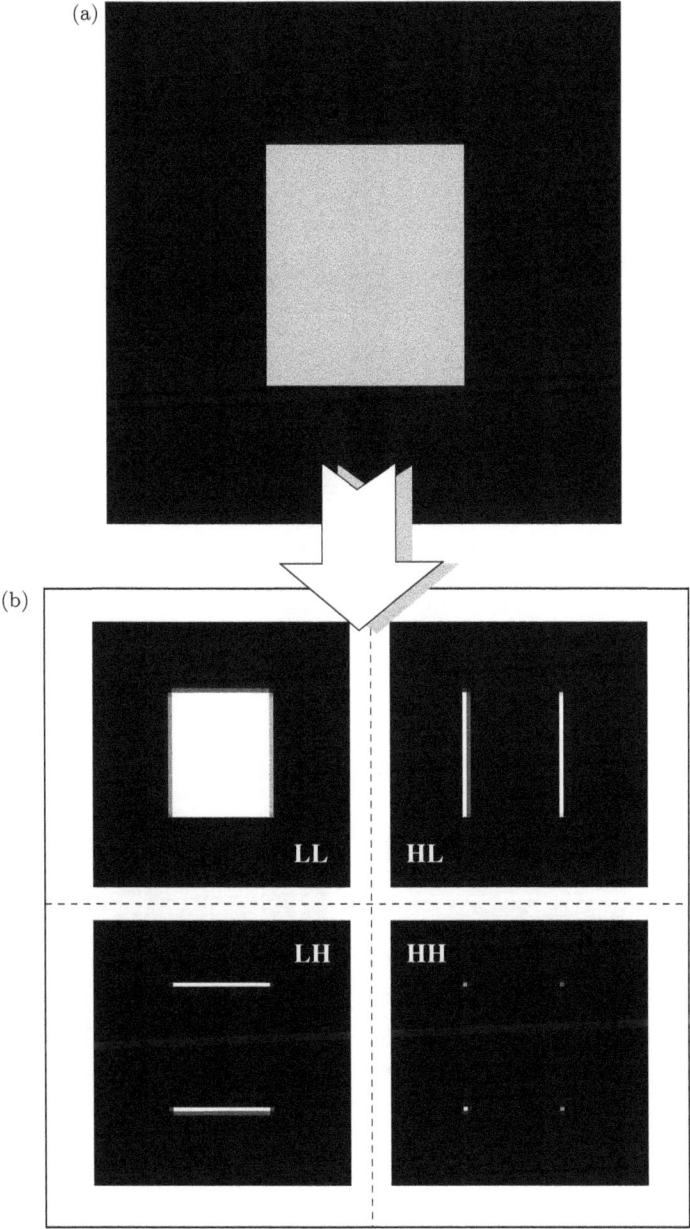

Fig. 11.2   A square image and its sub-images.

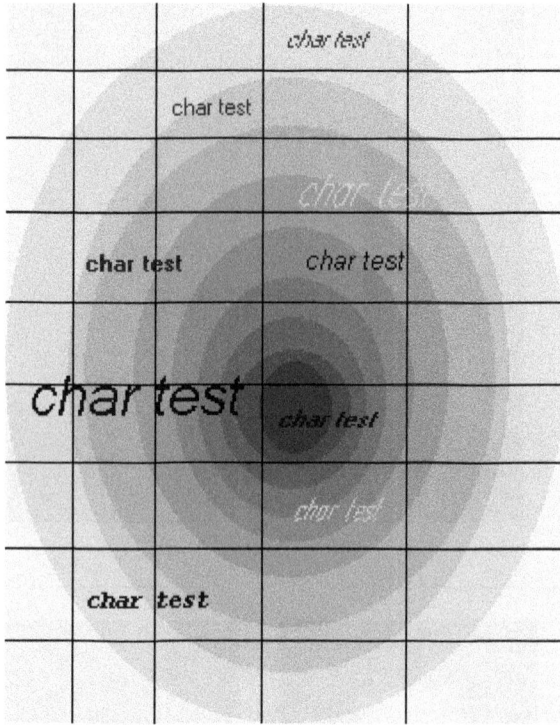

Fig. 11.3    Input image.

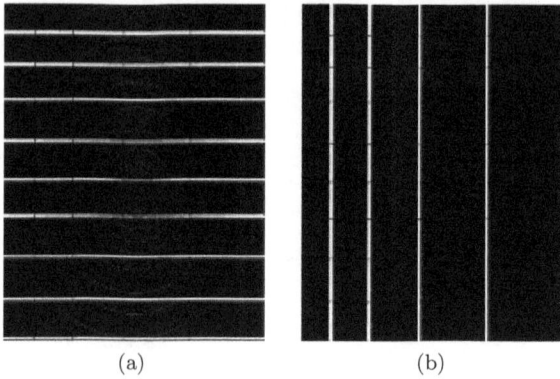

Fig. 11.4    LH (a) and HL (b) sub-images.

(a)

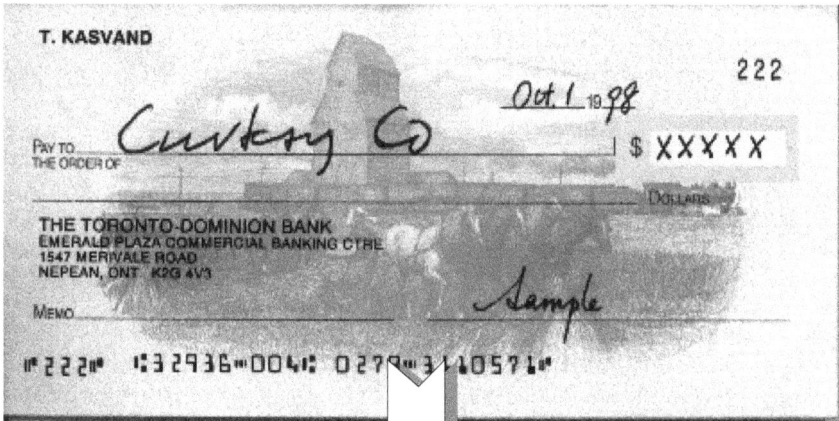

(b)

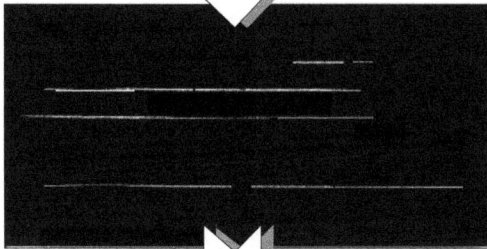

(c)

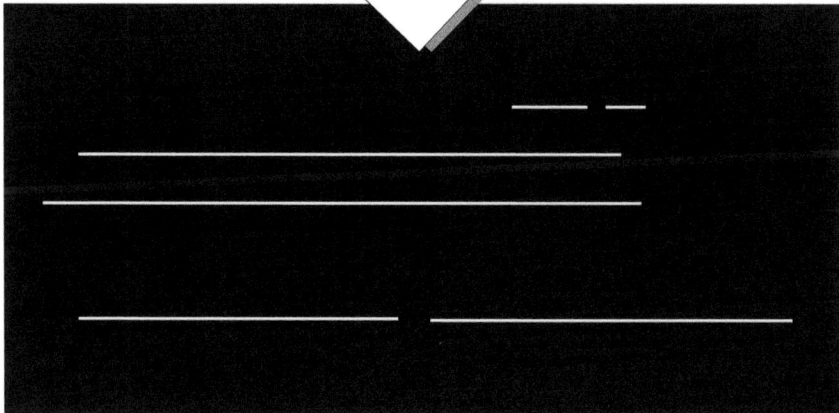

Fig. 11.5    A bank cheque and its original and enhanced LH sub-images. (a) The original check image. (b) The LH sub-image of the check. (c) The enhanced image.

## 11.3 Experiments

Using sub-images of wavelet, experiments have been conducted to process financial documents by a personal computer system. The document is entered into the system by an optical scanner and converted to a digital image. A document can be digitized into a color image, monochrome gray scale image, or binary image according to different requirements and applications. In our experiments, an HP scanner is employed to capture the image of the document. The resolution of digitization in our experiments can vary over a range of 200–300 DPI. All documents are converted to grayscale images in the experiments. We have examined a variety of form documents originating from the United States, Canada, and China. They contain alpha-numerics, Chinese characters, some special symbols, and graphics. Both simple financial documents such as Canadian bank cheques and complex documents such, as Canadian Federal tax return forms and deposit forms of Chinese banks have been tried by our new method. Due to the page limit in this chapter, only a few examples will be presented:

(1) cheque of a Canadian bank (TD Bank) (Fig. 11.5);
(2) portion of a federal tax return form (Fig. 11.6);
(3) deposit form of Chinese bank (Fig. 11.8).

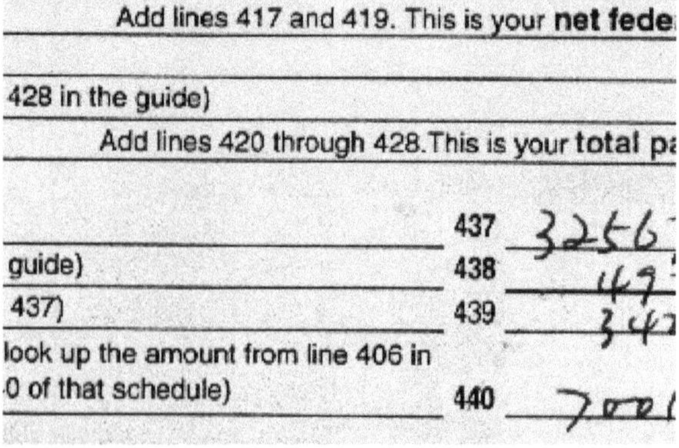
Allowable credit 414
Add lines 410, 412, and 414 416
line 406; if negative, enter "0")
the amount from line 406 in Table B. If you are using
Add lines 417 and 419. This is your **net fede**
428 in the guide)
Add lines 420 through 428.This is your **total p**
437
guide) 438
437) 439
look up the amount from line 406 in
0 of that schedule) 440

Fig. 11.6 Portion of a federal tax return form.

Table 11.1.   Values of $h_N(n)$'s, $N = 2 - 5$.

| $h_N(n)$ | | | | $h_N(n)$ | |
|---|---|---|---|---|---|
| $N = 2$ | $n = 0$ | 0.482 962 913 145 | $N = 3$ | $n = 0$ | 0.332 670 552 950 |
| | 1 | 0.836 516 303 738 | | 1 | 0.806 891 509 311 |
| | 2 | 0.224 143 868 042 | | 2 | 0.459 877 502 118 |
| | 3 | −0.129 409 522 551 | | 3 | −0.135 011 020 010 |
| | | | | 4 | −0.085 441 273 882 |
| | | | | 5 | 0.035 226 291 882 |
| $N = 4$ | $n = 0$ | 0.230 377 813 309 | $N = 5$ | $n = 0$ | 0.160 102 397 974 |
| | 1 | 0.714 846 570 553 | | 1 | 0.603 829 269 797 |
| | 2 | 0.630 880 767 930 | | 2 | 0.724 308 528 438 |
| | 3 | −0.027 983 769 417 | | 3 | 0.138 428 145 901 |
| | 4 | −0.187 034 811 719 | | 4 | −0.242 294 887 066 |
| | 5 | 0.030 841 381 836 | | 5 | −0.032 244 869 585 |
| | 6 | 0.032 883 011 667 | | 6 | 0.077 571 493 840 |
| | 7 | −0.010 597 401 785 | | 7 | −0.006 241 490 213 |
| | | | | 8 | −0.012 580 751 999 |
| | | | | 9 | 0.003 335 725 285 |

Table 11.2.   Values of $h_N(n)$'s, $N = 6-7$.

| $h_N(n)$ | | | | $h_N(n)$ | |
|---|---|---|---|---|---|
| $N = 6$ | $n = 0$ | 0.111 540 743 350 | $N = 7$ | $n = 0$ | 0.077 852 054 085 |
| | 1 | 0.494 623 890 398 | | 1 | 0.396 539 319 482 |
| | 2 | 0.751 133 908 021 | | 2 | 0.729 132 090 846 |
| | 3 | 0.315 250 351 709 | | 3 | 0.469 782 287 405 |
| | 4 | −0.226 264 693 965 | | 4 | -0.143 906 003 929 |
| | 5 | −0.129 766 867 567 | | 5 | -0.224 036 184 994 |
| | 6 | 0.097 501 605 587 | | 6 | 0.071 309 219 267 |
| | 7 | 0.027 522 865 530 | | 7 | 0.080 612 609 151 |
| | 8 | −0.031 582 039 318 | | 8 | -0.038 029 936 935 |
| | 9 | 0.000 553 842 201 | | 9 | -0.016 574 541 631 |
| | 10 | 0.004 777 257 511 | | 10 | 0.012 550 998 556 |
| | 11 | −0.001 077 301 085 | | 11 | 0.000 429 577 937 |
| | | | | 12 | -0.001 801 640 704 |
| | | | | 13 | 0.000 353 713 800 |

In these experiments, the 2D MRA algorithm has been applied. The compactly supported orthonormal wavelets used in our study have been chosen from Daubechies [1988], and the values of $h_k$'s are listed in Table 11.1–11.4.

Table 11.3.    Values of $h_N(n)$'s, $N = 8-9$.

| $h_N(n)$ | | | | $h_N(n)$ | | |
|---|---|---|---|---|---|---|
| $N = 8$ | $n = 0$ | 0.054 415 842 243 | $N = 9$ | $n = 0$ | 0.038 077 947 364 |
| | 1 | 0.312 871 590 914 | | 1 | 0.243 834 674 613 |
| | 2 | 0.675 630 736 297 | | 2 | 0.604 823 123 690 |
| | 3 | 0.585 354 683 654 | | 3 | 0.657 288 078 051 |
| | 4 | −0.015 829 105 256 | | 4 | 0.133 197 385 825 |
| | 5 | −0.284 015 542 962 | | 5 | −0.293 273 783 279 |
| | 6 | 0.000 472 484 574 | | 6 | −0.096 840 783 223 |
| | 7 | 0.128 747 426 620 | | 7 | 0.148 540 749 338 |
| | 8 | −0.017 369 301 002 | | 8 | 0.030 725 681 479 |
| | 9 | −0.044 088 253 931 | | 9 | −0.067 832 829 061 |
| | 10 | 0.013 981 027 917 | | 10 | 0.000 250 947 115 |
| | 11 | 0.008 746 094 047 | | 11 | 0.022 361 662 124 |
| | 12 | −0.004 870 352 993 | | 12 | −0.004 723 204 758 |
| | 13 | −0.000 391 740 373 | | 13 | −0.004 281 503 682 |
| | 14 | 0.000 675 449 406 | | 14 | 0.001 847 646 883 |
| | 15 | −0.000 117 476 784 | | 15 | 0.000 230 385 764 |
| | | | | 16 | −0.000 253 963 189 |
| | | | | 17 | 0.000 039 347 320 |

Table 11.4.    Values of $h_N(n)$'s, $N = 10$.

| $h_N(n)$ | | | $h_N(n)$ | | |
|---|---|---|---|---|---|
| $N = 10$ | $n = 0$ | 0.026 670 057 901 | $N = 10$ | $n = 10$ | −0.029 457 536 822 |
| | 1 | 0.188 176 800 078 | | 11 | 0.033 212 674 059 |
| | 2 | 0.527 201 188 932 | | 12 | 0.003 606 553 567 |
| | 3 | 0.688 459 039 454 | | 13 | −0.010 733 157 483 |
| | 4 | 0.281 172 343 661 | | 14 | 0.001 395 351 747 |
| | 5 | −0.249 846 424 327 | | 15 | 0.001 992 405 295 |
| | 6 | −0.195 946 274 377 | | 16 | −0.000 685 856 695 |
| | 7 | 0.127 369 340 336 | | 17 | −0.000 116 466 855 |
| | 8 | 0.093 057 364 604 | | 18 | 0.000 093 588 670 |
| | 9 | −0.071 394 147 166 | | 19 | −0.000 013 264 203 |

For the Canadian Bank cheque shown in Fig. 11.5(a), there exists a gray-level background on it. To remove the gray-level background, the special properties of the sub-images can be used. Precisely, the gray-level background has been removed in both the LH sub-image and the HL sub-image.

The LH sub-image results from a filter which allows lower frequencies to pass through along the horizontal direction and higher frequencies along the vertical direction. That is an "enhancing" effect on the vertical and a

(a)

(b)

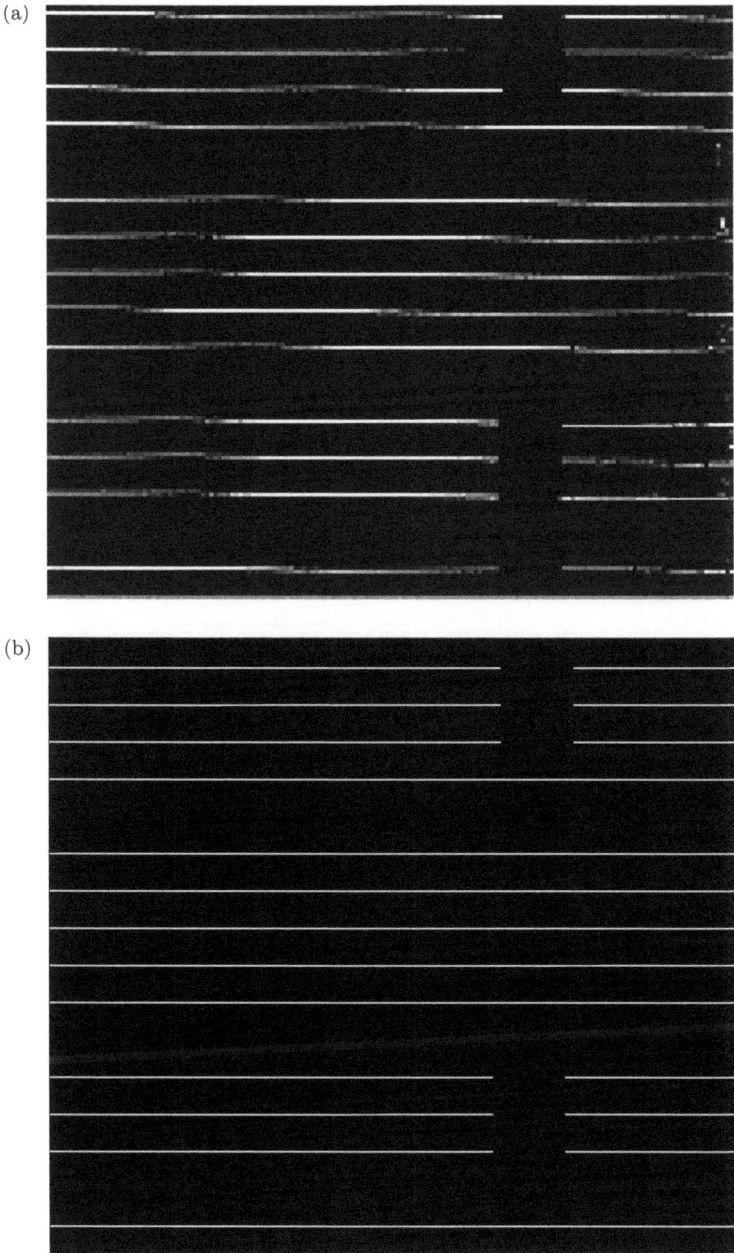

Fig. 11.7   (a) LH sub-image of the portion of a Federal tax return form shown in Fig. 11.6 and (b) the enhanced image.

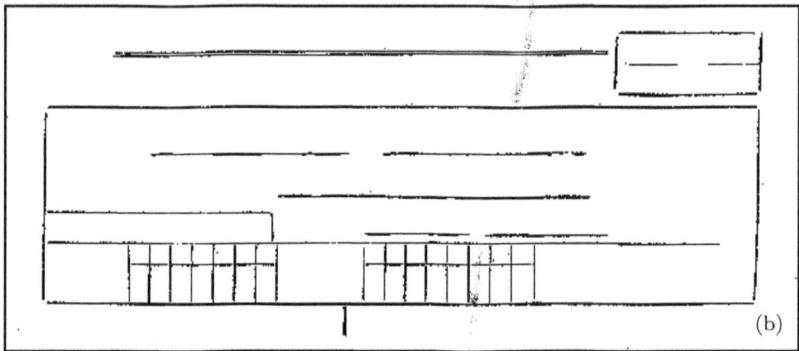

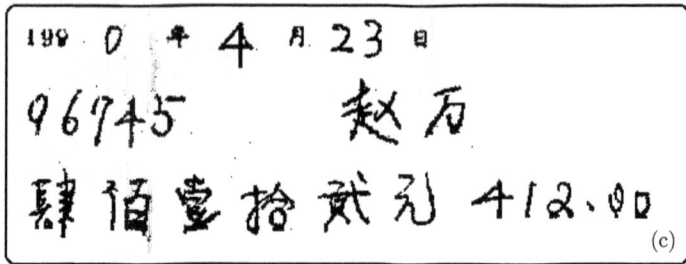

Fig. 11.8   (a) An image of the deposit form of a Chinese bank; (b) the reference lines detected from (a); (c) the item images extracted from (a) according to the reference lines.

"smoothing" effect on the horizontal. As a result, only horizontal lines remain in the LH sub-image. The HL sub-image is opposite to the LH one. The high-frequency spectrum can pass along the horizontal direction, while low-frequency spectrum along the vertical direction. That is an "enhancing" effect on the horizontal and a "smoothing" effect on the vertical. Thus, only vertical lines remain in the HL sub-image.

In our experiments, the special property of the LH sub-image has been used. To produce the LH sub-image, the 2D MRA algorithm and compactly supported orthonormal wavelets have been applied to the documents. The LH sub-image of Fig. 11.5(a) has been obtained and shown in Fig. 11.5(b). Since only horizontal lines remain in the LH sub-image, reference lines, which guide the writer to fill data in the proper location, can be extracted.

A point worthy of mention is that the number of pixels contained in any sub-image is one-fourth of that of the original document image. Thus, the size of LH sub-image shown in Fig. 11.5(b) is one-fourth of that of the original cheque image shown in Fig. 11.5(a). To achieve a high-quality image and correctly map the filled data with reference lines, the LH sub-image has been processed by regular enhancement and smoothing techniques and has been scaled up to the same size as the original image. The result can be found in Fig. 11.5(c).

For the form shown in Fig. 11.6, there are altogether 20 horizontal lines and many text strings in it. The result of extracting the pre-printed reference lines from the LH sub-image of the wavelet is presented in Fig. 11.7.

Note that the extracted lines in both Fig. 11.5(a) and Fig. 11.7(b) are gray levels, since the document images to be processed in our method are grayscale images.

The last example is a deposit form of a Chinese bank, as shown in Fig. 11.8, which is rather complicated. The original image of the Chinese bank deposit form is given in Fig. 11.8(a). As applying wavelet decomposition algorithm, the reference lines are detected and shown in Fig. 11.8(b). Based on these reference lines, the images of some items which contain useful information are extracted, and the results are displayed in Fig. 11.8(c).

# Chapter 12

# Chinese Character Processing with B-Spline Wavelet Transform

Chinese character recognition is a very significant branch of pattern recognition, and Chinese character processing is an important technology within it. Chinese character processing is to operate and modify Chinese characters including generation, storage, display, printing, transferring, and geometric transformation with the modern computer.

In this chapter, several algorithms for Chinese character processing are studied, based on cubic B-spline wavelet transform:

(1) compression of Chinese characters,
(2) arbitrary enlargement of the typeface size of Chinese characters,
(3) generation of Chinese-type style.

The outline of Chinese character processing by wavelet transform presented in this chapter is graphically illustrated in Fig. 12.1. It consists of three stages: (1) pre-processing, (2) wavelet transform, and (3) objective processing. They are presented in the following:

**(1) Pre-processing:** The original character is entered into a computer system by scanning, and it is further undergone by a contour extraction operator. Thus, a Chinese character is converted to one or more contours, which can be considered to be curves. Each curve is represented by an array of coordinate points, $Q_i(i = 1, 2, \ldots, M)$. It is then sent to the next stage.

**(2) Wavelet transform:** After the pre-processing stage, a Chinese character is already represented by its contours (curves), and each of them is formed by an array of coordinate points $Q_i(i = 1, 2, \ldots, M)$. In this stage, the cubic B-spline function is employed to interpolate such a parameter curve. Suppose $S^n(t)$ indicates the interpolation curve, which is produced

Input
Character

Pre-Processing

Contour of
Character

Wavelet
Transform

Wavelet
Coefficients

Compression     Enlargement     Generation

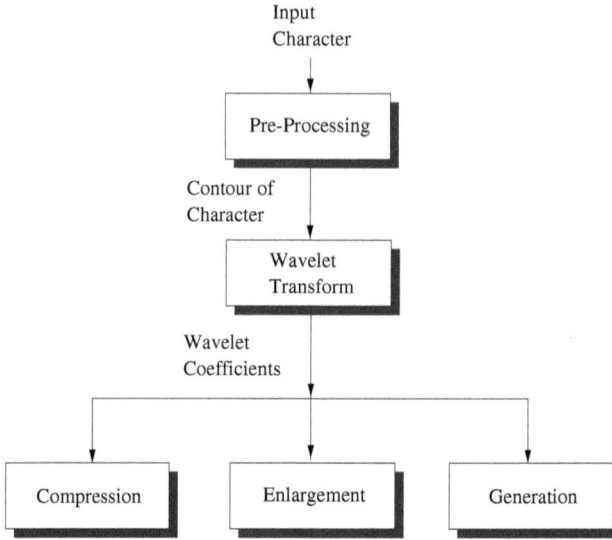

Fig. 12.1   Procedure of Chinese character processing with wavelet transform.

in accordance with the cubic B-spline function and passes through these coordinate points.

The interpolation curve $S^n(t)$ using cubic B-spline can be written as

$$S^n(t) = (s_0^n(t), s_1^n(t), \ldots, s_{M-2}^n(t)), \tag{12.1}$$

where $n$ stands for the level number of decompositions in wavelet analysis, which is presented later. $s_i^n(t)$ denotes the $i$th interpolation sub-curve with cubic B-spline and can be expressed by

$$s_i^n(t) = \sum_{l=1}^{4} c_{i+l-1}^n \cdot N_{l,3}(t), \quad 0 \le i \le M-2, \ 0 \le t \le 1, \tag{12.2}$$

where $N_{l,3}(t)$ is a basis function of the cubic B-spline interpolation, which means polynomial spline function with equally spaced sample knots. $C^n = (c_0^n, c_1^n, \ldots, c_{M+1}^n,)$ is a coefficient sequence, which is called the control point of the cubic B-spline interpolation curve. These control points correspond to the coordinate points (sometimes, those points are called vertices of the eigen polygon of B-spline).

The relationship between the coordinate points and the control points is

$$\frac{(c_{i-1}^n + 4c_i^n + c_{i+1}^n)}{6} = Q_i, \quad i = 1, 2, \ldots, M, \tag{12.3}$$

there are $M$ equations in (12.3), but there are $M + 2$ unknown variables. Hence, two boundary conditions are needed to supply. For the interpolation curve with cubic B-spline, we have the following at the end points:

$$c_0^n = c_1^n, \quad c_{M+1}^n = c_M^n.$$

Consequently, we can obtain a linear equation as follows:

$$\begin{bmatrix} 1 & -1 & 0 & 0 & 0 & 0 \\ \frac{1}{6} & \frac{4}{6} & \frac{1}{6} & 0 & 0 & 0 \\ \cdots & \cdots & \cdots & \cdots & \cdots & \cdots \\ 0 & 0 & 0 & \frac{1}{6} & \frac{4}{6} & \frac{1}{6} \\ 0 & 0 & 0 & 0 & -1 & 1 \end{bmatrix} \cdot \begin{bmatrix} c_0^n \\ c_1^n \\ \cdots \\ c_M^n \\ c_{M+1}^n \end{bmatrix} = \begin{bmatrix} 0 \\ Q_1 \\ \cdots \\ Q_M \\ 0 \end{bmatrix}. \tag{12.4}$$

According to the above analysis, the following is clear:

- Owing to (12.2) and (12.4), we can find the relationship among the coordinate points $Q_i$, sub-curves $s_i^n$, and control points $c_i^n$.
- A cubic B-spline curve constructed by $C^n = (c_0^n, c_1^n, \ldots, c_{M+1}^n,)$ passes through the coordinate points $Q_i(i = 1, 2, \ldots, M)$.
- By (12.2), four control points produce a sub-curve. $M$ coordinate points express $M - 1$ sub-curves, i.e., a curve needs $M + 2$ control points.
- The coordinate points can be converted to the control points by solving Eq. (12.4).

Next, the multiresolution analysis in wavelet theory is applied to describe a curve $S^n(t)$. In this way, the control points, by which the curve is formed, are transformed with wavelet function. For effectively decomposing a curve, the number of sub-curves is usually 2 to the power $n$, i.e., curve $S^n(t)$ should have $2^n + 3$ control points. When the control points are less than $2^n + 3$, the coordinate points must be extended as well as the control points. The performance of the extension can be periodical spinning out or adding zeros or re-sampling. For a Chinese character, which is composed of several closed curves, the periodic extension is usually utilized.

When the number of control points is equal to $2^n + 3$, after the first decomposition by wavelet transform, the sequence of the control points $C^n$ becomes $C^{n-1}$ and $D^{n-1}$. Thereafter, $C^{n-1}$ is decomposed into $C^{n-2}$ and

$D^{n-2}$ again. This process continues, as a total number of $n$ decompositions, finally, we arrive at $C^0$ and $D^0$. This process can be illustrated as follows:

$$C^n = D^{n-1} \oplus C^{n-1}$$
$$\Downarrow$$
$$(D^{n-2} \oplus C^{n-2})$$
$$\Downarrow$$
$$(D^{n-3} \oplus C^{n-3})$$
$$\Downarrow$$
$$\cdots$$
$$\Downarrow$$
$$(D^0 \oplus C^0).$$

Consequently, it is clear that the sequence of the control point $C^n$ can be represented by wavelet coefficients:

$$C^n \Longrightarrow C^{n-1}, C^{n-2}, \ldots, C^1, C^0,$$
$$D^{n-1}, D^{n-2}, \ldots, D^1, D^0.$$

Finally, we obtain

$$C^n = D^{n-1} \oplus D^{n-2} \oplus \cdots \oplus D^1 \oplus D^0 \oplus C^0.$$

The curve $S^n$ can be decomposed as

$$S^n \Longrightarrow S^{n-1}, S^{n-2}, \ldots, S^1, S^0.$$

An example of this procedure is shown in Fig. 12.2; here we know $n = 4$. Therefore, four wavelet decomposition layers occur. Owing to Fig. 12.2, we can further find the relationship between the sub-curves and the control points as well as the relationship between the wavelet coefficients and the sub-curves at the different layers. In this example, we obtain

$$C^4 \Longrightarrow C^3, C^2, C^1, C^0,$$
$$D^3, D^2, D^1, D^0,$$
$$S^4 \Longrightarrow S^3, S^2, S^1, S^0,$$

and

$$C^4 = D^3 \oplus D^2 \oplus D^1 \oplus D^0 \oplus C^0,$$

$$C^4: \quad S^4_0\, S^4_1\, S^4_2\, S^4_3\, S^4_4\, S^4_5\, S^4_6\, S^4_7\, S^4_8\, S^4_9\, S^4_{10}\, S^4_{11}\, S^4_{12}\, S^4_{13}\, S^4_{14}\, S^4_{15}$$
$$c^4_0\, c^4_1\, c^4_2\, c^4_3\, c^4_4\, c^4_5\, c^4_6\, c^4_7\, c^4_8\, c^4_9\, c^4_{10}\, c^4_{11}\, c^4_{12}\, c^4_{14}\, c^4_{15}\, c^4_{16}\, c^4_{17}\, c^4_{18}$$

$$C^3: \quad S^3_0\, S^3_1\, S^3_2\, S^3_3\, S^3_4\, S^3_5\, S^3_6\, S^3_7 \qquad d^3_0\, d^3_1\, d^3_2\, d^3_3\, d^3_4\, d^3_5\, d^3_6\, d^3_7\ D^3$$
$$c^3_0\, c^3_1\, c^3_2\, c^3_3\, c^3_4\, c^3_5\, c^3_6\, c^3_7\, c^3_8\, c^3_9\, c^3_{10}$$

$$C^2: \quad S^2_0\, S^2_1\, S^2_2\, S^2_3 \qquad d^2_0\, d^2_1\, d^2_2\, d^2_3\ D^2$$
$$c^2_0\, c^2_1\, c^2_2\, c^2_3\, c^2_4\, c^2_5\, c^2_6$$

$$C^1: \quad S^1_0\, S^1_1 \qquad d^1_0\, d^1_1\ D^1$$
$$c^1_0\, c^1_1\, c^1_2\, c^1_3\, c^1_4$$

$$C^0: \quad S^0_0 \qquad d^0_0\ D^0$$
$$c^0_0\, c^0_1\, c^0_2\, c^0_3$$

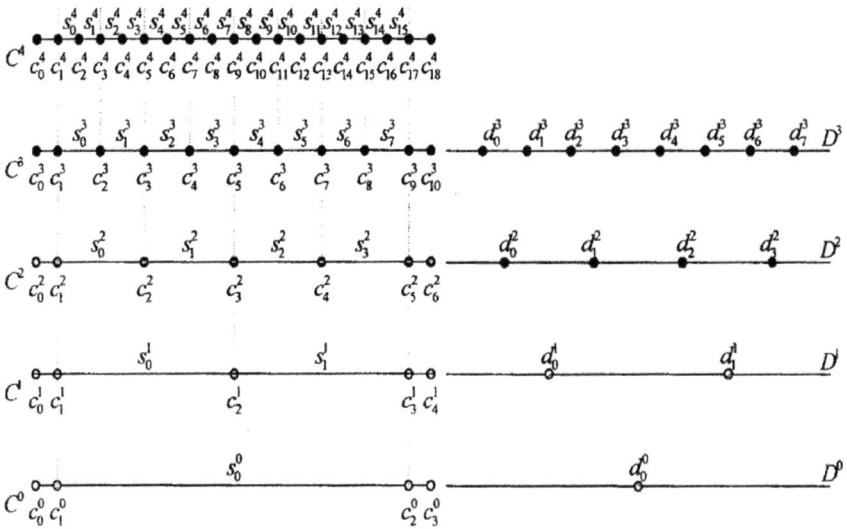

Fig. 12.2   Decomposition of cubic B-spline curve.

where the highest frequency in $C^3$ must be greater than that in $C^2$. In turn, the highest frequency in $C^2$ must be greater than that in $C^1$, etc. In general, we have

$$F_{\max}(C^k) > F_{\max}(C^{k-1}),$$

where $F_{\max}(C^k)$ denotes the highest frequency in $C^k$. The same thing appears in $D^i$ and $S^i$. Therefore,

$$F_{\max}(D^k) > F_{\max}(D^{k-1}), \quad F_{\max}(S^k) > F_{\max}(S^{k-1}).$$

If we consider the curve to be a 1D signal, this implies that $S^k$ (or $C^k$ or $D^k$) contains more particulars than $S^{k-1}$ (or $C^{k-1}$ or $D^{k-1}$) has.

**(3) Objective processing:** In the stage of the wavelet transform, a curve is decomposed into several layers, and the details of the curve at each layer can be described by wavelet coefficients. Therefore, by adequately choosing and processing these wavelet coefficients, the different objective processing of the Chinese character can be performed. In this book, compression of characters, arbitrary scale of the typeface, and generation of Chinese type styles are introduced using the cubic B-spline wavelet transform. In the rest of this chapter, we emphasize these objective operations and implementation.

## 12.1 Compression of Chinese Character

Compression of Chinese characters is a very important aspect of Chinese information processing with computers. It can benefit by reducing the amount of character storage memory, speeding up the processing, decreasing the cost, etc.

In the first three stages in Fig. 12.1, each Chinese character is converted into several contours (closed curves) by the process of extracting strokes and detecting edges, and the individual curve is estimated by a cubic B-spline expression, which can be represented by a sequence of control points. The control points of the B-spline curve are transformed into the different layers of details $D^{n-1}, D^{n-2}, \ldots, D^1, D^0$ plus $C^0$ or the different layers of control points $C^{n-1}, C^{n-2}, \ldots, C^1, C^0$ by wavelet transform. Therefore, each Chinese character can be described with wavelet transform coefficients (or control points) at different decomposition layers. According to the characteristics of wavelet transform coefficients (or control points), two approaches to compressing Chinese characters are present in this section: (1) the global approach and (2) the local approach.

### 12.1.1 *Algorithm 1 (Global Approach)*

The method of this algorithm is similar to image compression by wavelet transform. The basic idea is that, under a certain accuracy or error range, we consider the entire curve to delete some details $D^i$ and reserve fewer coefficients on the basis of the influence of different layer details on the curve. More precisely, in the result of the wavelet decomposition, a threshold $J$ is chosen so that the certain accuracy is satisfied, and thereafter, based on this threshold, some wavelet transform coefficients are removed, and some are kept. This can be described as follows:

$$D^n = \underbrace{D^{n-1} \oplus D^{n-2} \oplus \cdots \oplus D^{J+1}}_{\text{Deleted}} \oplus \underbrace{D^J \oplus D^{J-1} \oplus \cdots \oplus D^1 \oplus D^0 \oplus C^0}_{\text{Kept}}$$

$$\approx \underbrace{D^J \oplus D^{J-1} \oplus \cdots \oplus D^1 \oplus D^0 \oplus C^0}_{N}.$$

In this algorithm, a method of the non-fixed length code is used in encoding these remained details as follows:

- Give a quantification level $L$ and the maximum absolute value (max) in the residual details.
- Calculate the quantified value $x$ of the residual details by the equation $x = [(L * V_D)/\max + 0.5]$, where $V_D$ stands for the value of the residual details and $[.]$ denotes obtaining the integer value.
- In encoding the quantified value $x$, define two expressions $l = [\log_2(|x|)]$ and $y = (|x| - 2^l)$ except the value $x$ equating zero. There are two situations to be discussed. If $x = 0$, the code is represented by one byte. If $x < 0$ or $x > 0$, the $x$ code is composed of three parts, such as $\underbrace{0, 0, \ldots, 0, 1}_{l+1}, \text{flag}, \underbrace{(y)_2}_{l}$. Here, the first part shows the code's unique feature, the second part flag is a positive/negative flag, and the final part is a binary value with $l$ representing the quantified value $x$.

As the encoding method mentioned above, some examples are listed in the Table 12.1. At the same time, the compressed data in the Algorithm record information as follows:

- First, record the length of the original control points and the number of decomposition layers $N$.
- Then, store the quantification level $L$ and the value max.

Table 12.1.   Non-fixed length code.

| Quantified value | Positive value code | Negative value code |
|---|---|---|
| 0 | 1 | 1 |
| 1 | 010 | 011 |
| 2 | 00100 | 00110 |
| 3 | 00101 | 00111 |
| 4 | 0001000 | 0001100 |
| 5 | 0001001 | 0001101 |
| ... | ... | ... |
| 8 | 000010000 | 000011000 |
| ... | ... | ... |
| 15 | 000010111 | 000011111 |
| ... | ... | ... |
| $x$ | $\underbrace{0,0,\ldots,0,1,0,}_{l+1}\underbrace{(y)_2}_{l}$ | $\underbrace{0,0,\ldots,0,1,1,}_{l+1}\underbrace{(y)_2}_{l}$ |

- Finally, record the code of the details at different layers and the low-frequency components of the curve $C^0$.

## 12.1.2  *Algorithm 2 (Local Approach)*

The second algorithm is a specific one for the curves. The control points of a curve are decomposed to $C^{n-1}, C^{n-2}, \ldots, C^0$ by the wavelet transform, which approximates the original curve $S^n$. Those points at the different decomposition layers reflect the characteristics of the sub-curve approximation. The control points in $C^k$ are more close to the original curve than those in $C^{k-1}$. In a cubic B-spline expression, four control points describe a sub-curve. Suppose that at the $j$th layer, $s_i^j$ is a sub-curve, which is determined by four control points $c_i^j, c_{i+1}^j, c_{i+2}^j, c_{i+3}^j (0 \le j \le (n-1), 0 \le i \le 2^j)$, the sub-curve $s_i^j$ corresponds to exactly two (more detailed) curves $s_{2i}^{j+1}$ and $s_{2i+1}^{j+1}$ at the $(j+1)$th layer. Consequently, in this case, the number of control points are added up to 5 from 4. In Fig. 12.2, the sub-curve $s_0^2$ at the second layer corresponds to two curves both $s_0^3$ and $s_1^3$ at the third layer, i.e., the curve $s_0^2$ is the approximation of two curves both $s_0^3$ and $s_1^3$. As the layers increase, the approximation to the original curve becomes better. Hence, under a certain error, the entire original curve can be estimated by many sub-curves established by control points separately at the different layers. The steps of the algorithm are listed as follows:

**Step 1:** Given an error, the approximation process starts from the lowest layer;

**Step 2:** At the $j$th layer, the error between the approximated sub-curves and the original ones is checked at each segment;

**Step 3:** If the error is less than the given error in a certain segment, then the control points of this segment remain, otherwise, the control points at the $(j+1)$th layer are chosen, and we repeat the second step, and so on, until the whole curve can be described with control points at the different layers.

In this algorithm, the compressed data are organized as follows:

- The first byte records the length of curve $N$, i.e., how many segments (sub-curves) on the entire curve have been divided in the approximation.
- The successive bits are used to store the position information of the control points at each decomposition layer. For example, the approximated

curve described by control points of $k$th segment at the $j$th layer, $k$ is the position information. According to incremental change of the position information, the information code of each layer is performed below: (1) To compute the difference value of the position information $k$; (2) to code the difference value according to the non-fixed length code (Table 12.1).

- The last few bits are utilized to quantize and code the control points of each decomposition layer.

### 12.1.3 *Experiments*

The process of our experiment is listed as follows:

(1) Extract the contours (closed curves) from the bitmap image of a Chinese character.
(2) Check the number of coordinate points (or control points). The number of points is usually 2 to the power $n$, exactly, curve $S^n(t)$ should have $2^n + 3$ control points. When the control points are less than $2^n + 3$, the coordinate points must be extended as well as the control points. The performance of the extension can be periodical spinning out or adding zeros or re-sampling.
(3) Convert each contour into a cubic B-spline curve.
(4) Apply the wavelet transform to control points $C^n$, and the wavelet transform coefficients are produced.
(5) Choose adequate coefficients according to the above algorithms.
(6) Organize compression data by the above approaches.

In this experiment, $120 \times 120$ bitmap images of characters (1,800 bytes) with four typeface fonts (Song, Fang Song, Kai, and Hei) are chosen. In Algorithm 1, the quantification level $L$ mainly affects the compression ratio. In the second algorithm, the given error also has an enormous effect on compression ratio. Table 12.2 shows the compression results of Algorithms 1 and 2 as well as the traditional methods. In this table, the quantization level is in the first algorithm is 32 or 64. The given error range in the second algorithm is 1 or 2 pixels. Examples of the compression of Chinese typeface are illustrated in Figs. 12.3 and 12.4. The compression result using the first algorithm is shown in Fig. 12.3, and that using the second algorithm is displayed in Fig. 12.4.

Table 12.2.　Results of Algorithms 1 and 2 as well as the traditional methods.

| Font (1,800) | CR (Algorithm 1) L = 32 | CR (Algorithm 1) L = 64 | CR (Algorithm 2) $\varepsilon = 2$ |
|---|---|---|---|
| Songti | 6.59 | 6.12 | 8.45 |
| F-songti | 7.28 | 6.41 | 9.20 |
| Kaishu | 5.46 | 4.85 | 7.69 |
| Black | 6.29 | 5.47 | 7.79 |
| Average | 6.41 | 5.71 | 8.28 |

| Font (1,800) | CR (Algorithm 2) $\varepsilon = 1$ | Method of white and black | Method of outline link-code |
|---|---|---|---|
| Songti | 5.96 | 2.25 | 3.50 |
| F-songti | 6.45 | 1.93 | 2.61 |
| Kaishu | 5.90 | 2.03 | 3.74 |
| Black | 5.69 | 2.67 | 4.83 |
| Average | 6.00 | 2.22 | 3.67 |

*Note*: CR = Compression ratio.

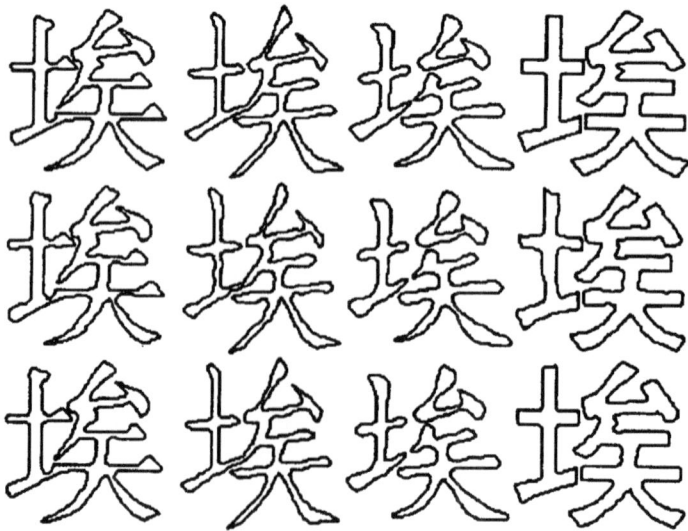

Fig. 12.3　The compressed results using the first algorithm: The first row is an original Chinese character with four fonts; the second row shows the compression result with the quantization level $L = 32$; the third one is the result with quantization level $L = 64$.

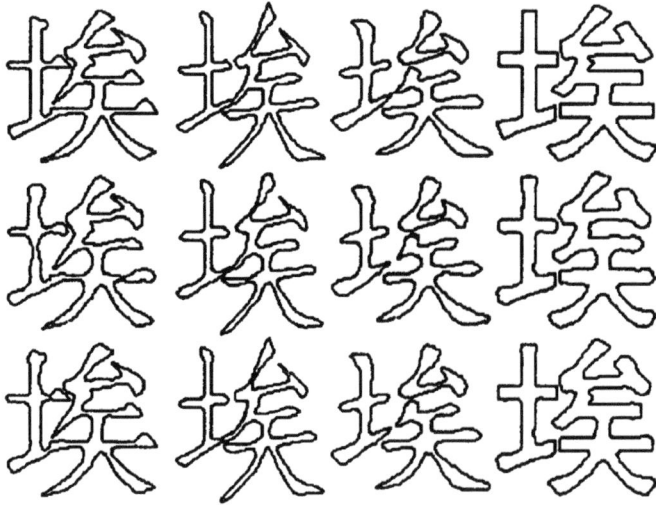

Fig. 12.4    The compressed results using the second algorithm: The first row is an original Chinese character with four fonts; the second row shows the compression result one with the error $\varepsilon = 2$; the third one is the result with the error $\varepsilon = 1$.

## 12.2   Enlargement of Type Size with Arbitrary Scale Based on Wavelet Transform

A font has a particular size and style of type. The size modification of Chinese characters, including the reduction and enlargement, is one of the most useful technologies in Chinese character processing systems. Compared with the font zooming out, the font enlargement is more difficult to be realized. Recently, some effective methods, such as the homogeneous coordinate and the logical equation interpolation, have been proposed to deal with the enlargement of the type size of Chinese characters with arbitrary (smooth) scale. This section aims at a new method for font enlargement with arbitrary scale. In this way, a bitmap image is converted to one or more contours, then the arbitrary scale enlargement is performed to these contours based on wavelet transform.

### 12.2.1   *Algorithms*

In practice application, when a curve (line) is displayed or printed on the output device, it is formed by a finite amount of unit points, such as pixels or dots. For a certain curve, if an amount of unit points is fixed, the higher the resolution of the output device, the smaller the curve is displayed or printed, and vice versa. Thus, under the same resolution, for zooming in on a curve, it is necessary to increase the amount of unit points by adding some pixels to the original curve. Several techniques, such as interpolation and complement, can be used to perform this task.

According to wavelet transform theory, we know that the wavelet reconstruction is performed by interpolation. Is it possible to enlarge a curve by a wavelet reconstruction? The answer is definite. In order to do so, we should keep the distance between two neighboring control points on a curve unchangeable. Otherwise, the length of the reconstructed curve will not be changed.

Let $C^n$ denote the control points of an original curve and $C^{n+1}$ be that of its enlarged curve. Based on the interpolation theory, we have known that

$$C^{n+1} = P^{n+1} \cdot C^n + Q^{n+1} \cdot D^n, \tag{12.5}$$

where $P^{n+1}$ and $Q^{n+1}$ are two filters. The particulars, $D^n$, contained on the original curve can be viewed as zero, therefore, the control points of the new curve are

$$C^{n+1} = P^{n+1} \cdot C^n. \tag{12.6}$$

Equation (12.6) can be considered to be a procedure of interpolation by wavelet reconstruction based on the cubic B-spline function. According to the wavelet reconstruction, the control points on a curve are doubled after the interpolation. In this way, an original curve can be enlarged at any $2^j$ times, where $j$ is a positive integer. It means that the scale of the enlargement is always a multiple of 2. A question is how the Chinese type size can be enlarged with arbitrary scale. We solve this question in this section, where two algorithms are proposed:

- The first one is used to perform the enlargement of a curve with arbitrary scale by a cubic B-spline wavelet transform.
- The second algorithm is employed to accomplish the arbitrary enlargement of Chinese type size using the first algorithm associated with other techniques.

**(1) Algorithm 1:** Let $S^j$ and $S^{j+1}$ represent two enlarged curves, the sizes of which are $2^j$ and $2^{j+1}$ times the original curve, respectively. Our object is to find a curve $S^{j+t}(0 \leq t \leq 1)$ between $S^j$ and $S^{j+1}$, as shown in Fig. 12.5. It can be seen that, after wavelet reconstruction, a control point

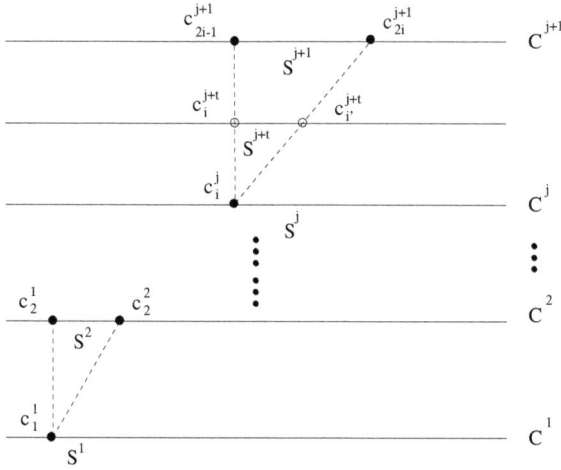

Fig. 12.5 Curve $S^{j+t}$ is produced by the different wavelet transform layers.

$c_i^j$ on the curve $S^j$ ($1 \leq i \leq 2^j + 1$) corresponds to two control points $c_{2i-1}^{j+1}$ and $c_{2i}^{j+1}$ on the curve $S^{j+1}$, thus, one point becomes two. Note that, in this section, the control points can be considered to be the coordinate points. For example, in Fig. 12.5, $c_1^1$ on $S^1$ corresponds to $c_1^2$ and $c_2^2$ on $S^2$. If we draw two lines from $c_i^j$ to $c_{2i-1}^{j+1}$ and $c_{2i}^{j+1}$, respectively, we can obtain two cross points $c_i^{j+t}$ and $c_{i'}^{j+t}$ on the curve $S^{j+t}$, which can be computed with line equations as follows:

$$c_i^{j+t} = c_i^j + t \cdot (c_{2i-1}^{j+1} - c_i^j)$$
$$= (1-t) \cdot c_i^j + t \cdot c_{2i-1}^{j+1}, \qquad (12.7)$$

$$c_{i'}^{j+t} = c_i^j + t \cdot (c_{2i}^{j+1} - c_i^j)$$
$$= (1-t) \cdot c_i^j + t \cdot c_{2i}^{j+1}. \qquad (12.8)$$

Note that, we should consider a special case, i.e., the distance between the control points $c_i^{j+t}$ and $c_{i'}^{j+t}$ is less than a unit point (a pixel). In this situation, we should combine these two control points into a single one.

Consequently, the control points $C^{j+t}$ on the curve $S^{j+t}$ can be obtained by the control points $C^j$ and $C^{j+1}$ on the curves $S^j$ and $S^{j+1}$ and can be written by

$$C^{j+t} = C^j + (C^{j+1} - C^j) \cdot t$$
$$= (1-t) \cdot C^j + t \cdot C^{j+1}. \qquad (12.9)$$

It is clear that two special cases also should be considered: (1) $S^{j+t} \to S^j$ when $t \to 0$; (2) by contrast, $S^{j+t} \to S^{j+1}$ when $t \to 1$.

Finally, we have the following algorithm for the enlargement of a curve with arbitrary scale.

---

**Algorithm 1 (Enlargement of Curves).**

**Step 1:** For a given scale $d$, compute $j$ and $t$ by $2^j \leq d \leq 2^{j+1}$ and $t = (d - 2^j)/2^j$.

**Step 2:** According to formula (12.6), enlarge the original curve to $2^j$ and $2^{j+1}$ times with the wavelet reconstruction based on the cubic B-spline function.

**Step 3:** Obtain $C^{j+t}$ by formula (12.9).

---

**(2) Algorithm 2:** A Chinese character is composed of a set of contours (closed curves). Therefore, the enlargement of the entire character can be performed by treating each closed curve using Algorithm 1 separately. In practice, it is found that the jaggy and staircase phenomenon emerges on the contour when a curve is directly enlarged without any preprocessing. An example of the character with such a jaggy and staircase phenomenon can be found in Fig. 12.6(e). To avoid this phenomenon, it is necessary to remove the jaggy and staircase before we enlarge a curve. We have known that the decomposition transform functions as a filter, which can eliminate the jaggy and staircase phenomenon. For this reason, the wavelet decomposition with the cubic B-spline function is employed to perform this task.

In conclusion, we have the following algorithm for the enlargement of a Chinese type size with an arbitrary scale based on the cubic B-spline wavelet transform.

---

**Algorithm 2 (Enlargement of Chinese type Size).**

**Step 1:** Divide a Chinese character into a set of connected domains in terms of some regular techniques of image processing.

**Step 2:** Extract the edge of each connected domain by the edge detection algorithm, which has been presented in Chapter 11.

**Step 3:** Trace the contour of each edge, and as a result, a set of closed curves (contours) can be obtained for the Chinese character.

---

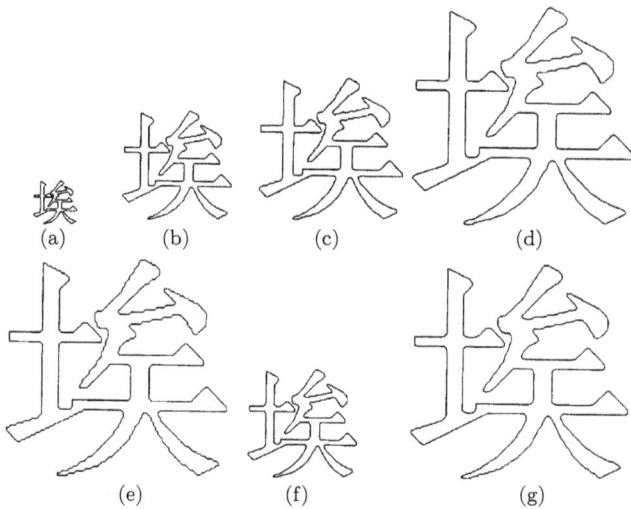

Fig. 12.6   Results and comparison of enlarged fonts: (a) Original font, (b) enlarged two times, (c) enlarged three times, (d) enlarged four times, (e) directly enlarged four times without step 5 in Algorithm 2, (f) enlarged two times by traditional interpolation, and (g) enlarged four times by traditional interpolation.

**Step 4:** Perform the pre-processing of each curve with wavelet decomposition, first, to decompose $C^n$ into $C^{n-1}$ by

$$C^n = D^{n-1} \oplus C^{n-1}.$$

Let $D^{n-1} = 0$, which means we remove the particulars (high-frequency components). Then, the preprocessing can be implemented by reconstructing $C^n$ with formula (12.6).

**Step 5:** Enlarge each curve with a given scale by Algorithm 1, which has been described in this section.

**Step 6:** After enlarging the contours, fill the domains, which are enclosed by the contours.

**Step 7:** Repeat Steps 4–6, until all contours have been processed.

## 12.2.2 *Experiments*

In our experiments, $100 \times 100$ bitmap images of Chinese characters are selected to be processed using both Algorithm 2 and a traditional algorithm, which is an interpolation algorithm with a Bezier curve equation, respectively. The original image of a character is scanned into a computer system, and its contour can be extracted by the edge detection technique and contour tracing algorithm. The contour points of an original curve are chosen as its control points $C^n$. The results of our experiments are illustrated in Fig. 12.6, in that (a) is an original font, (b)–(d) are the results, where the original font is enlarged by two $\sim$ four times with Algorithm 2, respectively, (e) is the font which is directly enlarged from the original one by four times with Algorithm 2 without the preprocessing (Step 4) and (f)–(g) are the resulting fonts, where the original font is enlarged by two and four times, respectively, with the traditional interpolation. Clearly, the curves enlarged by Algorithm 2 are smoother than the curves enlarged by the traditional interpolation. On the other hand, the distortion produced by Algorithm 2 is less than that produced by the traditional method, especially for the corner or short line in the characters.

Figure 12.7 shows other experiments, where the handwritten Chinese characters are enlarged by the algorithm described above in this section.

In this section, we apply the cubic B-spline wavelet transform to implement the enlargement of Chinese characters with arbitrary scales. Its advantages can be concluded as follows:

(1) The algorithm is very simple and easy to be implemented. (2) The computational complexity of the algorithm is O(N). (3) The distortion of amplified font is very light because it is only related to that of the original image itself.

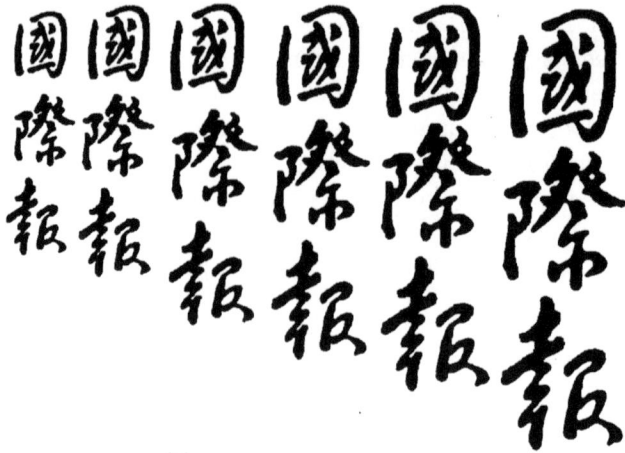

Fig. 12.7 Chinese handwritings are enlarged with arbitrary scales by the algorithms described in this section.

## 12.3 Generation of Chinese Type Style Based on Wavelet Transform

With the increasing requirement of the Chinese computing processed by the computer including the Chinese character recognition and Chinese press, the generation of type style has become another important application of the Chinese character processing systems. In this section, a new approach, which applies the cubic B-spline wavelet transform to the generation of the Chinese type style is discussed.

Usually, a new type style can be derived by modifying the structural features of an existing font. For instance, changing the length and the width of its strokes can produce a new type style. The structural features of a font correspond to the details at various wavelet decomposition layers when each contour of a character is decomposed by a wavelet transform. With the cubic B-spline wavelet transform, which is viewed as a "mathematical microscope," the details of a font at different layers can be effectively extracted. For a given font, modifying its structural features can be performed by changing its details in some layers, or composing its details with that of others. In this way, three steps are involved:

- the re-sample of the spline curves of the Chinese character,
- the wavelet transform of each curve,
- the modification or composition of the original curves to produce a new type style.

Two approaches are proposed in this section:

- The first one is to generate a new font by modifying the structural features of an original one.
- The second one is to generate a new font by composing the structural features of several existing ones.

### 12.3.1 *Modification*

Several algorithms, which can perform the modifications of the type styles, are discussed in this section.

**Algorithm 1.** When a cubic B-spline curve is utilized to fit the contour of a font with $M$ coordinate points, the number of the control points should be $M + 2$. Further, if $M \neq 2^n + 1$, the number of the control points of a cubic B-spline curve needs to be extended to $2^n + 3$. On the other hand, when two or more fonts are composed to form a new one, it is necessary to ensure that they have the same topological structure, e.g., their length, direction, start, and endpoint. In order to extend the control points with the required length and hold the structure unchangeable, it is necessary to re-sample the B-spline curve as follows:

**Step 1:** Given $M$ coordinate points of a curve, to find a cubic B-spline curve $F^n(u)$, which passes through these coordinate points,

$$F^n(u) = (f_1^n(u), f_2^n(u), \ldots, f_M^n(u)),$$

where $f_i^n(u)$ is the $i^{th}$ B-spline sub-curve, it can be written as

$$f_i^n(u) = \sum_{j=1}^{4} c_{i+j-2}^n N_{j,3}(u) \quad i = 1, 2, \ldots, M, u \in [0, 1], \quad (12.10)$$

where $N_{j,3}(u)$ are the base functions of the cubic B-spline. $c_i^n$ are the control points of the cubic B-spline curve, corresponding to the coordinate points of the curve.

**Step 2:** Select the decomposition levels $n$, in accordance with the length of the curve, $L$, satisfying $M \leq L$ and $L = 2^n - 1$.

**Step 3:** Re-sample $F^n(u)$ with a new sample interval of $\Delta t = \frac{M}{2^n - 1}$ producing a set of new coordinate points $Q_i' (i = 1, 2, \ldots, M')$, $M' = L + 1$.

**Step 4:** Owing to formula (12.10), we can obtain the extended control points $c_{i'-1}^n (i' = 1, 2, \ldots, M')$ with the length of $2^n + 3$. We have

$$F^n(u) = (f_1^n(u), f_2^n(u), \ldots, f_{M'}^n(u)).$$

The wavelet base we used here is the cubic B-spline wavelet, which can interpolate a function with equal intervals. Further, since the contour of the character is a closed curve, its start point is the same as its end point, that is, $c_0^n = c_1^n = c_{M'+2}^n = c_{M'+3}^n$. Therefore, only $2^n - 1$ points in curve $F^n(u)$ need to be re-sampled.

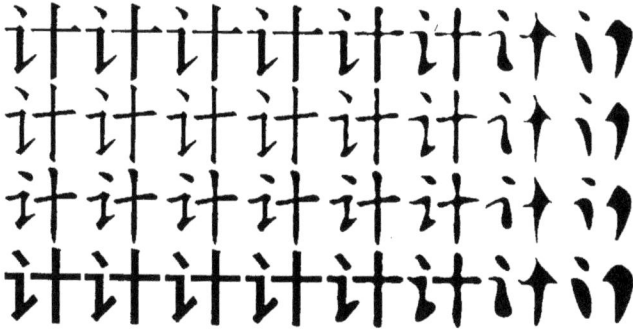

Fig. 12.8   Example 1.

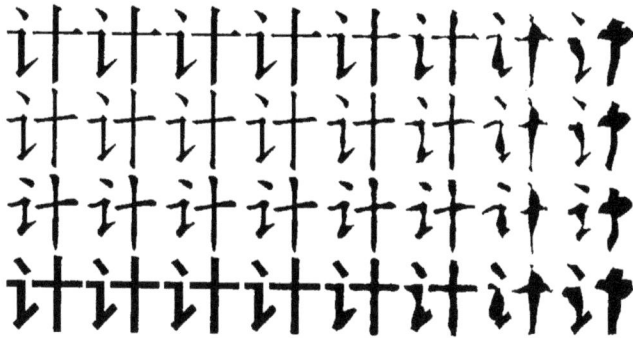

Fig. 12.9   Example 2.

Based on the re-sampled B-spline curve, the wavelet transform can be employed to modify the Chinese type styles. We have three stages to do so:

**Step 1:** Apply wavelet transform to the re-sampled B-spline curve, which has the length of $2^n + 3$.

**Step 2:** Process the details at some wavelet decomposition layers.

**Step 3:** Reconstruct a new type style by the wavelet reconstruction algorithm.

Two examples are shown in Figs. 12.8 and 12.9, in which four different fonts of a Chinese character are decomposed into seven layers by wavelet transform.

In Fig. 12.8, the modifications are conducted as follows:

- The detail $D^{n-1}$ is removed from the first layer.
- The details $D^{n-1}$ and $D^{n-2}$ are deleted from the second layer.

- The details $D^{n-1}$, $D^{n-2}$, and $D^{n-3}$ are discarded from the third layer.
- ...
- All of the details $D^{n-1}, D^{n-2}, \ldots, D^{n-7}$ are no longer kept at the seventh layer.

The results of the above modifications are illustrated from left to right in Fig. 12.8.

The modifications in Fig. 12.9 are carried out as follows:

- Only the detail $D^{n-1}$ is removed from the first layer.
- Only the detail $D^{n-2}$ is deleted from the second layer.
- Only the detail $D^{n-3}$ is discarded from the third layer.
- ...
- Only the detail $D^{n-7}$ is no longer kept at the seventh layer.

The results of the above modification are illustrated from left to right in Fig. 12.9. From these, it is easy to see that the details contained at different layers affect the features of the type style variously. The deeper the layer, the greater the effect of the details will be.

---

**Algorithm 2 (Smoothing of Curves).** Suppose an objective curve $r(t)$ contains $m = 2^j + 3$ control points: $C^j = (c_1^j, c_2^j, \ldots, c_m^j)$. According to the least-squared error, an approximated curve with $m' = 2^{j'} + 3$ control points can be obtained ($j' < j$ is a non-negative integer). From the previous section, we have known

$$C^{n-1} = A^n \cdot C^n. \tag{12.11}$$

Therefore, the control points $C'$ of the approximated curve can be expressed as

$$C' = A^{j'+1} A^{j'+2} \cdots A^j C^j. \tag{12.12}$$

A remarkable property of the multiresolution curve is its discrete nature, i.e., we can use $K$ control points to construct an approximated curve efficiently at the $j$th layer. In this way, $K$ can be any of the integers 4, 5, 7, 11, or $2^j + 3$, and $j$ can be any integer. In practice, we can also define another non-integer-layer curve $\gamma^{j+\mu}(t)$, $\mu \in R$ and $0 \le \mu \le 1$, which can be achieved by two curves, $\gamma^j(t)$ and $\gamma^{j+1}(t)$, at the neighboring integer layers, as follows:

$$\gamma^{j+\mu}(t) = (1 - \mu)\gamma^j(t) + \mu\gamma^{j+1}(t)$$
$$= (1 - \mu)\Phi^j(t)C^j + \mu\Phi^{j+1}(t)C^{j+1}. \tag{12.13}$$

This non-integer-layer curve can benefit to smooth the curve at any continuous scale. We can continuously edit any segment on this curve, which vary curve from the smoothest form (with only four control points) to its highest resolution (with $m > 4$ control points).

Suppose curve $\gamma(t)$ is to be smoothed, the concrete procedure to smooth it is presented in the following:

**Step 1:** Decompose the curve $\gamma(t)$ into a sequence of basic control points (including four basic points) and a group of details using wavelet transform.

**Step 2:** Choose the corresponding details of wavelet transform coefficients so that the requirement of the smoothness is satisfied (note that the lesser the value of $j$, the better the smoothness). Fix some chosen details from one layer or more layers to be zero, generating new details $D'^j$.

**Step 3:** Obtain a new curve $\gamma'(t)$ by wavelet reconstruction using changed details $D'^j$ and $C^0$.

---

**Algorithm 3 (Edition of the Shape and Details of the Curve).**
For this algorithm, we will only give some conclusions. The detailed discussion and inference can be found in the work of Stollnitz *et al.* [1996]. A curve, which has $C^J$ control points, can be described by wavelet coefficients $C^0$ and $D^0, D^1, \ldots, D^{J-1}$, or the control points $C^0, C^1, \ldots, C^{J-1}$ at the different layers. Two methods can be used to edit such a curve:

(1) In the first method, the global shape of the curve is changed, while the details of it are kept.
(2) The second method is just opposite the above method. The details of the curve are only altered, while the elementary shape of the curve is nearly the same as the original one.

We discuss these methods as follows:

**Method 1:** The basic idea of this method is as follows: first, some of $C^j$ at some layers are modified, then the original details $D^0, D^1, \ldots, D^{J-1}, (0 < j < J)$ are added to the modified $C^j$, and finally the global shape of the whole curve can be changed.

We shall describe the modified control points by writing $\hat{C}^j = C^j + \Delta C^j$, where $C^j$ and $\hat{C}^j$ denote the original points and modified ones, respectively. The $\Delta C^j$ indicates the difference between them. At the same time, some modification appears in $C^J$ through reconstruction, i.e.,

$$\hat{C}^J = C^J + \Delta C^J. \tag{12.14}$$

Here the $j$ value mainly affects the above modification. Precisely, the less the $j$ value, the wider the influenced range of the control points furthermore, the stronger the effect on the whole curve. Otherwise, the closer to $J$ the $j$ value is, the more narrow the influenced range of the control points is, at the same time, the less the effect on the whole curve is. This method can be extended to the non-integer layers. Suppose a non-integer-layer curve $\gamma^{j+\mu}$ satisfies (12.13), and it contains control points $C^{j+\mu}$, we have

$$\gamma^{j+\mu}(t) = \Phi^{j+1}(t) \cdot C^{j+\mu}. \tag{12.15}$$

If we modify a particular point $c_i^{j+\mu}$ in curve $C^{j+\mu}$, the positions of the neighboring points of $c_i^{j+\mu}$ will be changed. The size of the influenced range and degree is in inverse proportion to $\mu$:

- If $\mu$ closes to zero, then all control points at the $j + \mu$ layer will be simultaneously moved. In other words, every point at the $j$ layer will be edited.
- If $\mu$ approximates 1, the neighboring points will not be moved. Only a single point at the $j + 1$ layer will be edited.

Let the $\Delta C^{j+\mu}$ be the values of the modifications at the $j + \mu$ non-integer layer. Here the $\Delta c_i^{j+\mu}$, which is one of the $i$th point in $\Delta C^{j+\mu}$, is selected by the user. The $\Delta C^{j+\mu}$ can be considered as two parts, namely, $\Delta C^j$ and $\Delta D^j$ at the $j$th layer. We can define $\Delta D^j = B^{j+1} \cdot \Delta C^{j+1}$. Therefore, the modified $\Delta C^{j+\mu}$ of the whole curve is represented as follows:

$$\Delta C^J = P^J P^{J-1} \cdots P^{j+2}(P^{j+1} \Delta C^j + Q^{j+1} \Delta D^j). \tag{12.16}$$

Furthermore, from (12.13), (12.15), and $\Phi^{j-1}(u) = \Phi^j(u) \cdot P^j$, we obtain

$$\Delta C^{j+\mu} = (1 - \mu)P^{j+1} \cdot \Delta C^j + \mu \Delta C^{j+1}. \tag{12.17}$$

Considering the filter equation,

$$C^j = P^j \cdot C^{j-1} + Q^j \cdot D^{j-1},$$

where $P^j$ and $Q^j$ are metrics of the filters. By replacing $\Delta C^{j+1}$ with both $\Delta C^j$ and $\Delta D^j$, we can rewrite (12.17) as

$$\Delta C^{j+\mu} = P^{j+1} \cdot \Delta C^j + \mu Q^{j+1} \Delta D^j. \tag{12.18}$$

From (12.18), it is clear that the modified control points at the non-integer layer can be described by two components: (1) the modified control points at the lower layer and (2) the modified details at the lower layer. The entire modified curve can be determined by both (12.16) and (12.14). If $\Delta C^{j+\mu}$ is known, it is quite difficult to calculate directly $\Delta C^j$ and $\Delta D^j$ from (12.18). In practice, we can denote $\Delta C^{j+\mu} := (0,\ldots,\Delta c_i^{j+\mu},0,\ldots,0)^T$ and define

$$\Delta C^j = (1-\mu)A^{j+1} \cdot \Delta C^{j+\mu},$$

$$\Delta D^j = \mu B^{j+1} \cdot \Delta C^{j+\mu}. \tag{12.19}$$

The main editing procedure can be implemented as follows:

**Step 1:** Determine editing vector $C^{j+\mu} = (0,\ldots,\Delta c_i^{j+\mu},0,\ldots,0)^T$.
**Step 2:** Compute corresponding $\Delta C^j$ and $\Delta D^j$, according to (12.19).
**Step 3:** Obtain the bias of the curve, according to (12.16).
**Step 4:** Compose the new curve, according to (12.14).

**Method 2:** This method is just the opposite process to that described above. In Method 1, the global shape of the curve is altered, but the detail of it can be kept. However, in this method, the details of the curve are changed, and the basic shape of it is nearly the same as the original curve. The basic idea of this method is that we only modify the details $D^j, D^{j+1},\ldots,D^{J-1},(0 < j < J)$ and preserve the low-resolution components of the curve, $C^0, C^1,\ldots,C^j,(0 < j < J)$. Here the original details $D^j, D^{j+1},\ldots,D^{J-1},(0 < j < J)$ are replaced by a group of new details $\hat{D}^j, \hat{D}^{j+1},\ldots,\hat{D}^{J-1},(0 < j < J)$. This editing method needs to establish a library, which stores various content about curve details. They are samples of the standard curves, such as folding line, spiral and snake line. When the object curve is to be edited, we can perform the edition by the following steps:

**Step 1:** Choose some standard samples from the library, in accordance with the requirement of the edition.
**Step 2:** Decompose these curves by the wavelet transform.
**Step 3:** Extract the details, $D_n^k$, from one or several layers. They can represent the details of the shape efficiently, and we have $\hat{D}_i^j = \zeta(D_k^k,l,k)$.
**Step 4:** Add these details to the original low-frequency coefficients, to obtain the designed curve, i.e., $C^J = P^J \cdot C^{J-1} + Q^J \cdot \hat{D}^{J-1}$.

Fig. 12.10   Results of modifying the typeface of the Chinese character "ji": (a) Original character, where 1, 2, and 3 describe three closed curves, which can be transformed with wavelet transform; (b)–(d) the result of the local edition.

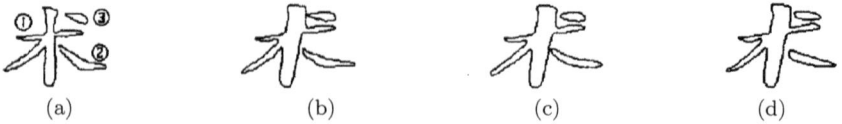

Fig. 12.11   Results of modifying the typeface of the Chinese character "shu": (a) Original character, where 1, 2, and 3 describe three closed curves, which can be transformed with wavelet transform; (b)–(d) the result of the local edition.

$180 \times 80$ bitmap images of Chinese characters are chosen as the basic samples in this experiment. The coordinate $x$ and $y$ in the contour of a character is considered to be the control points, i.e., $C_x^n$ and $C_y^n$. According to the above method, we can process Chinese characters by the local edition. Figures 12.10 and 12.11 show the experimental results.

In Figs. 12.10 and 12.11, each character consists of three contours, which are labeled by the numbers 1, 2, and 3 as shown in the figure. The result of the local edition is displayed in Fig. 12.10 using the following procedure:

- For label one, the point $c_i^3$ at the third layer is tensioned in both $x$ and $y$ directions.
- For label two, the point $c_i^4$ at the fourth layer is tensioned.
- For label three, the point $c_3^7$ at the seventh layer is tensioned.

Due to the extent of the modification, there is little difference among (b)–(d) in Fig. 12.10.

For the Chinese character in Fig. 12.11, the following modification is done:

- For label one, the control point $c_i^8$ at the eighth layer is tensioned toward the directions of down-left and up-right. So an italic type style is obtained. The results is shown in Fig. 12.11(b).
- For label two, the high-frequency component (details) will be removed. The result is shown in Fig. 12.11(c).

- The character is modified in accordance with the above two operations. Figure 12.11(d) illuminates the result.

## 12.3.2 *Composition*

Another way of creating new fonts is to compose the details of two or more existing fonts with different ratios. The new type style, which contains a mixture of details, can be restored by wavelet reconstruction. The composition algorithm based on the cubic B-spline wavelet transform is presented as follows:

**Step 1:** Divide each Chinese character into many separate connected domains (strokes). Then, extract the edge of each connected domain to produce a contour, and thereafter process each contour by Steps 2–6.

**Step 2:** Represent a contour by a cubic B-spline curve with $2^n + 3$ control points by re-sample, where $n$ is the number of the decomposition levels.

**Step 3:** Apply wavelet decomposition to control points $C^n$ with the cubic B-spline wavelet transform so that the detail information $D^0, D^1, \ldots, D^{n-1}$ at different layers and the control points $C^0$ at the lowest layer can be obtained.

**Step 4:** Compose the details of $m$ ($m \geq 1$) fonts at certain layers, and the detail information can be found as follows:

$$D^j_{New} = \sum_{i=1}^{m} r_i D^j_i,$$

where $D^j_i$ ($i = 1, 2, \ldots, m$) are the font details at the $j$th layer and $r_i$ ($i = 1, 2, \ldots, m$) are ratio factors, $0 \leq r_i \leq 1$ ($i = 1, 2, \ldots, m$).

**Step 5:** Reconstruct a new set of control points $C^n_{New}$ with new details $D^j_{New}$ and a set of control points $C^j$ at the $j$th layer by wavelet reconstruction. The length of $C^n_{New}$ is $L = 2^n + 3$.

**Step 6:** Re-sample the B-spline curve $F^n_{New}(u)$ formed by control points $C^n_{New}$. It is said to convert the B-spline curve $F^n_{New}(u)$ to a curve (contour) with the length of $M$ again. Here, the sample interval is $\Delta't = \frac{2^n - 1}{N}$.

**Step 7:** Fill in all the connected domains surrounded by the contours. The new Chinese type style is completed.

Two examples can be found in Figs. 12.12 and 12.13. Four Chinese type styles are presented in these examples, namely, Song (the first row in

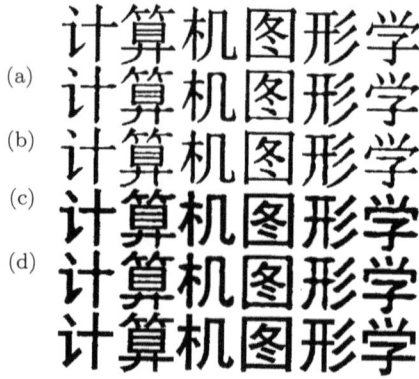

Fig. 12.12    Results of combining two typefaces. The first row and the last row are original typefaces.

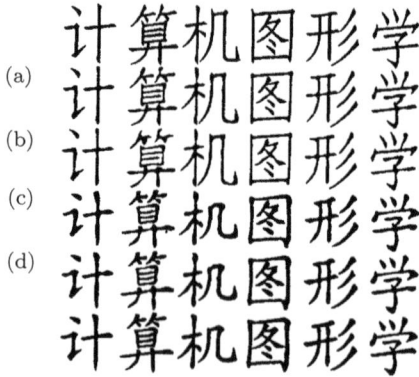

Fig. 12.13    Results of combining several typefaces. The first row and the last row are original typefaces.

Fig. 12.12), Hei (the last row in Fig. 12.12), Fang Song (the first row in Fig. 12.13), and Kai (the last row in Fig. 12.13). The size of the images is $200 \times 200$.

In Fig. 12.12, we have the following:

- Characters in groups (A) and (B) are obtained by removing the details of Song and Hei at the 1–4th layers, respectively.
- Ones in (C) and (D) are the resulting fonts, which are composed by the type style of Song and Hei.

In Fig. 12.13, we have the following:

- Characters in group (A) are the composition results of three fonts, namely, Fang Song, Kai, and Song, with ratios $r_1 = 0.2$, $r_2 = 0.4$, $r_3 = 0.4$, and the control points $C^{n-4}$ at the lowest decomposition layer is in the style of Fang Song.
- Ones in group (B) are the composition results of the four fonts with ratios $r_1 = 0.1$ (Fang Song), $r_2 = 0.3$ (Kai), $r_3 = 0.3$ (Song), $r_4 = 0.3$ (Hei), and $C^{n-4}$ is in Fang Song.
- Group (C) are the composition results of the four fonts with ratios $r_1 = 0.1$ (Kai), $r_2 = 0.3$ (Fang Song), $r_3 = 0.3$ (Song), $r_4 = 0.3$ (Hei), and $C^{n-4}$ is in the font of Kai.
- The type styles in group (D) are the composition results of three fonts with ratios $r_1 = 0.2$ (Kai), $r_2 = 0.4$ (Fang Song), $r_3 = 0.4$ (Song), and $C^{n-4}$ is in the type style of Kai.

# Chapter 13

# Classifier Design Based on Orthogonal Wavelet Series

Conceptually speaking, the methodology of statistical pattern recognition draws on classic Bayesian statistical decision theory. It essentially deals with two categories of problems; one is concerned with the identification and extraction of significant features for describing features and another with the design of classifiers for discriminating and recognizing patterns. In this chapter, we provide an in-depth examination of the classifier design problem. First, we present an overview of the fundamentals of pattern classifier design. In so doing, our emphasis will be on *minimum average-loss classifier design* and *minimum error-probability classifier design*. Next, we specifically describe and discuss the use of orthogonal wavelet series in classifier design.

## 13.1 Fundamentals

Without loss of generality, we shall only consider two-class pattern recognition problems. That is to say, the complete set of patterns to be classified, $\Omega$, is composed of patterns of two classes, $\Omega_1$ and $\Omega_2$. For instance, $\Omega$ may contain samples of human cells, and $\Omega_1$ and $\Omega_2$ may correspond to non-cancer and cancer cells, respectively. In other words, $\Omega_1 \cup \Omega_2 = \Omega$ and $\Omega_1 \cap \Omega_2 = \emptyset$. If based on some medical statistics, we observe that the priori probabilities of not having and having cancers in a certain region are $P_1$ and $P_2$, respectively, then we also know that the probabilities of $\Omega_1$ and $\Omega_2$ occurrences will be $P_1$ and $P_2$, and furthermore, $P_1 + P_2 = 1$. In order to classify these samples (e.g., to identify cancer cells), it is a common approach that we select a set of features to form a feature vector, $X$, which may be symbolically expressed as follows:

$$X = (X_1, X_2, \ldots, X_n), \tag{13.1}$$

where $X$ denotes an $n$-dimensional random vector that defines the following mapping:

$$(\Omega, \mathcal{F}, P) \xrightarrow{X(\omega)} (\mathbb{R}^n, \mathcal{B}^n, P_{X^{-1}}), \qquad (13.2)$$

where $\Omega$ and $\mathbb{R}^n$ correspond to the domain and range of $X$. $\mathbb{R}^n$ is an $n$-dimensional Euclidean space. $\mathcal{F}$ denotes a $\sigma$-field constructed by the subsets of $\Omega$. $P$ denotes a probability measure defined over $\mathcal{F}$ and determined by priori probabilities. $\mathcal{B}^n$ denotes a Borel $\sigma$-field in $\mathbb{R}^n$. $P_{X^{-1}}$ is a probability measure derived from random variable $X$ in feature vector space $(\mathbb{R}^n, \mathcal{B}^n)$; whose probability density function is denoted as $p(x)$, $x = (x_1, x_2, \ldots, x_n) \in \mathbb{R}^n$. Suppose that $p(x \mid j)$, the conditional probability density function of each pattern $\Omega_j$ is given. Then we can have the following expression:

$$p(x) = P_1 p(x \mid 1) + P_2 p(x \mid 2). \qquad (13.3)$$

In what follows, we shall introduce the notion of decision function.

**Definition 13.1.** Function $d(\omega)$ is called a decision function if and only if it satisfies the following conditions: $d(\omega)$ is defined in domain $\Omega$, and its corresponding values are given in set $\{1, 2\}$. In other words,

$$(\Omega, \mathcal{F}, P) \xrightarrow{d(\omega)} \{1, 2\}. \qquad (13.4)$$

In addition, $d^{-1}(\{i\}) \in \mathcal{F}$, $i = 1, 2$, where $d^{-1}(\cdot)$ denotes the inverse of $d(\cdot)$ and $\{i\}$ denotes a single-number set.

From Definition 13.1, we note that $d(\omega)$ is a discrete random variable. When $d(\omega) = i$, we say that $\omega$ belongs to the $i$th pattern, $\Omega_i$.

Note that here we make a decision on whether or not a pattern belongs to a specific class depending on the value of the feature vector. In other words, we have to consider a Borel measurable function $c(x)$ in feature vector space $\mathbb{R}^n$, whose value is given by $\{1, 2\}$:

$$(\mathbb{R}^n, \mathcal{B}^n) \xrightarrow{c(x)} \{1, 2\}. \qquad (13.5)$$

The above-mentioned decision function $d(\omega)$ is merely the composite of random feature vector $X$ and the Borel measurable function $c(x)$, i.e.,

$$d(\omega) = c(X(\omega)). \qquad (13.6)$$

When $x$, the observed sample value of the feature vector for pattern $\omega$, satisfies $c(x) = i$, we can arrive at the conclusion that pattern $\omega \in \Omega_i$. In this respect, we usually also refer to the medium function $c(x)$ as a decision function.

It can be noted that since classifications based on $c(x)$ are in essence statistical decisions, they are inevitably subject to statistical errors. In real-life applications, misclassifications can cause damages of varying degrees. For instance, in character recognition, it is sometimes possible to misclassify the letter $c$ into the letter $d$ and vice versa. In both cases, the damages caused may not seem to be as serious as misclassifying non-cancer cells into cancer ones in cancer diagnosis. The misclassification of cells could make patients devastated, and they may spend a fortune on their medication. The situation can be even worse if we misclassify cancer cells into non-cancer ones. In such a case, we may delay necessary medical treatments for the patients. As a result of this, the patients may lose their lives. Owing to the above considerations, it is important to explicitly define a loss function in order to reflect the degree of damage caused by misclassification. Before we formally define a loss function, let us first introduce the notion of a class function as follows.

**Definition 13.2.** $J(\omega)$ is called a class function if and only if it satisfies the following:

$$(\Omega, \mathcal{F}, P) \xrightarrow{J(\omega)} \{1, 2\}, \tag{13.7}$$

where

$$J(\omega) = \begin{cases} 1 & \text{if } \omega \in \Omega_1, \\ 2 & \text{if } \omega \in \Omega_2, \end{cases}$$

$J(\omega)$ is a discrete random variable.

We can now give the formal definition of a loss function based on those of decision function $d(\omega)$ and class function $J(\omega)$.

**Definition 13.3.** Let bivariate function $L(i, j)$ be defined as follows:

$$L(i, j) = c_{ij}, \ i, j \in \{1, 2\}, \ c_{ij} \in (-\infty, +\infty).$$

If we compose $L(i, j)$ with the random vector made of decision function $d(\omega)$ and class function $J(\omega)$, i.e., $(d(\omega), J(\omega))$, we will have composite random variable $L(d(\omega), J(\omega))$. This variable is referred to as a loss function. $c_{ij}$ is referred to as the loss caused by misclassifying pattern $\Omega_j$ into pattern $\Omega_i$.

Since $d(\omega)$ is the composite of medium function $c(x)$ and feature vector $x(\omega)$, we can write the following:

$$L(d(\omega), J(\omega)) = L(c(X(\omega)), J(\omega)). \tag{13.8}$$

In this way, we can also view $L(c(X), J)$ as the composite of function $L(c(\cdot), \cdot)$ and random vector $(X, J)$. The joint distribution of $(x, J)$ can be given as follows:

$$p(x, j) = P_j p(x \mid j) \quad j = 1, 2. \tag{13.9}$$

Based on the above interpretation of loss function, it becomes quite convenient to calculate an average loss.

**Definition 13.4.** The mathematical expectation of $L(c(x), J)$ is referred to as an average loss, which can be written as

$$R = E[L(c(X), J)]$$

$$= \sum_{j=1}^{2} \int_{\mathbb{R}^n} L(c(x), J) p(x, j) dx$$

$$= \sum_{j=1}^{2} \int_{\mathbb{R}^n} L(c(x), J) P_j p(x \mid j) dx.$$

From the above definition, it can be noted that given the priori probability of a pattern, $P_j$, and the conditional probability, $p(x \mid j)$, selecting different decision functions $c(x)$ can result in different average losses. Practically speaking, we hope to have the average loss as little as possible. Therefore, we are particularly interested in the problem of how to find $c_0(x)$ such that the average loss is the minimum. In what follows, we shall attempt to provide a detailed solution to this problem.

For ease of description, we shall consider only two-class pattern recognition problems, and assume that priori probability $P_i$, conditional probability density function $p(x \mid j)$, and the value of a lose function, $c_{ij}$, are known.

## 13.2 Minimum Average Loss Classifier Design

In statistical pattern recognition, since the feature vector for a pattern is a random variable, there is always a probability of committing errors no matter which decision scheme we choose to apply in classifications. To put it more accurately, the problem that we shall focus on here is how to compare the strengths and weakness of various decision methods from the point of view of certain statistical criteria and thereafter under such criteria to find the best solution to our problem.

In order to make our discussions more concise, apart from the assumption that $P_j$, $p(x \mid j)$, and $c_{ij}$ are known, we further assume that the damage caused by misclassifications is greater than that by correct classifications, that is, $c_{12} > c_{22}$ and $c_{21} > c_{11}$. Under these assumptions, we attempt to find the optimal decision function, $c_0(x)$, such that the criterion of minimizing $R = E[L(c(X), J)]$ is satisfied:

$$R_0 = E[L(c_0(X), J)]$$
$$= \min_{c(x)} E[L(c(X), J)],$$

where decision function $c_0(x)$ is referred to as the minimum average loss decision function.

The decision function $c(x)$ is essentially a Borel measurable function whose value is given by $\{1, 2\}$, i.e.,

$$(\mathbb{R}^n, \mathcal{B}^n) \xrightarrow{c(x)} \{1, 2\}. \tag{13.10}$$

Now, let $B_i = c^{-1}(\{i\})$, $i = 1, 2$, where $c^{-1}(\cdot)$ denotes the inverse of $c(\cdot)$. Thus, we can have $B_1 \cap B_2 = \emptyset$, $B_1 \cup B_2 = \mathbb{R}^n$. As a result, we can reduce the problem of finding decision function $c(x)$ to that of decomposing the complete pattern feature vector space, $\mathbb{R}^n$, into two non-intersecting Borel measurable sets, $B_1$ and $B_2$. $B_1$ and $B_2$ are called decision regions; when $x \in B_i$, we define $c(x) = i$. Once decision region $B_i$ is determined, the average loss can be rewritten as follows:

$$R = \sum_{j=1}^{2} \sum_{i=1}^{2} \int_{B_i} L(c(x), J) P_j p(x \mid j) dx$$

$$= \sum_{j=1}^{2} \sum_{i=1}^{2} \int_{B_i} c_{ij} P_j p(x \mid j) dx$$

$$= c_{11} P_1 \int_{B_1} p(x \mid 1) dx + c_{12} P_2 \int_{B_1} p(x \mid 2) dx$$

$$+ c_{21} P_1 \int_{B_2} p(x \mid 1) dx + c_{22} P_2 \int_{B_2} p(x \mid 2) dx.$$

Since

$$\int_{B_2} p(x \mid j) dx = 1 - \int_{B_1} p(x \mid j) dx \quad j = 1, 2, \tag{13.11}$$

it is obvious that

$$R = c_{21}P_1 + c_{22}P_2 + \int_{B_1} \{[P_2(c_{12} - c_{22})p(x \mid 2)]$$

$$-[P_1(c_{21} - c_{11})p(x \mid 1)]\}dx.$$

Note that in the above expression, $c_{21}P_1 + c_{22}P_2$ is a constant, and $P_j$, $p(x \mid j)$, and $c_{ij}$ are given. The only thing changeable in the integral term is the region of integral, $B_1$. In other words, we can only change $B_1$ in order to change the average loss $R$. At the same time, we may also observe that two integrands $P_2(c_{12} - c_{22})p(x \mid 2)$ and $P_1(c_{21} - c_{11})p(x \mid 1)$ are both non-negative. If we are to change $B_1$ in order to find the minimum average loss, we must realize that $B_1$ will include only those feature vectors $x$ at which integrands $P_2(c_{12} - c_{22})p(x \mid 2)$ is smaller than $P_1(c_{21} - c_{11})p(x \mid 1)$. That is to say, the selected $B_1$ and the corresponding $B_2$ $(B_2 = \mathbb{R}^n - B_1)$ will allow to establish the following relationships:

$$P_2(c_{12} - c_{22})p(x \mid 2) < P_1(c_{21} - c_{11})p(x \mid 1) \quad \text{when } x \in B_1,$$

$$P_2(c_{12} - c_{22})p(x \mid 2) \geq P_1(c_{21} - c_{11})p(x \mid 1) \quad \text{when } x \in B_2.$$

Based on the above, we can readily have the following decision rules: Let $x$ be the feature vector of a certain sample, $\omega$, thus:

$$\text{if } \frac{p(x \mid 2)}{p(x \mid 1)} < \frac{P_1(c_{21} - c_{11})}{P_2(c_{12} - c_{22})}, \quad \text{then } x \in B_1, \text{ hence } \omega \in \Omega_1,$$

$$\text{if } \frac{p(x \mid 2)}{p(x \mid 1)} \geq \frac{P_1(c_{21} - c_{11})}{P_2(c_{12} - c_{22})}, \quad \text{then } x \in B_2, \text{ hence } \omega \in \Omega_2. \quad (13.12)$$

In rule (13.12), $\frac{p(x \mid 2)}{p(x \mid 1)}$ and $\frac{P_1(c_{21} - c_{11})}{P_2(c_{12} - c_{22})}$ are referred to as the likelihood ratio and the decision threshold, respectively. The selected decision regions, $B_1$ and $B_2$, together with the above decision rule will enable us to design a minimum average loss classifier.

## 13.3  Minimum Error-Probability Classifier Design

Recall that in the above discussion on minimum average loss classifiers, we assume that the values of lose function $c(x)$ satisfy the following:

$$c_{11} = c_{22} = 0, \quad c_{21} = c_{12} = 1.$$

From the decision rule of minimum average loss classifiers, we know that if $x$ denotes the value of a feature vector for some pattern $\omega$, then:

$$\text{if } \frac{p(x \mid 2)}{p(x \mid 1)} < \frac{P_1}{P_2}, \quad \text{then } x \in B_1, \text{ hence } \omega \in \Omega_1,$$

$$\text{if } \frac{p(x \mid 2)}{p(x \mid 1)} \geq \frac{P_1}{P_2}, \quad \text{then } x \in B_2, \text{ hence } \omega \in \Omega_2.$$

$$(13.13)$$

The above decision rule can be rewritten as follows:

$$\begin{aligned} \text{if} \quad & P_2 p(x \mid 2) < P_1 p(x \mid 1), \quad \text{then } x \in B_1, \text{ hence } \omega \in \Omega_1, \\ \text{if} \quad & P_2 p(x \mid 2) \geq P_1 p(x \mid 1), \quad \text{then } x \in B_2, \text{ hence } \omega \in \Omega_2. \end{aligned} \quad (13.14)$$

As pointed out earlier, in statistical pattern recognition, feature vector $x$ of a certain sample is a random vector, and hence decision function $d(c(X))$ is also a random function. Therefore, it cannot be guaranteed that all classifications based on such a decision function are correct. To evaluate the effectiveness and features of a certain decision rule, we need to consider the probability of misclassifying one pattern into another. In the problems involving two classes of patterns, the feature vectors of patterns $\Omega_1$ and $\Omega_2$ are to be classified into regions $B_1$ and $B_2$, respectively. In such a case, two types of error may be committed: (1) misclassification of samples from pattern $\Omega_1$ as pattern $\Omega_2$ and (2) misclassification of samples from pattern $\Omega_2$ as pattern $\Omega_1$. The probability of total error will be equal to the sum of the probabilities of the two errors:

$$R = \int_{B_2} P_1 p(x \mid 1) dx + \int_{B_1} P_2 p(x \mid 2) dx. \quad (13.15)$$

In order to give a graphical illustration of the above-mentioned probabilities, let us suppose that feature vector $x$ of a certain sample is a 1D random variable satisfying a normal distribution. Figure 13.1 shows the illustration of the error probabilities in the two-pattern problems.

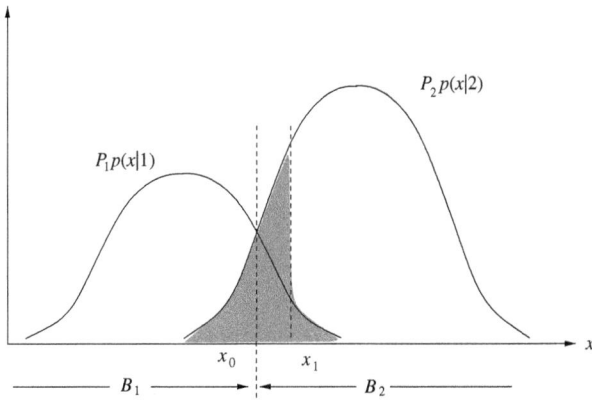

Fig. 13.1   The error probabilities in the two-pattern problems.

From Fig. 13.1, we readily note that dividing point $x_0$ between decision regions $B_1$ and $B_2$ satisfies the following:

$$P_1 p(x_0 \mid 1) = P_2 p(x_0 \mid 2). \qquad (13.16)$$

Also, we know that if the dividing point (i.e., decision threshold) is shifted to an arbitrary point, $x_1$, the total error probability will consequently be increased. This can be confirmed from the following calculations:

$$\int_{B_1} P_2 p(x \mid 2) dx + \int_{B_2} P_1 p(x \mid 1) dx$$

$$= \int_{-\infty}^{x_0} P_2 p(x \mid 2) dx + \int_{x_0}^{+\infty} P_1 p(x \mid 1) dx$$

$$\leq \int_{-\infty}^{x_1} P_2 p(x \mid 2) dx + \int_{x_1}^{+\infty} P_1 p(x \mid 1) dx. \qquad (13.17)$$

Therefore, decision rule (13.14) has the minimum error probability. It is because of this reason that people often refer to the classifiers built using decision rule minimum error probability classifiers.

Based on the above discussion, we may note that a minimum error-probability classifier is a special case of a minimum average lose classifier. To be more specific, a minimum average lose classifier is called a minimum error-probability classifier when lose function $c(x)$ satisfies $c_{11} = c_{22} = 0$ and $c_{12} = c_{21} = 1$.

In addition, it should be pointed out that in the above discussion on classifier design, we have assumed that priori probability $P_i$ and conditional probability density function $p(x \mid j)$ are both given. The two probabilities are used in decision rules (13.13) and (13.14) for the minimum average loss and the minimum error probability classifiers. However, in real-life applications, it may not be the case that both $P_i$ and $p(x \mid j)$ are known beforehand. In this book, we shall not deal with the cases of unknown priori probability. Interested readers may find discussions about such cases from other sources. As far as the problems of unknown $p(x \mid j)$ are concerned, we shall discuss how to estimate $p(x \mid j)$ by applying a method of orthogonal series approximation from the theory of statistical non-parametric estimation. In particular, we shall discuss how to use orthogonal wavelet series in estimating conditional probability density function $p(x \mid j)$.

## 13.4  Probability Density Estimation Based on Orthogonal Wavelet Series

In the preceding sections, we have assumed that conditional probability density function $p(x \mid j)$ is known. From such an assumption, we have shown how to design classifiers. In the case where $p(x \mid j)$ is unknown, we have to estimate $p(x \mid j)$ based on a set of sample feature vectors, $X_1, X_2, \ldots, X_N$. Hence, the estimation of $p(x \mid j)$ is the same as the estimation of general probability density function $p(x)$. In order to simplify the notations, in what follows, we shall deal only with the non-parametric estimation of general density function $p(x)$.

### 13.4.1  *Kernel Estimation of a Density Function*

In density function estimation, the easiest as well as most commonly used method is the Histogram method. In this method, we use a series of points, $\cdots < a_{-1} < a_0 < a_1 < \cdots$ to subdivide a real domain into a set of disjoint intervals $(a_i, a_{i+1})$. At each interval, a probability value can be estimated based on the following calculation:

$$\# \left( \{j; 1 \leq j \leq, a_i \leq X_j < a_{i+1}\} \right) / N, \qquad (13.18)$$

where $\#(A)$ returns the numbers of elements in set $A$. Thus, density function $p(x)$ in $[a_i, a_{i+1})$ can be estimated as follows:

$$\# \left( \{j; 1 \leq j \leq, a_i \leq X_j < a_{i+1}\} \right) / N(a_{i+1} - a_i). \qquad (13.19)$$

Use of the above-mentioned method can be dated back to as early as the seventh century. In the middle of the twentieth century, statisticians made several significant progresses in the area of non-parametric estimation of density functions. The most pioneering contributions during that time proposed and developed an important density estimator known as *kernel estimator*. In this way, for each $x$, we can construct a small region $[x - h_N, x + h_N)$ and then use the following to provide an estimate:

$$\tilde{p}_N(x) = \# \left( \{j; 1 \leq j \leq N, x - h_N \leq X_j < x + h_N\} \right) / 2N h_N, \qquad (13.20)$$

where $\tilde{p}_N(x)$ denotes an estimate. $h_N$ is a pre-defined positive constant related to $N$.

The kernel estimation method can be viewed as an improvement over the earlier-mentioned histogram method. Now, let us define a uniform probability density function, $\tilde{K}(x)$, over $[-1, 1)$, as follows:

$$\tilde{K}(x) = \frac{1}{2} I_{[-1,1)}(x)$$

$$= \begin{cases} \frac{1}{2} & -1 \leq x < 1, \\ 0 & \text{otherwise.} \end{cases} \tag{13.21}$$

Using $\tilde{K}(x)$, we can rewrite Eq. (13.20) as follows:

$$\tilde{p}_N(x) = \frac{1}{N h_N} \sum_{i=1}^{N} \tilde{K}\left(\frac{x - X_i}{h_N}\right). \tag{13.22}$$

As a matter of fact, $\tilde{K}(x)$ can be changed to any density functions or even other general functions. Thus, we can readily give the following definition.

**Definition 13.5.** Let $\tilde{K}(y)$ be a Borel measurable function in a 1D Euclidean space. Thus, we call

$$\tilde{p}_N(x) = \frac{1}{N h_N} \sum_{i=1}^{N} \tilde{K}\left(\frac{x - x_i}{h_N}\right), \tag{13.23}$$

a kernel estimator of probability density function $p(x)$ for pattern feature vector $X$. $\tilde{K}(y)$ is a kernel function and $h_N$ is a window width.

From Eq. (13.23), we can observe the following geometric interpretation for the estimate of $p(x)$, $\tilde{p}_N(x)$: For each sample $X_i$, we construct step function $\frac{1}{h_N}\tilde{K}(\frac{x-X_i}{h_N})$ of window width $2h_N$. The average for $N$ such step functions will become the estimate of density function $p(x)$. Note that $\lim_{N \to \infty} h_N = 0$; i.e. when the number of samples, $N$, increases, the step function is gradually becoming an impulse function.

Since the pioneering work on kernel estimation, statisticians have further studied the characteristics of large sample size in density function kernel estimation and examined the asymptotically unbiasedness, mean square consistency, asymptotical properties of mean square error, and uniformly convergence in probability. Many interesting theoretical results have been obtained from such efforts. Nevertheless, when they apply the large-sample-size kernel estimation in solving practical problems, they immediately face a difficulty, that is, how to determine window width $h_N$ given the actual number of samples, so that the estimation error of $\tilde{p}_N(x)$ with respect to $p(x)$ is smaller than a threshold.

## 13.4.2 *Orthogonal Series Probability Density Estimators*

As an alternative to the above-mentioned density function kernel estimation, we can also use the orthogonal series of a function in the $L^2(\mathbb{R}^n)$ space as an asymptotical estimator for density function $p(x)$.

Since most probability density functions are square integrable, i.e., $p(x) \in L^2(\mathbb{R}^n)$, it is possible to expand $p(x)$ using the orthogonal basis of $L^2(\mathbb{R}^n)$. For the sake of brevity, here we shall consider square integrable function space $L^2(\mathbb{R})$ on 1D real space $\mathbb{R}$.

Let $\{\phi_j(x)\}$ be an orthonormal basis (ONB) in $L^2(\mathbb{R})$ and $\hat{p}(x)$ denote an estimator of density function $p(x)$. Since $p(x) \in L^2(\mathbb{R})$, we have

$$p(x) = \sum_{j=-\infty}^{+\infty} c_j \phi_j(x). \tag{13.24}$$

Generally speaking, we can use

$$\hat{p}(x) = \sum_{j=1}^{m} c_j \phi_j(x), \tag{13.25}$$

that is, part of the orthogonal series summation for $p(x)$, $\sum_{j=1}^{m} c_j \phi_j(x)$, to approximate $p(x)$. In such a case, the mean square error of estimation becomes

$$\gamma_e = \int_{-\infty}^{+\infty} |p(x) - \hat{p}(x)|^2 dx$$

$$= \int_{-\infty}^{+\infty} \left| p(x) - \sum_{j=1}^{m} c_j \phi_j(x). \right|^2 dx. \tag{13.26}$$

The necessary condition for minimizing the above mean square error can be stated as follows:

$$\frac{\partial \gamma_e}{\partial c_k} = 0, \quad k = 1, 2, \ldots, m. \tag{13.27}$$

That is,

$$\int_{-\infty}^{+\infty} 2(p(x) - \sum_{j=1}^{m} c_j \phi_j(x))\phi_k(x)dx = 0. \tag{13.28}$$

Hence,

$$\sum_{j=1}^{m} c_j \int_{-\infty}^{+\infty} \phi_j(x)\phi_k(x)dx = \int_{-\infty}^{+\infty} \phi_k(x)p(x)dx. \tag{13.29}$$

Since $\{\phi_j(x)\}$ is an orthogonal system, the above expression is essentially the following:

$$c_k = \int_{-\infty}^{+\infty} \phi_k(x)p(x)dx. \tag{13.30}$$

The right-hand side of Eq. (13.30), as the mathematical expectation of random variable $\phi(x)$ can be estimated using the mean value of $N$ samples. Hence, we can write

$$\hat{c}_k = \frac{1}{N}\sum_{i=1}^{N} \phi_k(X_i). \tag{13.31}$$

Therefore,

$$\hat{p}(x) = \sum_{j=1}^{m} \hat{c}_j\phi_j(x). \tag{13.32}$$

There exist many orthonormal bases in $L^2(\mathbb{R})$. The commonly used ones include Hermite orthogonal system, Laguerre orthogonal system, and Legendre orthogonal system. No matter which system is chosen in asymptotical series expression $\sum_{j=1}^{m} \hat{c}_j\phi_j(x)$ to approximate density function $p(x)$, we are inevitably facing the next difficult problem. As we know, the quality of estimator $\hat{p}(x)$ is closely related to the term number, $m$, in the base function of $\phi_j(x)$. Our problem here is how to determine $m$ according to the number of samples, $N$, such that $\hat{p}(x) = \sum_{j=1}^{m} \hat{c}_j\phi_j(x)$ can best represent $p(x)$ — if so, any pattern classifier based on $\hat{p}(x)$ will be able to function effectively. Theoretically speaking, there is no explicit decision rule for determining such an $m$. The best way to do so is through empirical experimentation.

### 13.4.3 *Orthogonal Wavelet Series Density Estimators*

In order to derive, from a theoretical point of view, a theorem about the large sample size in the case of orthogonal wavelet series density estimation, we first of all introduce notions of slowly increasing generalized functions space and Sobolev space.

**Definition 13.6.** Rapidly decreasing function space $S$ is composed of functions, $C^{\infty}(\mathbb{R}^n)$, that satisfy the following condition:

$$\sup_{\mathbb{R}^n} |x^{\alpha}D^{\beta}\theta(x)| < \infty, \quad \forall \alpha, \beta \in N^n. \tag{13.33}$$

$C^{\infty}(\mathbb{R}^n)$ denotes infinitely differentiable real functions in $\mathbb{R}^n$.

From the above definition, it can be noted that any function in space $S$ is a $C^\infty$ function that will approach 0 when $\mid x \mid \to \infty$ at a speed faster than any power of $\frac{1}{|x|}$. Due to this reason, $S$ is normally referred to as rapidly decreasing function space. In $S$, it is possible to obtain several countable semi norms:

$$\gamma_{\alpha,\beta}(\theta) = \sup_{\mathbb{R}^n}|x^\alpha D^\beta(\theta)|, \quad \alpha, \beta \in N^n. \tag{13.34}$$

Thus, space $S$ becomes a linear topological space, which means when $\theta_v \to \theta$, for any exponents $\alpha, \beta$, we have

$$\lim_{v \to \infty} x^\alpha D^\beta \left(\theta_v(x) - \theta(x)\right) = 0, \quad \text{uniformly } x \in \mathbb{R}^n. \tag{13.35}$$

**Definition 13.7.** The space composed of all continuous linear functionals on rapidly decreasing function space $S$ is called slowly increasing generalized function space or tempered distribution space, denoted by $S'$.

In addition, we use $S_\gamma$ to denote the space composed of $C^\infty(\mathbb{R}^n)$ functions that satisfy the following conditions:

$$\sup_{\mathbb{R}^n}|x^\alpha D^\beta \theta(x)| < \infty, \quad \forall \alpha, \beta \in N^n, \ \mid \beta \mid \leq \gamma. \tag{13.36}$$

**Definition 13.8.** Sobolev space $H^S(\mathbb{R}^n)$ is defined as follows:

$$H^S(\mathbb{R}^n) = \{u(x); u(x) \in S', (1+ \mid \omega \mid^2)^{S/2}\hat{u}(\omega) \in L^2\}. \tag{13.37}$$

In Sobolev space, with Hermite inner product,

$$(u, v)_S = \frac{1}{(2\pi)^n} \int (1+ \mid \omega \mid^2)^S \hat{u}(\omega)\hat{v}(\omega)d\omega, \tag{13.38}$$

it is possible to show that this inner product can make $H^S(\mathbb{R}^n)$ become Hilbert space.

In what follows, we shall introduce the notion of reproducing kernel Hilbert space. The multiresolution analysis $\{V_m\}$ in wavelet analysis theory is in fact a sequence in reproducing kernel Hilbert space.

**Definition 13.9.** Let a bivariate function $\mathcal{L}(x, y) : \mathbb{R} \times \mathbb{R} \to \mathbb{R}$ be symmetric and non-negative. It is known that there exists a unique Hilbert space, $H(\mathbb{R})$, such that $\forall x \in \mathbb{R}, \mathcal{L}(x, \cdot) \in H(\mathbb{R})$. Furthermore, $\forall g(y) \in H(\mathbb{R})$, the following holds:

$$(g(\cdot), \mathcal{L}(x, \cdot))_{H(\mathbb{R})} = g(x). \tag{13.39}$$

The space $H(\mathbb{R})$ is called a reproducing kernel Hilbert space (RKHS), and bivariate function $\mathcal{L}(x, y)$ is called a reproducing kernel (RK) for $H(\mathbb{R})$.

In fact, each space, $V_m$, in MRA is a reproducing kernel Hilbert space, and the reproducing kernel, $\mathcal{L}(x, y)$, of $V_0$ can be written as follows:

$$\mathcal{L}(x, y) = \sum_{-\infty}^{+\infty} \phi(x - n)\phi(y - n), \tag{13.40}$$

where $\phi(x)$ is a scaling function. The reproducing kernel, $\mathcal{L}_m(x, y)$, of $V_m$ is given in the following:

$$\mathcal{L}_m(x, y) = 2^m \mathcal{L}(2^m x, 2^m y). \tag{13.41}$$

In order to study the asymptotical properties of reproducing kernel $\mathcal{L}_m(x, y)$, we need to further introduce the property, $Z_\lambda$, of scaling function $\phi(x)$.

**Definition 13.10.** Let scaling function $\phi(x) \in S_r$. $\phi(x)$ is said to satisfy property $Z_\lambda$ if and only if it satisfies the following:

(i)  $\hat{\phi}(\omega) = 1 + O(|\omega|^\lambda)$   as $\omega \to 0,$ $\tag{13.42}$

(ii)  $Z\phi(x, \omega) = e^{-i\omega x}(1 + O(|\omega|^\lambda))$   uniformly as $\omega \to 0,$

$$\tag{13.43}$$

where $Z\phi(x, \omega) \overset{\triangle}{=} \sum_{k=-\infty}^{+\infty} e^{-i\omega k}\phi(x - k)$ is called the Zak transform of scaling function $\phi(x)$.

Having introduced the $Z_\lambda$ property for scaling function $\phi(x)$, in what follows, we state a related theorem without giving its proof.

**Theorem 13.1.** *Let scaling function $\phi(x) \in S_r$, and for a certain $\lambda > 0$ satisfying the $Z_\lambda$, $\mathcal{L}_m(x, y)$ is a reproducing kernel for space $V_m$. Thus, we have:*

$$\|\mathcal{L}_m(x, \cdot) - \delta(x - \cdot)\|_{-\alpha} = O(2^{-m\lambda})  \text{ uniformly for } y \in \mathbb{R}, \tag{13.44}$$

*where $\| \cdot \|_{-\alpha}$ is a Sobolev norm and $\alpha > \lambda + \frac{1}{2}$.*

With the above preparation, we can now address the issue of how to derive a probability density estimate based on orthogonal wavelet series.

As we have mentioned earlier, common density function $p(x) \in L^2(\mathbb{R})$. From the multiresolution theory in wavelet analysis, it is know that

$$L^2(\mathbb{R}) = \overline{\bigcup_m V_m}. \tag{13.45}$$

Let $p_m(x)$ denote the orthogonal project of $p(x)$ in space $V_m$. Thus,

$$(L^2) \lim_{m \to \infty} p_m(x) = p(x), \tag{13.46}$$

where

$$p_m(x) = \sum_{n=-\infty}^{+\infty} a_{mn} 2^{m/2} \phi(2^m x - n). \tag{13.47}$$

The minimum mean square error estimator of $p_m(x)$ will be written as follows:

$$\hat{p}_m(x) = \sum_{n=-\infty}^{+\infty} \hat{a}_{mn} 2^{m/2} \phi(2^m x - n), \tag{13.48}$$

where

$$\hat{a}_{mn} = \frac{1}{N} \sum_{i=1}^{N} 2^{m/2} \phi(2^m X_i - n). \tag{13.49}$$

Thus, we can have the following:

$$\begin{aligned}
\hat{p}_m(x) &= \sum_{n=-\infty}^{+\infty} \left[ \frac{1}{N} \sum_{i=1}^{N} 2^{m/2} \phi(2^m X_i - n) \right] 2^{m/2} \phi(2^m x - n) \\
&= \frac{1}{N} \sum_{i=1}^{N} \sum_{n=-\infty}^{+\infty} 2^m \phi(2^m x - n) \phi(2^m X_i - n) \\
&= \frac{1}{N} \sum_{i=1}^{N} \mathcal{L}_m(x, X_i). 
\end{aligned} \tag{13.50}$$

Comparing (13.50) with (13.22), we note that orthogonal wavelet series density estimator $\hat{p}_m(x)$ and kernel estimator $\tilde{p}_N(x)$ are quite similar. From the geometrical point of view, $\mathcal{L}_m(x, X_i)$ is an impulse function scaled from $\mathcal{L}(x, X_i)$. The mean of $N$ impulse functions corresponds to the density estimator, $\hat{p}_m(x)$.

The following theorem further indicates that when scaling function $\phi(x)$ and unknown density function $p(x)$ satisfy certain specific properties, orthogonal wavelet series density estimator $\hat{p}_m(x)$ will converge to $p(x)$.

**Theorem 13.2.** *Let scaling function $\phi(x) \in S_r$, and for a certain $\lambda \geq 1$, it satisfies property $Z_\lambda$. Let $X$ be a continuous bounded density function random variable and $X_1, X_2, \ldots, X_N$ be $N$ independent identically distributed samples of $X$. Thus, if $p(x) \in H^\alpha$, $\alpha > \lambda + \frac{1}{2}$, $m \approx lgN/(2\lambda + 1)lg2$, then:*

$$E|\hat{p}_m(x) - p(x)|^2 \leq O(2^{-2m\lambda}). \tag{13.51}$$

**Proof.** First, we have

$$E|\hat{p}_m(x) - p(x)|^2$$
$$= E|(\hat{p}_m(x) - p_m(x)) + (p_m(x) - p(x))|^2$$
$$= E|\hat{p}_m(x) - p_m(x)|^2 + E|p_m(x) - p(x)|^2$$
$$+ 2E[(\hat{p}_m(x) - p_m(x))(p_m(x) - p(x))].$$

Note that

$$E\hat{p}_m(x) = E\left(\frac{1}{N}\sum_{i=1}^{N}\mathcal{L}_m(x, X_i)\right)$$

$$= \frac{1}{N}\sum_{i=1}^{N}E\mathcal{L}_m(x, X_i)$$

$$= \int \mathcal{L}_m(x, y)p(y)dy$$

$$= p_m(x). \tag{13.52}$$

Hence,

$$E[(\hat{p}_m(x) - p_m(x))(p_m(x) - p(x))] = 0. \tag{13.53}$$

Therefore, we can have

$$E|\hat{p}_m(x) - p(x)|^2$$
$$= E|\hat{p}_m(x) - p_m(x)|^2 + |p_m(x) - p(x)|^2, \tag{13.54}$$

where

$$E|\hat{p}_m(x) - p_m(x)|^2 = E\left|\frac{1}{N}\sum_{i=1}^{N}[\mathcal{L}_m(x, X_i) - p_m(x)]\right|^2$$

$$= \frac{1}{N}\left[\int \mathcal{L}_m^2(x, y)p(y)dy - p_m^2(x)\right]$$

$$\leq \frac{1}{N}\int \mathcal{L}_m^2(x, y)p(y)dy$$

$$\leq \frac{1}{N}\|p(\cdot)\|_\infty \mathcal{L}_m(x,x)$$

$$= \frac{2^m}{N}\|p(\cdot)\|_\infty \mathcal{L}_m(2^m x, 2^m x)$$

$$= O\left(\frac{2^m}{N}\right), \tag{13.55}$$

where $\|p(\cdot)\|_\infty$ is a constant, $\mathcal{L}_m(2^m x, 2^m x)$ is an impulse function that has the same magnitude as $\mathcal{L}(x,x)$. If we let $m = O(lgN)$, then when $m \to \infty$, $\mathcal{L}(2^m x, 2^m x)$ will gradually become a point impulse. Furthermore, since $\lim_{m\to\infty} \frac{2^m}{N} = 0$, we have $\lim_{m\to\infty} E|\hat{p}_m(x) - p_m(x)|^2 = 0$. On the other hand,

$$|p_m(x) - p(x)| = \left|\int \mathcal{L}_m(x,y)p(y)dy - p(x)\right|$$

$$= \left|\int [\mathcal{L}_m(x,y) - \delta(x-y)]p(y)dy\right|$$

$$\leq \|\mathcal{L}_m(x,\cdot) - \delta(x-\cdot)\|_{-\alpha}\|p\|_\alpha. \tag{13.56}$$

Inequality (13.56) is a Schwarz inequality in the Sobolev space. From Theorem 13.1, we know that

$$\|\mathcal{L}_m(x,\cdot) - \delta(x-\cdot)\|_{-\alpha} = O(2^{-m\lambda}). \tag{13.57}$$

Therefore,

$$|p_m(x) - p(x)|^2 \leq O(2^{-2m\lambda}). \tag{13.58}$$

If we let $N = 2^{m(2\lambda+1)}$, we can write $m = lgN/(2\lambda+1)lg2 = O(lgN)$. Since $\frac{2^m}{N} = 2^{-2m\lambda}$, we have $\lim_{m\to\infty} \frac{2^m}{N} = \lim_{m\to\infty} 2^{-2m\lambda} = 0$.

Hence, based on the above derivations, we can arrive at the following expression:

$$E|\hat{p}_m(x) - p(x)|^2 \leq O(2^{-2m\lambda}). \tag{13.59}$$

This concludes our proof. □

Let us now recall the orthogonal wavelet series density estimator, $\hat{p}_m(x)$,

$$\hat{p}_m(x) = \frac{1}{N}\sum_{I=1}^{N}\mathcal{L}_m(x,X_i). \tag{13.60}$$

Based on Theorem 13.2, we know that if the number of samples $X_i$ for feature vector $X$ is given, as denoted by $N$, $m$ can be rewritten as $m \approx lgN/(2\lambda+1)lg2$. Once $m$ is known, the window width of impulse function

$\mathcal{L}_m(x, X_i)$ can accordingly be determined. In other words, Theorem 13.2 provides a criterion for determining the window width of impulse function $\mathcal{L}_m(x, X_i)$ from the number of samples, $N$, which is exactly what the kernel estimator mentioned earlier is lacking of.

From the above discussions, we can note that the orthogonal wavelet series estimator differs from the kernel estimator and the traditional orthogonal series density estimator. Its basic idea shares some similarities to that of the traditional orthogonal series density estimator. However, it also satisfies several key properties of kernel estimator and exhibits some additional features. Generally speaking, the orthogonal wavelet series density estimator represents a new non-parametric way of estimating density functions, which has a great potential for practical applications. For instance, in pattern classifier design, sometimes, the probability density function, $p(x)$, of a certain feature vector may not be available. In such a case, we can readily replace $p(x)$ with $\hat{p}_m(x)$ using the above-described orthogonal wavelet series density estimator and thus effectively design the classifiers.

# PART 4

# Chapter 14

# Deep Learning-Based Texture Classification by Scattering Transform with Wavelet

Wavelet transform has been widely used in texture classification. Texture refers to the visual patterns or structures that repeat in an image, and they can be fine-grained, coarse, regular, or irregular. Texture classification involves dividing an image or image regions into different texture categories or identifying a specific texture type. Wavelet can decompose a signal or image into frequency components at different scales. This decomposition process can extract local texture features from the image and represent them as a set of wavelet coefficients. The wavelet coefficients represent the detailed information of the image at different scales and orientations. These coefficients can be used for texture classification tasks.

This chapter introduces a fusion-based architecture that uses multiple wavelets to increase the energy of scattering coefficients at higher levels via a further step of applying a linear operator followed by a nonlinear operator. Experimental results show that the proposed fusion scattering transform, which integrates scattering transforms with two different (Morlet and Shannon) wavelets, improves the error of classification in comparison with the original scattering transform. This hierarchical structure of deep networks can build a suitable representation of different texture categories [Dadashnialehi *et al.*, 2017].

## 14.1 Texture Classification by Scattering Transforms

The architecture of a scattering transform is shown in Fig. 14.1. In this transform, a wavelet-modulus operator $|W_m|$ followed by a low-pass filter produces scattering coefficients $S_m x$ at each layer $m$ (where

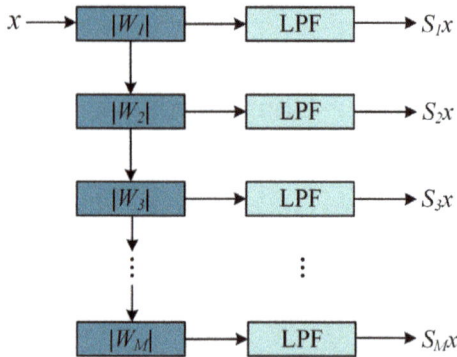

Fig. 14.1   The architecture of a scattering transform (LPF is the abbreviation of low-pass filter).

$m \geq 1$) of the network. A scattering transform $\Phi$ is locally translation-invariant, non-expansive, and stable to small deformations [Mallat, 2012; Bruna and Mallat, 2013]. The transform satisfies the Lipschitz condition, which guarantees that the distance between the transformation of data and the transformation of a slightly deformed version of the same data in the new compressed space is bounded by the size of deformation. Considering a diffeomorphism $\tau$, translation-invariant operator $\Phi$ is said to be Lipschitz continuous if there exists $C$ such that for all $\tau$ we have

$$\|\Phi(x) - \Phi(L_\tau x)\| \leq C\|x\|\|\tau\|, \tag{14.1}$$

the scattering transform satisfies the above Lipschitz condition and computes an invariant representation with a cascade of wavelet decomposition (linear convolution operator), complex modulus (nonlinear operator), and a local averaging [Mallat, 2012; Sifre and Mallat, 2013].

Given a specific number of training examples per texture class, the goal in a texture classification problem is to learn the value of a function $g(x)$ that predicts the class of a new texture image $x$. A well-known solution for the calculation of the function $g(x)$ is to consider its neighbor data values and use an interpolation method (e.g., nearest-neighbor interpolation) to determine the value of the function at each new $x$. While this approach probably works effectively for low-dimensional data, it is not guaranteed to perform well for this application. Texture images commonly have high resolutions and are made up of a huge number of (typically millions of) pixels. The texture classification is a problem of high-dimensional data classification, and the intrinsic problem in dealing with high-dimensional data is that the existence of such close neighbors is highly unlikely.

The Euclidian distance measure between two texture images that belong to the same class can be very large due to slight deformations. As such, it is not a useful similarity measure and does not provide suitable information for the classification purpose. However, the strategy used in the architecture of the scattering transform is an effective way to address this issue. First, the Lipschitz condition in Eq. (14.1) guarantees that the distance between an image and its slightly deformed version always has an upper bound (i.e., representations of images that belong to a certain class would stay in the vicinity of each other in the transformed data space). As such, the scattering transform effectively reduces the volume of the data space, ensuring that essential information for recognition is preserved. Second, to separate different classes from each other in the transformed space, the scattering transform uses the same strategy that is implemented in other forms of deep learning networks (i.e., to increase the dimensionality of data even further, see bottom of Fig. 14.2). As such, the architecture of the scattering transform also effectively uses the generalization capability of linear classifiers by increasing the dimensionality of data through linear convolutions with several rotated and dilated versions of (2D) mother wavelets.

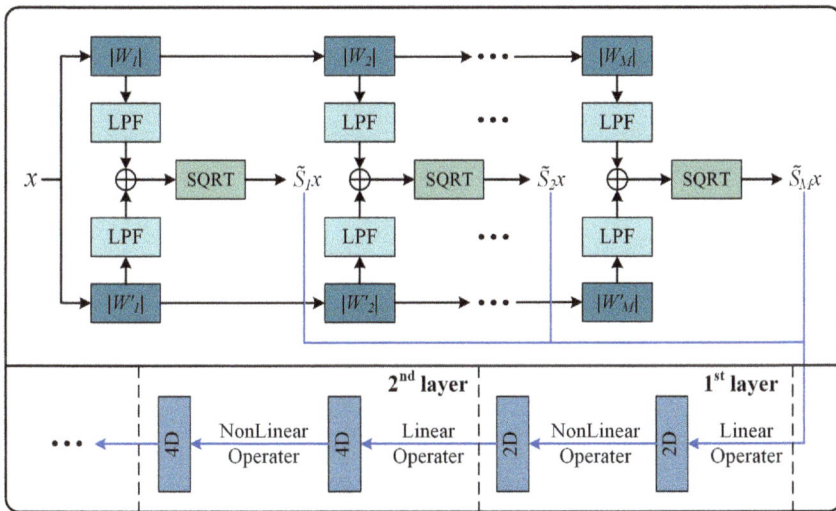

Fig. 14.2   Overview of the multi-wavelet fusion of scattering transforms. Top: Modified scattering transform based on the fusion of two different (Morlet and Shannon) wavelets (SQRT is the abbreviation of the square root operator). Bottom: Hierarchical structure of deep networks for building a suitable representation of a signal. The dimensionality of the data space is increased by using linear convolutions.

As it was mentioned earlier, the energy of coefficients in a scattering transform is spread across different layers. However, the energy rapidly vanishes at higher layers of the network. For example, 98% of the energy of scattering coefficients is distributed in layers 0, 1, and 2 in Caltech101 image database. As such, for classification purposes, the scattering network depth is typically limited to two layers [Bruna and Mallat, 2011]. The following section introduces a fusion-based architecture that increases the energy of scattering coefficients at higher levels by using multiple wavelets. The experiments show that the introduced method improves texture classification accuracy when using two different wavelets.

## 14.2 Multi-Wavelet Fusion of Scattering Transforms

Data fusion concept is based on the idea of synergistically using multiple sources of information to provide more reliable and accurate understanding of a given phenomenon. Fusion-based techniques are potentially advantageous in that they can effectively exploit complementary and redundant information to achieve a better inference [Luo *et al.*, 2002]. Data fusion techniques have found applications in many fields, including navigation, robotics, measurement science, and vehicular technology [Hoseinnezhad and Bab-Hadiashar, 2006; Dadashnialehi *et al.*, 2013].

A block diagram of the introduced fusion-based architecture to improve the distinguishability of data using scattering networks with two different wavelets is shown in Fig. 14.2. As it was mentioned earlier, a scattering transform can be considered as a deep convolutional network that cascades linear and nonlinear operators in multiple layers. Similarly, the fusion-based architecture shown in the top of Fig. 14.2 integrates two scattering networks (with distinct wavelets) by cascading a linear operation (addition) followed by a nonlinear operation (square root).

Each scattering transform uses a specific mother wavelet that has a different shape compared to that used in its counterpart scattering network. Wavelet transforms, unlike Fourier transforms, exploit a wide variety of basis functions to decompose a given signal. This enables a more targeted feature extraction by using application-specific shapes for the basis functions. For example, Morlet wavelets are closely related to Gabor wavelets, which are derived from Gaussian functions modulated by complex exponentials. The 2D Morlet wavelet is defined by

$$\psi(u, v) = e^{\frac{u^2 + s^2 v^2}{2\sigma^2}} \left( K - e^{iu\xi} \right), \tag{14.2}$$

where $\sigma$ is the spread of the Gaussian envelop, $\xi$ is the frequency of the oscil-latory exponential, $s$ is the eccentricity of the elliptical Gaussian envelop, and $K$ is a constant. On the other hand, the 2D Shannon wavelet uses an entirely different basis function that is defined as

$$\dot{\psi}(u, v) = e^{2i\pi(u+v)} 1_{1>|u|>1/2, 1>|u|>1/2}, \tag{14.3}$$

both Morelt and Shannon wavelets capture high frequency information (e.g., lines and edges.) that can be redundant. However, the experiments show that each wavelet captures some features (complementary informa-tion) more effectively compared to the other wavelet. It should also be noted that the Morlet wavelet usually produces a scattering coefficient with better distinguishability.

Incorporation of different wavelets in the fusion-based architecture shown in Fig. 14.2 enables the integration of complementary information captured by different wavelet shapes in a single framework. This allows the extraction of features that are difficult to capture using just one wavelet.

Although wavelets provide unique information depending on their shape, the information provided by different wavelets can also be redun-dant. This redundancy improves the overall accuracy of classification results through the reinforcement of important features. Integration of redun-dant and complementary information using multiple wavelets increases the energy of scattering coefficients at higher levels of a fusion-based scattering transform, alleviates the issue of low energy at higher layers, and improves the results of classifications.

## 14.3 Experimental Results and Discussions

To show that scattering transforms that exploit different mother wavelets can produce redundant and complementary information, the scattering coefficients (at level two) were calculated for a simple example (the square shape shown in Fig. 14.3(a) [Dadashnialehi *et al.*, 2017]) using Morlet and Shannon wavelets and are shown in Fig. 14.3(b) [Dadashnialehi *et al.*, 2017] and Fig. 14.3(c) [Dadashnialehi *et al.*, 2017], respectively. The figure shows that both Morelt and Shannon wavelets have identified horizontal and verti-cal edges (redundant information); however, the use of the Shannon wavelet has produced stronger scattering coefficients at the corners of the square (complementary information).

To demonstrate the effectiveness of the introduced fusion-based archi-tecture in real-world texture classification problems, the introduced method has conducted a number of experiments on a well-known texture dataset

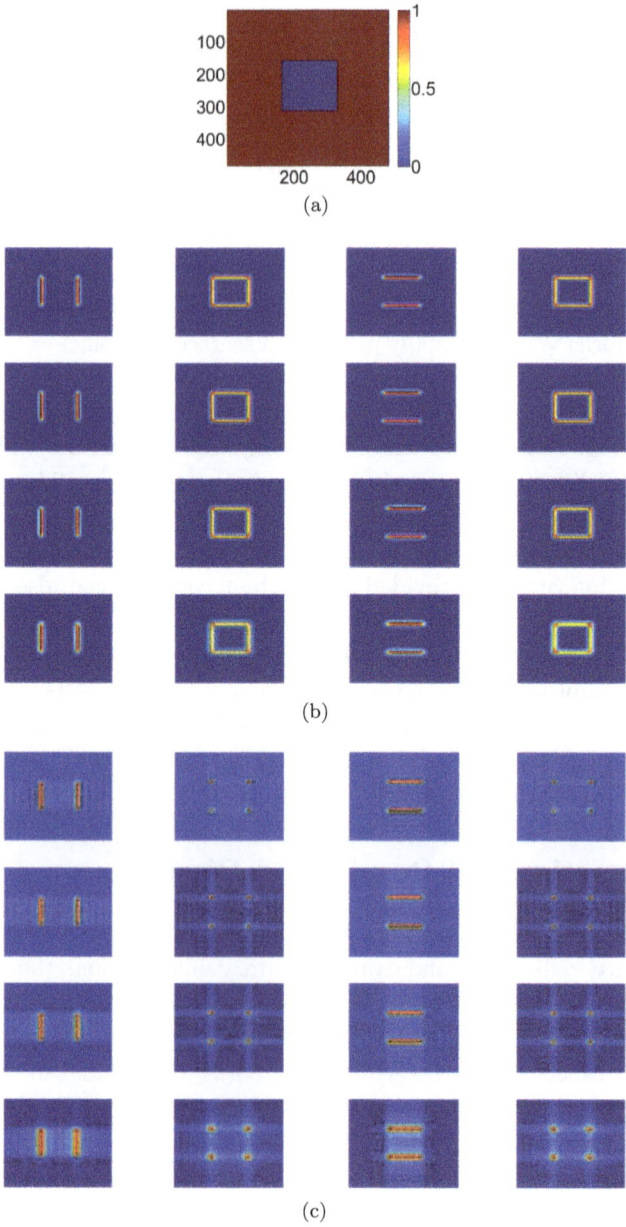

Fig. 14.3   (a) Image of a square; (b) scattering coefficients (at level two) calculated for the square (shown in part (a)) using Morlet wavelet; (c) scattering coefficients using Shannon wavelet.

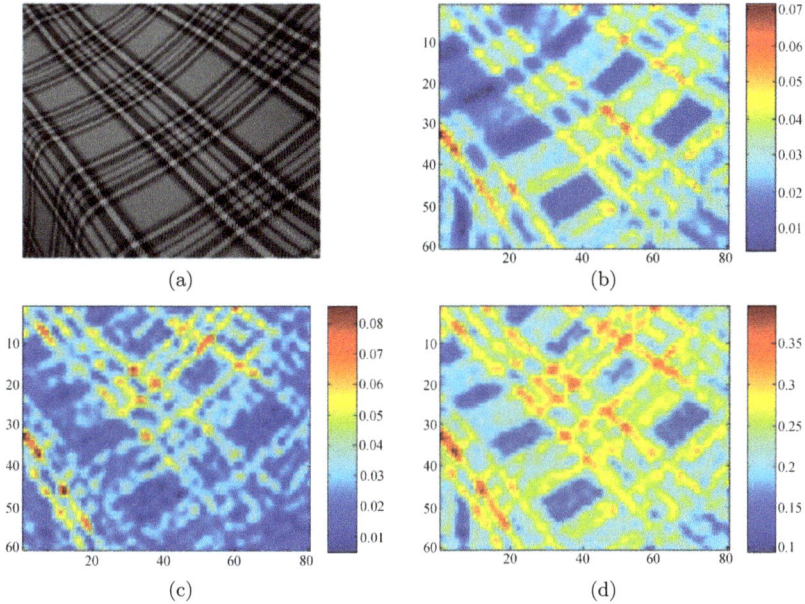

Fig. 14.4   (a) Sample image from UIUC texture dataset; (b) a sample of scattering coefficients at level 3 calculated by original scattering transform using Morlet wavelet; (c) scattering coefficients corresponding to part (b), calculated by original scattering transform using Shannon wavelet; (d) fusion of scattering transforms using Morlet and Shannon wavelets. The energy of scattering coefficients is effectively increased using the introduced fusion-based method.

called UIUC [Lazebnik *et al.*, 2005]. The database includes several classes of textures with 40 images for each texture class. A sample image of a texture class from this dataset is shown in Fig. 14.4(a) [Dadashnialehi *et al.*, 2017].

Figure 14.4(b) [Dadashnialehi *et al.*, 2017] shows a sample of scattering coefficients for this image at level 3, which is calculated by the original scattering transform using the Morlet wavelet. The corresponding scattering coefficients calculated by the Shannon wavelet are shown in Fig. 14.4(c) [Dadashnialehi *et al.*, 2017]. Comparison between these two figures shows that the Morlet wavelet outperforms the Shannon wavelet in producing a scattering coefficient with better distinguishability.

The scattering coefficients calculated by the fusion of scattering transforms using both Morlet and Shannon wavelets are shown in Fig. 14.4(d) [Dadashnialehi *et al.*, 2017]. The figure shows that the number of scattering coefficients with higher energy is effectively increased, which leads to a better distinguishability when compared to using a single wavelet. Results of the experiments also confirm the effectiveness of the introduced fusion method because the classification outcome for the fusion-based method is better than that of the original scattering transform.

To show that the increases in the energy of scattering coefficients lead to better classification outcomes, the method calculates the correct classification rates for the introduced fusion-based method and compares those with their counterparts calculated by the original scattering transform. First, the UIUC database is rearranged into six subsets, each subset consisting of $n$ number of classes, where $n$ changes between 2 and 7. For example, six classes $(n = 6)$ of the UIUC database are used in the classification of the fifth subset. The original classification code for scattering transforms uses an affine space classifier for the classification purpose, and a modified version of the original code based on the fusion of scattering transforms using two wavelets (as shown in Figs. 14.5 and 14.6) are used to perform classification on each of the above six subsets. The data in each subset were split in half so that each class was partitioned into the training and testing

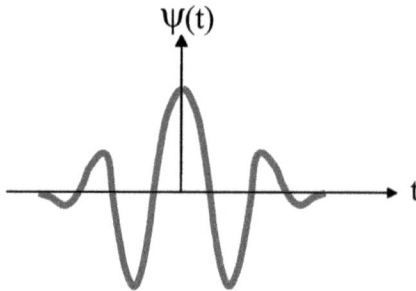

Fig. 14.5     Morlet wavelet.

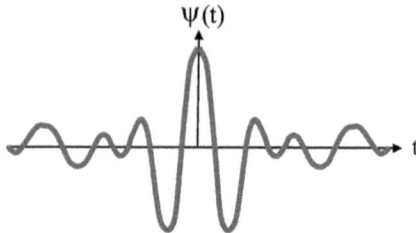

Fig. 14.6     Shannon wavelet.

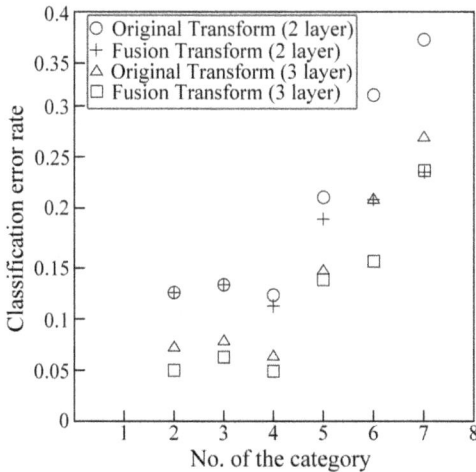

Fig. 14.7 Calculated classification errors for deep scattering networks (using scatnet 0.2 software) using one (Morlet) and fusion of two (Morlet and Shannon) wavelets at levels 2 and 3.

sets with equal size. The classification errors on all subsets are shown in Fig. 14.7 for networks with two and three layers.

Figure 14.7 shows that, for a fixed number of layers, the classification error for the fusion-based method is always less than or equal to that of the original scattering transform. In addition, the figure shows that the classification error for the fusion-based method at level 2 is comparable or better compared to that of the original scattering transform at level 3, particularly better results were achieved when classification was conducted on higher numbers of classes (e.g., $n = 7$).

For a fixed number of layers, the fusion-based method requires more computation compared to the original scattering transform. However, the experimental results show that the computational time for the fusion-based (using two wavelets) method at level $M$ is slightly less than that of the original scattering transform at level $M + 1$, while it improves the correct classification rate when classification was conducted on higher numbers of classes. Taking that into consideration, the trade-offs between the computational requirements and the improvements in the texture classification error can be compared for each method in Fig. 14.7. The figure shows that the introduced fusion-based method results in comparable or smaller classification error at one layer less than that of the original scattering transform, which is due to the effective increased energy of scattering coefficients.

# Chapter 15

# An Approach to Image Classification by Deep Learning-Wavelet Architecture

The problem addressed in this chapter is feature extraction and classification of images. As a solution, this chapter proposed a deep wavelet network (DWN) architecture based on the wavelet network (WN) and the stacked auto encoders (AEs). In this chapter, the deep learning based on neural networks is shifted to deep learning based on wavelet networks. The latter doesn't change the general form of the deep learning based on the neural network but it is a novel method that shows the process of feature extraction and explains the system of image classification.

Specifically, the DWN in this chapter constructs a WN for each image in all classes by using the best contribution algorithm (BCA). Then the score of each wavelet used in all WNs is calculated. The global wavelet network (GWN) is created using the wavelets with the highest scores. DWN uses the GWN to create AE for each class in the dataset. Finally, the hidden layers of all self-encoders are used as the first hidden layer of the deep network by stitching them together as a vertical stitch. In the classification phase, this chapter uses a linear classifier like the Softmax classifier [Blel *et al.*, 2022].

## 15.1 Recent Work

In a previous study, the objective was the combination of two notions: wavelet network and deep learning. The proposed algorithm [Hassairi *et al.*, 2015, 2020; Said *et al.*, 2016; Bouallégue *et al.*, 2017; Jemel *et al.*, 2020] provides the creation of a deep stacked wavelet auto encoder (DSWAE) to supervise feature extraction for pattern classification. This architecture is formed according to the following steps:

**Step 1:** Creation of a WN for each element of class from the dataset.

**Step 2:** Calculation of the scores of all wavelets taking their position and their numbers of appearance in all the WN.

**Step 3:** Creation of the GWN using the best wavelet contribution for a class.

**Step 4:** Transformation of the GWN to an AE.

**Step 5:** Building a series of AE to get the DSWAE.

The proposed DSWAE was examined with three different datasets (the Columbia Object Image Library (COIL-100) [Nene *et al.*, 1996], Arabic printed text image (APTI) [Slimane *et al.*, 2013], and ImageNet [Hassairi *et al.*, 2016]).

Despite the performance gained from this architecture in terms of classification rate, the only inconvenience remains the use of a DSWAE for each class of the database. This architecture consists of the classification of one class versus all the other classes (they consider that the dataset is composed of two classes: Class 1 is the class that they want to classify. Class 2 contains all the other classes of the dataset). This problem is addressed in the contribution.

## 15.2 The Proposed Approach

In this section, an improved pattern recognition method is presented.

### 15.2.1 *System Overview*

This algorithm contains two main stages: the training stage and classification stage. The training phase is the main stage of the contribution and it includes many sub-steps that illustrate the overall architecture of the system (Fig. 15.1) [Blel *et al.*, 2022].

Where: Class 1, Class 2, Class 3, ... and Class $N$ are the classes that we desire to classify.

1. the construction of a WN for each image in all classes using the BCA,
2. score calculation of each wavelet used in all WN,
3. the creation of the GWN using wavelets that have the highest score,
4. the production of the AE with wavelets used in GWN,
5. the formation of the deep DWN with a linear classifier in the last layer.

The various strides of the methodology are itemized in the accompanying sections.

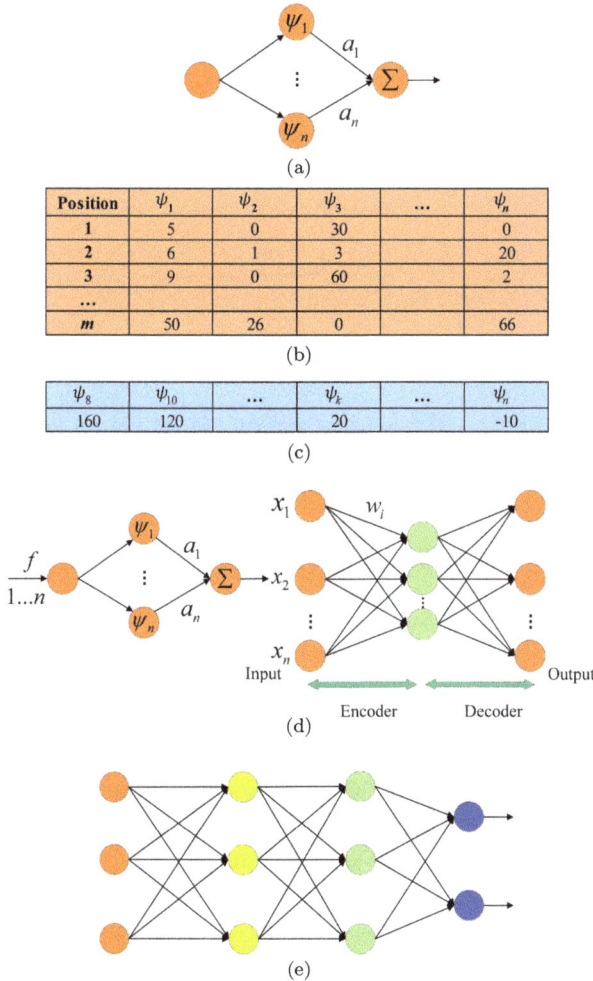

Fig. 15.1   The overall architecture of the system: (a) Construction WN; (b) Wavelets for Class 1; (c) Top scoring; (d) Create the wavelet AE for Class 1; (e) DWN with linear classifier.

## 15.2.2   *The Construction of a WN*

The first step is to create a WN for each element in the class, applying the BCA [Jemai *et al.*, 2010; Zaied *et al.*, 2011] (Fig. 15.2).

The output signal of the WN is evaluated by the following equation:

$$\hat{f} = \sum_{i=1}^{n} a_i \psi_i, \tag{15.1}$$

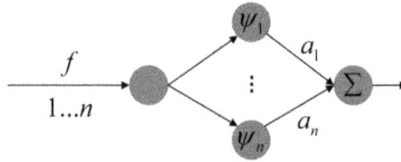

Fig. 15.2   Wavelet network for one image of a class.

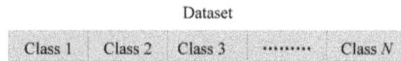

Fig. 15.3   An example of the division of the dataset into $N$ classes: Class 1, Class 2, Class 3, ..., Class $N$.

where $a_i$ is the coefficient of $\psi_i$. $n$ represents the number of wavelets in the hidden layer of WN.

### 15.2.3   *Wavelet Score Calculation to Elaborate the GWN*

In this section, the technique of computing the wavelet scores to approximate the GWN of a category is explained. Approximating the GWN of a category is obtained. The dataset will then be considered containing $N$ categories (Fig. 15.3).

The algorithm allows the classification of all classes in the database. So, let's consider the dataset that consists of $N$ classes. Then, we need to calculate the score of each wavelet used in all WNs of Class 1. For that reason, we need to collect all wavelets of each signal in the form of a table in Fig. 15.4 to obtain the position of apparition of each wavelet in each WN for each signal.

From this point, Blel *et al.* [2022] had the idea to compute the number of occurrences of every wavelet in each position within the WN for all classes separately (Table 15.1).

After that, the method of calculating the coefficients of the wavelets is described. The coefficient of $\psi_i$ of Class 1 is obtained by the sum of all the appearance values of $\psi_i$ which are multiplied by the $((m - j) + 1)$ in each position $j$. So this global coefficient in Class 1 is obtained according to this equation:

$$\psi_i GalobCoef_{Class\ 1} = \sum_{j=1}^{m} V_j * ((m - j) + 1), \qquad (15.2)$$

where $V_j$ is the number of apparitions of $\psi_i$ in position $j$.

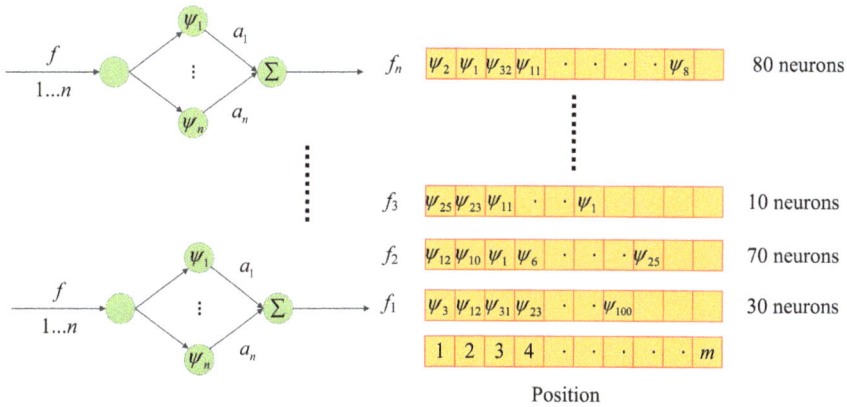

Fig. 15.4    All WNs for Class 1.

Table 15.1. Apparition numbers of wavelets of Class 1.

| Position | $\psi_1$ | $\psi_2$ | $\psi_3$ | $\psi_4$ | $\ldots$ | $\psi_n$ |
|----------|------|------|------|------|-----|------|
| 1 | 5 | 0 | 30 | 100 | | 0 |
| 2 | 6 | 1 | 3 | 0 | | 20 |
| 3 | 9 | 0 | 60 | 3 | | 0 |
| 4 | 0 | 22 | 0 | 35 | | 80 |
| $\ldots$ | | | | | | |
| $m$ | 50 | 26 | 0 | 0 | | 66 |

Then the same operation is applied to $\psi_i$ of Class 2, Class 3, ..., Class $N$ to get its importance in Class 2, Class 3, ..., Class $N$. Here is the equation of the calculation of the score:

$$\psi_i GalobCoef_{\text{Class 2}} = \sum_{j=1}^{m} V_j * ((m-j)+1), \qquad (15.3)$$

$$\psi_i GlobCoef_{\text{Class 3}} = \sum_{j=1}^{m} V_j * ((m-j)+1), \qquad (15.4)$$

$$\psi_i GlobCoef_{\text{Class } N} = \sum_{j=1}^{m} V_j * ((m-j)+1). \qquad (15.5)$$

Finally, the global score of each wavelet is obtained according to the following equation:

$$GlabalCoeff_{\psi_i} = \psi_1 GlabCoef_{Class\,1} -$$
$$\psi_1 GlobCoef_{Class\,2} - \cdots - \psi_i GlabCoef_{Class\,N}, \tag{15.6}$$

the wavelets are then sorted according to their global coefficient from the biggest to the smallest. For wavelet selection, we can use classical wavelets, such as Haar, Mexican hat, Gaussian, and so on.

### 15.2.4 *Creation of the AE to get the DWN*

In this section Blel *et al.* [2022] produce an AE (Fig. 15.5) for each class in the dataset. For that reason, the wavelets are used in connection with both encoding and decoding of the AE and not inside the neurons. The wavelet AE utilized is biorthogonal: They are composed of two arrangements of wavelets made by a mother wavelet $\psi$ and a double wavelet $\tilde{\psi}$.

$$\langle \psi_i, \tilde{\psi}_j \rangle = \begin{cases} 0 & i \neq j, \\ 1 & i = j. \end{cases} \tag{15.7}$$

Given this:

$$\begin{aligned} f &= x_1, x_2, x_3, \ldots, x_n, \\ \hat{f} &= \hat{x}_1, \hat{x}_2, \hat{x}_3, \ldots, \hat{x}_n, \\ \psi_1 &= w_{1i}, w_{2i}, w_{3i}, \ldots, w_{ni}, \\ \tilde{\psi}_1 &= \tilde{w}_{1i}, \tilde{w}_{2i}, \tilde{w}_{3i}, \ldots, \tilde{w}_{ni}. \end{aligned} \tag{15.8}$$

As well as,

$$a_i = \langle f, \psi_i \rangle \Rightarrow a_i = \sum_{j=1}^{n} w_{ji} x_j. \tag{15.9}$$

Therefore, the output signal of the AE is obtained by the following equation:

$$\hat{x}_i = \sum_{j=1}^{n} \tilde{w}_{ji} a_j. \tag{15.10}$$

In this stage, an AE for each class from the dataset is created (Fig. 15.6(a)). To build the DWN, all hidden layers of all AEs are combined to construct the first hidden layer of the deep network (Fig. 15.6(c)). From this first hidden layer of the DWN, a series of AE is applied to create further hidden layers for the learning phase (Fig. 15.6(b)).

Finally, in the classification phase, Blel *et al.* [2022] use a linear classifier similar to the Softmax classifier [AUEB *et al.*, 2016]. The frequently used

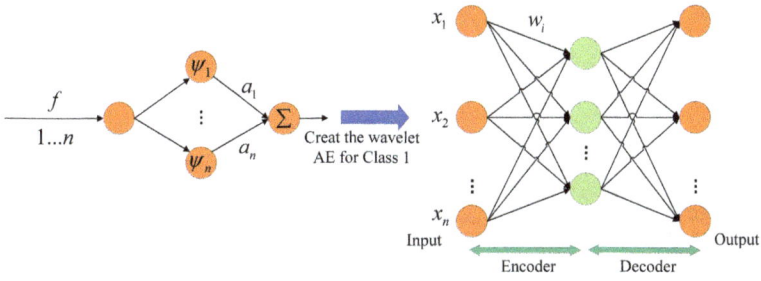

Fig. 15.5 A wavelet AE for a class.

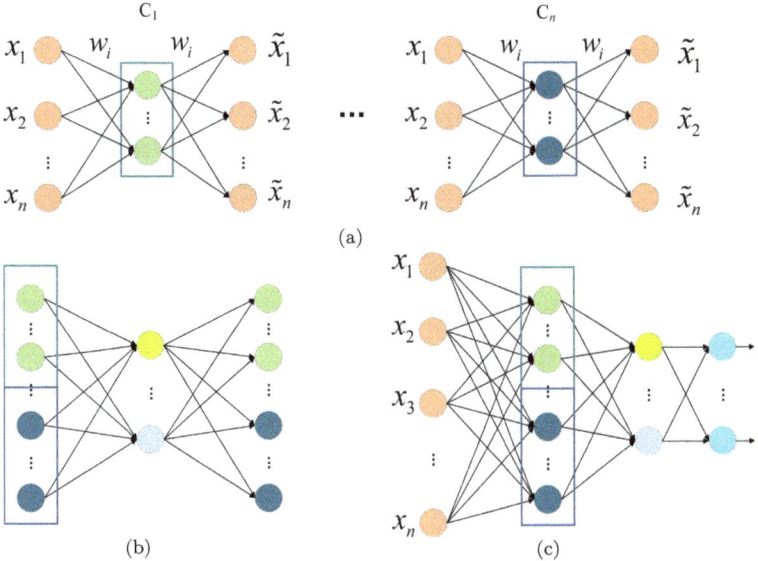

Fig. 15.6 DWN steps creation: (a) An AE for each class; (b) the Second AE; (c) the DWN architecture.

Softmax function is the activation function applied to the last layer of the DWN. In order to greatly improve the performance of the method, Blel *et al.* [2022] used a backpropagation algorithm for fine-tuning.

## 15.3 The Experimental Results

In the next part of this chapter, some results are presented and the experimental testing of the proposed pattern classification method are explained in detail. All results were done on COIL-100.

Fig. 15.7    The COIL-100 dataset.

COIL-100 is a database that includes color images collected from 100 objects. The objects are placed on a motorized turntable on a black background with a 360-degree rotation in order to vary the pose of the objects compared to a fixed color camera. Images of the objects were taken at exposure intervals of 5 degrees (Fig. 15.7).

In the methodology of this chapter, images from each category (2/3) of the dataset were used in the training phase and the remaining (1/3) images were used in the classification testing phase. The network itself initializes the weights based on the data.

Peak signal-to-noise ratio (PSNR) metric is used to compare the performance of user-aided and manual approaches. PSNR is most commonly used to measure the quality of reconstruction of images [Vasavi and Rao, 2015].

Table 15.2. PSNR results of Class 1.

| Input | | | |
|---|---|---|---|
| PSNR | 62,1199 | 60,3787 | 59,5136 |

Table 15.3. PSNR results of Class 2.

| Input | | | |
|---|---|---|---|
| PSNR | 48,7831 | 38,6872 | 40,5936 |

Table 15.4. PSNR results of Class 3.

| Input | | | |
|---|---|---|---|
| PSNR | 58,7453 | 56,8655 | 56,2427 |

Table 15.5. PSNR results of Class 4.

| Input | | |
|---|---|---|
| PSNR | 46,9255 | 41,8629 |

The four classes of images randomly use the PSNR metric. The PSNR values for images in categories 1, 2, 3, and 4 are shown in Tables 15.2–15.5 in that order.

So, the PSNR is good for all classes. Classes 1–4 have nearly the same wavelets. For this reason, we can explicate the performance of the signal quality in all classes. Moreover, the rate obtained by the PSNR of Class 1 is better than that in Classes 2–4. It is obvious that the AE models just the class it represents from the PSNR values.

After that, a series of AEs is applied to learn the features of images. We obtain the DWN with two hidden layers and a Softmax classifier in the last layer. The fine-tuning helps get a good classification rate. Thus, this rate

Table 15.6.    Image classification rate of different methods.

| Methods | Classification rates (%) |
|---------|--------------------------|
| DWN | 97,60 |
| MARM | 96,83 |
| DPL | 81,77 |
| SRRS | 89,85 |

gives an idea about the efficiency and the evaluation of the performance of the approach (Table 15.6) when compared to other approaches. The classification test of the algorithm shows very good performance with a rate that reaches $97,60\%$. We find, from Table 15.6, that the approach provided a better classification rate of the MARM [Wang *et al.*, 2020], the DPL [Wang *et al.*, 2017], and the SRRS [Li and Fu, 2015] tested on the COIL-100 dataset. This solution is encouraged to be applied to large databases, such as ImageNet.

## Chapter 16

# Brain Tumor Identification Based on Wavelet and CNN-LSTM Deep Learning

The wavelet transform is a widely utilized mathematical tool in the domains of signal processing and image processing. In the segmentation of brain images, the wavelet transform can extract information at various frequencies, aiding in the division of brain images into distinct regions and structures. Wavelet transforms are employed in brain image segmentation tasks to extract distinguishing features of various brain tissues. Various tissue types exhibit distinct frequency and spatial distribution properties in the wavelet domain. These features can be extracted for subsequent segmentation processes through statistical analysis of wavelet coefficients and the application of image processing techniques.

The application of wavelet transform in brain image segmentation can facilitate the extraction of feature information and the segmentation of brain tissues using appropriate algorithms. This method can enhance the accuracy and robustness of segmentation while providing valuable support for diagnosing and treating brain diseases. This chapter introduces the application of wavelets in image feature extraction with the example of a hybrid deep learning-based brain tumor classification and segmentation by stationary wavelet packet transform and adaptive kernel fuzzy $C$-means clustering. A hybrid convolution neural network-long short-term memory (CNN-LSTM) deep learning is used to improve tumor classification accuracy [Devi *et al.*, 2022].

## 16.1 Literature Review

This section highlighted some of the most recent studies on deep learning-based brain tumor classification and segmentation.

Sultan *et al.* [2019] suggested a convolutional neural network (CNN) founded on data left (DL) to categorize distinct brain cancers utilizing two public datasets. This outcome demonstrated the model's capacity to achieve the goal of tumor multi classification. Spectral clustering was used by Maruthamuthu *et al.* [2020] to separate the tissues of brain tumors using magnetic resonance images (MRIs) to create high-standard clusters. This clustering approach has the disadvantage of requiring a large amount of data to create dense similarity matrices. The author proposed brain tumor segmentation as a solution to this problem. This method outperformed existing clustering algorithms using competitive dice score values from MRI images for edema and tumor core (TC) segmentation.

To develop the precise brain tumor classification strategy, Gumaei *et al.* [2019] recommended combining a hybrid feature extraction method with a regularized extreme learning machine. The tumor kind was classified using a regularized extreme learning machine (RELM). The submitted work was more efficient than previous methodologies, according to the findings. For picture reduction, Mallick *et al.* [2019] utilized a deep wavelet auto encoder (DWA). It was a mix of the auto encoder, a fundamental feature reduction property, and the wavelet transform, an image decomposition technique. A DWA-DNN image classifier was used to create a brain imaging dataset.

A CNN-based U-net has been proposed by Naser and Deen [2020]. It was beneficial for segmenting the tumor, and transfer learning relies on a Vgg16 pre-trained convolutional base, and a fully connected classifier was created to grade the tumor. Kumar and Mankame [2020] introduced the dolphin echolocation-based sine cosine algorithm (Dolphin-SCA) deep learning method, which is based on deep CNN. It comes in handy when we need to make quick categorization judgments and enhance accuracy. Dolphin-SCA and the fuzzy deformable fusion model were used to complete the segmentation procedure. Feature extraction was carried out using power local differential privacy (LDP) and statistical characteristics (skewness, mean, and variance). The collected features were used in the deep convolution neural network (D CNN) to perform the brain tumor classification using Dolphin-SCA (as the training method).

A hybrid clustering approach was introduced by Maheswari *et al.* [2021]. The $k$-means algorithm and the fuzzy $C$-means algorithm have been used to segment brain tumors. The method has been tested on synthetic and real-time datasets. When compared to other procedures, this strategy was more accurate. The architecture for segmenting tumors was started by Zhou *et al.* [2020]. The 3D deep supervision approach was used to improve the training. Chen *et al.* [2020] used a DCNN to automatically portion brain tumors. Deep convolutional symmetric neural network (DCSNN) achieves this by casting symmetric masks in several layers. Its improved segmentation networks are based on DCNN. The dice similarity coefficient (DSC) metric was used to calculate the results.

An automatic brain tumor segmentation was presented by Bal *et al.* [2022]. Shape-based topological characteristics and rough-fuzzy *C*-means were employed by the author in this study. In the rough-fuzzy *C*-means (RFCM) approach, fuzzy membership handled the overlapping division. The upper and lower bounds of rough datasets are in charge of resolving the dataset's uncertainty. Crisp approximation in RFCM and fuzzy border were critical in brain tumor segmentation utilizing MRI. The approach was used to pick the first centroid by the author.

According to a thorough review of the literature, the suggested model is built on a hybrid deep learning method. From the above literature survey, the problems identified are as follows for enhancing the accuracy of the segmentation, various region of interest (ROI) features should be extracted with the effective feature extraction technique. The existing system should not provide an effective technique to identify the tumor at the subregions, such as core region and enhanced region. The major drawback of the existing system is it consumes more processing time. The performance of segmentation and classification should be enhanced.

## 16.2 Hybrid Deep Learning-Based Brain Tumor Classification and Segmentation

The brain tumor is the most lethal illness in adults because it is caused by an abnormal mass of cells that develops fast and disrupts organ function. Diagnostics, growth prediction, and therapy of brain tumors all benefit from automatic segmentation and categorization of medical pictures. The suggested method's main goal is to align and improve the strategy for tumor detection using brain MRI segmentation. This work introduces a deep learning-based technique for autonomously classifying, localizing, and segmenting brain tumors. The research's major goal is to use an effective approach to segment an MRI picture. As indicated in the picture, the suggested method is divided into five categories: preprocessing, feature selection, feature extraction, segmentation, and classification, as shown in Fig. 16.1.

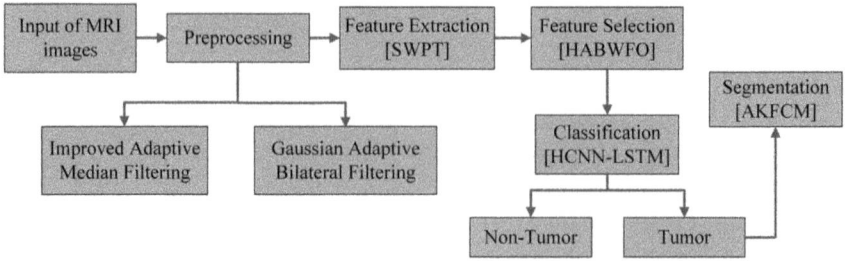

Fig. 16.1   The schematic block diagram of the proposed methodology.

Initially, we preprocessed the brain MRI for creating the input as fit for segmentation. Next, the preprocessed output is given to the input of feature extraction. The suggested approach named stationary wavelet packet transform (SWPT) is utilized for feature extraction. Then, a hybrid adaptive black widow optimization with moth flame optimization (HABWMFO) is utilized for the optimal feature selection. The feature values are supplied into the clustering algorithm for segmentation after picking the best features. For segmentation, the adaptive kernel fuzzy $C$-means clustering technique (AKFCM) is used. Following segmentation, characteristics such as texture and certain numerical features will be established on this large feature set, and the final classification will be performed. Deep learning is used to classify brain tumors using a hybrid convolutional neural network with long short-term memory (CNN-LSTM). The accuracy of segmentation is used to estimate the process of the recommended approach.

## 16.2.1  *Preprocessing*

In this chapter, a preprocessing step is performed in order to improve image quality. To lessen the noise and blurring impact, several preprocessing techniques are applied. The various procedures determine the higher-quality measurements. This preprocessing is done using an enhanced adaptive median and adaptive bilateral filter in this suggested method. The processed pictures are then taken to the next step, where feature extraction is performed in preparation for further processing.

### 16.2.1.1  *Improved Adaptive Median Filter*

This filter is used to remove the salt and pepper noise from the actual image while maintaining the image's clarity. If $W_{\max}$ denoted the maximum window size, the mean square displacement (MSD) of the threshold is represented by thalamus (TH). The non-pollution point's closeness pixels are represented as $Z_k$.

(1) Evaluate the current pixel's minimum and maximum values $\mathcal{F}_{\min}$ and $\mathcal{F}_{\max}$ with $A_{ij}$.
(2) If $\mathcal{F}_{\min} < \mathcal{F}_{ij} < \mathcal{F}_{\max}$ has been changed, the signal point value $P(i,j)$ would be the signal significance different; proceed to Step 5 as well as add a non-pollution point from $A_{ij}$ into $\Omega$, i.e., $Z_k \in A_{ij}$ and $\mathcal{F}_{\min} < \mathcal{F}_{ij} < \mathcal{F}_{\max}$, then $\Omega = \Omega \cup \{Z_k\}$.
(3) If $w < w_{\max}$, set $w = w + 2$ and then move through Step 1.
(4) If $w \geq w_{\max}$ is true, make sure the collection $\Omega$ isn't blank. If $\Omega$ is blank, leave $\mathcal{F}_{ij}$ alone and proceed to Step 5; otherwise, use the following equation to calculate $\mathcal{F}_{ij}$'s MSD using the set $\Omega$:

$$\mathrm{MSD} = \min_{z_k \in \Omega} \left\{ |\mathcal{F}_{ij} - \mathcal{F}_{z_k}| \right\}. \tag{16.1}$$

The absolute value operation is defined as $|.|$ in this case, if MSD $\leq$ TH is true, its signal point is $\mathcal{P}(i,j)$, its pixel value is $\mathcal{F}_{ij}$, and Step 5 is performed; otherwise, the noise point is at $\mathcal{P}(i,j)$, the set $\Omega$ median value is $\mathcal{F}_{\mathrm{median}}$, and Step 5 is performed.
(5) The software will end after all of the pixels have been processed; otherwise, you must slide the windows continually and return to Step 1.

### 16.2.1.2 *Adaptive Bilateral Filter*

Once the guiding $g$ and noise filtering input $I$ are comparable, the BF deteriorates dramatically. As a result, the Gaussian adaptive bilateral filter (GABF) proposes that once a low-pass filtering process is applied to offer range kernel guidance, $g$ and $I$ are non-identical. To construct a low-pass guiding image, the Gaussian blur method, which correlates to low-pass filtering, is first applied to a provided picture $I$. To achieve this, a cumulative sum of pixels at nearby locations with a weight descending from either the center position $I$ is employed. The process can be briefly expressed by the following equation:

$$F(i) = \sum_j W_{i,j}^g(g)F_j. \tag{16.2}$$

The filter kernel is defined as follows:

$$W_{i,j}^g(g) = \frac{1}{K_i^r} e^{\left(-\frac{\|i-j\|^2}{\sigma_s^2}\right)}. \tag{16.3}$$

As the number grows, more averaging is done, and the resulting image loses its substance, sharp edges, and abrupt discontinuities.

For noise filtering inputs, it must be ensured that $g$ and $I$ are non-identical, furthermore the Gaussian-adaptive bilateral kernel is

$$W_{i,j}^{GBF}(I,\bar{g}) = \frac{1}{K_i} e^{\left(-\frac{\|i-j\|^2}{\sigma_s^2}\right)} e^{\left(-\frac{\|F-\bar{j}_j\|^2}{\sigma_r^2}\right)}, \tag{16.4}$$

where $\bar{g}$ is the denoted low-pass guidance. As a result, the GABF's filtering output $I$ is assessed by

$$F(i) = \sum_j W_{i,j}^{GBF}(\mathcal{J},\bar{g})\mathcal{F}_j. \tag{16.5}$$

## 16.2.2 *Feature Extraction*

Wavelet packet transform (WPT) is a variant of discrete wavelet packet (DWT) in which the filtering process is used to decompose both approximations and detailed sub-bands while still decimating the filter outputs. Down sampling is a crucial computational step in WPT. The SWPT, on the other hand, is executed without down-sampling, thus all of the elements in the coefficients are preserved throughout all decomposition levels. For each level $j$, SWPT coefficients may be calculated by

$$S_{j+1,2n}(t) = \sqrt{2}\sum_k H_{j+1}(k)S_{j,n}(2t-k), \tag{16.6}$$

$$S_{j+1,2n+1}(t) = \sqrt{2}\sum_k L_{j+1}(k)S_{j,n}(2t-k) \tag{16.7}$$

where $n$ denoted the node number. The frequency resolution of the SWPT at each level is

$$F_r = \frac{F_s}{2^{j+1}}. \tag{16.8}$$

SWPT coefficient's frequency bandwidth is $\left[\frac{nF_s}{2^{j+1}}, \frac{(n+1)F_s}{2^{j+1}}\right]$.

## 16.2.3 *Feature Selection*

If all available features are included in the testing set, the feature vector will be greater than the feature size. As feature sizes get greater, they will no longer fit in memory, posing a scalability issue. As a result, an optimal feature selection approach is necessary to solve this challenge. The smallest amount of features selected from a huge feature collection is the problem's optimization. Picking characteristics has the key advantage of allowing you to select the best contender characteristics for enhancing the accuracy of classification while minimizing communication complexity, resource consumption, and internal storage requirements. The most important features are chosen at this stage so that the client can grasp the relationship between functionality and classes.

To identify the optimal function for a classification issue, this research presents a hybrid of the adaptive black widow optimization algorithm and moth flame (ABWO-MF) optimization. The choice of ABWO-MF was made in order to improve MF's convergence speed. ABWO-MF, the suggested method, controls an overall population space that ABWO and MF share. To increase population diversity, premature convergence is to be prevented, and convergence speed needs to be accelerated with the moth flame optimization algorithm. This algorithm enables a better convergence speed, therefore, the performance and results can be obtained as most effective. Here, the section mainly covers an ABWO-MF where the hybridization of the ABWO algorithm and moth flame is discussed in detail. In moth flame, it can gradually increase population diversity against premature convergence. In addition, it effectively enables the local optimum. The following section provides the mathematical models of the initial population, procreate, cannibalism, mutation, and convergence.

### 16.2.3.1 *Initial Population*

A basic black widow optimization (BWO) algorithm has four different stages: a random population initialization stage, a procreation stage that outcomes in some of the most promising solution sustaining, a three-layered Cannibalism stage that resembles the exploratory phase, and a mutation stage that allows for effective exploitation of optimum solutions. In addition to expanding the core BWO algorithms to handle two-dimensional challenges, the proposed ABWO features two major structural alterations. A chaotic start-up phase and a moth flame approach are both used to assist the algorithm's convergence capabilities.

*Chaotic initialization*: Instead of using a typical random number generator, a logistic chaotic map is employed to initialize the solution space:

$$X_{i,j} = (X_j^{\min} + (X_j^{\max} - X_j^{\min}) \, \text{ch}(j)), \tag{16.9}$$

where $I$ is the number of separate applicants in the solution space and $j$ is the magnitude of the issue. The potential solution populations $X_{i,j}$ are utilized for minimizing or maximizing the following objective function:

$$\text{RMSE} = \frac{1}{n} \sum_{i=1}^{n} w_i \left( t_i - \widehat{t_i} \right)^2, \tag{16.10}$$

where $N$ denotes the number of samples, $t_i$ represents the true sample value, and $\widehat{t_i}$ corresponds to the predictive value.

### 16.2.3.2 *Procreate*

Each pair is independent in the group where they parallelly act for mating to produce a new generation. As discussed, they individually process mating in the web from other spiders. Therefore, in the real-time process, they produced $10K$ eggs approximately. However, the fittest spider or strongest spider in the web only survives. In this algorithm, an array is considered for the reproduction process, and this array-based reproduction is carried out until a widow array with random numbers is available. Then, $\mu$ is denoted for creating an offspring based on the following equations:

$$y_1 = \mu \times x_1 + (1 - \mu) \times x_2, \tag{16.11}$$

$$y_2 = \mu \times x_2 + (1 - \mu) \times x_1, \tag{16.12}$$

where $x_1$ and $x_2$ represent the parents, $y_1$ and $y_2$ represent the offspring, $i$ and $j$ can be represented in the range of 1 to $N$, and $\mu$ can be determined in the random range of 0 and 1.

### 16.2.3.3 *Cannibalism*

Cannibalism can be executed in three types, such as sexual cannibalism, child cannibalism, and sibling cannibalism. In sexual cannibalism, the male spider is eaten by the female while mating or after mating. Here, fitness value is highly considered in this process. The second cannibalism is that children eat their parents based on their fitness value to determine whether the spiderlings are weak or strong. Likewise, sibling cannibalism is spider eats its sibling if it is weak. In this algorithm, the cannibalism rate is determined for computing the survivor rate.

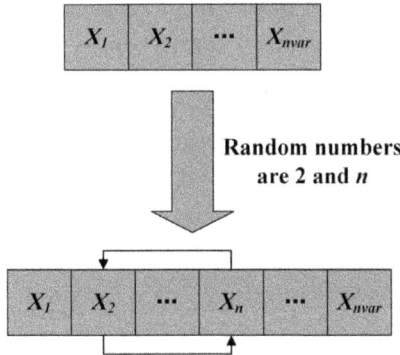

Fig. 16.2   The schematic block diagram of the proposed methodology.

#### 16.2.3.4 *Mutation*

The mutation process is executed based on the random selection process for producing the population by selecting the mute pop number. In Fig. 16.2, the two elements in the array at random are to be exchanged based on the chosen solution. The mutation rate is used to calculate mute pop.

#### 16.2.3.5 *Convergence*

To increase population diversity, premature convergence is to be prevented, convergence speed needs to be accelerated with the MF algorithm. The moth-fly has the unique properties of increasing population diversity sequentially. This process helps come out from the local optimum. Moreover, this method is efficient for the exploration and exploitation process using moth flame optimization (MFO).

Here, the convergence process of the modified MF algorithm is incorporated based on applying Eq. (16.13). The mathematical formulation of this process is as follows:

$$X_i^{i+1} = X_i^t + u \, \text{sign} \, [r \, \text{and} - 0.5] \oplus \text{Levy}(\beta), \tag{16.13}$$

where $X_i^t$ signifies the $i$th solution vector or moth, $X_i$ denotes the number of iterations, $t, u$ denotes the random parameter for uniform distribution, $\oplus$ describes the dot product (entry-wise multiplications), and rand is the random initialization range of $[0, 1]$.

The representation is presented with the symbol [rand 0.5], with just three possible values being 0, 1, and −1. Similarly, the $u$ sign [rand 0.5] is combined in Eq. (16.13) and the moth flame may do the random walk.

With the introduction of modified MF in ABWO, the local minima can be minimized and the global search capacity can be increased.

Moreover, the MF process mainly relies on random walks where step length helps determine the step in its process, and the Levy distribution process is confirmed by the jumps process. The mathematical operation of this process is as follows:

$$\text{Levy}(\beta)\mu = t^{-1-\beta}, \quad (0 \le \beta \le 2), \tag{16.14}$$

$$\text{Levy}(\beta)\frac{\theta \times \mu}{|v|^{1/\beta}}, \tag{16.15}$$

where $\mu$ and $v$ denote the standard normal distributions, $\Gamma$ denotes the standard Gamma Function, $\beta = 1.5$, and $\Phi$ is defined as follows:

$$\phi = \left(\Gamma(1+\beta) \times \sin(\pi \times \beta/2) \mid \Gamma\left(((1+\beta)/2) \times \beta \times 2^{(\beta-1/2)}\right)\right)^{1/\beta}. \tag{16.16}$$

Here, the global search ability of the proposed algorithm is improved by the incorporation of a random walk with levy-flight to eliminate the weakness of the MFO algorithm.

With this process, the local minimum process can be mitigated to generate a successful result, especially multi model benchmark functions and unimodal functions.

### 16.2.4 *Segmentation by AKFCM Algorithm*

The feature values are supplied into the clustering procedure for segmentation once the features have been chosen. The adaptive kernel fuzzy $C$-means algorithm is used in this proposed methodology to execute an effective clustering procedure, which may deliver superior performance than other techniques. The clustering centroid $x_i$ must be chosen carefully for the best segmentation results. To improve the segmentation effect, keep updating the cluster center $x$, $I$ and clustering with the new centroid until you have the best clustering classification results. The membership function linked with the cluster centroid $x$ is updated on a regular basis since it is altered. The following is a description of the fuzzy clustering technique for picture segmentation:

**Step 1:** Initialize cluster centroid $x_i$, the convergence requirement $\varepsilon$, the total of the neighborhood pixel there in filtering frame $D_k$, as well as the cluster quantity $c$.

**Step 2:** For both the filtering image and the adaptive weighting means, use the equation to calculate the adaptive weighting means:

$$\bar{y}_i = \frac{\sum_{j=1}^{D_k} m_j \times y_i}{\sum_{j=1}^{D_k} m_j}. \tag{16.17}$$

**Step 3:** Considering the equation, find the optimal membership function:

$$\bar{z}_{ik} = \frac{\frac{\alpha}{S_r} \sum_{r \in R_k} \bar{z}_{ir} - \sum_{i=1}^{c} (1 - K(y_k, x_i) + \alpha)}{(1 - K(y_k, x_i) + \alpha)}. \tag{16.18}$$

**Step 4:** To use the optimal membership function as well as the clustering centroid, optimize this objective function with the following equation:

$$O_{\text{AKFCM}} = \sum_{i=1}^{c} \sum_{k=1}^{D} (K(y_k, x_i) \bar{z}_{ik})^p (1 - K(y_k, x_i))$$

$$+ \frac{\alpha}{S_r} \sum_{i=1}^{c} \times \sum_{k=1}^{D} (K(y_k, x_i) \bar{z}_{ik})^p \sum_{r \in R_k} (1 - K(y_k, x_i)). \tag{16.19}$$

**Step 5:** Determine the absolute value. $|x_{i+1} - x_i|$, iteration is ended until to $\max_i |x_{i+1} - x_i| < \varepsilon$, return to Step 3 if necessary.

**Step 6:** The best membership degree rules are used to complete the clustering segmentation.

## 16.2.5 *Classification Using Hybrid CNN-LSTM*

The input, hidden, and output layers make up the three layers of the CNN model. A convolutional layer is commonly given the input of a 3D array, height, and weight, and also the number of connections reflect the dimensions:

$$a^1(i, h) = (w_h^1 \times x)(i) = \sum_{j=-\infty}^{\infty} w_h^1(j) \times (i - j). \tag{16.20}$$

The hidden level defines a convolutional layer, a pooling layer, and a fully connected layer. The convolutional layer is responsible for automatically featuring extraction from various parts of the extraction of information images. Neurons in one convolution layer are only linked to a tiny portion of the neurons in the next layer. As a result, the convolutional operation is performed by the filter using a shared weight matrix. The pooling layer sums together all of the values in the pooling window. With a max-pooling operation, which chooses the maximum value from each sub-area of the preceding layer, this transformation reduces the size of the input layer. Furthermore, this layer lowers the learning process's computing cost and handles any overfitting difficulties. In the level that is concealed $l = 2, \ldots, L$, the characteristic pattern of the input $f^{l-1} \epsilon R^{1 \times N_{l-1}} \times M_{l-1}$, where $1 \times N_{l-1} \times M_{l-1}$ is the size of the preceding convolution's output features with $N_{l-1} = N_{l-2} - k + 1$, is constrained by a collection of $M_1$ features $w_h^1 \epsilon R^{1 \times k \times N_{l-1}}$, $h = 1, \ldots, M_1$, to create a feature map $a^1 \times R^{1 \times N_{l-1} \times M_{l-1}}$ as follows:

$$a^1(i, h) = \left( w_h^1 \times x \right)(i) = \sum_{j=-\infty}^{\infty} \sum_{m=1}^{M_{l-1}} w_h^1(j, m) f^{l-1}(i - j, m). \qquad (16.21)$$

To create a complete output, a fully connected layer flattens and combines the high-level retrieved properties learned by the convolution layer. Many memory blocks are connected through layers in a long short-term memory (LSTM) network. Each layer has a set of recurrently connected memory cells as well as three multiplicative units: input, forget, and output gates. The LSTM can predict the variable $y = (y_1, y_2, \ldots, y_{t-1}, y_t)$ by making the gates more modern (output gate $y_t$, input gate $i_t$, and forget gate $f_t$) within a memory cell $c_t$, from moment to time $t = 1$ to $T$. The following are the mathematical formulae for the LSTM:

$$i_t = \sigma \left( w_i x_t + R_i h_{t-1} + b_i \right), \qquad (16.22)$$

$$f_t = \sigma \left( w_f x_t + R_f h_{t-1} + b_f \right), \qquad (16.23)$$

$$y_t = \sigma \left( w_y x_t + R_y h_{t-1} + b_y \right), \qquad (16.24)$$

$$c_t = f_t c_{t-1} + i_t c_b \qquad (16.25)$$

$$c_t = \sigma \left( w_c x_t + R_c h_{t-1} + b_c \right), \qquad (16.26)$$

$$h_t = y_t \sigma \left( c_t \right), \qquad (16.27)$$

where $x_t$ is the input vector, $w_i, w_f$, and $w_y$ are the weight matrix from the input, forget, and output gates to the input, respectively, $R_i, R_f$, and $R_y$ are the weight matrix from the input, forget, and output gates to the input, respectively, and $b_i, b_f$, and $b_y$ seem to be the gate bias vectors for the input, forget, and output gates, correspondingly, $c_{t-1}$. The preceding cell and its output vector are $c_{t-1}$ and $h_{t-1}$ and $h_t$, respectively, and the output vector is $h_t$.

## 16.3 Results and Discussions

The validation of the suggested approach is discussed in this part using performance matrices, such as precision, accuracy, loss, and $F$-measure. The experimental procedure is carried out using MathWorks' MATLAB program. A hybrid deep-learning algorithm is used to identify brain tumors. To select and categorize the features, the convolutional neural network is linked with long short-term memory (CNN LSTM). The median filter is utilized for pre processing, while ABWO-MF is employed for optimization. The continuity equation should be used to generate the values of the factors underlying in order to measure the performance of the recommended strategy.

### 16.3.1 *Performance Metrics*

The section contains the suggested model's measurements, some of which have been evaluated and verified. The study of the performance metrics employed in the scheduled and current evaluations is presented mathematically as follows:

**Accuracy:** The ratio of accuracy to precision is defined as several true patterns to the sum of all patterns. The given formula is described for accuracy:

$$\text{Accuracy} = \frac{\text{TP} + \text{TN}}{\text{TP} + \text{FN} + \text{FP} + \text{TN}}. \tag{16.28}$$

**Precision:** Precision is the ratio predicted based on positive values to the overall predicted positive values. The formula for predicting the precision is

$$\text{Precision} = \frac{\text{TP}}{\text{TP} + \text{FP}}. \tag{16.29}$$

**Recall:**   Recall measure is utilized for obtaining the true positive and the false negative values. The recall is also known as sensitivity which is shown in the following equation:

$$\text{Recall} = \frac{\text{TP}}{\text{TP} + \text{FN}}. \qquad (16.30)$$

**F-Score:**   F-score is a mutual combination of precision and recall value that helps compute the score. With this observation, the weighted average of recall and precision can be obtained. Here, 1 is the best value and 0 is the worst score. The following is the equation of $F_{\text{Score}}$:

$$F_{\text{Score}} = 2 \cdot \frac{\text{Precision} \times \text{recall}}{\text{Precision} + \text{recall}}. \qquad (16.31)$$

### 16.3.2   *Performance Analysis*

All of the aforementioned statistical statistics are used to demonstrate that the proposed model outperforms current models.

In Table 16.1, 0.9785 represents accuracy, 0.9538 represents precision, 0.9781 represents recall, 0.96 represents $F$-measure, and 0.7840 represents the suggested method's loss. The artificial neural network (ANN) approach has an accuracy value of 0.8316, a precision rating of 0.7963, a recall rate of 0.7792, an $F$-measure value of 0.83, and a loss rate of 0.9325. For the support vector machine (SVM) technique, overall $F$-measure, precision, recall, accuracy, and loss value, are 0.7225, 0.7212, 0.7356, 0.73 and 1.2566, respectively. Furthermore, the accuracy is 0.8641, the precision is 0.8365, the recall is 0.8254, the $F$-measure is 0.86, and the loss is 1.6522 with the deep neural network (DNN) approach.

Figure 16.4 shows the accuracy result in terms of epoch with 100 iterations. Moreover, the square symbol line indicates the accuracy. Therefore,

Table 16.1.   Performance comparison of the proposed and existing methods.

| Methods | Performance of classifier | | | | |
| | Accuracy | Precision | Recall | F-measure | Loss |
|---|---|---|---|---|---|
| DNN | 0.8641 | 0.8365 | 0.8254 | 0.86 | 1.6522 |
| SVM | 0.7225 | 0.7212 | 0.7356 | 0.73 | 1.2566 |
| ANN | 0.8316 | 0.7963 | 0.7792 | 0.83 | 0.9325 |
| Proposed | 0.9785 | 0.9538 | 0.9781 | 0.96 | 0.7840 |

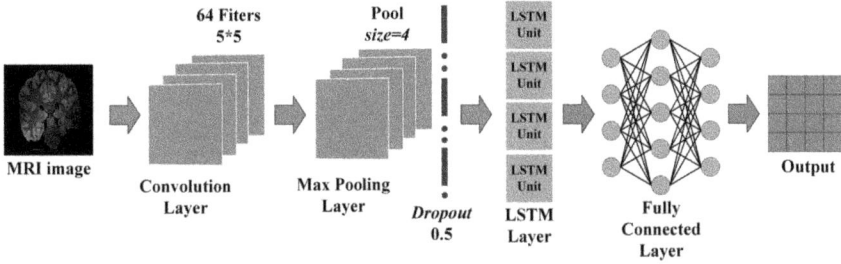

Fig. 16.3   The execution process of the CNN-LSTM model.

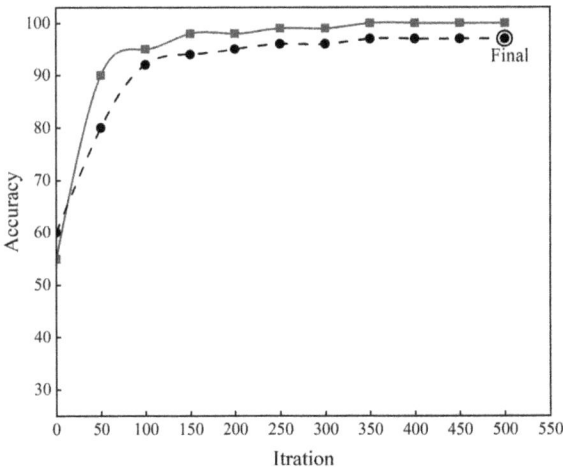

Fig. 16.4   Accuracy-based results based on the graphical representation.

the performance of the proposed can be trained and validated for successful evaluation process. Thus, the performance shows high efficiency for proposed. The proposed model is obtained from an open-source internet area.

The obtained outcomes of the training and validation procedure for the supplied dataset are graphically displayed in the following photos to highlight the clarity of the recommended brain tumor-based CNN LSTM technique identification. The proposed and existing techniques' performance characteristics, such as precision, recall, $F$-measure, accuracy, and loss, are compared in the graphs here.

According to the data, the recommended technique has a high detection rate. The accompanying graph demonstrates the great accuracy of CNN

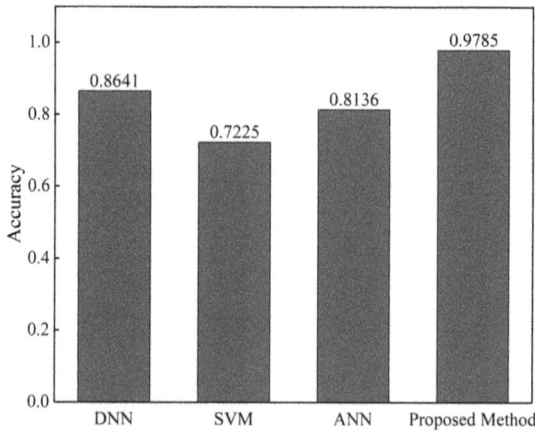

Fig. 16.5    Proposed and existing methods' accuracy.

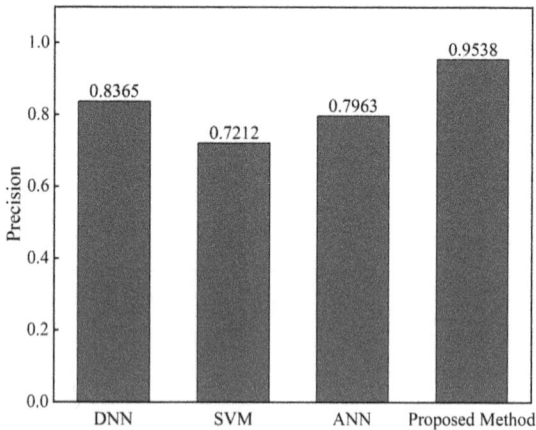

Fig. 16.6    Precision.

LSTM-based brain tumor identification. When the number of iterations is increased, for example, accuracy improves steadily. The precision for proposed and present algorithms like ANN, SVM, and DNN is shown in Fig. 16.5 This recommended strategy has a higher accuracy rating of 0.9785 than the other current solutions. The ANN has an accuracy of 0.8316, the SVM has an accuracy of 0.7225, and the DNN has an accuracy of 0.8641. The proposed technique is more precise than the other technique.

Figure 16.6 shows the precision of proposed and present algorithms like ANN, SVM, and DNN. The accuracy value of 0.9538 for the recommended

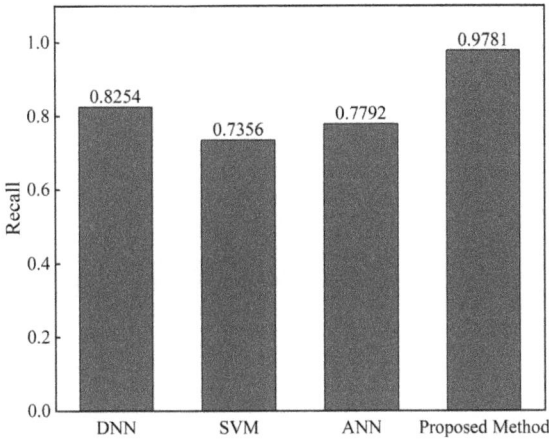

Fig. 16.7   Recall.

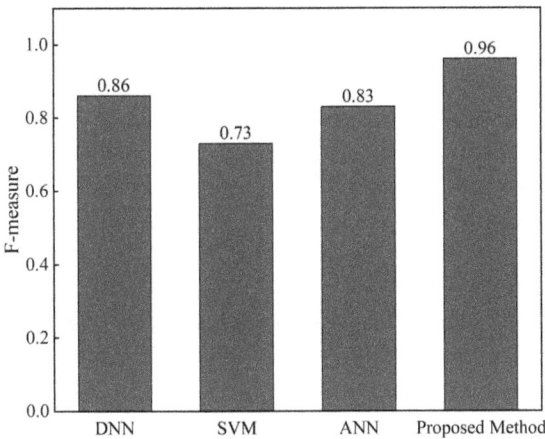

Fig. 16.8   *F*-measure.

method is better than the other current methods. The accuracy of ANN is 0.7963, the accuracy of SVM is 0.7212, and the accuracy of DNN is 0.8365. The suggested strategy, according to the analysis, is more accurate than the other ways.

Figure 16.7 shows how proposed and present algorithms like ANN, SVM, and DNN compare in terms of accuracy. The recommended strategy has a higher accuracy value of 0.9781 than the other current solutions. ANN's accuracy is 0.7792, SVM's accuracy is 0.7356, and DNN's accuracy is 0.8254.

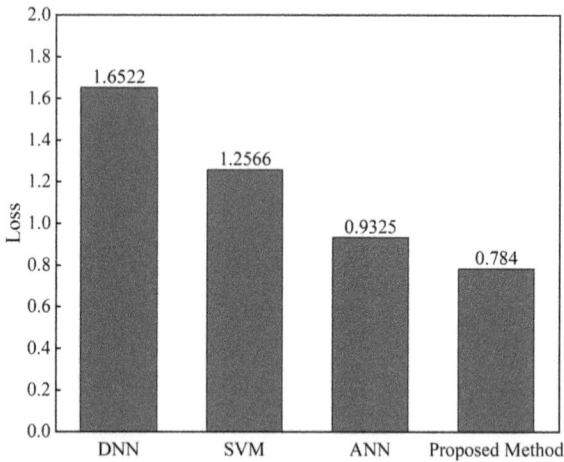

Fig. 16.9　Loss.

The recommended technique, according to the study, is more precise than other techniques.

The efficiency of proposed and present algorithms such as ANN, SVM, and DNN is shown in Fig. 16.8. The recommended strategy has a higher accuracy value of 0.96 than the other current solutions. ANN has an accuracy of 0.83, SVM has an accuracy of 0.73, and DNN has an accuracy of 0.86. The recommended technique, according to the study, is more precise than the other techniques.

The suggested attribute selection with the CNN-LSTM model results in a loss with the smallest detection error, as seen in the above image. The detection error of the suggested solution diminishes when the iteration is increased. As a consequence, the proposed system is found to increase accuracy and error function. Figure 16.9 compares the loss of the proposed technique to that of current methods, like ANN, SVM, and DNN. The loss value for the suggested technique is 0.7840, 0.9325 for the ANN, 1.2566 for SVM, and 1.6522 for DNN.

# Chapter 17

# Speech Enhancement Method Combining Wavelet and Deep Learning

Noise can have an impact on the performance of today's automatic speech recognition systems, hearing aids, and other applications, so it is essential to use speech enhancement techniques to improve the quality of speech. For this reason, many speech enhancement techniques have been proposed and implemented over the past decades to meet a large number of possible conditions and applications. Recently, deep learning-based approaches have surpassed previous methods in many aspects of speech enhancement, and hybrid approaches are considered as a possibility to extend the capabilities of individual methods, thus increasing their application capabilities. In this chapter, we evaluate a hybrid method that combines wavelet transform and deep learning [Gutiérrez-Muñoz and Coto-Jiménez, 2022]. Numerous experiments were conducted to select the appropriate wavelets and train the neural network, and the experiments evaluate whether the hybrid approach is beneficial for speech enhancement tasks under several types and levels of noise, providing useful information for future implementations. This chapter consists of the following four sections.

## 17.1 Related Work

This section focuses on the hybrid approaches to speech denoising and previous experiences with wavelet transform presented in the literature. The application of deep neural networks as an isolated algorithm for this purpose has been reported in a number of publications and reviewed recently in the works of Azarang and Kehtarnavaz [2020] and Lun *et al.* [2012].

The wavelet-denoising-based references usually specify the problem of the threshold in the wavelet functions and measure the signal-to-noise ratio (SNR) before and after the application of the functions. For example, in a recent report presented in the work of Tan *et al.* [2020], four threshold selection methods were applied using sym4 and db8 wavelets. Some authors provide experimental validation for different noisy signals and proved that the denoising method of the speech signal based on wavelet analysis is effective in enhancing noisy speech signals.

A two-stage wavelet approach was developed in the work of Balaji *et al.* [2021], first by estimating the speech presence probability and then by removing the coefficients of the noise floor. Results in speech degraded with pink and white noise from an SNR of 0 to an SNR of −8 surpassed several classical algorithms.

A hybrid of deep learning and the vector Wiener filter was presented recently in the work of Bahadur *et al.* [2021], showing benefits from the combined application of algorithms. Other than the deep learning-based hybrid approach, contemplating harmonic regeneration noise reduction and a comb filter was reported in the work of Lun and Hsung [2010] and validated using also subjective measurements. Another two-stage estimation algorithm based on wavelets was proposed in the work of Bahoura and Rouat [2006], as a previous stage to more traditional algorithms, such as the Wiener filter and minimum mean-squared estimation (MMSE).

One implementation of the wavelet transform for enhancing noisy signals in ranges of the SNR from −10 to SNR 10, with a great variety of natural and artificial noises, was presented in the work of Bouzid *et al.* [2016]. The success of the proposal was observed, especially for lower SNR levels.

Hybrid approaches that combine wavelets and other techniques for speech enhancement are also part of the proposals presented in the literature. For example, a combination of wavelets and a modified version of principal component analysis (PCA) was presented in the work of Ram and Mohanty [2019]. The results showed relevant noise reduction for several kinds of artificial and natural noises and a lower signal distortion without introducing artifacts.

In terms of wavelets and deep learning hybrid approaches, some recent experiences were explored in the work of Mihov *et al.* [2009], by applying the wavelet transform for the decomposition of the signal, and in a second stage, the radial basis function network (RBFN). The performance of the proposal was described as excellent by the authors, using objective measures, such as the segmental signal-to-noise ratio (SegSNR) and perceptual evaluation of speech quality (PESQ).

This chapter takes advantage of the application of wavelets as presented in the work of Tan *et al.* [2020], with a hybrid approach similar to those of Ram and Mohanty [Mihov *et al.*, 2009], but with the incorporation of initialized LSTM networks using transfer learning.

## 17.2 Problem Statement

The purpose of speech enhancement of a noisy signal is to estimate the uncorrupted speech signal from the degraded speech signal. Several speech-denoising algorithms estimate the characteristics of the noise from silent segments in the speech utterances or by mapping the signal into new domains, such as with the wavelet transform.

This section considered segments of noisy $\mathbf{y}_t$ and clean $\mathbf{s}_t$ speech to compare the enhancement using wavelets, deep learning, and both methods in cascade. Chui [1992b] stated the enhancing process using wavelets can be summarized as follows: Given $W(\cdot)$ and $W^{-1}(\cdot)$, the forward and inverse wavelet transform operators, and $D(\cdot, \lambda)$, the denoising operator with threshold $\lambda$, the process is performed using the following three steps:

- Transform $y(t)$ using a wavelet: $Y = W(y(t))$.
- Obtain the denoised version using the threshold, in the wavelet domain: $Z = D(Y, \lambda)$.
- Transform the denoised version into the time domain: $\tilde{s}_1 = W^{-1}(Z)$.

On the other hand, the enhancement using ANNs is performed by learning a mapping function $f$ between the spectrum of $\mathbf{y}_t$ and $\mathbf{s}_t$ with the criteria

$$\min \sum_{t=1}^{T} \|\mathbf{s}_t - f(\mathbf{y}_t)\|^2, \tag{17.1}$$

where $f$ is approximated using a recurrent neural network, which outputs a version of the denoised signal $\tilde{s}_2$ after the training process.

In the hybrid approach, the first step of wavelet denoising provides $\tilde{\mathbf{s}}_t$ to the neural networks, which are trained with the criteria:

$$\min \sum_{t=1}^{T} \|\mathbf{s}_t - f(\tilde{\mathbf{s}}_t)\|^2, \tag{17.2}$$

with the purpose of obtaining $\tilde{s}$, a better approximation of $s_t$ than $\tilde{s}_1$ and $\tilde{s}_2$.

## 17.3  Materials and Methods

In this section, the main techniques and procedures to establish the experimental setup to evaluate the proposed hybrid approach are presented.

### 17.3.1  *Wavelets*

Wavelets are a class of functions that have been successfully applied in the discrimination of data from noise data, emulating a filter. The wavelet transform uses an infinite set of functions of different scales and at different locations to map a signal into a new domain, the wavelet domain [Chavan and Mastorakis, 2010].

It has become an alternative to the Fourier transform and can be related to similar families of function transformations but with a particular interest in the scale or resolution of the signals.

In the continuous-time domain, a wavelet transform of a function $f(t)$ is defined as

$$\text{CWT}_{\Psi} f(a,b) = W_f(a,b) = |a|^{\frac{1}{2}} \int_{-\infty}^{\infty} f(t) \Psi \left( \frac{t-b}{a} \right) dt, \qquad (17.3)$$

where $a \neq 0$ and $a, b$ are real numbers that represent dilating and translating coefficients. The function $\Psi(t)$ is called the mother wavelet and requires the property of having a zero net area. There is a variety of mother wavelet functions, for example: Haar, Symlet, and Ricker, among many others. Different values of $a$ and $b$ provide variants of scales and shifts of the mother wavelet, as shown in Fig. 17.1.

The fundamental idea behind wavelets is to analyze the functions according to the scale [Priyadarshani *et al.*, 2016], representing them as a combination of time-shifted and scaled representations of the mother wavelet. For the selection of the best mother wavelet for a particular application, an experimental approach needs to be implemented [Al-Qazzaz *et al.*, 2014]. For example, in the case of electroencephalogram (EEG) signals, more than 40 mother functions were tested in the work of Gargour *et al.* [2009], to determine the Symlet wavelet of order 9 as the best option for that problem.

The wavelet transform provides coefficients related to the similarity of the signal with the mother function. A detailed mathematical description of wavelets can be found in the works of Mallat [1999], Taswell [2000] and Donoho and Johnstone [1994].

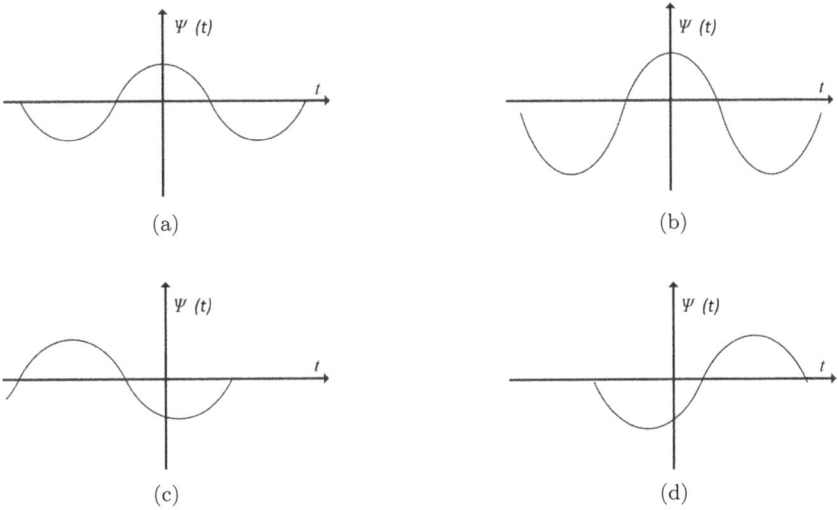

Fig. 17.1   Different scales and shifts of the Ricker wavelet, also known as the "Mexican hat" wavelet.

The application of wavelets for denoising signals using thresholding emerged in the 1990s from the works [Donoho, 1995; Oktar *et al.*, 2016]. The threshold can be of two types: soft thresholding and hard thresholding, and the idea is to reduce the magnitude or completely remove the coefficients in the wavelet domain.

The process of denoising using this approach can be described using the following steps [Verma and Verma, 2012]:

(1) Apply the wavelet transform to the noisy signal to obtain the wavelet coefficients.
(2) Apply the thresholding function and procedure to obtain new wavelet coefficients.
(3) Reconstruct the signal by inverse-transforming the coefficients after the threshold.

Valencia *et al.* [2016], stated that wavelet denoising gives good results in enhancing noisy speech in the case of white Gaussian noise. Wavelet denoising is considered a non-parametric method. The choice of the mother wavelet function determines the final waveform shape and has an important role in the quality of the denoising process.

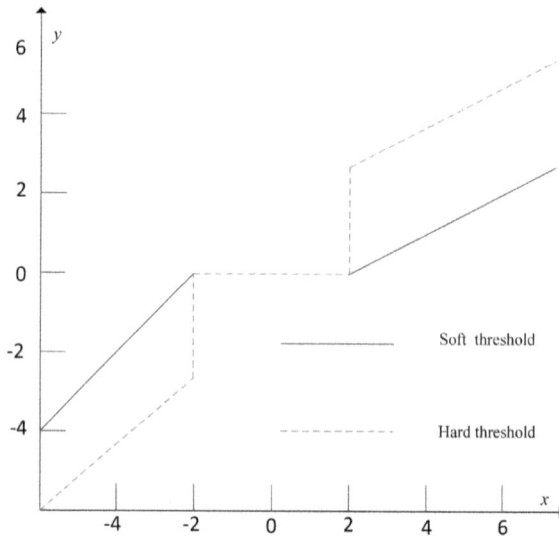

Fig. 17.2    Illustration of hard and soft thresholding for wavelet coefficients.

### 17.3.1.1  *Thresholding*

The threshold process affects the magnitude or the amount of coefficients in the wavelet domain. The two most popular approaches are hard
thresholding and soft thresholding. In the first type, hard thresholding,
the coefficients, whose absolute values are lower than $\lambda$, are set to zero.
Soft thresholding performs a similar operation but also shrinks the non-
zero coefficients. The two types of thresholding are illustrated in Fig. 17.2.

   To implement the thresholds, several estimation methods are available
in the literature. Four of the well-known standard threshold estimation
methods are as follows [Schimmack and Mercorelli, 2018, 2019]:

(1) **Minimax criterion:** In statistics, the estimators face the problem of
    estimating a deterministic parameter from observations. The minimax
    method minimizes the cost of the estimator in the worst case. For the
    case of threshold selection, the principle is applied by assimilating the
    denoised signal to the estimator of the unknown regression function.
    This way, the threshold can be expressed as

$$\lambda = \begin{cases} \sigma\left(0.336 + 0.1829\log_2 N\right) & N > 32, \\ 0 & N < 32, \end{cases} \tag{17.4}$$

where $\sigma = \text{median}\left(\frac{|w|}{0.6745}\right)$ and $\omega$ is the wavelet coefficient vector of length $N$.

(2) **Sqtwolog criterion:** The threshold is calculated using the following equation:

$$\lambda_j = \sigma_j\sqrt{2\log(N_j)}, \tag{17.5}$$

where $\sigma_j$ is the median absolute deviation (MAD) and $N_j$ is the length of the noisy signal at the $j$th scale.

(3) **Rigrsure:** The soft threshold can be expressed as

$$\lambda = \sigma\sqrt{\omega_b}, \tag{17.6}$$

where $\omega_b$ is the $b$th squared wavelet coefficient chosen from a vector consisting of the squared values of the wavelet coefficients and $\sigma$ is the standard deviation.

(4) **Hersure:** The threshold combines Sqtwolog and Rigrsure, given the property that the Rigrsure threshold does not perform well at a low SNR. In such a case, the Sqtwolog method gives better threshold estimation. If the estimation from Sqtwolog is $\lambda_1$ and from Rigrsure is $\lambda_2$, then Hersure uses

$$\lambda = \begin{cases} \lambda_1 & A > B, \\ \min(\lambda_1, \lambda_2) & A \geq B, \end{cases} \tag{17.7}$$

where given the length of the wavelet coefficient $N$ and $s$, the sum of squared wavelet coefficients, the values of $A$ and $B$ are calculated as

$$A = \frac{s - N}{N}, \tag{17.8}$$

$$B = (\log_2 N)^{\frac{3}{2}}\sqrt{N}. \tag{17.9}$$

### 17.3.1.2 *No Thresholding Alternative*

Research on the implementation of wavelet denoising without using a threshold can be found the works of Goodfellow *et al.* [2016] and Abiodun *et al.* [2018]. This approach considers using a functional analysis method based on the entropy of the signal, and this algorithm takes advantage of a constructive structural property of the wavelet tree with respect to a defined seminorm; it consists of searching for minima for the low-frequency domain and other minima for the high-frequency domain.

## 17.3.2 *Deep Learning*

Deep learning is a subset of machine learning techniques that allows computers to process information in terms of a hierarchy of concepts [LeCun *et al.*, 2015]. Typically, deep learning is based on artificial neural networks, which are known for their capacity as universal function approximations with good properties of self-learning, adaptivity, and advancement in input to output mapping. With this capacity, computers can learn complex operations and functions by building them out of simpler ones.

Previous to the development of deep learning techniques and algorithms, other approaches were almost unable to process natural data in their raw form. For this reason, the application of pattern recognition or machine learning systems requires domain expertise to understand the problems, obtain the best descriptors, and apply the techniques using feature vectors that encompass the descriptors [Waseem *et al.*, 2021].

The most common form of deep learning is by composing layers of neural network units. The first level receives the data (in some cases, raw data), and subsequent layers perform other transformations. After several layers of this process, significantly complex functions can be learned. The feedforward deep neural network, or multi layer perceptron (MLP) with more than three hidden layers, is a typical example of a deep learning algorithm. The architecture of an MLP organized with multiple units, inputs, outputs, and weights can be represented graphically [Purwins *et al.*, 2019], as shown in Fig. 17.3.

The purpose of the training process of a deep neural network is to approximate some mapping function $f(x; \theta)$, where $x$ are inputs and $\theta$ the networks' parameters, such as the value of the connections between units and the hyperparameters of learning (e.g., the learning rate and the bias). One of the most relevant aspects of deep learning is that the parameter $\theta$ is learned from data using a general-purpose learning procedure. This way, deep learning has shown its capability to solve problems that have resisted previous attempts in the artificial intelligence community.

The success of deep learning in speech recognition and image classification in 2012 is often cited as the leading result of the renaissance of deep learning, using architectures and approaches, such as deep feedforward neural networks, convolutional neural networks (CNNs), and LSTM.

One of the most important architectures of deep neural networks applied to signal processing is autoencoders. Autoencoders are designed to reconstruct or denoise input signals. For this reason, the output presents the

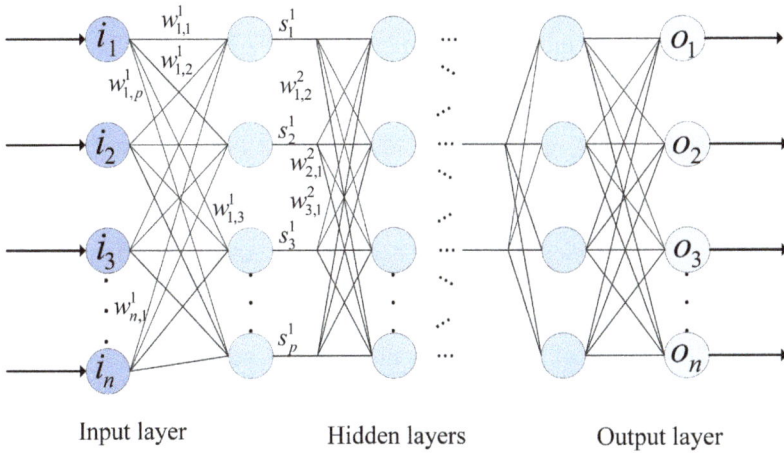

Fig. 17.3   Illustration of an MLP. Information flows from inputs to outputs through connections between unit $i$ and unit $j$ denoted as $w_j^i$. In each node, outputs $s_k^i$ are produced and propagated toward the outputs $o_m$ of the network. Hidden layers may differ in the number of units. Each layer performs a function from the inputs (or the outputs of the previous layer) to the outputs, using activation functions defined in each unit and the value of the weight of the connections. For example, a network with three layers defines functions $f_1(\cdot)$, $f_2(\cdot)$, $f_3(\cdot)$, and this way, the whole network performs a function from inputs $x$ to the outputs defined as $f(x) = f_3\left(f_2\left(f_1(x)\right)\right)$.

same dimensionality as the inputs. Thus, autoencoders consist of encoding layers and decoding layers. The first stage removes redundant information in the input, while decoding layers reverse the process. With proper training, pairs of noisy/clean parameters can be presented to the autoencoder, and the approximation function gives denoising properties to the network.

A massive amount of data is often required, given the huge amount of parameters and hyperparameters of autoencoders and the deep networks in general. Furthermore, the recent advances in machine parallelism, such as cloud computing and GPUs, are of great importance to perform the training procedures in a short time [Westhausen and Meyer, 2020].

From this experience, Gutiérrez-Muñoz and Coto-Jiménez [2022] selected the recent implementation of the stacked dual signal transformation LSTM network (DTLN). This implementation combines a short-time Fourier transform (STFT) and a pre-trained stacked network. This combination enables the DTLN approach to extract information from magnitude spectra and incorporate phase information, providing state-of-the-art performance.

The DTLN has 128 units in each of its LSTM layers. The networks' inputs correspond to information extracted from a frame size of 32 ms and a shift of 8 ms, using an FFT of size 512. An internal convolutional layer with 256 filters to create the learned feature representation is also included. During training, 25% of dropout is implemented between the LSTM layers. The optimization algorithm applied to update the network weights was Adam, first presented in the work of Mercorelli [2017], using a learning rate of $10 \times e^{-3}$. This implementation is capable of real-time processing, showing state-of-the-art performance. Its architecture combines LSTM, dropout, and convolutional layers, resulting in a total of 986753 trainable parameters. Further details of the implementation can be found in the work of Kominek and Black [2004].

### 17.3.3 *Proposed System*

In order to test the proposed method, the first step is to generate a dataset of noisy speech with both natural and artificial noise at several SNR levels. This procedure establishes parallel data of clean and noisy speech and allows the comparison of speech quality before and after the application of the denoising procedures.

This chapter is on the combination of wavelet-based and deep-learning-based speech denoising, with the purpose of comparing the performance of both separately and analyzing the suitability of both in a two-stage approach. In the case of wavelet-based denoising, the following four steps were applied in an extensive experimentation, according to the description presented in Section 17.3.1:

(1) Select a suitable mother wavelet.
(2) Transform each speech signal using the mother wavelet.
(3) Select the appropriate threshold to remove the noise.
(4) Apply the inverse wavelet transform to obtain the denoised signal.

There is a variety of criteria that can be used to choose the mother wavelet, such as the ones presented in the work of Rix *et al.* [2001]. In this case, an experimental approach was implemented, following a process of trial and error with commonly used wavelet families for speech denoising, such as Daubechies, Symlet, and biorthogonal. Different wavelets from each family (using common ranges) were tested using objective measures, and the wavelets with the best results in each case were tested again; finally, the wavelet with the best results among the wavelet families was selected.

For the application of deep-learning-based denoising, the procedure can be summarized in the following steps:

(1) *Select one architecture of the network.* In the experiments Gutiérrez-Muñoz and Coto-Jiménez [2022] used the stacked dualsignal transformation LSTM network architecture presented in the work of Kominek and Black [2004]. The architecture was based on two LSTM layers followed by a fully connected (FC) layer.
(2) Train the deep neural network with pairs of noisy and clean speech at the inputs and at the outputs. For the case of the hybrid approach, the outputs of the wavelet denoising were used as the inputs of the neural network, which were re-trained completely using pairs of wavelet-based denoising and clear speech.
(3) Establish a stop criterion for the training procedure.

As in the case of wavelets, objective measures can be applied to validate the benefits of the deep neural networks in each noise type and level. With the purpose of performing a proper comparison, the same amount of epochs for training the deep neural networks was used for both (noisy, clean) and (wavelet-denoised, clean) procedures. Additionally, for the sake of completeness in the experiments, we also considered a two-stage approach with the application of wavelet denoising to the results of the deep-learning-based denoising procedure. This experimental approach can be summarized in four possibilities to implement and compare, as illustrated in Fig. 17.4.

### 17.3.4  *Experimental Setup*

In this section, a detailed description of the data and the evaluation process is presented.

#### 17.3.4.1  *Dataset*

In order to test the hybrid proposal based on wavelets and deep learning, Gutiérrez-Muñoz and Coto-Jiménez [2022] chose the CMU ARCTIC databases, constructed at the Language Technologies Institute at Carnegie Mellon University. The recordings and their transcriptions are freely available in the work of Rix *et al.* [2002]. The dataset consists of more than 1100 recorded sentences, selected from Project Gutenberg's copyright-free texts. The recordings were sampled at 16 KHz in WAV format.

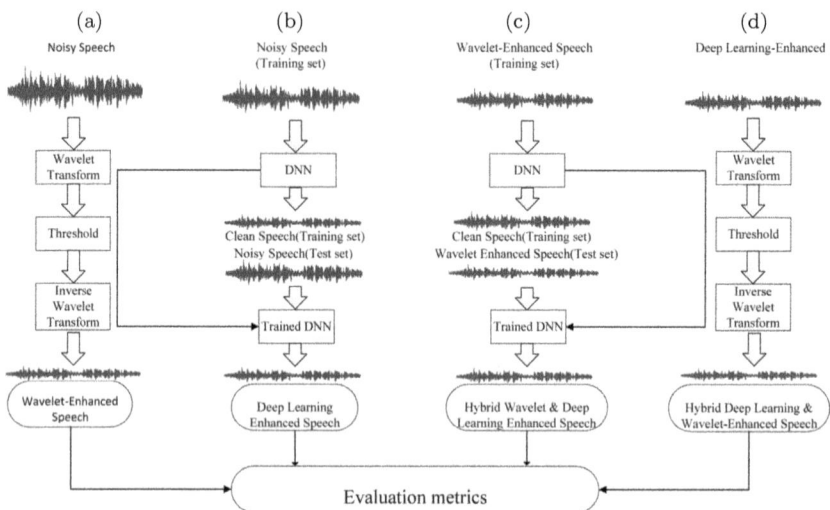

Fig. 17.4 The four implementations for experimental setup: (a) Wavelet enhancement; (b) deep learning enhancement; (c) wavelet + deep learning enhancement; (d) deep learning + wavelet enhancement.

Four native English speakers recorded each sentence, and their corresponding files were labeled as bdl (male), slt (female), clb (female), and rms (male). For the experiments, Gutiérrez-Muñoz and Coto-Jiménez [2022] chose the slt voice and defined the training, validation, and test sets according to the common criteria of the data available: 70%, 20%, and 10%, respectively.

### 17.3.4.2 Noise

To compare the capacity of the four cases contemplated in our proposal, the database was degraded with additive noise of three types: two artificially generated noises (white and pink) and one natural noise (babble). To cover a wide range of conditions, five levels of SNR ratios were considered for each case. This gives the following dataset:

The whole set of voices to compare can be listed as

- Clean, as the dataset described in the previous section.
- The same dataset degraded with additive white noise added at five SNR levels: SNR −10, SNR −5, SNR 0, SNR 5, and SNR 10.

- The clean dataset degraded with additive pink noise added at five SNR levels: SNR −10, SNR −5, SNR 0, SNR 5, and SNR 10.
- The clean dataset degraded with additive babble noise added at five SNR levels: SNR −10, SNR −5, SNR 0, SNR 5, and SNR 10.

### 17.3.4.3 *Evaluation*

The evaluation metrics defined for our experiments were based on measures commonly applied in noise reduction and speech enhancement, namely, PESQ, and frequency domain SegSNR.

The first measure is based on a psychoacoustic model to predict the subjective quality of speech, according to ITU-T recommendation P.862.ITU. Results are given in the interval $[0.5, 4.5]$, where 4.5 corresponds to a perfect signal reconstruction [Wang *et al.*, 2021; Gnanamanickam *et al.*, 2021].

The second measure is frame based, calculated by averaging the SNR estimates at each frame, using the following equation:

$$\text{SegSNR}_f = \frac{10}{N} \sum_{i=1}^{N} \log \left[ \frac{\sum_{j=0}^{L-1} S^2(i,j)}{\sum_{j=0}^{L-1} (S(i,j) - X(i,j))^2} \right], \tag{17.10}$$

where $X(i,j)$ is the Fourier transform coefficient of frame $i$ and $S(i,j)$ is the coefficient for the processed speech. $N$ is the number of frames and $L$ is the number of frequency bins. The values of this measure are given in the interval $[-20, 35]$ dB.

Additionally, Gutiérrez-Muñoz and Coto-Jiménez [2022] present waveforms and spectrogram visual inspection to illustrate the result of the different approaches.

## 17.4  Results and Discussion

This study examines five signal-to-noise levels, three noise types, and two objective measurements. A sample visualization of the different waveforms involved in this study is shown in Figs. 17.5–17.9.

The objective measures of PESQ and SegSNR are reported as the mean of 50 measures calculated on the test set. To select the mother wavelet and the threshold, extensive experimentation was conducted. For every case reported in the results, more than 20 possibilities were tested. The most successful mother wavelets were db1 and db2.

For the deep learning and hybrid approaches involving neural networks, the stop criterion was defined as the number of epochs. The same number of

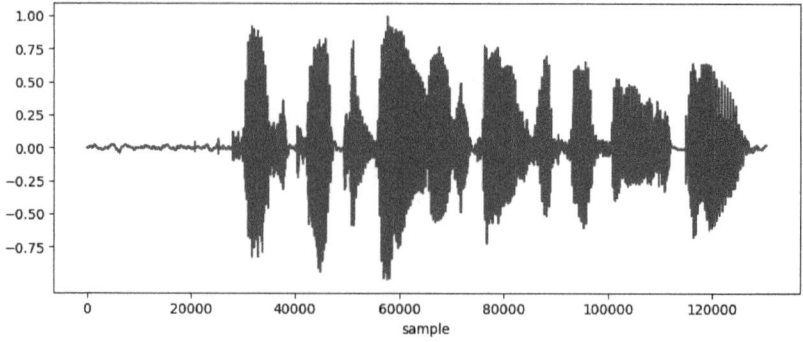

Fig. 17.5    Clean utterance.

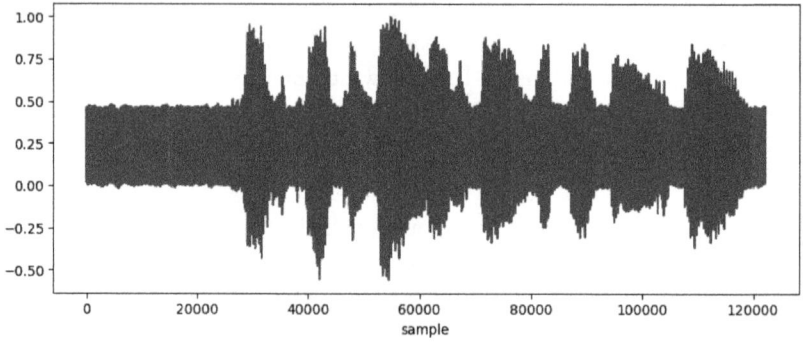

Fig. 17.6    The same previous utterance degraded with white noise at SNR-5.

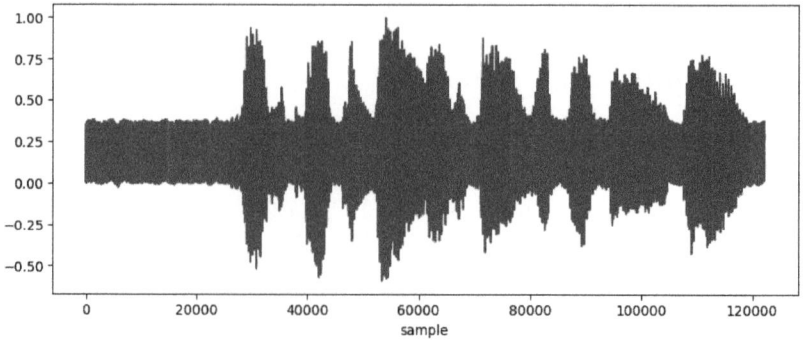

Fig. 17.7    The previous degraded utterance enhanced with wavelets.

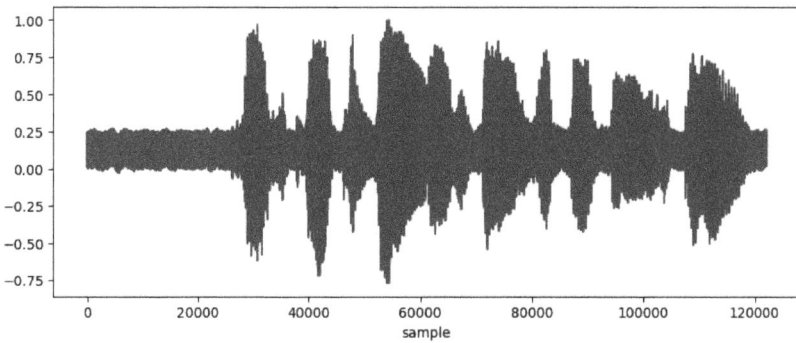

Fig. 17.8   The noise-degraded utterance enhanced using deep learning.

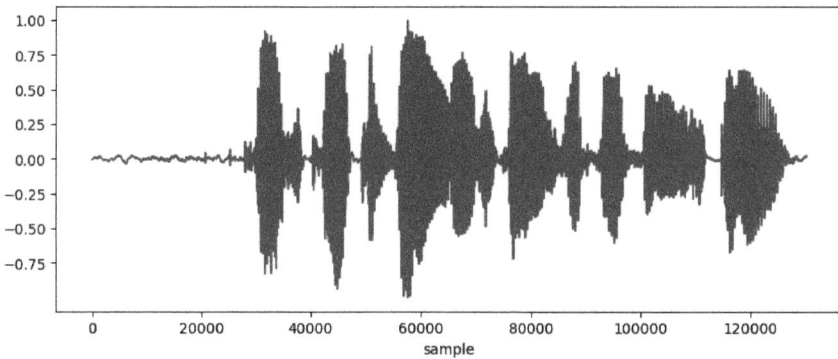

Fig. 17.9   The noise-degraded utterance enhanced using a hybrid of wavelets + deep learning.

epochs used for the training of the networks from the noisy to clean signal was replicated in the hybrid proposals, for the sake of comparison.

The results for the PESQ measure and the babble noise degradation and filtering are presented in Table 17.1. The first relevant result can be observed in the small benefit that was measured for the case of the wavelet enhancement. A better performance than those of the wavelets was obtained with the deep learning enhancement for every SNR level of babble noise. In terms of the hybrid combination of wavelets and deep learning, the deep neural networks as a second stage achieved an improvement in three of the five cases of PESQ. Furthermore, an increase in SegSNR was measured in

Table 17.1.   Babble noise PESQ.

| SNR | Noisy | Wavelets | DL | Wavelets + DL | DL + wavelets |
|---|---|---|---|---|---|
| −10 | 0.44 | 0.49 | **0.53** | 0.51 | 0.52 |
| −5 | 0.53 | 0.54 | 0.95 | **1.43** | 0.95 |
| 0 | 0.82 | 0.83 | 1.85 | **1.86** | 1.85 |
| 5 | 1.32 | 1.32 | **2.20** | 2.16 | **2.20** |
| 10 | 1.94 | 1.94 | 2.42 | **2.53** | 2.43 |

*Note*: The higher values represent better results. In bold is the best result for each SNR level.

Table 17.2.   Babble noise SegSNR.

| SNR | Noisy | Wavelets | DL | Wavelets + DL | DL + wavelets |
|---|---|---|---|---|---|
| −10 | −15.74 | −15.72 | **−0.98** | -0.99 | −0.94 |
| −5 | −10.75 | −10.74 | **0.76** | 0.69 | 0.621 |
| 0 | −5.80 | −5.82 | 4.90 | 4.94 | 4.62 |
| 5 | −0.98 | −1.04 | **6.38** | 6.02 | 5.92 |
| 10 | 3.60 | 3.43 | 7.12 | **7.58** | 6.45 |

*Note*: The higher values represent better results. In bold is the best result for each SNR level.

two of the five cases with the same hybrid combination, as presented in Table 17.2.

For all the cases of the SegSNR measure with babble noise, it was observed that the wavelet transform did not represent significant improvements in the results. This can explain why none of the hybrid approaches performed better than deep learning alone, with the exception of two cases.

The benefits of the hybrid approaches were consistently better for the case of pink noise. The results for PESQ and the different levels of this type of noise are presented in Table 17.3. For this measure, the hybrid approach of wavelets + deep learning gave better results in four of the five noise levels. These results are important because the application of wavelets did not improve any of the cases in terms of the SNR, but, as shown in Table 17.4, an increase in SegSNR was consistent.

Such results can be interpreted in terms of improvements incorporated into the signals with the application of wavelets, which did not improve the perceptual quality of the speech sounds, but the mapping of the noisy signals into a different version is beneficial for the enhancement using deep learning.

Table 17.3.  Pink noise PESQ.

| SNR | Noisy | Wavelets | DL | Wavelets + DL | DL + wavelets |
|---|---|---|---|---|---|
| −10 | 0.16 | 0.04 | 1.27 | **1.29** | 1.26 |
| −5 | 0.46 | 0.42 | 1.50 | **1.54** | 1.49 |
| 0 | 0.83 | 0.83 | 1.65 | **1.74** | 1.63 |
| 5 | 1.39 | 1.39 | **2.14** | 2.13 | 2.13 |
| 10 | 1.99 | 1.99 | 2.31 | **2.32** | 2.30 |

*Note*: The higher values represent better results. In bold is the best result for each SNR level.

Table 17.4.  Pink noise SegSNR.

| SNR | Noisy | Wavelets | DL | Wavelets + DL | DL + wavelets |
|---|---|---|---|---|---|
| −10 | −15.11 | −9.98 | 4.26 | 4.51 | 4.36 |
| −5 | −10.14 | −5.11 | 4.95 | 5.23 | 5.09 |
| 0 | −5.22 | −5.11 | 5.05 | 5.65 | 5.15 |
| 5 | −0.43 | −0.42 | 7.31 | 7.22 | 7.16 |
| 10 | 4.08 | 3.92 | 7.57 | 7.53 | 7.12 |

*Note*: The higher values represent better results. In bold is the best result for each SNR level.

Table 17.5.  White noise PESQ.

| SNR | Noisy | Wavelets | DL | Wavelets + DL | DL + wavelets |
|---|---|---|---|---|---|
| −10 | 0.28 | 0.11 | 1.34 | **1.36** | 1.34 |
| −5 | 0.58 | 0.56 | 1.67 | **1.75** | 1.65 |
| 0 | 0.94 | 0.94 | 1.76 | **1.81** | 1.75 |
| 5 | 1.43 | 1.43 | 1.92 | **1.92** | 1.90 |
| 10 | 1.95 | 1.94 | 2.23 | **2.44** | 2.20 |

*Note*: The higher values represent better results. In bold is the best result for each SNR level.

The best results of the hybrid approach of wavelets + deep learning were obtained for the case of white noise. Table 17.5 shows the results of the PESQ measure. For all the noise SNR levels, such a hybrid approach gave the best results, even when the first stage of wavelet enhancement did not improve the quality of the signal. However, in a similar way to the previous case, Table 17.6 shows how the wavelets improved the SegSNR in all the cases of white noise degradation.

The improvements on the SegSNR measure with the hybrid approach were consistent also at all SNR levels of white noise. For this case, it is also

Table 17.6.   White noise SegSNR.

| SNR | Noisy | Wavelets | DL | Wavelets + DL | DL + wavelets |
|---|---|---|---|---|---|
| −10 | −15.74 | −12.77 | 2.83 | **3.64** | 2.90 |
| −5 | −10.77 | −7.84 | 5.50 | **6.40** | 5.70 |
| 0 | −5.84 | −3.03 | 7.71 | **8.61** | 7.85 |
| 5 | −1.03 | 1.49 | 9.54 | **10.21** | 9.66 |
| 10 | 3.54 | 5.51 | 11.29 | **11.51** | 11.34[1] |

*Note*: In bold is the best result for each SNR level. The higher values represent better results.

significant that the hybrid combination of deep learning and wavelets as a second stage also surpassed the results of deep learning.

For all types of noise, the application of wavelets as the second stage of enhancement did not represent any relevant benefit in terms of PESQ, in comparison to the deep learning approach. From a visual inspection of the spectrograms in Figs. 17.5–17.9, it seems that the application of wavelets introduced some patterns at the higher frequencies and blurred some relevant information at those bands as well.

This kind of result may explain also why wavelets did not improve significantly the PESQ of the noisy utterances but helped improve the SegSNR (in particular in the case of white noise). Especially for the case of white noise, the wavelets as an intermediate representation of the signals seemed to represent advantages for the application of the deep learning enhancement.

The results of this work may represent similar benefits to recent proposals of combining wavelets and deep learning. In this case, deep learning was applied as a second stage of enhancement (and as a first stage prior to wavelets). In that work, relevant benefits in terms of improving the SNR were found.

Other hybrid or cascade approaches have been tested recently in similar domains, for example, speech emotion recognition. The results of this study may represent an opportunity to develop more hybrid approaches, where the benefit of each stage can be analyzed separately, in a similar way to image enhancement, where different algorithms to enhance aspects such as noise, blur, and compression have been applied separately, using a cascade approach.

In this chapter, Azarang and Kehtarnavaz [2020] and Lun *et al.* [2012] have examined the hybrid combination of wavelets and deep learning for enhancing speech signals. To achieve this goal, Azarang and Kehtarnavaz [2020] and Lun *et al.* [2012] conducted extensive experiments to select the appropriate parameters for the wavelet transform and trained over 40 deep neural networks to determine whether the combination of wavelets and deep learning, either as the first or second stage, improves speech signal enhancement in the presence of various types of noise. To ensure a proper comparison, Azarang and Kehtarnavaz [2020] and Lun *et al.* [2012] imposed certain constraints on the experiments, such as limiting the number of epochs during training to match the hybrid and deep learning cases.

The results indicate that the hybrid approach of wavelet enhancement as the first stage and deep learning as the second stage is particularly effective for white noise, as it outperforms both the noisy signal and the wavelet and deep learning approaches separately. However, for Babble noise, which is more complex and irregular than synthetic white and pink noise, the hybrid approach produced mixed results, with some improvements in SNR levels. In this case, applying deep learning alone yielded better outcomes. For pink noise, the hybrid approach produced better results than the separate algorithms for higher levels of noise. When the SNR was as low as 5 or 10, deep learning performed better.

The wavelet denoising successfully enhanced the signals in terms of SegSNR (except for babble noise, where improvements were almost negligible), but some spectrogram artifacts may explain why the benefits were not measurable in terms of PESQ. Nevertheless, the output obtained with the wavelet enhancement represents a better input to deep neural networks than the noisy signals. The advantages of combining deep learning with other algorithms are well documented in the scientific literature, and future research may explore the specific benefits of each algorithm to establish optimized hybrid applications for particular noise types and levels. This study provides several research opportunities, such as analyzing delays in real-time applications using the hybrid proposal to determine the feasibility of implementation in specific hardware and integrated software. Additionally, new scenarios, such as far-field speech enhancement for use in video conferencing or multi speaker enhancement, can be examined in terms of hybrid approaches to select the most effective simple or hybrid algorithms.

# List of Symbols

- $\mathbb{R}$: the set of all the real numbers;
- $\mathbb{N}$: the set of all the natural numbers;
- $\mathbb{Z}$: the set of all the integers;
- $\mathbb{Z}^+$: the set of all the negative integers;
- $\mathbb{R}^d$: $d$-dimensional Euclidean space;
- $\mathbb{T}$: the quotient group $\mathbb{R}/2\pi\mathbb{Z}$;
- $\mathrm{Re}(z)$: the real part $x$ of complex number $z = x + iy$;
- $\mathrm{Im}(z)$: the imaginary part $y$ of complex number $z = x + iy$;
- For sets $A$ and $B$, $A\backslash B := \{x | x \in A \text{ and } x \notin B\}$;
- a.e.: almost everywhere;
- Let $M$ be a matrix, its transposed matrix is denoted by $M^t$. If $M$ is invertible, its inverse matrix is denoted by $M^{-1}$ and its determinant is denoted by $det(M)$;
- $L^p$, $L^p(\mathbb{R})$ ($1 \leq p < \infty$): the space of all the $p$-power integrable functions, i.e.,

$$L^p := L^p(\mathbb{R}) := \left\{ f \ \middle| \ \int_{\mathbb{R}} |f(x)|^p dx < \infty \right\}.$$

- $L^p$, $L^p(\mathbb{T})$ ($1 \leq p < \infty$): the space of all the $p$-power integrable, $2\pi$-periodic functions, i.e.,

$$L^p := L^p(\mathbb{T}) := \left\{ f \ \middle| \ \int_{\mathbb{T}} |f(x)|^p dx < \infty \right\}.$$

- $l^p$, $l^p(\mathbb{Z})$ ($1 \leq p < \infty$): the space of all the $p$-power sumable sequences, i.e.,

$$l^p := l^p(\mathbb{Z}) := \left\{ \{c_k\} \ \middle| \ \sum_{k\in\mathbb{Z}} |c_k|^p < \infty \right\}.$$

- The *Fourier transform* of $f \in L^1(\mathbb{R})$ is defined by

$$\hat{f}(\xi) := \int_{\mathbb{R}} f(x)e^{-ix\cdot\xi}dx,$$

and the *Inverse Fourier* is defined by

$$\check{f} = \frac{1}{2\pi}\int_{\mathbb{R}} f(x)e^{ix\cdot\xi}dx.$$

- The *inner product* of $f$, $g \in L^2(\mathbb{R})$ is defined by

$$\langle f, g \rangle := \int_{\mathbb{R}} f(x)\bar{g}(x)dx,$$

and the following equality holds always:

$$\langle f, g \rangle = \frac{1}{2\pi}\langle \hat{f}, \hat{g} \rangle.$$

- The *Fourier coefficients* of $f \in L^1(\mathbb{T})$ are defined by

$$c_k(f) := \int_{\mathbb{T}} f(x)e^{-ik\cdot x}dx.$$

It is always true that $f(x) \sim \sum_{k\in\mathbb{Z}} c_k(f)e^{ik\cdot x}$.
- Let $\phi(x)$ be a function defined on $\mathbb{R}$, denote

$$\phi_{j,k}(x) := 2^{j/2}\phi(2^j x - k) \quad (\forall j \in \mathbb{Z},\ k \in \mathbb{Z}).$$

- The support of a complex sequence $\{c_k\}_{k\in\mathbb{Z}}$ is defined by

$$\mathrm{supp}\{c_k\}_{k\in\mathbb{Z}} := \{k \in \mathbb{Z} \mid c_k \neq 0\}.$$

# Bibliography

Abiodun, O. I., Jantan, A., Omolara, A. E., Dada, K. V., Mohamed, N. A., and Arshad, H. (2018). State-of-the-art in artificial neural network applications: A survey. *Heliyon*, 4(11):e00938.

Aghajan, H. K. and Kailath, T. (1994). SLIDE: Subspace-based line detection. *IEEE Transactions on Pattern Analysis and Machine Intelligence*, 16(11):1057–1073.

Al-Qazzaz, N. K., Ali, S., Ahmad, S. A., Islam, M. S., and Ariff, M. I. (2014). Selection of mother wavelets thresholding methods in denoising multi-channel EEG signals during working memory task. In *2014 IEEE Conference on Biomedical Engineering and Sciences (IECBES)*, pp. 214–219. IEEE.

Arbelaez, P., Maire, M., Fowlkes, C., and Malik, J. (2010). Contour detection and hierarchical image segmentation. *IEEE Transactions on Pattern Analysis and Machine Intelligence*, 33(5):898–916.

AUEB, T. R. *et al.* (2016). One-vs-each approximation to softmax for scalable estimation of probabilities. *Advances in Neural Information Processing Systems*, 29.

Auslander, L., Kailath, T., and Mitter, S., editors (1990). *Signal Processing I: Signal Processing Theory*. Springer-Verlag, New York.

Azarang, A. and Kehtarnavaz, N. (2020). A review of multi-objective deep learning speech denoising methods. *Speech Communication*, 122:1–10.

Bahadur, I., Kumar, S., and Agarwal, P. (2021). Performance measurement of a hybrid speech enhancement technique. *International Journal of Speech Technology*, 24:665–677.

Bahoura, M. and Rouat, J. (2006). Wavelet speech enhancement based on time-scale adaptation. *Speech Communication*, 48(12):1620–1637.

Bal, A., Banerjee, M., Chakrabarti, A., and Sharma, P. (2022). MRI brain tumor segmentation and analysis using rough-fuzzy c-means and shape based properties. *Journal of King Saud University-Computer and Information Sciences*, 34(2):115–133.

Balaji, V., Sathiya Priya, J., Dinesh Kumar, J., and Karthi, S. (2021). Radial basis function neural network based speech enhancement system using SLANT-LET transform through hybrid vector wiener filter. In *Inventive Communication and Computational Technologies: Proceedings of ICICCT 2020*, pp. 711–723. Springer.

Bengio, Y. *et al.* (2009). Learning deep architectures for AI. *Foundations and trends®  in Machine Learning*, 2(1):1–127.

Beylkin, G., Coifman, R., Daubechies, I., Mallat, S., Meyer, Y., Raphael, L., and Ruskai, B. (1991). *Wavelets and Their Applications*. Burlington, MA: Jones and Bartlett.

Blel, D., Hassairi, S., and Ejbali, R. (2022). Wavelet network-based deep learning system for image classification. In *Fourteenth International Conference on Machine Vision (ICMV 2021)*, vol. 12084, pp. 377–385. SPIE.

Blum, H. (1967). A transformation for extracting new descriptors of shape. In W. Wathen-Dunn, editor, Models for the Perception of Speech and Visual Form. The MIT Press, Cambridge, MA.

Bouallégue, A., Hassairi, S., Ejbali, R., and Zaied, M. (2017). Learning deep wavelet networks for recognition system of Arabic words. In *International Joint Conference SOCO 16-CISIS 16-ICEUTE 16: San Sebastián, Spain, October 19th–21st, 2016 Proceedings 11*, pp. 498–507. Springer.

Bouzid, A., Ellouze, N., *et al.* (2016). Speech enhancement based on wavelet packet of an improved principal component analysis. *Computer Speech & Language*, 35:58–72.

Bow, S. T. (1992). *Pattern Recognition and Image Preprocessing*. New York: Marcel-Dekker.

Brady, M. (1983). Criteria for representation of shape. In J. Beck and B. Hope and A. Rosenfeld, editors, *Human and Machine Vision*. New York: Academic Press.

Brady, M. and Asada, H. (1993). Smoothed local symmetries and their implementation. *International Journal of Robotics Research*, 15:973–981.

Bruna, J. and Mallat, S. (2011). Classification with scattering operators. In *CVPR 2011*, pp. 1561–1566. IEEE.

Bruna, J. and Mallat, S. (2013). Invariant scattering convolution networks. *IEEE Transactions on Pattern Analysis and Machine Intelligence*, 35(8):1872–1886.

Candes, E. J., Wakin, M. B., and Boyd, S. P. (2008). Enhancing sparsity by reweighted? 1 minimization. *Journal of Fourier Analysis and Applications*, 14:877–905.

Canny, J. (1986). A computational approach to edge detection. *IEEE Transactions on Pattern Analysis and Machine Intelligence*, 8(6):679–698.

Chambolle, A., DeVore, R. A., Lee, N. Y., and Lucier, B. J. (1998). Nonlinear wavelet image processing: Variational problems, compression, and noise removal through wavelet shrinkage. *IEEE Transactions on Image Processing*, 7(3):319–335.

Chang, H. S. and Yan, H. (1999). Analysis of stroke structures of handwritten Chinese characters. *IEEE Transactions on Systems, Man, and Cybernetics (B)*, 29:47–61.

Chaohao, G. (1992). *Mathematics Dictionary*. Shanghai, Shanghai Dictionary Press, China.

Chavan, M. S. and Mastorakis, N. (2010). Studies on implementation of Harr and daubechies wavelet for denoising of speech signal. *International Journal of Circuits, Systems and Signal Processing*, 4(3):83–96.

Chen, C., Lee, J., and Sun, Y. (1995). Wavelet transformation for gray-level corner detection. *Pattern Recognition*, 28(6):853–861.

Chen, G. and Yang, Y. H. H. (1995). Edge detection by regularized cubic B-spline fitting. *IEEE Transactions on Systems, Man, and Cybernetics*, 25(4): 636–643.

Chen, H., Qin, Z., Ding, Y., Tian, L., and Qin, Z. (2020). Brain tumor segmentation with deep convolutional symmetric neural network. *Neurocomputing*, 392:305–313.

Chen, H., Zhang, Y., Kalra, M. K., Lin, F., Chen, Y., Liao, P., Zhou, J., and Wang, G. (2017). Low-dose CT with a residual encoder-decoder convolutional neural network. *IEEE Transactions on Medical Imaging*, 36(12): 2524–2535.

Chuang, G. C. H. and Kuo, C. C. J. (1996). Wavelet descriptor of planar curves: Theory and applications. *IEEE Transactions on Image Processing*, 5(1): 56–70.

Chui, C. K. (1992a). *An Introduction to Wavelets*. Boston, MA: Academic Press.

Chui, C. K. (1992b). *An Introduction to Wavelets*, vol. 1. Academic Press.

Combettes, P. L. (1998). Convex multiresolution analysis. *IEEE Transactions on Pattern Analysis and Machine Intelligence*, 20(12):1308–1318.

Combettes, P. L. and Pesquet, J. C. (2004). Wavelet-constrained image restoration. *International Journal of Wavelets, Multiresolution and Information Processing*, 2(4):371–389.

Connell, J. H. and Brady, M. (1987). Generating and generalizing models of visual objects. *Artificial Intelligence*, 31:159–183.

Cotter, F. and Kingsbury, N. (2018). Deep learning in the wavelet domain.

Dadashnialehi, A., Bab-Hadiashar, A., Cao, Z., and Kapoor, A. (2013). Enhanced ABS for in-wheel electric vehicles using data fusion. In *2013 IEEE Intelligent Vehicles Symposium (IV)*, pp. 702–707. IEEE.

Dadashnialehi, A., Bab-Hadiashar, A., and Hoseinnezhad, R. (2017). Deep learning for texture classification via multi-wavelet fusion of scattering transforms. In *2017 IEEE International Conference on Mechatronics (ICM)*, pp. 324–329. IEEE.

Daubechies, I. (1988). Orthonormal bases of compactly supported wavelets. *Communications on Pure & Applied Mathematics*, 41(7):909–996.

Daubechies, I. (1990). Wavelet transform, time-frequency localization and signal analysis. *IEEE Transactions on Information Theory*, 36:961–1005.

Daubechies, I. (1992). *Ten Lectures on Wavelets*, vol. 61. CBMS — Conference Lecture Notes, SIAM, Philadelphia.

Daugman, J. (2003). Demodulation by complex-valued wavelets for stochastic pattern recognition. *International Journal of Wavelets, Multiresolution and Information Processing*, 1(1):1–317.

de Wouwer, G., Scheunders, P., Livens, S., and Dyck, D. V. (1999a). Wavelet correlation signatures for color texture characterization. *Pattern Recognition*, 32:443–451.

de Wouwer, G. V., Schenuders, P., and Dyck, D. V. (1999b). Statistical texture characterization from discrete wavelet representation. *IEEE Transactions on Image Processing*, 8:592–598.

Deng, L., Yu, D., *et al.* (2014). Deep learning: Methods and applications. *Foundations and Trends®️ in Signal Processing*, 7(3–4):197–387.

Deng, W. and Lyengar, S. S. (1996). A new probability relaxation scheme and its application to edge detection. *IEEE Transactions on Pattern Analysis and Machine Intelligence*, 18(4):432–443.

Devi, R. S., Perumal, B., and Rajasekaran, M. P. (2022). A hybrid deep learning based brain tumor classification and segmentation by stationary wavelet packet transform and adaptive kernel fuzzy c means clustering. *Advances in Engineering Software*, 170:103146.

Donoho, D. L. (1995). De-noising by soft-thresholding. *IEEE Transactions on Information Theory*, 41(3):613–627.

Donoho, D. L. and Johnstone, I. M. (1994). Ideal spatial adaptation by wavelet shrinkage. *Biometrika*, 81(3):425–455.

Edgar, G. A. (1990). *Measure, Topology, and Fractal Geometry*. New York: Springer-Verlag.

El-Khamy, S. E., Hadhoud, M. M., Dessouky, M. I., Salam, B. M., and El-Samie, F. E. A. (2006). Wavelet fusion: A tool to break the limits on LMMSE image super-resolution. *International Journal of Wavelets, Multiresolution and Information Processing*, 4(1):105–118.

Falconer, K. L. (1985). *The Geometry of Fractal Sets*. Cambridge University Press.

Falconer, K. L. (1990). *Fractal Geometry: Mathematical Foundation and Applications*. New York: Wiley.

Gabor, D. (1946). Theory of communication. *Journal of the Institution of Electrical Engineers*, 93:429–457.

Gargour, C., Gabrea, M., Ramachandran, V., and Lina, J.-M. (2009). A short introduction to wavelets and their applications. *IEEE Circuits and Systems Magazine*, 9(2):57–68.

Ge, Y. and Fitzpatrick, J. M. (1996). On the generation of skeletons from discrete Euclidean distance maps. *IEEE Transactions on Pattern Analysis and Machine Intelligence*, 18:1055–1066.

Gilbarg, D. and Trudinger, N. S. (1977). *Elliptic Partial Differential Equations of Second Order*. Springer, Berlin, New York.

Girshick, R., Donahue, J., Darrell, T., and Malik, J. (2014). Rich feature hierarchies for accurate object detection and semantic segmentation. In *Proceedings of the IEEE Conference on Computer Vision and Pattern Recognition*, pp. 580–587.

Gnanamanickam, J., Natarajan, Y., and Sri Preethaa, K. R. (2021). A hybrid speech enhancement algorithm for voice assistance application. *Sensors*, 21(21):7025.

Goodfellow, I., Bengio, Y., and Courville, A. (2016). *Deep Learning*. MIT Press.

Grossmann, A. and Morlet, J. (1984). Decomposition of Hardy function into square integrable wavelets of constant shape. *SIAM Journal of Mathematical Analysis*, 15:723–736.

Grossmann, A. and Morlet, J. (1985). Decomposition of functions into wavelets of constant shape, and related transforms. In Streit, L., editor, *Lecture on Recent Results*, Mathematics and Physics. Singapore: World Scientific Publishing.

Grossmann, A., Morlet, J., and Paul, T. (1985). Transforms associated to square integrable group representations I. General results. *Journal of Mathemtical Physics*, 26:2473–2479.

Gumaei, A., Hassan, M. M., Hassan, M. R., Alelaiwi, A., and Fortino, G. (2019). A hybrid feature extraction method with regularized extreme learning machine for brain tumor classification. *IEEE Access*, 7:36266–36273.

Gutiérrez-Muñoz, M. and Coto-Jiménez, M. (2022). An experimental study on speech enhancement based on a combination of wavelets and deep learning. *Computation*, 10(6):102.

Haley, G. M. and Manjunath, B. S. (1999). Rotation-invariant texture classification using a complete space-frequency model. *IEEE Transactions on Image Processing*, 8(2):255–269.

Hassairi, S., Ejbali, R., and Zaied, M. (2015). Supervised image classification using deep convolutional wavelets network. In *2015 IEEE 27th International Conference on Tools with Artificial Intelligence (ICTAI)*, pp. 265–271. IEEE.

Hassairi, S., Ejbali, R., and Zaied, M. (2016). Sparse wavelet auto-encoders for image classification. In *2016 International Conference on Digital Image Computing: Techniques and Applications (DICTA)*, pp. 1–6. IEEE.

Hassairi, S., Jemel, I., Ejbali, R., and Zaied, M. (2020). Sparse representation of images using substitution of wavelet by patches. In *Twelfth International Conference on Machine Vision (ICMV 2019)*, vol. 11433, pp. 402–409. SPIE.

He, K., Zhang, X., Ren, S., and Sun, J. (2016). Deep residual learning for image recognition. In *Proceedings of the IEEE Conference on Computer Vision and Pattern Recognition*, pp. 770–778.

Hochreiter, S. and Schmidhuber, J. (1997). Long short-term memory. *Neural Computation*, 9(8):1735–1780.

Horn, B. K. P. and Brooks, M. J. (1989). *Shape from Shading*. MIT Press, Cambridge, MA.

Hoseinnezhad, R. and Bab-Hadiashar, A. (2006). Fusion of redundant information in brake-by-wire systems using a fuzzy voter. *Journal of Advanced Information Fusion*, 1(1):52–62.

Hsieh, J.-W., Liao, H.-M., Ko, M.-T., and Fan, K.-C. (1995). Wavelet-based shape form shading. *Graphical Models and Image Processing*, 57(4):343–362.

IEEE (1992). Special issue on wavelet transforms and multiresolution signal analysis. *IEEE Transactions on Information Theory*, 38(2):529–924.

IEEE (1993). Special issue on wavelets and signal processing. *IEEE Transactions on Signal Processing*, 41(12):3213–3600.

Jain, P. and Merchant, S. N. (2004). Wavelet-based multiresolution histogram for fast image retrieval. *International Journal of Wavelets, Multiresolution and Information Processing*, 2(1):59–73.

Janssen, R. D. T. (1997). Interpretation of maps: From bottom-up to model-based. In H. Bunke and P. S. P. Wang, editors, Handbook of Character Recognition and Document Image Analysis. Singapore: World Scientific.

Jemai, O., Zaied, M., Amar, C. B., and Alimi, A. M. (2010). FBWN: An architecture of fast beta wavelet networks for image classification. In *The 2010 International Joint Conference on Neural Networks (IJCNN)*, pp. 1–8. IEEE.

Jemel, I., Hassairi, S., Ejbali, R., and Zaied, M. (2020). Deep stacked sparse auto-encoder based on patches for image classification. In *Twelfth International Conference on Machine Vision (ICMV 2019)*, vol. 11433, pp. 468–474. SPIE.

Kingma, D. P. and Ba, J. (2014). Adam: A method for stochastic optimization. arXiv preprint arXiv:1412.6980.

Kingsbury, N. (2001). Complex wavelets for shift invariant analysis and filtering of signals. *Applied and Computational Harmonic Analysis*, 10(3):234–253.

Kominek, J. and Black, A. W. (2004). The CMU arctic speech databases. In *Fifth ISCA Workshop on Speech Synthesis*.

Kouzani, A. Z. and Ong, S. H. (2003). Lighting-effects classification in facial images using wavelet packets transform. *International Journal of Wavelets, Multiresolution and Information Processing*, 1(2):199–215.

Kovesi, P. (1995). Image features from phase congruency. Technical report, University of Western Australia, Robotics and Vision Group, Australia.

Kovesi, P. (1997). Symmetry and asymmetry from local phase. In *Tenth Australian Joint Convergence on Artificial Intelligence*, pp. 2–4.

Ksantini, R., Ziou, D., Dubeau, F., and Harinarayan, P. (2006). Image retrieval based on region separation and multiresolution analysis. *International Journal of Wavelets, Multiresolution and Information Processing*, 4(1):147–175.

Kubo, M., Aghbari, Z., and Makinouchi, A. (2003). Content-based image retrieval technique using wavelet-based shift and brightness invariant edge feature. *International Journal of Wavelets, Multiresolution and Information Processing*, 1(2):163–178.

Kumar, S. and Kumar, D. K. (2005). Visual hand gestures classification using wavelet transforms and moment based features. *International Journal of Wavelets, Multiresolution and Information Processing*, 3(1):79–101.

Kumar, S., Kumar, D. K., Sharma, A., and McLachlan, N. (2003). Visual hand gestures classification using wavelet transforms. *International Journal of Wavelets, Multiresolution and Information Processing*, 1(4):373–392.

Kumar, S. and Mankame, D. P. (2020). Optimization driven deep convolution neural network for brain tumor classification. *Biocybernetics and Biomedical Engineering*, 40(3):1190–1204.

Kunte, R. S. and Samuel, R. D. S. (2007). Wavelet descriptors for recognition of basic symbols in printed Kannada text. *International Journal of Wavelets, Multiresolution and Information Processing*, 5(2):351–367.

Lai, J. H., Yuen, P. C., and Feng, G. C. (1999). Spectroface: A Fourier-based approach for human face recognition. In *Proceedings of ICMI '99*, pp. VI115–120.

Lam, L., Lee, S. W., and Suen, C. Y. (1992). Thinning methodologies — A comprehensive survey. *IEEE Transactions on Pattern Analysis and Machine Intelligence*, 14:869–885.

Law, T., Iton, H., and Seki, H. (1996). Image filtering, edge detection, and edge tracing using fuzzy reasoning. *IEEE Transactions on Pattern Analysis and Machine Intelligence*, 18(5):481–491.

Lazebnik, S., Schmid, C., and Ponce, J. (2005). A sparse texture representation using local affine regions. *IEEE Transactions on Pattern Analysis and Machine Intelligence*, 27(8):1265–1278.

LeCun, Y., Bengio, Y., and Hinton, G. (2015). Deep learning. *Nature*, 521(7553):436–444.

LeCun, Y., Bottou, L., Bengio, Y., and Haffner, P. (1998). Gradient-based learning applied to document recognition. *Proceedings of the IEEE*, 86(11): 2278–2324.

Lee, S. W., Kim, C. H., Ma, H., and Tang, Y. Y. (1996). Multiresolution recognition of unconstrained handwritten numerals with wavelet transform and multilayer cluster neural network. *Pattern Recognition*, 29:1953–1961.

Leyton, M. (1988). *Symmetry, Causality, Mind*. Cambridge, MA: The MIT Press.

Li, H. (2006). Wavelet-based weighted average and human vision system image fusion. *International Journal of Wavelets, Multiresolution and Information Processing*, 4(1):97–103.

Li, S. (2008). Multisensor remote sensing image fusion using stationary wavelet transform: Effects of basis and decomposition level. *International Journal of Wavelets, Multiresolution and Information Processing*, 6(1):37–50.

Li, S. and Fu, Y. (2015). Learning robust and discriminative subspace with low-rank constraints. *IEEE Transactions on Neural Networks and Learning Systems*, 27(11):2160–2173.

Liang, K. H., Chang, F., Tan, T. M., and Hwang, W. L. (1999). Multiresolution Hadamard representation and its application to document image analysis. In *Proceedings of The Second International Conference on Multimodel Interface (ICMI'99)*, pp. V1–V6, Hong Kong.

Liang, K. H. and Tjahjadi, T. (2006). Adaptive scale fixing for multiscale texture segmentation. *IEEE Transactions on Image Processing*, 15(1):249–256.

Liao, Z. and Tang, Y. Y. (2005). Signal denoising using wavelets and block hidden Markov model. *International Journal of Pattern Recognition and Artificial Intelligence*, 19(5):681–700.

Long, R. (1995). *High-Dimensional Wavelet Analysis*. World Books Co. Pte. Ltd, Beijing.

Lun, D. P.-K. and Hsung, T.-C. (2010). Improved wavelet based a-priori snr estimation for speech enhancement. In *Proceedings of 2010 IEEE International Symposium on Circuits and Systems*, pp. 2382–2385. IEEE.

Lun, D. P.-K., Shen, T.-W., Hsung, T.-C., and Ho, D. K. (2012). Wavelet based speech presence probability estimator for speech enhancement. *Digital Signal Processing*, 22(6):1161–1173.

Luo, R. C., Yih, C.-C., and Su, K. L. (2002). Multisensor fusion and integration: Approaches, applications, and future research directions. *IEEE Sensors journal*, 2(2):107–119.

Ma, H., Xi, D. H., Mao, X. G., Tang, Y. Y., and Suen, C. Y. (1995). Document analysis by fractal signatures. In Spitz, A. L. and Dengel, A., editors, *Document Analysis Systems*. World Scientific Publishing Co. Pte, Ltd., Singapore.

Maheswari, K., Balamurugan, A., Malathi, P., and Ramkumar, S. (2021). Hybrid clustering algorithm for an efficient brain tumor segmentation. *Materials Today: Proceedings*, 37:3002–3006.

Mallat, S. (1989a). Multiresolution approximations and wavelet orthonormal bases of $L^2(R)$. *Transactions on American Mathematical Society*, 315: 69–87.

Mallat, S. (1989b). A theory of multiresolution signal decomposition: The wavelet representation. *IEEE Transactions on Pattern Analysis and Machine Intelligence*, 11:674–693.

Mallat, S. (1998). *Wavelet Tour of Signal Processing*. Academic Press, San Diego, USA.

Mallat, S. (1999). *A Wavelet Tour of Signal Processing*. Elsevier.

Mallat, S. (2012). Group invariant scattering. *Communications on Pure and Applied Mathematics*, 65(10):1331–1398.

Mallat, S. and Hwang, W. L. (1992). Singularity detection and processing with wavelets. *IEEE Transactions on Information Theory*, 38:617–643.

Mallat, S. and Zhong, S. (1992). Characterization of signals from multiscale edges. *IEEE Transactions on Pattern Analysis and Machine Intelligence*, 14(7):710–732.

Mallat, S. G. (1989c). Multifrequency channel decompositions of images and wavelet models. *IEEE Transactions on Acoustics Speech and Signal Processing*, 37(12):2091–2110.

Mallick, P. K., Ryu, S. H., Satapathy, S. K., Mishra, S., Nguyen, G. N., and Tiwari, P. (2019). Brain MRI image classification for cancer detection using deep wavelet autoencoder-based deep neural network. *IEEE Access*, 7: 46278–46287.

Mandelbrot, B. B. (1982). *The Fractal Geometry of Nature*. New York: Freeman.

Marr, D. and Hildreth, E. C. (1980). Theory of edge detection. In *Proceedings of Royal Society*, pp. 187–217, London.

Martin, D., Fowlkes, C., Tal, D., and Malik, J. (2001). A database of human segmented natural images and its application to evaluating segmentation algorithms and measuring ecological statistics. In *Proceedings Eighth IEEE International Conference on Computer Vision. ICCV 2001*, vol. 2, pp. 416–423. IEEE.

Maruthamuthu, A. *et al.* (2020). Brain tumour segmentation from MRI using superpixels based spectral clustering. *Journal of King Saud University-Computer and Information Sciences*, 32(10):1182–1193.

Matalas, L., Benjamin, R., and Kitney, R. (1997). An edge detection technique using the facet model and parameterized relaxation labeling. *IEEE Transactions on Pattern Analysis and Machine Intelligence*, 19(4):328–341.

Mercorelli, P. (2017). A fault detection and data reconciliation algorithm in technical processes with the help of Haar wavelets packets. *Algorithms*, 10(1):13.

Meyer, Y. (1990). *Ondelettes et Operateurs*, volumes I, II. Paris: Hermann.

Michelucci, U. (2022). An introduction to autoencoders. arXiv preprint arXiv:2201.03898.

Mihov, S. G., Ivanov, R. M., and Popov, A. N. (2009). Denoising speech signals by wavelet transform. *Annual Journal of Electronics*, (6):2–5.

Moghaddam, H. A., Khajoie, T., Rouhi, A. H., and Tarzjan, M. S. (2005). Wavelet correlogram: A new approach for image indexing and retrieval. *Pattern Recognition*, 38(12):2506–2518.

Morlet, J., Arens, G., Fourgeau, E., and Giard, D. (1982a). Wave propagation and sampling theory—Part 1: Complex signal and scattering in multilayered media. *Geophysics*, 47:203–221.

Morlet, J., Arens, G., Fourgeau, E., and Giard, D. (1982b). Wave propagation and sampling theory—Part 2: Sampling theory and complex waves. *Geophysics*, 47:222–236.

Muneeswaran, K., Ganesan, L., Arumugam, S., and Harinarayan, P. (2005). A novel approach combing Gabor wavelet transforms and moments for texture segmentation. *International Journal of Wavelets, Multiresolution and Information Processing*, 3(4):559–572.

Munkres, J. R. (1975). *Topology, A First Course*. Upper Saddle River, NJ: Prentice-Hall Inc.

Murtagh, F. and Starck, J. L. (1998). Pattern clustering based on noise modeling in wavelet space. *Pattern Recognition*, 31:847–855.

Naser, M. A. and Deen, M. J. (2020). Brain tumor segmentation and grading of lower-grade glioma using deep learning in MRI images. *Computers in Biology and Medicine*, 121:103758.

Nastar, C. and Ayache, N. (1996). Frequency-based non-rigid motion analysis. *IEEE Transactions on Pattern Analysis and Machine Intelligence*, 18(11).

Nene, S. A., Nayar, S. K., Murase, H., *et al.* (1996). Columbia object image library (coil-20).

Ogniewicz, R. L. and Kubler, O. (1995). Hierarchic Voronoi skeletons. *Pattern Recognition*, 28:343–359.

Oktar, M. A., Nibouche, M., and Baltaci, Y. (2016). Denoising speech by notch filter and wavelet thresholding in real time. In *2016 24th Signal Processing and Communication Application Conference (SIU)*, pp. 813–816. IEEE.

Osher, S., Burger, M., Goldfarb, D., Xu, J., and Yin, W. (2005). An iterative regularization method for total variation-based image restoration. *Multiscale Modeling & Simulation*, 4(2):460–489.

O'Toole, A., Abdi, H., Deffenbacher, K., and Valentin, D. (1993). Low-dimensional representation of faces in higher dimensions of the face space. *Journal Optical Society of America A.*, 10(3):405–411.

Pavlidis, T. (1986). Vectorizer and feature extractor for document recognition. *Computer Vision, Graphics, Image Processing*, 35:111–127.

Priyadarshani, N., Marsland, S., Castro, I., and Punchihewa, A. (2016). Birdsong denoising using wavelets. *PLOS one*, 11(1):e0146790.

Procter, P. (1993). *Longman English-Chinese Dictionary of Contemporary English*. Hong Kong: Longman Asia Limited.

Purwins, H., Li, B., Virtanen, T., Schlüter, J., Chang, S.-Y., and Sainath, T. (2019). Deep learning for audio signal processing. *IEEE Journal of Selected Topics in Signal Processing*, 13(2):206–219.

Ram, R. and Mohanty, M. N. (2019). Use of radial basis function network with discrete wavelet transform for speech enhancement. *International Journal of Computational Vision and Robotics*, 9(2):207–223.

Ramzi, Z., Michalewicz, K., Starck, J.-L., Moreau, T., and Ciuciu, P. (2023). Wavelets in the deep learning era. *Journal of Mathematical Imaging and Vision*, 65(1):240–251.

Rix, A. W., Beerends, J. G., Hollier, M. P., and Hekstra, A. P. (2001). Perceptual evaluation of speech quality (PESQ) — A new method for speech quality assessment of telephone networks and codecs. In *2001 IEEE International Conference on Acoustics, Speech, and Signal Processing. Proceedings (Cat. No. 01CH37221)*, vol. 2, pp. 749–752. IEEE.

Rix, A. W., Hollier, M. P., Hekstra, A. P., and Beerends, J. G. (2002). Perceptual evaluation of speech quality (PESQ) the new ITU standard for end-to-end speech quality assessment part I–time-delay compensation. *Journal of the Audio Engineering Society*, 50(10):755–764.

Rom, H. and Medioni, G. (1984). Hierarchical decomposition and axial shape description. *IEEE Transactions on Pattern Analysis and Machine Intelligence*, 3:36–61.

Ronneberger, O., Fischer, P., and Brox, T. (2015). U-net: Convolutional networks for biomedical image segmentation. In *Medical Image Computing and Computer-Assisted Intervention–MICCAI 2015: 18th International Conference, Munich, Germany, October 5–9, 2015, Proceedings, Part III 18*, pp. 234–241. Springer.

Rudin, W. (1973). *Functional Analysis*. New York: McGraw-Hill, Inc.

Rudin, W. (1974). *Real and Complex Analysis*, 2nd edition. McGraw-Hill Book Company.

Said, S., Jemai, O., Hassairi, S., Ejbali, R., Zaied, M., and Amar, C. B. (2016). Deep wavelet network for image classification. In *2016 IEEE International Conference on Systems, Man, and Cybernetics (SMC)*, pp. 000922–000927. IEEE.

Saint-Marc, P., Rom, H., and Medioni, G. (1993). B-spline contour representation and symmetry detection. *IEEE Transactions on Pattern Analysis and Machine Intelligence*, 15:1191–1197.

Schimmack, M. and Mercorelli, P. (2018). An on-line orthogonal wavelet denoising algorithm for high-resolution surface scans. *Journal of the Franklin Institute*, 355(18):9245–9270.

Schimmack, M. and Mercorelli, P. (2019). A structural property of the wavelet packet transform method to localise incoherency of a signal. *Journal of the Franklin Institute*, 356(16):10123–10137.

Schmidhuber, J. (2015). Deep learning. *Scholarpedia*, 10(11):32832.

Selesnick, I. W., Baraniuk, R. G., and Kingsbury, N. C. (2005). The dual-tree complex wavelet transform. *IEEE Signal Processing Magazine*, 22(6):123–151.

Shankar, B. U., Meher, S. K., and Chosh, A. (2007). Neuro-wavelet classifier for multispectral remote sensing images. *International Journal of Wavelets, Multiresolution and Information Processing*, 5(4):589–611.

Sharnia, A., Kumart, D. K., and Kumar, S. (2004). Wavelet directional histograms of the spatio-temporal templates of human gestures. *International Journal of Wavelets, Multiresolution and Information Processing*, 2(3): 283–298.

Shen, D. and Ip, H. H. S. (1999). Discriminative wavelet shape descriptors for recognition of 2-D pattern. *Pattern Recognition*, 32:151–165.

Shensa, M. J. (1992). The discrete wavelets: Wedding the atrous and Mallat algorithms. *IEEE Transactions on Signal Processing*, 40:2464–2482.

Shi, X., Chen, Z., Wang, H., Yeung, D.-Y., Wong, W.-K., and Woo, W.-C. (2015). Convolutional LSTM network: A machine learning approach for precipitation nowcasting. *Advances in Neural Information Processing Systems*, 28.

Shiri, F. M., Perumal, T., Mustapha, N., and Mohamed, R. (2023). A comprehensive overview and comparative analysis on deep learning models: CNN, RNN, LSTM, GRU. arXiv preprint arXiv:2305.17473.

Sifre, L. and Mallat, S. (2013). Rotation, scaling and deformation invariant scattering for texture discrimination. In *Proceedings of the IEEE Conference on Computer Vision and Pattern Recognition*, pp. 1233–1240.

Sirovich, L. and Kirby, M. (1987). Low-dimensional procedure for the characterization of human faces. *Journal of Optical Society of America A.*, 4(3):519–524.

Slimane, F., Kanoun, S., Hennebert, J., Alimi, A. M., and Ingold, R. (2013). A study on font-family and font-size recognition applied to Arabic word images at ultra-low resolution. *Pattern Recognition Letters*, 34(2):209–218.

Smeulders, A. W. M., Worring, M., Santini, S., Gupta, A., and Jain, R. (2000). Content-based image retrieval at the end of early years. *IEEE Transactions on Pattern Analysis and Machine Intelligence*, 22:1349–13805.

Smith, R. W. (1987). Computer processing of line images: A survey. *Pattern Recognition*, 20:7–15.

Special-Issue-Digital-Library (1996). *IEEE Transactions on Pattern Analysis and Machine Intelligence*, 18.

SPIE (1994). Special issue on wavelet applications. In Szu, H. H., editor, *Proceedings of SPIE 2242*.

Starck, J. L., Bijaoui, A., and Murtagh, F. (1995). Multiresolution support applied to image filtering and deconvolution. *Graphical Models Image Processing*, 57:420–431.

Stollnitz, E. J., Derose, T. D., and Salesin, D. H. (1996). *Wavelets for Computer Graphics: Theory and Applications*. Morgan Kaufmann Publishers.

Sultan, H. H., Salem, N. M., and Al-Atabany, W. (2019). Multi-classification of brain tumor images using deep neural network. *IEEE Access*, 7:69215–69225.

Swets, D. L. and Weng, J. (1996). Using discriminant eigenfeatures for image retrieval. *IEEE Transactions on Pattern Analysis and Machine Intelligence*, 18(8):831–836.

Tan, L., Chen, Y., and Wu, F. (2020). Research on speech signal denoising algorithm based on wavelet analysis. In *Journal of Physics: Conference Series*, 1627:012027.

Tang, Y. Y., Cheng, H. D., and Suen, C. Y. (1991). Transformation-ring-projection (TRP) algorithm and its VLSI implementation. *International Journal of Pattern Recognition and Artificial Intelligence*, 5(1 and 2): 25–56.

Tang, Y. Y., Li, B., Ma, H., and Liu, J. (1998a). Ring-projection-wavelet-fractal signatures: A novel approach to feature extraction. *IEEE Transactions on Circuits and Systems II.*

Tang, Y. Y., Lihua, Q. S., Yang, and Feng, L. (1998b). Two-dimensional overlap-save method in handwriting recognition. In *Proc. of 6th International Workshop on Frontiers in Handwriting Recognition(IWFHR'98)*, pp. 627–633, Taejon, Korea.

Tang, Y. Y., Liu, J., Ma, H., and Li, B. (1996a). Two-dimensional wavelet transform in document analysis. In *Proceedings of 1st International Conference on Multimodal Interface*, pp. 274–279, Beijing, China.

Tang, Y. Y., Liu, J., Ma, H., and Li, B. F. (1999). Wavelet orthonormal decomposition for extracting features in pattern recognition. *International Journal of Pattern Recognition and Artificial Intelligence*, 13(6):803–831.

Tang, Y. Y. and Ma, H. (2000). Classifier design based on orthogonal wavelet series. Technical report, Hong Kong Baptist.

Tang, Y. Y., Ma, H., Li, B., and Liu, J. (1996b). Character recognition based on doubechies wavelet. In *Proceedings of 1st International Conference on Multimodal Interface*, pp. 215–220, Beijing, China.

Tang, Y. Y., Ma, H., Liu, J., Li, B., and Xi, D. (1997a). Multiresolution analysis in extraction of reference lines from documents with graylevel background. *IEEE Transactions on Pattern Analysis and Machine Intelligence*, 19: 921–926.

Tang, Y. Y., Ma, H., Xi, D., Cheng, Y., and Suen, C. Y. (1995a). Extraction of reference lines from document with grey-level background using sub-image of wavelets. In *Proceedings of 3-rd International Conference on Document Analysis and Recognition*, pp. 571–574, Montreal, Canada.

Tang, Y. Y., Ma, H., Xi, D., Cheng, Y., and Suen, C. Y. (1995b). A new approach to document analysis based on modified fractal signature. In *Proceedings of 3-rd International Conference on Document Analysis and Recognition*, pp. 567–570, Montreal, Canada.

Tang, Y. Y., Ma, H., Xi, D., Mao, X., and Suen, C. Y. (1997b). Modified fractal signature (MFS): A new approach to document analysis for automatic knowledge acquisition. *IEEE Transactions on Knowledge and Data Engineering*, 9(5):747–762.

Tang, Y. Y., Suen, C. Y., and Yan, C. D. (1994). Document processing for automatic knowledge acquisition. *IEEE Transactions on Knowledge and Data Engineering*, 6(1):3–21.

Tang, Y. Y., Yan, C. D., Cheriet, M., and Suen, C. Y. (1995c). Financial document processing based on staff line and description language. *IEEE Transactions on Systems, Man, and Cybernetics*, 25(5):738–754.

Tang, Y. Y., Yang, L., and Liu, J. (2000). Characterization of Dirac-structure edges with wavelet transform. *IEEE Transactions on Systems, Man, and Cybernetics (B)*, 30(1):93–109.

Tang, Y. Y., Yang, L. H., and Feng, L. (1998c). Characterization and detection of edges by Lipschitz exponent and MASW wavelet transform. In *Proceedings of the 14th International Conference on Pattern Recognition*, pp. 1572–1574, Brisbane, Australia.

Tang, Y. Y., Yang, L. H., and Feng, L. (1998d). Contour detection of handwriting by modular-angle-separated wavelets. In *Proceedings of the 6-th International Workshop on Frontiers of Handwriting Recognition (IWFHR-VI)*, pp. 357–366, Taejon, Korea.

Tang, Y. Y., Yang, L. H., Feng, L., and Liu, J. (1998e). Scale-independent wavelet algorithm for detecting step-structure edges. Technical report, Hong Kong Baptist University.

Tang, Y. Y., Yang, L. H., and Liu, J. (1997c). Wavelet-based edge detection in Chinese document. In *Proceedings of the 17th International Conference on Computer Processing of Oriental Languages*, vol. 1, pp. 333–336.

Tang, Y. Y. and You, X. G. (2003). Skeletonization of ribbon-like shapes based on a new wavelet function. *IEEE Transactions on Pattern Analysis and Machine Intelligence*, 25:1118–1133.

Taswell, C. (2000). The what, how, and why of wavelet shrinkage denoising. *Computing in Science & Engineering*, 2(3):12–19.

Thune, M., Olstad, B., and Thune, N. (1997). Edge detection in noisy data using finite mixture distribute analysis. *Pattern Recognition*, 30(5):685–699.

Tieng, Q. M. and Boles, W. W. (1997a). Recognition of 2D object contours using the wavelet transform zero-crossing representation. *IEEE Transactions on Pattern Analysis and Machine Intelligence*, 19:910–916.

Tieng, Q. M. and Boles, W. W. (1997b). Wavelet-based affine invariant representation: A tool for recognizing planar objects in 3D space. *IEEE Transactions on Pattern Analysis and Machine Intelligence*, 19:846–857.

Tombre, K. (1998). Analysis of engineering drawings: State of the art and challenges, In K. Tombre and A. K. Chhabra, editors, Graphics Recognition: Algorithm and Systems, vol. 1389 of *Lecture Notes in Computer Science*. Berlin: Springer-Verlag.

Tou, J. T. and Gonzalez, R. C. (1974). *Pattern Recognition Principles*. London: Addison-Wesley.

Turk, M. and Pentland, A. (1991). Eigenfaces for recognition. *Journal of Cognitive Neuroscience*, 3(1):71–86.

Unser, M., Aldroubi, A., and Eden, M. (1992). On the asymptotic convergence of B-spline wavelets to Gabor functions. *IEEE Transactions on Information Theory*, 38:864–872.

Unser, M., Aldroubi, A., and Eden, M. (1993). A family of polynomial spline wavelets transforms. *Signal Processing*, 30:141–162.

Valencia, D., Orejuela, D., Salazar, J., and Valencia, J. (2016). Comparison analysis between rigrsure, sqtwolog, heursure and minimaxi techniques using hard and soft thresholding methods. In *2016 XXI Symposium on Signal Processing, Images and Artificial Vision (STSIVA)*, pp. 1–5. IEEE.

Vasavi, S. and Rao, V. S. (2015). Metadata based object detection and classification using key frame extraction method. *Journal of Image and Graphics*, 3(2):90–95.

Venkatesh, S. and Owens, R. (1990). On the classification of image features. *Pattern Recognition Letters*, 11:339–349.

Verma, N. and Verma, A. (2012). Performance analysis of wavelet thresholding methods in denoising of audio signals of some Indian musical instruments. *International Journal of Engineering Science Technology*, 4(5):2040–2045.

Wang, L., Zheng, W., Ma, X., and Lin, S. (2021). Denoising speech based on deep learning and wavelet decomposition. *Scientific Programming*, 2021:1–10.

Wang, S., Ding, Z., and Fu, Y. (2017). Feature selection guided auto-encoder. In *Proceedings of the AAAI Conference on Artificial Intelligence*, vol. 31.

Wang, X., Yang, L. T., Song, L., Wang, H., Ren, L., and Deen, M. J. (2020). A tensor-based multiattributes visual feature recognition method for industrial intelligence. *IEEE Transactions on Industrial Informatics*, 17(3):2231–2241.

Waseem, M., Lin, Z., Liu, S., Jinai, Z., Rizwan, M., and Sajjad, I. A. (2021). Optimal bra based electric demand prediction strategy considering instance-based learning of the forecast factors. *International Transactions on Electrical Energy Systems*, 31(9):e12967.

Westhausen, N. L. and Meyer, B. T. (2020). Dual-signal transformation LSTM network for real-time noise suppression. arXiv preprint arXiv:2005.07551.

Weston, J., Chopra, S., and Bordes, A. (2014). Memory networks. *arXiv preprint arXiv:1410.3916.*

Wunsch, P. and Laine, A. F. (1995). Wavelet descriptors for multiresolution recognition of handprinted characters. *Pattern Recognition*, 28(8):1237–1249.

Xiao, H., Li, L., Liu, Q., Zhu, X., and Zhang, Q. (2023). Transformers in medical image segmentation: A review. *Biomedical Signal Processing and Control*, 84:104791.

Xin, W., Liu, R., Liu, Y., Chen, Y., Yu, W., and Miao, Q. (2023). Transformer for skeleton-based action recognition: A review of recent advances. *Neurocomputing*.

Yang, F., Wang, Z., and Yu, Y. L. (1998). Chinese typeface generation and composition using B-Spline wavelet transform. In *Proceedings of SPIE, Wavelet Applications V, Orlando, FL, 1998*, pp. 616–620.

Yang, L., Suen, C. Y., and Tang, Y. Y. (2003a). A width-invariant property of curves based on wavelet transform with a novel wavelet function. *IEEE Transactions on Systems, Man, and Cybernetics (B)*, 33(3):541–548.

Yang, L. H., Bui, T. D., and Suen, C. Y. (2003b). Image recognition based on nonlinear wavelet approximation. *International Journal of Wavelets, Multiresolution and Information Processing*, 1(2):151–161.

Yang, L. H., You, X., Haralick, R. M., Phillips, I. T., and Tang, Y. (2001). Characterization of Dirac edge with new wavelet transform. In *Proc. of 2th Int. Conf. Wavelets and its Application*, vol. 1, pp. 872–878, Hong Kong.

Yoon, S. H., Kim, J. H., Alexander, W. E., Park, S. M., and Sohn, K. H. (1998). An optimum solution for scale-invariant object recognition based on the multi-resolution approximation. *Pattern Recognition*, 31:889–908.

You, X., Chen, Q., Fang, B., and Tang, Y. Y. (2006). Thinning character using modulus minima of wavelet transform. *International Journal of Pattern Recognition and Artificial Intelligence*, 20(3):361–376.

You, X. and Tang, Y. Y. (2007). Wavelet-based approach to character skeleton. *IEEE Transactions on Image Processing*, 16:1220–1231.

Young, R. K. (1993). *Wavelet Theory and Its Applications*. Boston, MA: Kluwer Academic Publishers.

Yuen, P. C., Dai, D. Q., and Feng, G. C. (1998). Wavelet-based PCA for human face recognition. *Proceeding of IEEE Southwest Symposium on Image Analysis and Interpretation*, pp. 223–228.

Zaied, M., Said, S., Jemai, O., and Amar, C. B. (2011). A novel approach for face recognition based on fast learning algorithm and wavelet network theory. *International Journal of Wavelets, Multiresolution and Information Processing*, 9(06):923–945.

Zhang, G. (1986). *Lecture of Functional Analysis*, vol. 1. Beijing University Press.

Zhang, K., Zuo, W., Chen, Y., Meng, D., and Zhang, L. (2017). Beyond a Gaussian denoiser: Residual learning of deep CNN for image denoising. *IEEE Transactions on Image Processing*, 26(7):3142–3155.

Zhou, T., Li, Q., Lu, H., Cheng, Q., and Zhang, X. (2023). GAN review: Models and medical image fusion applications. *Information Fusion*, 91:134–148.

Zhou, Z., He, Z., Shi, M., Du, J., and Chen, D. (2020). 3d dense connectivity network with atrous convolutional feature pyramid for brain tumor segmentation in magnetic resonance imaging of human heads. *Computers in Biology and Medicine*, 121:103766.

Zou, J. J. and Yan, H. (2001). Skeletonization of ribbon-like shapes based on regularity and singularity analyses. *IEEE Transactions on Systems, Man, and Cybernetics (B)*, 31(3):401–407.

# Index

9 789811 284045